Graduate Texts in Mathematics 227

Editorial Board
S. Axler   K.A. Ribet

T0207442

Graduate Texts in Mathematics

(continued after index)

Ezra Miller
Bernd Sturmfels

Combinatorial
Commutative Algebra

With 102 Figures

 Springer

Ezra Miller
School of Mathematics
University of Minnesota
Minneapolis, MN 55455
USA
ezra@math.umn.edu

Bernd Sturmfels
Department of Mathematics
University of California at Berkeley
Berkeley, CA 94720-3840
USA
bernd@math.berkeley.edu

Editorial Board
S. Axler
Mathematics Department
San Francisco State University
San Francisco, CA 94132
USA
axler@sfsu.edu

K. A. Ribet
Mathematics Department
University of California, Berkeley
Berkeley, CA 94720-3840
USA
ribet@math.berkeley.edu

Mathematics Subject Classification (2000): 13-01, 05-01

Library of Congress Control Number: 2005923441

ISBN-10: 0-387-23707-0   Printed on acid-free paper.
ISBN-13: 978-0387-23707-7

© 2005 Springer Science+Business Media, Inc.
All rights reserved. This work may not be translated or copied in whole or in part without
the written permission of the publisher (Springer Science+Business Media, Inc., 233 Spring
Street, New York, NY 10013, USA), except for brief excerpts in connection with reviews or
scholarly analysis. Use in connection with any form of information storage and retrieval, electronic adaptation, computer software, or by similar or dissimilar methodology now known or
hereafter developed is forbidden.
The use in this publication of trade names, trademarks, service marks, and similar terms,
even if they are not identified as such, is not to be taken as an expression of opinion as to
whether or not they are subject to proprietary rights.

Printed in the United States of America.   (EB)

9 8 7 6 5 4 3 2 1

springeronline.com

To Elen and Hyungsook

Preface

The last decade has seen a number of exciting developments at the intersection of commutative algebra with combinatorics. New methods have evolved out of an influx of ideas from such diverse areas as polyhedral geometry, theoretical physics, representation theory, homological algebra, symplectic geometry, graph theory, integer programming, symbolic computation, and statistics. The purpose of this volume is to provide a self-contained introduction to some of the resulting combinatorial techniques for dealing with polynomial rings, semigroup rings, and determinantal rings. Our exposition mainly concerns combinatorially defined ideals and their quotients, with a focus on numerical invariants and resolutions, especially under gradings more refined than the standard integer grading.

This project started at the COCOA summer school in Torino, Italy, in June 1999. The eight lectures on monomial ideals given there by Bernd Sturmfels were later written up by Ezra Miller and David Perkinson and published in [MP01]. We felt it would be nice to add more material and turn the COCOA notes into a real book. What you hold in your hand is the result, with Part I being a direct outgrowth of the COCOA notes.

Combinatorial commutative algebra is a broad area of mathematics, and one can cover but a small selection of the possible topics in a single book. Our choices were motivated by our research interests and by our desire to reach a wide audience of students and researchers in neighboring fields. Numerous references, mostly confined to the Notes ending each chapter, point the reader to closely related topics that we were unable to cover.

A milestone in the development of combinatorial commutative algebra was the 1983 book by Richard Stanley [Sta96]. That book, now in its second edition, is still an excellent source. We have made an attempt to complement and build on the material covered by Stanley. Another boon to the subject came with the arrival in 1995 of the book by Bruns and Herzog [BH98], also now in its second edition. The middle part of that book, on "Classes of Cohen–Macaulay rings", follows a progression of three chapters on combinatorially defined algebras, from Stanley–Reisner rings through semigroup rings to determinantal rings. Our treatment elaborates on these three themes. The influence of [BH98] can be seen in the subdivision of our book into three parts, following the same organizational principle.

We frequently refer to two other textbooks in the same Springer series as ours, namely Eisenbud's book on commutative algebra [Eis95] and Ziegler's book on convex polytopes [Zie95]. Students will find it useful to place these two books next to ours on their shelves. Other books in the GTM series that contain useful material related to combinatorial commutative algebra are [BB04], [Eis04], [EH00], [Ewa96], [Grü03], [Har77], [MacL98], and [Rot88].

There are two other fine books that offer an introduction to combinatorial commutative algebra from a perspective different than ours, namely the ones by Hibi [Hib92] and Villarreal [Vil01]. Many readers of our book will enjoy learning more about computational commutative algebra as they go along; for this we recommend the books by Cox, Little, and O'Shea [CLO98], Greuel and Pfister [GP02], Kreuzer and Robbiano [KR00], Schenck [Sch03], Sturmfels [Stu96], and Vasconcelos [Vas98]. Additional material can be found in the proceedings volumes [EGM98] and [AGHSS04].

Drafts of this book have been used for graduate courses taught by Victor Reiner at the University of Minnesota and by the authors at UC Berkeley. In our experience, covering all 18 chapters would require a full-year course, either two semesters or three quarters (one for each of Part I, Part II, and Part III). For a first introduction, we view Chapter 1 and Chapters 3–8 as being essential. However, we recommend that this material be supplemented with a choice of one or two of the remaining chapters, to get a feel for a specific application of the theory presented in Chapters 7 and 8. Topics that stand alone well for this purpose are Chapter 2 (which could, of course, be presented earlier), Chapter 9, Chapter 10, Chapter 11, Chapter 14, and Chapter 18. We have also observed success in covering Chapter 12 with only the barest introduction to injective modules from Chapter 11, although Chapters 11 and 12 work even more coherently as a pair. Other two-chapter sequences include Chapters 11 and 13 or Chapters 15 and 16. Although the latter pair forms a satisfying end, it becomes even more so as a triplet with Chapter 17. Advanced courses could begin with Chapters 7 and 8 and continue with the rest of Part II, or instead continue with Part III.

In general, we assume knowledge of commutative algebra (graded rings, free resolutions, Gröbner bases, and so on) at a level on par with the undergraduate textbook of Cox, Little, and O'Shea [CLO97], supplemented with a little bit of simplicial topology and polyhedral geometry. Although these prerequisites are fairly modest, the mix of topics calls for considerable mathematical maturity. Also, more will be gained from some of the later chapters with additional background in homological algebra or algebraic geometry. For the former, this is particularly true of Chapters 11 and 13, whereas for the latter, we are referring to Chapter 10 and Chapters 15–18. Often we work with algebraic groups, which we describe explicitly by saying what form the matrices have (such as "block lower-triangular"). All of our arguments that use algebraic groups are grounded firmly in the transparent linear algebra that they represent. Typical conclusions reached using algebraic geometry are the smoothness and irreducibility of orbits. Typical uses of homological algebra include statements that certain operations (on resolutions, for example) are well-defined independent of the choices made.

Each chapter begins with an overview and ends with Notes on references and pointers to the literature. Theorems are, for the most part, attributed only in the Notes. When an exercise is based on a specific source, that source is credited in the Notes. For the few exercises used in the proofs of theorems in the main body of the text, solutions to the nonroutine ones are referenced in the Notes. The References list the pages on which each source is cited. The mathematical notation throughout the book is kept as consistent as possible, making the glossary of notation particularly handy, although some of our standard symbols occasionally moonlight for brief periods in nonstandard ways, when we run out of letters. Cross-references have the form "Item aa.bb" if the item is number bb in Chapter aa. Finally, despite our best efforts, errors are sure to have kept themselves safely hidden from our view. Please do let us know about all the bugs you may discover.

In August 2003, a group of students and postdocs ran a seminar at Berkeley covering topics from all 18 chapters. They read the manuscript carefully and provided numerous comments and improvements. We wish to express our sincere gratitude to the following participants for their help: Matthias Beck, Carlos D'Andrea, Mike Develin, Nicholas Eriksson, Daniel Giaimo, Martin Guest, Christopher Hillar, Serkan Hoşten, Lionel Levine, Edwin O'Shea, Julian Pfeifle, Bobby Poon, Nicholas Proudfoot, Brian Rothbach, Nirit Sandman, David Speyer, Seth Sullivant, Lauren Williams, Alexander Woo, and Alexander Yong. Additional comments and help were provided by David Cox, Alicia Dickenstein, Jesus De Loera, Joseph Gubeladze, Mikhail Kapranov, Diane Maclagan, Raymond Hemmecke, Bjarke Roune, Olivier Ruatta, and Günter Ziegler. Special thanks are due to Victor Reiner, for the many improvements he contributed, including a number of exercises and corrections of proofs. We also thank our coauthors Dave Bayer, Mark Haiman, David Helm, Allen Knutson, Misha Kogan, Laura Matusevich, Isabella Novik, Irena Peeva, David Perkinson, Sorin Popescu, Alexander Postnikov, Mark Shimozono, Uli Walther, and Kohji Yanagawa, from whom we have learned so much about combinatorial commutative algebra, and whose contributions form substantial parts of this book.

A number of organizations and nonmathematicians have made this book possible. Both authors had partial support from the National Science Foundation. Ezra Miller was a postdoctoral fellow at MSRI Berkeley in 2003. Bernd Sturmfels was supported by the Miller Institute at UC Berkeley in 2000–2001, and as a Hewlett–Packard Research Professor at MSRI Berkeley in 2003–2004. Our editor, Ina Lindemann, kept us on track and helped us to finish at the right moment. Most of all, we thank our respective partners, Elen and Hyungsook, for their boundless encouragement and support.

Ezra Miller, Minneapolis, MN   `ezra@math.umn.edu`
Bernd Sturmfels, Berkeley, CA   `bernd@math.berkeley.edu`
12 May 2004

Contents

Part I

Monomial Ideals

Chapter 1

Squarefree monomial ideals

We begin by studying ideals in a polynomial ring $\Bbbk[x_1, \ldots, x_n]$ that are generated by squarefree monomials. Such ideals are also known as *Stanley–Reisner ideals*, and quotients by them are called *Stanley–Reisner rings*. The combinatorial nature of these algebraic objects stems from their intimate connections to simplicial topology. This chapter explores various enumerative and homological manifestations of these topological connections, including simplicial descriptions of *Hilbert series* and *Betti numbers*.

After describing the relation between simplicial complexes and squarefree monomial ideals, this chapter goes on to introduce the objects and notation surrounding both the algebra of general monomial ideals as well as the combinatorial topology of simplicial complexes. Section 1.2 defines what it means for a module over the polynomial ring $\Bbbk[x_1, \ldots, x_n]$ to be *graded by* $\mathbb{N}^n$ and what Hilbert series can look like in these gradings. In preparation for our discussion of Betti numbers in Section 1.5, we review simplicial homology and cohomology in Section 1.3 and free resolutions in Section 1.4. The latter section introduces *monomial matrices*, which allow us to write down $\mathbb{N}^n$-graded free resolutions explicitly.

1.1 Equivalent descriptions

Let $\Bbbk$ be a field and $S = \Bbbk[\mathbf{x}]$ the polynomial ring over $\Bbbk$ in n indeterminates $\mathbf{x} = x_1, \ldots, x_n$.

Definition 1.1 A **monomial** in $\Bbbk[\mathbf{x}]$ is a product $\mathbf{x}^{\mathbf{a}} = x_1^{a_1} x_2^{a_2} \cdots x_n^{a_n}$ for a vector $\mathbf{a} = (a_1, \ldots, a_n) \in \mathbb{N}^n$ of nonnegative integers. An ideal $I \subseteq \Bbbk[\mathbf{x}]$ is called a **monomial ideal** if it is generated by monomials.

As a vector space over $\Bbbk$, the polynomial ring S is a direct sum

$$S \;=\; \bigoplus_{\mathbf{a} \in \mathbb{N}^n} S_{\mathbf{a}},$$

where $S_{\mathbf{a}} = \Bbbk\{\mathbf{x}^{\mathbf{a}}\}$ is the vector subspace of S spanned by the monomial $\mathbf{x}^{\mathbf{a}}$. Since the product $S_{\mathbf{a}} \cdot S_{\mathbf{b}}$ of graded pieces equals the graded piece $S_{\mathbf{a}+\mathbf{b}}$ in degree $\mathbf{a} + \mathbf{b}$, we say that S is an $\mathbb{N}^n$-*graded* $\Bbbk$-algebra.

Monomial ideals are the $\mathbb{N}^n$-*graded ideals* of S, which means by definition that I can also be expressed as a direct sum, namely $I = \bigoplus_{\mathbf{x}^{\mathbf{a}} \in I} \Bbbk\{\mathbf{x}^{\mathbf{a}}\}$.

Lemma 1.2 *Every monomial ideal has a unique minimal set of monomial generators, and this set is finite.*

Proof. The Hilbert Basis Theorem says that every ideal in S is finitely generated. It implies that if I is a monomial ideal, then $I = \langle \mathbf{x}^{\mathbf{a}_1}, \ldots, \mathbf{x}^{\mathbf{a}_r} \rangle$. The direct sum condition means that a polynomial f lies inside I if and only if each term of f is divisible by one of the given generators $\mathbf{x}^{\mathbf{a}_i}$. $\quad\square$

Definition 1.3 A monomial $\mathbf{x}^{\mathbf{a}}$ is **squarefree** if every coordinate of $\mathbf{a}$ is 0 or 1. An ideal is **squarefree** if it is generated by squarefree monomials.

The information carried by squarefree monomial ideals can be characterized in many ways. The most combinatorial uses simplicial complexes.

Definition 1.4 An (abstract) **simplicial complex** Δ on the **vertex set** $\{1, \ldots, n\}$ is a collection of subsets called **faces** or **simplices**, closed under taking subsets; that is, if $\sigma \in \Delta$ is a face and $\tau \subseteq \sigma$, then $\tau \in \Delta$. A simplex $\sigma \in \Delta$ of cardinality $|\sigma| = i + 1$ has **dimension** i and is called an i-**face** of Δ. The **dimension** $\dim(\Delta)$ of Δ is the maximum of the dimensions of its faces, or it is $-\infty$ if $\Delta = \{\}$ is the **void complex**, which has no faces.

The empty set $\varnothing$ is the unique dimension -1 face in any simplicial complex Δ that is not the void complex $\{\}$. Thus the *irrelevant complex* $\{\varnothing\}$, whose unique face is the empty set, is to be distinguished from the void complex. The reason for this distinction will become clear when we introduce (co)homology as well as in numerous applications to monomial ideals.

We frequently identify $\{1, \ldots, n\}$ with the variables $\{x_1, \ldots, x_n\}$, as in our next example, or with $\{a, b, c, \ldots\}$, as in Example 1.8.

Example 1.5 The simplicial complex Δ on $\{1, 2, 3, 4, 5\}$ consisting of all subsets of the sets $\{1, 2, 3\}$, $\{2, 4\}$, $\{3, 4\}$, and $\{5\}$ is pictured below:

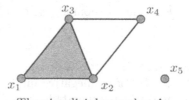

The simplicial complex Δ

Note that Δ is completely specified by its *facets*, or maximal faces, by definition of simplicial complex. ◇

Simplicial complexes determine squarefree monomial ideals. For notation, we identify each subset $\sigma \subseteq \{1, \ldots, n\}$ with its *squarefree vector* in $\{0,1\}^n$, which has entry 1 in the ith spot when $i \in \sigma$, and 0 in all other entries. This convention allows us to write $\mathbf{x}^\sigma = \prod_{i \in \sigma} x_i$.

Definition 1.6 The **Stanley–Reisner ideal** of the simplicial complex Δ is the squarefree monomial ideal

$$I_\Delta = \langle \mathbf{x}^\tau \mid \tau \notin \Delta \rangle$$

generated by monomials corresponding to **nonfaces** τ of Δ. The **Stanley–Reisner ring** of Δ is the quotient ring S/I_Δ.

There are two ways to present a squarefree monomial ideal: either by its generators or as an intersection of monomial prime ideals. These are generated by subsets of $\{x_1, \ldots, x_n\}$. For notation, we write

$$\mathfrak{m}^\tau = \langle x_i \mid i \in \tau \rangle$$

for the monomial prime ideal corresponding to τ. Frequently, τ will be the complement $\overline{\sigma} = \{1, \ldots, n\} \smallsetminus \sigma$ of some simplex σ.

Theorem 1.7 *The correspondence* $\Delta \rightsquigarrow I_\Delta$ *constitutes a bijection from simplicial complexes on vertices* $\{1, \ldots, n\}$ *to squarefree monomial ideals inside* $S = \Bbbk[x_1, \ldots, x_n]$. *Furthermore,*

$$I_\Delta = \bigcap_{\sigma \in \Delta} \mathfrak{m}^{\overline{\sigma}}.$$

Proof. By definition, the set of squarefree monomials that have nonzero images in the Stanley–Reisner ring S/I_Δ is precisely $\{\mathbf{x}^\sigma \mid \sigma \in \Delta\}$. This shows that the map $\Delta \rightsquigarrow I_\Delta$ is bijective. In order for $\mathbf{x}^\tau$ to lie in the intersection $\bigcap_{\sigma \in \Delta} \mathfrak{m}^{\overline{\sigma}}$, it is necessary and sufficient that τ share at least one element with $\overline{\sigma}$ for each face $\sigma \in \Delta$. Equivalently, τ must be contained in no face of Δ; that is, τ must be a nonface of Δ. □

Example 1.8 The simplicial complex $\Delta = $ from Example 1.5, after replacing the variables $\{x_1, x_2, x_3, x_4, x_5\}$ by $\{a, b, c, d, e\}$, has Stanley–Reisner ideal

$$I_\Delta = \langle d, e \rangle \quad \cap \quad \langle a, b, e \rangle \quad \cap \quad \langle a, c, e \rangle \cap \langle a, b, c, d \rangle$$
$$= \langle ad, ae, bcd, be, ce, de \rangle.$$

This expresses I_Δ via its prime decomposition and its minimal generators. Above each prime component is drawn the corresponding facet of Δ.   ◇

Remark 1.9 Because of the expression of Stanley–Reisner ideals I_Δ as intersections in Theorem 1.7, they are also in bijection with unions of coordinate subspaces in the vector space $\Bbbk^n$, or equivalently, unions of coordinate subspaces in the projective space $\mathbb{P}_{\Bbbk}^{n-1}$. A little bit of caution is warranted here: if $\Bbbk$ is finite, it is not true that I_Δ equals the ideal of polynomials vanishing on the corresponding collection of coordinate subspaces; in fact, this vanishing ideal will not be a monomial ideal! On the other hand, when $\Bbbk$ is infinite, the Zariski correspondence between radical ideals and algebraic sets does induce the bijection between squarefree monomial ideals and their zero sets, which are unions of coordinate subspaces. (The *zero set* inside $\Bbbk^n$ of an ideal I in $\Bbbk[\mathbf{x}]$ is the set of points $(\alpha_1, \ldots, \alpha_n) \in \Bbbk^n$ such that $f(\alpha_1, \ldots, \alpha_n) = 0$ for every polynomial $f \in I$.)

1.2 Hilbert series

Even if the goal is to study monomial ideals, it is necessary to consider graded modules more general than ideals.

Definition 1.10 An S-module M is $\mathbb{N}^n$**-graded** if $M = \bigoplus_{\mathbf{b} \in \mathbb{N}^n} M_\mathbf{b}$ and $\mathbf{x}^\mathbf{a} M_\mathbf{b} \subseteq M_{\mathbf{a}+\mathbf{b}}$. If the vector space dimension $\dim_\Bbbk(M_\mathbf{a})$ is finite for all $\mathbf{a} \in \mathbb{N}^n$, then the formal power series

$$H(M; \mathbf{x}) \quad = \quad \sum_{\mathbf{a} \in \mathbb{N}^n} \dim_\Bbbk(M_\mathbf{a}) \cdot \mathbf{x}^\mathbf{a}$$

is the **finely graded** or $\mathbb{N}^n$**-graded Hilbert series** of M. Setting $x_i = t$ for all i yields the ($\mathbb{Z}$**-graded** or **coarse**) **Hilbert series** $H(M; t, \ldots, t)$.

The ring of formal power series in which finely graded Hilbert series live is $\mathbb{Z}[[\mathbf{x}]] = \mathbb{Z}[[x_1, \ldots, x_n]]$. In this ring, each element $1 - x_i$ is invertible, the series $\frac{1}{1-x_i} = 1 + x_i + x_i^2 + \cdots$ being its inverse.

Example 1.11 The Hilbert series of S itself is the rational function

$$H(S; \mathbf{x}) \quad = \quad \prod_{i=1}^{n} \frac{1}{1 - x_i}$$

$$= \quad \text{sum of all monomials in } S.$$

Denote by $S(-\mathbf{a})$ the free module generated in degree $\mathbf{a}$, so $S(-\mathbf{a}) \cong \langle \mathbf{x}^\mathbf{a} \rangle$ as $\mathbb{N}^n$-graded modules. The Hilbert series

$$H(S(-\mathbf{a}); \mathbf{x}) \quad = \quad \frac{\mathbf{x}^\mathbf{a}}{\prod_{i=1}^{n}(1 - x_i)}$$

of such an $\mathbb{N}^n$-*graded translate* of S is just $\mathbf{x}^\mathbf{a} \cdot H(S; \mathbf{x})$.   ◇

In the rest of Part I, our primary examples of Hilbert series are

$$H(S/I; \mathbf{x}) \quad = \quad \text{sum of all monomials not in } I$$

for monomial ideals I. A running theme of Part I of this book is to analyze not so much the whole Hilbert series, but its numerator, as defined in Definition 1.12. (In fact, Parts II and III are frequently concerned with similar analyses of such numerators, for ideals in other gradings.)

Definition 1.12 If the Hilbert series of an $\mathbb{N}^n$-graded S-module M is expressed as a rational function $H(M; \mathbf{x}) = \mathcal{K}(M; \mathbf{x})/(1 - x_1) \cdots (1 - x_n)$, then its numerator $\mathcal{K}(M; \mathbf{x})$ is the K-**polynomial** of M.

We will eventually see in Corollary 4.20 (but see also Theorem 8.20) that the Hilbert series of every monomial quotient of S can in fact be expressed as a rational function as in Definition 1.12, and therefore every such quotient has a K-polynomial. That these K-polynomials are polynomials (as opposed to Laurent polynomials, say) is also proved in Corollary 4.20. Next we want to show that Stanley–Reisner rings S/I_Δ have K-polynomials by explicitly writing them down in terms of Δ.

Theorem 1.13 *The Stanley–Reisner ring S/I_Δ has the K-polynomial*

$$\mathcal{K}(S/I_\Delta; \mathbf{x}) \quad = \quad \sum_{\sigma \in \Delta} \left(\prod_{i \in \sigma} x_i \cdot \prod_{j \notin \sigma} (1 - x_j) \right).$$

Proof. The definition of I_Δ says which *squarefree* monomials are not in I_Δ. However, because the generators of I_Δ are themselves squarefree, a monomial $\mathbf{x}^{\mathbf{a}}$ lies outside I_Δ precisely when the squarefree monomial $\mathbf{x}^{\mathrm{supp}(\mathbf{a})}$ lies outside I_Δ, where $\mathrm{supp}(\mathbf{a}) = \{i \in \{1, \ldots, n\} \mid a_i \neq 0\}$ is the *support* of $\mathbf{a}$. Therefore

$$
\begin{aligned}
H(S/I_\Delta; x_1, \ldots, x_n) &= \sum \{\mathbf{x}^{\mathbf{a}} \mid \mathbf{a} \in \mathbb{N}^n \text{ and } \mathrm{supp}(\mathbf{a}) \in \Delta\} \\
&= \sum_{\sigma \in \Delta} \sum \{\mathbf{x}^{\mathbf{a}} \mid \mathbf{a} \in \mathbb{N}^n \text{ and } \mathrm{supp}(\mathbf{a}) = \sigma\} \\
&= \sum_{\sigma \in \Delta} \prod_{i \in \sigma} \frac{x_i}{1 - x_i},
\end{aligned}
$$

and the result holds after multiplying the summand for σ by $\prod_{j \notin \sigma} \frac{1 - x_j}{1 - x_j}$ to bring the terms over a common denominator of $(1 - x_1) \cdots (1 - x_n)$. $\square$

Example 1.14 Consider the simplicial complex Γ depicted in Fig. 1.1. (The reason for not calling it Δ is because we will compare Γ in Example 1.36 with the simplicial complex Δ of Examples 1.5 and 1.8.) The Stanley–Reisner ideal of Γ is

$$
\begin{aligned}
I_\Gamma &= \langle de, abe, ace, abcd \rangle \\
&= \langle a, d \rangle \cap \langle a, e \rangle \cap \langle b, c, d \rangle \cap \langle b, e \rangle \cap \langle c, e \rangle \cap \langle d, e \rangle,
\end{aligned}
$$

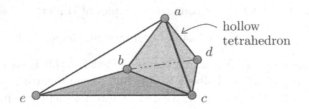

Figure 1.1: The simplicial complex Γ

and the Hilbert series of the quotient $\Bbbk[a,b,c,d,e]/I_\Gamma$ is

$$
1 + \frac{a}{1-a} + \frac{b}{1-b} + \frac{c}{1-c} + \frac{d}{1-d} + \frac{e}{1-e} + \frac{ab}{(1-a)(1-b)} + \frac{ac}{(1-a)(1-c)}
$$
$$
+ \frac{ad}{(1-a)(1-d)} + \frac{ae}{(1-a)(1-e)} + \frac{bc}{(1-b)(1-c)} + \frac{bd}{(1-b)(1-d)} + \frac{be}{(1-b)(1-e)}
$$
$$
+ \frac{cd}{(1-c)(1-d)} + \frac{ce}{(1-c)(1-e)} + \frac{abc}{(1-a)(1-b)(1-c)} + \frac{abd}{(1-a)(1-b)(1-d)}
$$
$$
+ \frac{acd}{(1-a)(1-c)(1-d)} + \frac{bcd}{(1-b)(1-c)(1-d)} + \frac{bce}{(1-b)(1-c)(1-e)}
$$
$$
= \frac{1 - abcd - abe - ace - de + abce + abde + acde}{(1-a)(1-b)(1-c)(1-d)(1-e)}.
$$

See Example 1.25 for a hint at a quick way to get this series. $\diamond$

The formula for the Hilbert series of S/I_Δ perhaps becomes a little neater when we coarsen to the $\mathbb{N}$-grading.

Corollary 1.15 *Letting f_i be the number of i-faces of Δ, we get*

$$
H(S/I_\Delta; t, \ldots, t) = \frac{1}{(1-t)^n} \sum_{i=0}^{d} f_{i-1} t^i (1-t)^{n-i},
$$

where $d = \dim(\Delta) + 1$.

Canceling $(1-t)^{n-d}$ from the sum and the denominator $(1-t)^n$ in Corollary 1.15, the numerator polynomial $h(t)$ on the right-hand side of

$$
\frac{1}{(1-t)^d} \sum_{i=0}^{d} f_{i-1} t^i (1-t)^{d-i} = \frac{h_0 + h_1 t + h_2 t^2 + \cdots + h_d t^d}{(1-t)^d}
$$

is called the *h-polynomial* of Δ. It and the *f-vector* $(f_{-1}, f_0, \ldots, f_{d-1})$ are, to some approximation, the subjects of a whole chapter of Stanley's book [Sta96]; we refer the reader there for further discussion of these topics.

1.3 Simplicial complexes and homology

Much of combinatorial commutative algebra is concerned with analyzing various homological constructions and invariants, and in particular, the manner in which they are governed by combinatorial data. Often, the analysis reduces to related (and hopefully easier) homological constructions purely in the realm of simplicial topology. We review the basics here, referring the reader to [Hat02], [Rot88], or [Mun84] for a full treatment.

Let Δ be a simplicial complex on $\{1, \ldots, n\}$. For each integer i, let $F_i(\Delta)$ be the set of i-dimensional faces of Δ, and let $\Bbbk^{F_i(\Delta)}$ be a vector space over $\Bbbk$ whose basis elements e_σ correspond to i-faces $\sigma \in F_i(\Delta)$.

Definition 1.16 The (**augmented** or **reduced**) **chain complex** of Δ over $\Bbbk$ is the complex $\widetilde{\mathcal{C}}.(\Delta; \Bbbk)$:

$$0 \longleftarrow \Bbbk^{F_{-1}(\Delta)} \xleftarrow{\partial_0} \cdots \longleftarrow \Bbbk^{F_{i-1}(\Delta)} \xleftarrow{\partial_i} \Bbbk^{F_i(\Delta)} \longleftarrow \cdots \xleftarrow{\partial_{n-1}} \Bbbk^{F_{n-1}(\Delta)} \longleftarrow 0.$$

The **boundary maps** ∂_i are defined by setting $\mathrm{sign}(j, \sigma) = (-1)^{r-1}$ if j is the rth element of the set $\sigma \subseteq \{1, \ldots, n\}$, written in increasing order, and

$$\partial_i(e_\sigma) \;\; = \;\; \sum_{j \in \sigma} \mathrm{sign}(j, \sigma)\, e_{\sigma \smallsetminus j}.$$

If $i < -1$ or $i > n - 1$, then $\Bbbk^{F_i(\Delta)} = 0$ and $\partial_i = 0$ by definition. The reader unfamiliar with simplicial complexes should make the routine check that $\partial_i \circ \partial_{i+1} = 0$. In other words, the image of the $(i+1)^{\text{st}}$ boundary map ∂_{i+1} lies inside the kernel of the ith boundary map ∂_i.

Definition 1.17 For each integer i, the $\Bbbk$-vector space

$$\widetilde{H}_i(\Delta; \Bbbk) \;\; = \;\; \ker(\partial_i)/\mathrm{im}(\partial_{i+1})$$

in **homological degree** i is the ith **reduced homology** of Δ over $\Bbbk$.

In particular, $\widetilde{H}_{n-1}(\Delta; \Bbbk) = \ker(\partial_{n-1})$, and when Δ is not the irrelevant complex $\{\varnothing\}$, we get also $\widetilde{H}_i(\Delta; \Bbbk) = 0$ for $i < 0$ or $i > n-1$. The irrelevant complex $\Delta = \{\varnothing\}$ has homology only in homological degree -1, where $\widetilde{H}_{-1}(\Delta; \Bbbk) \cong \Bbbk$. The dimension of the zeroth reduced homology $\widetilde{H}_0(\Delta; \Bbbk)$ as a $\Bbbk$-vector space is one less than the number of connected components of Δ. Elements of $\ker(\partial_i)$ are often called i-*cycles* and elements of $\mathrm{im}(\partial_{i+1})$ are often called i-*boundaries*.

Example 1.18 For Δ as in Example 1.5, we have

$$\begin{aligned}
F_2(\Delta) &= \{\{1, 2, 3\}\}, \\
F_1(\Delta) &= \{\{1, 2\}, \{1, 3\}, \{2, 3\}, \{2, 4\}, \{3, 4\}\}, \\
F_0(\Delta) &= \{\{1\}, \{2\}, \{3\}, \{4\}, \{5\}\}, \\
F_{-1}(\Delta) &= \{\varnothing\}.
\end{aligned}$$

Ordering the bases for $\Bbbk^{F_i(\Delta)}$ as suggested by the ordering of the faces listed above, the chain complex for Δ becomes

$$0 \longleftarrow \Bbbk \xleftarrow[\partial_0]{\begin{bmatrix} 1 & 1 & 1 & 1 & 1 \end{bmatrix}} \Bbbk^5 \xleftarrow[\partial_1]{\begin{bmatrix} -1 & -1 & 0 & 0 & 0 \\ 1 & 0 & -1 & -1 & 0 \\ 0 & 1 & 1 & 0 & -1 \\ 0 & 0 & 0 & 1 & 1 \\ 0 & 0 & 0 & 0 & 0 \end{bmatrix}} \Bbbk^5 \xleftarrow[\partial_2]{\begin{bmatrix} 1 \\ -1 \\ 1 \\ 0 \\ 0 \end{bmatrix}} \Bbbk \longleftarrow 0,$$

where vectors in $\Bbbk^{F_i(\Delta)}$ are viewed as columns of length $f_i = |F_i(\Delta)|$. For example, $\partial_2(e_{\{1,2,3\}}) = e_{\{2,3\}} - e_{\{1,3\}} + e_{\{1,2\}}$, which we identify with the vector $(1, -1, 1, 0, 0)$. The homomorphisms ∂_2 and ∂_0 both have rank 1 (that is, they are injective and surjective, respectively). Since the matrix ∂_1 has rank 3, we conclude that $\widetilde{H}_0(\Delta; \Bbbk) \cong \widetilde{H}_1(\Delta; \Bbbk) \cong \Bbbk$, and the other homology groups are 0. Geometrically, $\widetilde{H}_0(\Delta; \Bbbk)$ is nontrivial because Δ is disconnected, and $\widetilde{H}_1(\Delta; \Bbbk)$ is nontrivial because Δ contains a triangle that does not bound a face of Δ. $\diamond$

Remark 1.19 We would avoid making such a big deal about the difference between the irrelevant complex $\{\varnothing\}$ and the void complex $\{\}$ if it did not come up so much. Many of the formulas for Betti numbers, dimensions of local cohomology, and so on depend on the fact that $\widetilde{H}_i(\{\varnothing\}; \Bbbk)$ is nonzero for $i = -1$, whereas $\widetilde{H}_i(\{\}; \Bbbk) = 0$ for all i.

In some situations, the notion dual to homology arises more naturally. In what follows, we write $(\_)^*$ for vector space duality $\mathrm{Hom}_\Bbbk(\_, \Bbbk)$.

Definition 1.20 The **(reduced) cochain complex** of Δ over $\Bbbk$ is the vector space dual $\widetilde{C}^\bullet(\Delta; \Bbbk) = (\widetilde{C}_\bullet(\Delta; \Bbbk))^*$ of the chain complex, with **coboundary maps** $\partial^i = \partial_i^*$. For $i \in \mathbb{Z}$, the $\Bbbk$-vector space

$$\widetilde{H}^i(\Delta; \Bbbk) \;\; = \;\; \ker(\partial^{i+1})/\mathrm{im}(\partial^i)$$

is the ith **reduced cohomology** of Δ over $\Bbbk$.

Explicitly, let $\Bbbk^{F_i^*(\Delta)} = (\Bbbk^{F_i(\Delta)})^*$ have basis $F_i^*(\Delta) = \{e_\sigma^* \mid \sigma \in F_i(\Delta)\}$ dual to the basis of $\Bbbk^{F_i(\Delta)}$. Then

$$0 \longrightarrow \Bbbk^{F_{-1}^*(\Delta)} \xrightarrow{\partial^0} \cdots \longrightarrow \Bbbk^{F_{i-1}^*(\Delta)} \xrightarrow{\partial^i} \Bbbk^{F_i^*(\Delta)} \longrightarrow \cdots \xrightarrow{\partial^{n-1}} \Bbbk^{F_{n-1}^*(\Delta)} \longrightarrow 0$$

is the cochain complex $\widetilde{C}^\bullet(\Delta; \Bbbk)$ of Δ, where for an $(i-1)$-face σ,

$$\partial^i(e_\sigma^*) \;\; = \;\; \sum_{\substack{j \notin \sigma \\ j \cup \sigma \in \Delta}} \mathrm{sign}(j, \sigma \cup j)\, e_{\sigma \cup j}^*$$

is the transpose of ∂_i.

Since $\mathrm{Hom}_{\Bbbk}(\_, \Bbbk)$ takes exact sequences to exact sequences, there is a canonical isomorphism $\widetilde{H}^i(\Delta; \Bbbk) = \widetilde{H}_i(\Delta; \Bbbk)^*$. Elements of $\ker(\partial^{i+1})$ are called *i-cocycles* and elements of $\mathrm{im}(\partial^i)$ are called *i-coboundaries*.

Example 1.21 The cochain complex for Δ as in Example 1.18 is exactly the same as the chain complex there, except that the arrows should be reversed and the elements of the vector spaces should be considered as row vectors, with the matrices acting by multiplication on the right. The nonzero reduced cohomology of Δ is $\widetilde{H}^0(\Delta; \Bbbk) \cong \widetilde{H}^1(\Delta; \Bbbk) \cong \Bbbk$. $\diamond$

1.4 Monomial matrices

The central homological objects in Part I of this book, as well as in Chapter 9, are free resolutions. To begin, a *free S-module* of finite rank is a direct sum $F \cong S^r$ of copies of S, for some nonnegative integer r. In our combinatorial context, F will usually be $\mathbb{N}^n$-graded, which means that $F \cong S(-\mathbf{a}_1) \oplus \cdots \oplus S(-\mathbf{a}_r)$ for some vectors $\mathbf{a}_1, \ldots, \mathbf{a}_r \in \mathbb{N}^n$. A sequence

$$\mathcal{F}. : \quad 0 \longleftarrow F_0 \xleftarrow{\phi_1} F_1 \longleftarrow \cdots \longleftarrow F_{\ell-1} \xleftarrow{\phi_\ell} F_\ell \longleftarrow 0 \qquad (1.1)$$

of maps of free S-modules is a *complex* if $\phi_i \circ \phi_{i+1} = 0$ for all i. The complex is *exact* in homological degree i if $\ker(\phi_i) = \mathrm{im}(\phi_{i+1})$. When the free modules F_i are $\mathbb{N}^n$-graded, we require that each homomorphism ϕ_i be degree-preserving (or $\mathbb{N}^n$-*graded of degree* $\mathbf{0}$), so that it takes elements in F_i of degree $\mathbf{a} \in \mathbb{N}^n$ to degree $\mathbf{a}$ elements in F_{i-1}.

Definition 1.22 A complex $\mathcal{F}.$ as in (1.1) is a **free resolution** of a module M over $S = \Bbbk[x_1, \ldots, x_n]$ if $\mathcal{F}.$ is exact everywhere except in homological degree 0, where $M = F_0/\mathrm{im}(\phi_1)$. The image in F_i of the homomorphism ϕ_{i+1} is the ith **syzygy module** of M. The **length** of the resolution is the greatest homological degree of a nonzero module in the resolution; this equals ℓ in (1.1), assuming $F_\ell \neq 0$.

Often we *augment* the free resolution $\mathcal{F}.$ by placing $0 \longleftarrow M \xleftarrow{\phi_0} F_0$ at its left end instead, to make the complex exact everywhere.

The Hilbert Syzygy Theorem says that every module M over the polynomial ring S has a free resolution with length at most n. In cases that interest us here, $M = S/I$ is $\mathbb{N}^n$-graded, so it has an $\mathbb{N}^n$-graded free resolution. Indeed, the kernel of an $\mathbb{N}^n$-graded module map is $\mathbb{N}^n$-graded, so the syzygy modules—and hence the whole free resolution—of S/I are automatically $\mathbb{N}^n$-graded. Before giving examples, it would help to be able to write down maps between $\mathbb{N}^n$-graded free modules efficiently. To do this, we offer the following definition, in which the "$\succeq$" symbol is used to denote the partial order on $\mathbb{N}^n$ in which $\mathbf{a} \succeq \mathbf{b}$ if $a_i \geq b_i$ for all $i \in \{1, \ldots, n\}$.

Definition 1.23 A **monomial matrix** is an array of **scalar entries** λ_{qp} whose columns are labeled by **source degrees** $\mathbf{a}_p$, whose rows are labeled by **target degrees** $\mathbf{a}_q$, and whose entry $\lambda_{qp} \in \Bbbk$ is zero unless $\mathbf{a}_p \succeq \mathbf{a}_q$.

The general monomial matrix represents a map that looks like

$$
\begin{array}{c}
\cdots\ \mathbf{a}_p\ \cdots \\[4pt]
\mathbf{a}_q \ \begin{array}{c}\vdots \\ \ \\ \vdots\end{array}
\left[\ \ \lambda_{qp}\ \ \right]
\end{array}
$$

$$
\bigoplus_q S(-\mathbf{a}_q) \longleftarrow \bigoplus_p S(-\mathbf{a}_p).
$$

Sometimes we label the rows and columns with monomials $\mathbf{x}^{\mathbf{a}}$ instead of vectors $\mathbf{a}$. The scalar entry λ_{qp} indicates that the basis vector of $S(-\mathbf{a}_p)$ should map to an element that has coefficient λ_{qp} on the monomial that is $\mathbf{x}^{\mathbf{a}_p - \mathbf{a}_q}$ times the basis vector of $S(-\mathbf{a}_q)$. Observe that this monomial sits in degree $\mathbf{a}_p$, just like the basis vector of $S(-\mathbf{a}_p)$. The requirement $\mathbf{a}_p \succeq \mathbf{a}_q$ precisely guarantees that $\mathbf{x}^{\mathbf{a}_p - \mathbf{a}_q}$ has nonnegative exponents.

When the maps in a free resolution are written using monomial matrices, the top border row (source degrees $\mathbf{a}_p$) on a monomial matrix for ϕ_i equals the left border column (target degrees $\mathbf{a}_q$) on a monomial matrix for ϕ_{i+1}.

Each $\mathbb{N}^n$-graded free module can also be regarded as an ungraded free module, and most readers will have seen already matrices used for maps of (ungraded) free modules over arbitrary rings. In order to recover the more usual notation, simply replace each matrix entry λ_{qp} by $\mathbf{x}^{\mathbf{a}_p - \mathbf{a}_q} \lambda_{qp}$, and then forget the border row and column. Because of the conditions defining monomial matrices, $\mathbf{x}^{\mathbf{a}_p - \mathbf{a}_q} \lambda_{qp} \in S$ for all p and q.

Definition 1.24 A monomial matrix is **minimal** if $\lambda_{qp} = 0$ when $\mathbf{a}_p = \mathbf{a}_q$. A homomorphism of free modules, or a complex of such, is **minimal** if it can be written down with minimal monomial matrices.

Given that $\mathbb{N}^n$-graded free resolutions exist, it is not hard to show (by "pruning" the nonzero entries λ_{qp} for which $\mathbf{a}_p = \mathbf{a}_q$) that every finitely generated graded module possesses a minimal free resolution. In fact, minimal free resolutions are unique up to isomorphism. For more details on these issues, see Exercises 1.10 and 1.11; for a full treatment, see [Eis95, Theorem 20.2 and Exercise 20.1].

Minimal free resolutions are characterized by having scalar entry $\lambda_{qp} = 0$ whenever $\mathbf{a}_p = \mathbf{a}_q$ in any of their monomial matrices. If the monomial matrices are made ungraded as above, this simply means that the nonzero entries in the matrices are nonconstant monomials (with coefficients), so it agrees with the usual notion of minimality for $\mathbb{N}$-graded resolutions.

Example 1.25 Let Γ be the simplicial complex from Example 1.14. The Stanley–Reisner ring S/I_Γ has minimal free resolution

$$
0 \leftarrow S \xleftarrow{\;\;\begin{array}{c} de\ abe\ ace\ abcd \\ 1\begin{bmatrix} 1 & 1 & 1 & 1 \end{bmatrix} \end{array}\;\;} S^4 \xleftarrow{\;\;\begin{array}{c} \quad abce\ abde\ acde\ abcde \\ \begin{array}{c} de \\ abe \\ ace \\ abcd \end{array}\begin{bmatrix} 0 & -1 & -1 & -1 \\ 1 & 1 & 0 & 0 \\ -1 & 0 & 1 & 0 \\ 0 & 0 & 0 & 1 \end{bmatrix} \end{array}\;\;} S^4 \xleftarrow{\;\;\begin{array}{c} abcde \\ \begin{array}{c} abce \\ abde \\ acde \\ abcde \end{array}\begin{bmatrix} -1 \\ 1 \\ -1 \\ 0 \end{bmatrix} \end{array}\;\;} S \leftarrow 0
$$

with degrees below the free modules:

00000	00011	11101	11111
	11001	11011	
	10101	10111	
	11110	11111	

in which the maps are denoted by monomial matrices. We have used the more succinct monomial labels $\mathbf{x}^{\mathbf{a}_p}$ and $\mathbf{x}^{\mathbf{a}_q}$ instead of the vector labels $\mathbf{a}_p$ and $\mathbf{a}_q$. Below each free module is a list of the degrees in $\mathbb{N}^5$ of its generators. For an example of how to recover the usual matrix notation for maps of free S-modules, this free resolution can be written as

$$
0 \leftarrow S \xleftarrow{\begin{bmatrix} de & abe & ace & abcd \end{bmatrix}} S^4 \xleftarrow{\begin{bmatrix} 0 & -ab & -ac & -abc \\ c & d & 0 & 0 \\ -b & 0 & d & 0 \\ 0 & 0 & 0 & e \end{bmatrix}} S^4 \xleftarrow{\begin{bmatrix} -d \\ c \\ -b \\ 0 \end{bmatrix}} S \leftarrow 0,
$$

without the border entries and forgetting the grading.

As a preview to Chapter 4, the reader is invited to figure out how the labeled simplicial complex below corresponds to the above free resolution.

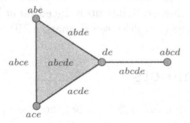

Hint: Compare the free resolution and the labeled simplicial complex with the numerator of the Hilbert series in Example 1.14.　　　　◇

Recall that in reduced chain complexes of simplicial complexes, the basis vectors are called e_σ for subsets $\sigma \subseteq \{1, \ldots, n\}$.

Definition 1.26 The **Koszul complex** is the complex $\mathbb{K}_\bullet$ of free modules given by monomial matrices as follows: in the reduced chain complex of the simplex consisting of all subsets of $\{1, \ldots, n\}$, label the column and the row corresponding to e_σ by σ itself (or $\mathbf{x}^\sigma$), and renumber the homological degrees so that the empty set $\varnothing$ sits in homological degree 0.

Example 1.27 The Koszul complex for $n = 3$ is

$$\mathbb{K}. : \quad 0 \longleftarrow S \xleftarrow{\; 1 \begin{bmatrix} x & y & z \\ 1 & 1 & 1 \end{bmatrix} \;} S^3 \xleftarrow{\; \begin{array}{c} \\ x \\ y \\ z \end{array}\begin{bmatrix} yz & xz & xy \\ 0 & 1 & 1 \\ 1 & 0 & -1 \\ -1 & -1 & 0 \end{bmatrix} \;} S^3 \xleftarrow{\; \begin{array}{c} \\ yz \\ xz \\ xy \end{array}\begin{bmatrix} xyz \\ 1 \\ -1 \\ 1 \end{bmatrix} \;} S \longleftarrow 0$$

after replacing the variables $\{x_1, x_2, x_3\}$ by $\{x, y, z\}$. $\diamond$

The method of proof for many statements about resolutions of monomial ideals is to determine what happens in each $\mathbb{N}^n$-graded degree of a complex of S-modules. To illustrate, we do this now for $\mathbb{K}.$ in some detail.

Proposition 1.28 *The Koszul complex $\mathbb{K}.$ is a minimal free resolution of $\Bbbk = S/\mathfrak{m}$ for the maximal ideal $\mathfrak{m} = \langle x_1, \ldots, x_n \rangle$.*

Proof. The essential observation is that a free module generated by 1_τ in squarefree degree τ is nonzero in squarefree degree σ precisely when $\tau \subseteq \sigma$ (equivalently, when $\mathbf{x}^\tau$ divides $\mathbf{x}^\sigma$). The only contribution to the degree $\mathbf{0}$ part of $\mathbb{K}.$, for example, comes from the free module corresponding to $\varnothing$, whose basis vector $1_\varnothing$ sits in degree $\mathbf{0}$.

More generally, for $\mathbf{b} \in \mathbb{N}^n$ with support σ, the degree $\mathbf{b}$ part $(\mathbb{K}.)_\mathbf{b}$ of the complex $\mathbb{K}.$ comes from those rows and columns labeled by faces of σ. In other words, we restrict $\mathbb{K}.$ to its degree $\mathbf{b}$ part by ignoring summands $S \cdot 1_\tau$ for which τ is not a face of σ. Therefore, $(\mathbb{K}.)_\mathbf{b}$ is, as a complex of $\Bbbk$-vector spaces, precisely equal to the reduced chain complex of the simplex σ! This explains why the homology of $\mathbb{K}.$ is just $\Bbbk$ in degree $\mathbf{0}$ and zero elsewhere: a simplex σ is contractible, so it has no reduced homology—that is, unless $\sigma = \{\varnothing\}$ is the irrelevant complex (see Remark 1.19). $\square$

1.5   Betti numbers

Since every free resolution of an $\mathbb{N}^n$-graded module M contains a minimal resolution as a subcomplex (Exercise 1.11), minimal resolutions of M are characterized by having the ranks of their free modules F_i all simultaneously minimized, among free resolutions (1.1) of M.

Definition 1.29 If the complex $\mathcal{F}.$ in (1.1) is a minimal free resolution of a finitely generated $\mathbb{N}^n$-graded module M and $F_i = \bigoplus_{\mathbf{a} \in \mathbb{N}^n} S(-\mathbf{a})^{\beta_{i,\mathbf{a}}}$, then the ith **Betti number** of M in degree $\mathbf{a}$ is the invariant $\beta_{i,\mathbf{a}} = \beta_{i,\mathbf{a}}(M)$.

There are other equivalent ways to describe the $\mathbb{N}^n$-graded Betti number $\beta_{i,\mathbf{a}}(M)$. For example, it measures the minimal number of generators required in degree $\mathbf{a}$ for any i^{th} syzygy module of M. A more natural (by which we mean functorial) characterization of Betti numbers uses tensor products and Tor, which we now review in some detail.

If M and N are $\mathbb{N}^n$-graded modules, then their tensor product $N \otimes_S M$ is $\mathbb{N}^n$-graded, with degree $\mathbf{c}$ component $(N \otimes_S M)_{\mathbf{c}}$ generated by all elements $f_{\mathbf{a}} \otimes g_{\mathbf{b}}$ such that $f_{\mathbf{a}} \in N_{\mathbf{a}}$ and $g_{\mathbf{b}} \in M_{\mathbf{b}}$ satisfy $\mathbf{a} + \mathbf{b} = \mathbf{c}$. For example, $S(-\mathbf{a}) \otimes_S M$ is a module denoted by $M(-\mathbf{a})$ and called the $\mathbb{N}^n$-*graded translate* of M by $\mathbf{a}$. Its degree $\mathbf{b}$ component is $M(-\mathbf{a})_{\mathbf{b}} = 1_{\mathbf{a}} \otimes M_{\mathbf{b}-\mathbf{a}}$, where $1_{\mathbf{a}}$ is a basis vector for $S(-\mathbf{a})$, so that $S \cdot 1_{\mathbf{a}} = S(-\mathbf{a})$. In particular, $S(-\mathbf{a}) \otimes_S \Bbbk$ is a copy $\Bbbk(-\mathbf{a})$ of the vector space $\Bbbk$ in degree $\mathbf{a} \in \mathbb{N}^n$.

Example 1.30 Tensoring the minimal free resolution in Example 1.25 with $\Bbbk = S/\mathfrak{m}$ yields a complex

$$0 \longleftarrow \Bbbk \longleftarrow \Bbbk^4 \longleftarrow \Bbbk^4 \longleftarrow \Bbbk \longleftarrow 0$$

00000	00011	11101	11111
	11001	11011	
	10101	10111	
	11110	11111	

of S-modules, each of which is a direct sum of translates of $\Bbbk$, and where all the maps are zero. The translation vectors, which are listed below each direct sum, are identified with the row labels to the right of the corresponding free module in Example 1.25, or the column labels to the left. $\diamond$

The modules $\mathrm{Tor}_i^S(M, N)$ are by definition calculated by applying $\_ \otimes N$ to a free resolution of M and taking homology [Wei94, Definition 2.6.4]. However, it is a general theorem from homological algebra (see [Wei94, Application 5.6.3] or do Exercise 1.12) that $\mathrm{Tor}_i^S(M, N)$ can also be calculated by applying $M \otimes \_$ to a free resolution of N and taking homology. When both M and N are $\mathbb{N}^n$-graded, we can choose the free resolutions to be $\mathbb{N}^n$-graded, so the Tor modules are also $\mathbb{N}^n$-graded.

Example 1.31 The homology of the complex in Example 1.30 is the complex itself, considered as a homologically and $\mathbb{N}^n$-graded module. By definition, this module is $\mathrm{Tor}_\bullet^S(S/I_\Gamma, \Bbbk)$. It agrees with the result of tensoring the Koszul complex with S/I_Γ, where again Γ is the simplicial complex from Examples 1.25 and 1.14. The reader is encouraged to check this explicitly, but we shall make this calculation abstractly in the proof of Corollary 5.12. $\diamond$

Now we can see that Betti numbers tell us the vector space dimensions of certain Tor modules.

Lemma 1.32 *The ith Betti number of an $\mathbb{N}^n$-graded module M in degree $\mathbf{a}$ equals the vector space dimension $\dim_{\Bbbk} \mathrm{Tor}_i^S(\Bbbk, M)_{\mathbf{a}}$.*

Proof. Tensoring a minimal free resolution of M with $\Bbbk = S/\mathfrak{m}$ turns all of the differentials ϕ_i into zero maps. $\square$

There is no general formula for the maps in a minimal free resolution of an arbitrary squarefree monomial ideal I_Δ. However, we can figure out what its Betti numbers are in terms of simplicial topology. More generally, we can get simplicial formulas for Betti numbers of quotients by arbitrary monomial ideals.

Definition 1.33 For a monomial ideal I and a degree $\mathbf{b} \in \mathbb{N}^n$, define

$$K^{\mathbf{b}}(I) \ = \ \{\text{squarefree vectors } \tau \mid \mathbf{x}^{\mathbf{b}-\tau} \in I\}$$

to be the **(upper) Koszul simplicial complex** of I in degree $\mathbf{b}$.

Theorem 1.34 *Given a vector* $\mathbf{b} \in \mathbb{N}^n$, *the Betti numbers of* I *and* S/I *in degree* $\mathbf{b}$ *can be expressed as*

$$\beta_{i,\mathbf{b}}(I) \ = \ \beta_{i+1,\mathbf{b}}(S/I) \ = \ \dim_{\Bbbk} \widetilde{H}_{i-1}(K^{\mathbf{b}}(I); \Bbbk).$$

Proof. For the first equality, use a minimal free resolution of I achieved by snipping off the copy of S occurring in homological degree 0 of a minimal free resolution of S/I. To equate $\beta_{i,\mathbf{b}}(I)$ with the dimension of the indicated homology, use Lemma 1.32 and Proposition 1.28 to write $\beta_{i,\mathbf{b}}(I)$ as the vector space dimension of the ith homology of the complex $\mathbb{K}. \otimes I$ in $\mathbb{N}^n$-graded degree $\mathbf{b}$. Then calculate this homology as follows.

Since I is a submodule of S, the complex in degree $\mathbf{b}$ of $\mathbb{K}. \otimes_S I$ is naturally a subcomplex of $(\mathbb{K}.)_{\mathbf{b}}$, which we saw in the proof of Proposition 1.28 is the reduced chain complex of the simplex with facet $\sigma = \mathrm{supp}(\mathbf{b})$. It suffices to identify which faces of σ contribute $\Bbbk$-basis vectors to $(\mathbb{K}.)_{\mathbf{b}}$.

The summand of $\mathbb{K}.$ corresponding to a squarefree vector τ is a free S-module of rank 1 generated in degree τ. Tensoring this summand with I yields $I(-\tau)$, which contributes a nonzero vector space to degree $\mathbf{b}$ if and only if I is nonzero in degree $\mathbf{b} - \tau$, which is equivalent to $\mathbf{x}^{\mathbf{b}-\tau} \in I$. $\square$

In the special case of squarefree ideals, the Koszul simplicial complexes have natural interpretations in terms of a simplicial complex closely related to Δ. In fact, the simplicial complex we are about to introduce is determined just as naturally from the data defining Δ as is Δ itself.

Definition 1.35 The squarefree **Alexander dual** of $I = \langle \mathbf{x}^{\sigma_1}, \ldots, \mathbf{x}^{\sigma_r} \rangle$ is

$$I^\star \ = \ \mathfrak{m}^{\sigma_1} \cap \cdots \cap \mathfrak{m}^{\sigma_r}.$$

If Δ is a simplicial complex and $I = I_\Delta$ its Stanley–Reisner ideal, then the simplicial complex $\Delta^\star$ **Alexander dual** to Δ is defined by $I_{\Delta^\star} = I_\Delta^\star$.

Example 1.36 The Stanley–Reisner ideals I_Δ and I_Γ from Examples 1.8 and 1.14 are Alexander dual; their generators and irreducible components are arranged to make this clear. $\diamond$

The following is a direct description of the Alexander dual simplicial complex. Recall that $\overline{\sigma} = \{1,\ldots,n\} \smallsetminus \sigma$ is the complement of σ in the vertex set.

Proposition 1.37 *If Δ is a simplicial complex, then its Alexander dual is $\Delta^{\star} = \{\overline{\tau} \mid \tau \notin \Delta\}$, consisting of the complements of the nonfaces of Δ.*

Proof. By Definition 1.6, $I_{\Delta} = \langle \mathbf{x}^{\tau} \mid \tau \notin \Delta \rangle$, so $I_{\Delta^{\star}} = \bigcap_{\tau \notin \Delta} \mathfrak{m}^{\tau}$ by Definition 1.35. However, this intersection equals $\bigcap_{\overline{\tau} \in \Delta^{\star}} \mathfrak{m}^{\tau}$ by Theorem 1.7, so we conclude that $\tau \notin \Delta$ if and only if $\overline{\tau} \in \Delta^{\star}$, as desired. $\qquad\square$

Specializing Theorem 1.34 to squarefree ideals requires one more notion.

Definition 1.38 The **link** of σ inside the simplicial complex Δ is

$$\mathrm{link}_{\Delta}(\sigma) \;=\; \{\tau \in \Delta \mid \tau \cup \sigma \in \Delta \text{ and } \tau \cap \sigma = \varnothing\},$$

the set of faces that are disjoint from σ but whose unions with σ lie in Δ.

Example 1.39 Consider the simplicial complex Γ from Examples 1.14 and 1.25, depicted in Fig. 1.1. The link of the vertex a in Γ consists of the vertex e along with all proper faces of the triangle $\{b, c, d\}$. The link of the vertex c in Γ is pure of dimension 1, its four facets being the three edges of the triangle $\{a, b, d\}$ plus the extra edge $\{b, e\}$ sticking out.

The simplicial complex $\mathrm{link}_{\Gamma}(e)$ consists of the vertex a along with the edge $\{b, c\}$ and its subsets. The link of the edge $\{b, c\}$ in Γ consists of the three remaining vertices: $\mathrm{link}_{\Gamma}(\{b, c\}) = \{\varnothing, a, d, e\}$. The link in Γ of the edge through a and e is the irrelevant complex: $\mathrm{link}_{\Gamma}(\{a, e\}) = \{\varnothing\}$. $\qquad\Diamond$

The next result is called the "dual version" of Hochster's formula because it gives Betti numbers of I_{Δ} by working with the Alexander dual complex $\Delta^{\star}$, and because it is dual to Hochster's original formulation, which we will see in Corollary 5.12.

Corollary 1.40 (Hochster's formula, dual version) *All nonzero Betti numbers of I_{Δ} and S/I_{Δ} lie in squarefree degrees σ, where*

$$\beta_{i,\sigma}(I_{\Delta}) \;=\; \beta_{i+1,\sigma}(S/I_{\Delta}) \;=\; \dim_{\Bbbk} \widetilde{H}_{i-1}(\mathrm{link}_{\Delta^{\star}}(\overline{\sigma}); \Bbbk).$$

Proof. For squarefree degrees, apply Theorem 1.34 by first checking that $K^{\sigma}(I_{\Delta}) = \mathrm{link}_{K^{1}(I_{\Delta})}(\overline{\sigma})$ and then verifying that $K^{1}(I_{\Delta}) = \Delta^{\star}$. Both of these claims are straightforward from the definitions and hence omitted. For degrees $\mathbf{b}$ with $b_i \geq 2$, the monomial $\mathbf{x}^{\mathbf{b}-(\tau \cup i)}$ lies in I_{Δ} if and only if $\mathbf{x}^{\mathbf{b}-\tau}$ does. This means that $K^{\mathbf{b}}(I_{\Delta})$ is a cone with vertex i. Cones, being contractible, have zero homology (see [Wei94, Section 1.5], for example). $\square$

We will have a lot more to say about Alexander duality in Chapter 5. The interested reader may even wish to skip directly to Sections 5.1, 5.2, and 5.5 (except the end), as these require no additional prerequisites.

Remark 1.41 Since we are working over a field $\Bbbk$, one may substitute reduced homology for reduced cohomology when calculating Betti numbers, since these have the same dimension.

Exercises

1.1 Let $n = 6$ and let Δ be the boundary of an octahedron.

(a) Determine I_Δ and $I_\Delta^\star$.
(b) Compute their respective Hilbert series.
(c) Compute their minimal free resolutions.
(d) Interpret the Betti numbers in part (c) in terms of simplicial homology.

1.2 Suppose that $\mathbf{x}^\mathbf{b}$ is not the least common multiple of some subset of the minimal monomial generators of I. Explain why $K^\mathbf{b}(I)$ is the cone over some subcomplex. Conclude that all nonzero Betti numbers of I occur in $\mathbb{N}^n$-graded degrees $\mathbf{b}$ for which $\mathbf{x}^\mathbf{b}$ equals a least common multiple of some minimal generators.

1.3 Fix a simplicial complex Δ. Exhibit a monomial ideal I and a degree $\mathbf{b}$ in $\mathbb{N}^n$ such that $\Delta = K^\mathbf{b}(I)$ is a Koszul simplicial complex. Is your ideal I squarefree?

1.4 Fix a set of monomials in $x_1, \ldots, x_n$, and let $I(\Bbbk)$ be the ideal they generate in $S = \Bbbk[x_1, \ldots, x_n]$, for varying fields $\Bbbk$.

(a) Can the $\mathbb{N}^n$-graded Hilbert series of $I(\Bbbk)$ depend on the characteristic of $\Bbbk$?
(b) Is the same true for Betti numbers instead of Hilbert series?
(c) Show that the Betti numbers of $S/I(\Bbbk)$ in homological degrees 0, 1, 2, and n are independent of $\Bbbk$.
(d) Prove that all Betti numbers of $S/I(\Bbbk)$ in homological degrees 0, 1, and n lie in distinct $\mathbb{N}^n$-graded degrees. Why is 2 not on this list? Give an example.

1.5 Let $\Bbbk = \mathbb{C}$ be the field of complex numbers. For each monomial $\mathbf{x}^\mathbf{a} \in \mathbb{C}[\mathbf{x}]$, the exponent vector $\mathbf{a}$ can be considered as a vector in $\mathbb{C}^n$. Show that $\mathbf{a}$ lies in the zero set of a Stanley–Reisner ideal I_Δ if and only if $\mathbf{x}^\mathbf{a}$ is nonzero in $\mathbb{C}[\mathbf{x}]/I_\Delta$.

1.6 For a monomial ideal $I = \langle m_1, \ldots, m_r \rangle$ and integers $t \geq 1$, the **Frobenius powers** of I are the ideals $I^{[t]} = \langle m_1^t, \ldots, m_r^t \rangle$. Given a simplicial complex Δ, write an expression for the K-polynomial of $S/I_\Delta^{[2]}$. What about $S/I_\Delta^{[3]}$? $S/I_\Delta^{[t]}$?

1.7 Is there a way to construct monomial matrices for a (minimal) free resolution of $I^{[t]}$ starting with monomial matrices for a (minimal) free resolution of I?

1.8 Let Δ be as in Examples 1.5 and 1.8. Use the links in Example 1.39 to compute as many nonzero Betti numbers of I_Δ as possible.

1.9 Which links in the simplicial complex Δ from Example 1.5 have nonzero homology? Verify your answer using Hochster's formula by comparing it to the Betti numbers of S/I_Γ that appear in Examples 1.25 and 1.30.

1.10 Suppose that ϕ is a nonminimal $\mathbb{N}^n$-graded homomorphism of free modules. Show that ϕ can be represented by a block diagonal monomial matrix Λ in which one of its blocks is a nonzero 1×1 matrix with equal row and column labels.

1.11 Using the fact that every $\mathbb{N}^n$-graded module M has a finite $\mathbb{N}^n$-graded free resolution, deduce from Exercise 1.10 that every $\mathbb{N}^n$-graded free resolution of M is the direct sum of a minimal free resolution of M and a free resolution of zero.

1.12 This exercise provides a direct proof that $\operatorname{Tor}_i^S(M, N) \cong \operatorname{Tor}_i^S(N, M)$. Let $\mathcal{F}$ and $\mathcal{G}$ be free resolutions of M and N, respectively, with differentials ϕ and ψ. Denote by $\mathcal{F} \otimes \mathcal{G}$ the free module $\bigoplus_{i,j} F_i \otimes G_j$, and think of the summands as lying in a rectangular array, with $F_i \otimes G_j$ in row i and column j.

(a) Explain why the *horizontal differential* $(-1)^i \otimes \psi$ on row i of $\mathcal{F} \otimes \mathcal{G}$, induced by ψ on $\mathcal{G}$ and multiplication by ± 1 on F_i, makes $F_i \otimes \mathcal{G}$ into a free resolution of $F_i \otimes N$. (The sign $(-1)^i$ is innocuous, but is needed for ∂, defined next.)

(b) Define a *total differential* ∂ on $\mathcal{F} \otimes \mathcal{G}$ by requiring that

$$\partial(f \otimes g) = \phi_i(f) \otimes g + (-1)^i f \otimes \psi_j(g)$$

for $f \in F_i$ and $g \in G_j$. Show that $\partial^2 = 0$, so we get a *total complex* $\operatorname{tot}(\mathcal{F} \otimes \mathcal{G})$ by setting $\operatorname{tot}(\mathcal{F} \otimes \mathcal{G})_k = \bigoplus_{i+j=k} F_i \otimes G_j$ in homological degree k.

(c) Prove that the map $\mathcal{F} \otimes \mathcal{G} \to \mathcal{F} \otimes N$ that kills $F_i \otimes G_j$ for $j > 0$ and maps $F_i \otimes G_0 \twoheadrightarrow F_i \otimes N$ induces a morphism $\operatorname{tot}(\mathcal{F} \otimes \mathcal{G}) \to \mathcal{F} \otimes N$ of *complexes*, where the ith differential on $\mathcal{F} \otimes N$ is the map $\phi_i \otimes 1$ induced by ϕ.

(d) Using the exactness of the horizontal differential, verify that the morphism $\operatorname{tot}(\mathcal{F} \otimes \mathcal{G}) \to \mathcal{F} \otimes N$ induces an isomorphism on homology. (The arguments for injectivity and surjectivity are each a diagram chase.)

(e) Deduce that the ith homology of $\operatorname{tot}(\mathcal{F} \otimes \mathcal{G})$ is isomorphic to $\operatorname{Tor}_i^S(M, N)$.

(f) Transpose the above argument, leaving the definition of $\operatorname{tot}(\mathcal{F} \otimes \mathcal{G})$ unchanged but replacing $(-1)^i \otimes \psi$ with the *vertical differential* $\phi \otimes 1$ on the jth column of $\mathcal{F} \otimes \mathcal{G}$, to deduce that $\operatorname{tot}(\mathcal{F} \otimes \mathcal{G})$ has j^{th} homology $\operatorname{Tor}_j^S(N, M)$.

(g) Conclude that $\operatorname{Tor}_i^S(M, N) \cong H_i(\operatorname{tot}(\mathcal{F} \otimes \mathcal{G})) \cong \operatorname{Tor}_i^S(N, M)$.

1.13 Let $m \leq n$ be positive integers, and $S = \Bbbk[x_1, \ldots, x_{m+n}]$. Setting $M = S/\langle x_{m+1}, \ldots, x_{m+n} \rangle$ and $N = S/\langle x_{n+1}, \ldots, x_{m+n} \rangle$, find the Hilbert series of the isomorphic modules $\operatorname{Tor}_i^S(M, N)$ and $\operatorname{Tor}_i^S(N, M)$. Which is easier to calculate? Write a succinct expression for the result of setting $x_i = q^i$ for all i in this series.

Notes

Stanley–Reisner rings and Stanley–Reisner ideals are sometimes called *face rings* and *face ideals*. Their importance in combinatorial commutative algebra cannot be overstated. Stanley's green book [Sta96] contains a wealth of information about them, including a number of important applications, such as Stanley's proof of the Upper Bound Theorem for face numbers of convex polytopes. We also recommend Chapter 5 of the book of Bruns and Herzog [BH98] and Hibi's book [Hib92] for more background on squarefree monomial ideals. The first two of these references contain versions of Hochster's formula, whose original form appeared in [Hoc77]; the form taken by Theorem 1.34 is that of [BCP99].

We have only presented the barest prerequisites in simplicial topology. The reader wishing a full introduction should consult [Hat02], [Mun84], or [Rot88].

Monomial matrices were introduced in [Mil00a] for the purpose of working efficiently with resolutions and Alexander duality. Monomial matrices will be convenient for the purpose of cellular resolutions in Chapters 4, 5, and 6. Other applications and generalizations will appear in the context of injective resolutions (Section 11.3) and local cohomology (Chapter 13).

The reader is encouraged to do explicit computations with the objects in this chapter, and indeed, in all of the chapters to come. Those who desire to compute numerous or complicated examples should employ a computer algebra system such as CoCoA, Macaulay2, or Singular [CoC, GS04, GPS01].

We included Exercise 1.12 because there seems to be no accessible proof of the symmetry of Tor in the literature. The proof outlined here shows that the natural map from the total complex of any bicomplex to its horizontal homology complex is an isomorphism on homology when the rows are resolutions (so their homology lies only in homological degree zero). This statement forms the crux of a great number of arguments producing isomorphisms arising in local cohomology and other parts of homological algebra. The argument given in Exercise 1.12 is the essence behind the spectral sequence method of deriving the same result. Those who desire to brush up on their abstract homological algebra should employ a textbook such as Mac Lane's classic [MacL95] or Weibel's book [Wei94].

Chapter 2

Borel-fixed monomial ideals

Squarefree monomial ideals occur mostly in combinatorial contexts. The ideals to be studied in this chapter, namely the Borel-fixed monomial ideals, have, in contrast, a more direct connection to algebraic geometry, where they arise as fixed points of a natural algebraic group action on the Hilbert scheme. The fact that we will not treat these schemes until Chapter 18 should not cause any worry—one need not know what the Hilbert scheme is to understand both the group action and its fixed points. After an introductory section concerning group actions on ideals, there are three main themes in this chapter: the construction of generic initial ideals, the minimal resolution of Borel-fixed ideals due to Eliahou–Kervaire, and the Bigatti–Hulett Theorem on extremal behavior of lexicographic segment ideals.

2.1 Group actions

Throughout this chapter, the ground field $\Bbbk$ is assumed to have characteristic 0, and all ideals of the polynomial ring $S = \Bbbk[x_1, \ldots, x_n]$ that we consider are homogeneous with respect to the *standard $\mathbb{Z}$-grading* (an $\mathbb{N}$-grading) given by $\deg(x_i) = 1$ for $i = 1, \ldots, n$. Consider the following inclusion of matrix groups:

$$
\begin{aligned}
GL_n(\Bbbk) &= \{\text{invertible } n \times n \text{ matrices}\} && \text{general linear group} \\
\cup \\
B_n(\Bbbk) &= \{\text{upper triangular matrices}\} && \text{Borel group} \\
\cup \\
T_n(\Bbbk) &= \{\text{diagonal matrices}\} && \text{algebraic torus group}
\end{aligned}
$$

The general linear group (and hence its subgroups) acts on the polynomial ring as follows. For an invertible matrix $g = (g_{ij}) \in GL_n(\Bbbk)$ and a polynomial $f = p(x_1, \ldots, x_n) \in S$, let g act on f by

$$g \cdot p = p(gx_1, \ldots, gx_n), \quad \text{where} \quad gx_i = \sum_{j=1}^{n} g_{ij} x_j.$$

Given an ideal $I \subset S$, we get a new ideal by applying g to every element of I:

$$g \cdot I = \{ g \cdot p \mid p \in I \}.$$

If I is an ideal with special combinatorial structure and the matrix g is fairly general, then passing from I to $g \cdot I$ will usually lead to a considerable increase in complexity. For a simple example, take $n = 4$ and let I be the principal ideal generated by the quadric $x_1 x_2 - x_3 x_4$. Then $g \cdot I$ is the principal ideal generated by

$$
\begin{aligned}
(g_{11}g_{21} &- g_{31}g_{41})x_1^2 + (g_{12}g_{22} - g_{32}g_{42})x_2^2 \\
&+ (g_{13}g_{23} - g_{33}g_{43})x_3^2 + (g_{14}g_{24} - g_{34}g_{44})x_4^2 \\
&+ (g_{11}g_{22} - g_{32}g_{41} + g_{12}g_{21} - g_{31}g_{42})x_1 x_2 \\
&+ (g_{13}g_{21} + g_{11}g_{23} - g_{33}g_{41} - g_{31}g_{43})x_1 x_3 \\
&+ (g_{14}g_{21} - g_{31}g_{44} - g_{34}g_{41} + g_{11}g_{24})x_1 x_4 \\
&+ (g_{12}g_{23} - g_{33}g_{42} + g_{13}g_{22} - g_{32}g_{43})x_2 x_3 \\
&+ (g_{14}g_{22} - g_{34}g_{42} - g_{32}g_{44} + g_{12}g_{24})x_2 x_4 \\
&+ (g_{13}g_{24} + g_{14}g_{23} - g_{34}g_{43} - g_{33}g_{44})x_3 x_4.
\end{aligned}
$$

We are interested in ideals I that are fixed under the actions of the three kinds of matrix groups. Let us start with the smallest of these three.

Proposition 2.1 *A nonzero ideal I inside S is fixed under the action of the torus $T_n(\Bbbk)$ if and only if I is a monomial ideal.*

Proof. Torus elements map each variable—and hence each monomial—to a multiple of itself, so monomial ideals are fixed by $T_n(\Bbbk)$. Conversely, let I be an arbitrary torus-fixed ideal, and suppose that $p = \sum c_{\mathbf{a}} \mathbf{x}^{\mathbf{a}}$ is a polynomial in I. Then $\mathbf{t} \cdot p = \sum c_{\mathbf{a}} \mathbf{t}^{\mathbf{a}} \mathbf{x}^{\mathbf{a}}$ is also in I, for every diagonal matrix $\mathbf{t} = \mathrm{diag}(t_1, \ldots, t_n)$. Let $\mathcal{T} = \{ \mathbf{t}^{(1)}, \ldots, \mathbf{t}^{(s)} \} \subset T_n(\Bbbk)$ be a generic set of diagonal matrices $\mathbf{t}^{(k)} = \mathrm{diag}(t_1^{(k)}, \ldots, t_n^{(k)})$, where the cardinality s equals the number of monomials with nonzero coefficient in p. For each monomial $\mathbf{x}^{\mathbf{a}}$ appearing in p and each diagonal matrix $\mathbf{t} \in \mathcal{T}$, there is a corresponding monomial $\mathbf{t}^{\mathbf{a}}$. Form the $s \times s$ matrix $(\mathbf{t}^{\mathbf{a}})$ whose columns are indexed by the monomials appearing in p and whose rows are indexed by $\mathcal{T}$. As a polynomial in the $n \cdot s$ symbols $\{ t_1^{(k)}, \ldots, t_n^{(k)} \mid k = 1, \ldots, s \}$, the determinant of $(\mathbf{t}^{\mathbf{a}})$ is nonzero, because all terms in the expansion are distinct. Hence $\det(\mathbf{t}^{\mathbf{a}}) \neq 0$, because $\mathcal{T}$ is generic. Multiplying the inverse of $(\mathbf{t}^{\mathbf{a}})$ with the column vector whose entries are the polynomials $\mathbf{t} \cdot p$ for $\mathbf{t} \in \mathcal{T}$ yields the column vector whose entries are precisely the terms $c_{\mathbf{a}} \mathbf{x}^{\mathbf{a}}$ appearing in p. We have therefore produced each term $c_{\mathbf{a}} \mathbf{x}^{\mathbf{a}}$ in p as a linear combination of polynomials $\mathbf{t} \cdot p \in I$. It follows that I is a monomial ideal. $\square$

Corollary 2.2 *A nonzero ideal I in S is fixed under the action of the general linear group $GL_n(\Bbbk)$ if and only if I is a power $\mathfrak{m}^d$ of the irrelevant maximal ideal $\mathfrak{m} = \langle x_1, \ldots, x_n \rangle$, for some positive integer d.*

Proof. The vector space of homogeneous polynomials of degree d is fixed by $GL_n(\Bbbk)$, and hence so is the ideal $\mathfrak{m}^d$ it generates. Conversely, suppose I is a $GL_n(\Bbbk)$-fixed ideal and that p is a nonzero polynomial in I of minimal degree, say d. For a general matrix g, the polynomial $g \cdot p$ contains all monomials of degree d in S. Since $g \cdot p$ is in I, and since I is a monomial ideal by Proposition 2.1, every monomial of degree d lies in I. But I contains no nonzero polynomial of degree strictly less than d, so $I = \mathfrak{m}^d$. $\square$

The characterization of monomial ideals in Proposition 2.1 is one of our motivations for having included a chapter on *toric varieties* later in this book: toric varieties are closures of T_n orbits. In representation theory and in the study of determinantal ideals in Part III, one is also often interested in actions of the Borel group B_n. Since B_n contains the torus T_n, and T_n-fixed ideals are monomial, every Borel-fixed ideal is necessarily a monomial ideal. Borel-fixed ideals enjoy the extra property that larger-indexed variables can be swapped for smaller ones without leaving the ideal.

Proposition 2.3 *The following are equivalent for a monomial ideal I.*

(i) I is Borel-fixed.

(ii) If $m \in I$ is any monomial divisible by x_j, then $m\frac{x_i}{x_j} \in I$ for $i < j$.

Proof. Suppose that I is a Borel-fixed ideal. Let $m \in I$ be any monomial divisible by x_j and consider any index $i < j$. Let g be the elementary matrix in $B_n(\Bbbk)$ that sends x_j to $x_j + x_i$ and that fixes all other variables. The polynomial $g \cdot m$ lies in $I = g \cdot I$, and the monomial mx_i/x_j appears in the expansion of $g \cdot m$. Since I is a monomial ideal, this implies that the monomial mx_i/x_j lies in I. We have proved the implication (i) $\Rightarrow$ (ii).

Suppose that condition (ii) holds for a monomial ideal I. Let m be any monomial in I and $g \in B_n(\Bbbk)$ any upper triangular matrix. Every monomial appearing in $g \cdot m$ can be obtained from the monomial m by a sequence of transformations as in (ii). All of these monomials lie in I. Hence $g \cdot m$ lies in I. Therefore condition (i) holds for I. $\square$

In checking whether a given ideal I is Borel-fixed, it suffices to verify condition (ii) for minimal generators m of the ideal I. Hence condition (ii) constitutes an explicit finite algorithm for checking whether I is Borel-fixed.

Example 2.4 Here is a typical Borel-fixed ideal in three variables:

$$I = \langle x_1^2, x_1 x_2, x_2^3, x_1 x_3^3 \rangle.$$

Each of the four generators satisfies condition (ii). The ideal I has the following unique irreducible decomposition (see Chapter 5.2 if these are unfamiliar), which is also a primary decomposition:

$$I = \langle x_1, x_2^3 \rangle \cap \langle x_1^2, x_2, x_3^3 \rangle.$$

The second irreducible component is not Borel-fixed. $\diamond$

The previous example is slightly surprising from the perspective of monomial primary decomposition. Torus-fixed ideals, namely monomial ideals, always admit decompositions as intersections of irreducible torus-fixed ideals; but the same statement does not hold for Borel-fixed ideals.

2.2   Generic initial ideals

This section serves mainly as motivation for studying Borel-fixed ideals, although it is also a convenient place to recall some fundamentals of Gröbner bases, which will be used sporadically throughout the book. The crucial point about Borel-fixed ideals is Theorem 2.9, which says that they arise naturally as initial ideals after generic changes of coordinates. Although this result and the existence of generic initial ideals are stated precisely, we refer the reader elsewhere for large parts of the proof. For a more detailed introduction to Gröbner bases, see [CLO97] or [Eis95, Chapter 15].

To find Gröbner bases, one must first fix a *term order* $<$ on the polynomial ring $S = \Bbbk[x_1, \ldots, x_n]$. By definition, $<$ is a total order on the monomials of S that is *multiplicative*, meaning that $\mathbf{x^b} < \mathbf{x^c}$ if and only if $\mathbf{x^{a+b}} < \mathbf{x^{a+c}}$, and *artinian*, meaning that $1 < \mathbf{x^a}$ for all nonunit monomials $\mathbf{x^a} \in S$. Unless stated otherwise, we assume that our chosen term order satisfies $x_1 > x_2 > \cdots > x_n$.

Given a polynomial $f = \sum_{\mathbf{a} \in \mathbb{N}^n} c_{\mathbf{a}} \mathbf{x^a}$, the monomial $\mathbf{x^a}$ that is largest under the term order $<$ among those whose coefficients are nonzero in p determines the *initial term* $\mathrm{in}_<(f) = c_{\mathbf{a}} \mathbf{x^a}$. When the term order has been fixed for the discussion, we sometimes write simply $\mathrm{in}(f)$. If I is an ideal in S, then the *initial ideal* of I,

$$\mathrm{in}(I) \;\; = \;\; \langle \mathrm{in}(f) \mid f \in I \rangle,$$

is generated by the set of initial terms of all polynomials in I.

Definition 2.5 Suppose that $I = \langle f_1, \ldots, f_r \rangle$. The set $\{f_1, \ldots, f_r\}$ of generators constitutes a **Gröbner basis** if the initial terms of $f_1, \ldots, f_r$ generate the initial ideal of I; that is, if $\mathrm{in}(I) = \langle \mathrm{in}(f_1), \ldots, \mathrm{in}(f_r) \rangle$.

Every ideal in S has a (finite) Gröbner basis for every term order, because $\mathrm{in}(I)$ is finitely generated by Hilbert's basis theorem. Note that there is no need to mention any ideals when we say, "The set $\{f_1, \ldots, f_r\}$

is a Gröbner basis," as the set must be a Gröbner basis for the ideal $I = \langle f_1, \ldots, f_r \rangle$ it generates. On the other hand, most ideals have many different Gröbner bases for a fixed term order. This uniqueness issue can be resolved by considering a *reduced* Gröbner basis $\{f_1, \ldots, f_r\}$, which means that $\text{in}(f_i)$ has coefficient 1 for each $i = 1, \ldots, r$, and that the only monomial appearing anywhere in $\{f_1, \ldots, f_r\}$ that is divisible by the initial term $\text{in}(f_i)$ is $\text{in}(f_i)$ itself; see Exercise 2.5.

In the proof of the next lemma, we will use a general tool due to Weispfenning [Wei92] for establishing finiteness results in Gröbner basis theory. Suppose that $\mathbf{y}$ is a set of variables different from $x_1, \ldots, x_n$, and let J be an ideal in $S[\mathbf{y}]$, which is the polynomial ring over $\Bbbk$ in the variables $\mathbf{x}$ and $\mathbf{y}$. Every $\Bbbk$-algebra homomorphism $\phi : \Bbbk[\mathbf{y}] \to \Bbbk$ determines a homomorphism $\phi_S : S[\mathbf{y}] \to S$ that sends the $\mathbf{y}$ variables to constants. The image $\phi_S(J)$ is an ideal in S. Given a fixed term order $<$ on S (not on $S[\mathbf{y}]$), Weispfenning proves that J has a *comprehensive Gröbner basis*, meaning a finite set $\mathcal{C}$ of polynomials $p(\mathbf{x}, \mathbf{y}) \in J$ such that for every homomorphism $\phi : \Bbbk[\mathbf{y}] \to \Bbbk$, the specialized set $\phi_S(\mathcal{C})$ is a Gröbner basis for the specialized ideal $\phi_S(J)$ in S with respect to the term order $<$.

Returning to group actions on S, every matrix $g \in GL_n(\Bbbk)$ determines the initial monomial ideal $\text{in}(g \cdot I)$. After fixing a term order, we call two matrices g and g' *equivalent* if

$$\text{in}(g \cdot I) = \text{in}(g' \cdot I).$$

The resulting partition of the group $GL_n(\Bbbk)$ into equivalence classes is a geometrically well-behaved stratification, as we shall now see.

To explain the geometry, we need a little terminology. Let $\mathbf{g} = (g_{ij})$ be an $n \times n$ matrix of indeterminates, so that the algebra $\Bbbk[\mathbf{g}]$ consists of (some of the) polynomial functions on $GL_n(\Bbbk)$. The term *Zariski closed set* inside of $GL_n(\Bbbk)$ or $\Bbbk^n$ refers to the zero set of an ideal in $\Bbbk[\mathbf{g}]$ or S. If V is a Zariski closed set, then a *Zariski open subset* of V refers to the complement of a Zariski closed subset of V.

Lemma 2.6 *For a fixed ideal I and term order $<$, the number of equivalence classes in $GL_n(\Bbbk)$ is finite. One of these classes is a nonempty Zariski open subset U inside of $GL_n(\Bbbk)$.*

Proof. Consider the polynomial ring $S[g_{11}, \ldots, g_{nn}] = \Bbbk[\mathbf{g}, \mathbf{x}]$ in $n^2 + n$ unknowns. Suppose that $p_1(\mathbf{x}), \ldots, p_r(\mathbf{x})$ are generators of the given ideal I in S. Let J be the ideal generated by the elements $\mathbf{g} \cdot p_1(\mathbf{x}), \ldots, \mathbf{g} \cdot p_r(\mathbf{x})$ in $\Bbbk[\mathbf{g}, \mathbf{x}]$, and fix a comprehensive Gröbner basis $\mathcal{C}$ for J.

The equivalence classes in $GL_n(\Bbbk)$ can be read off from the coefficients of the polynomials in $\mathcal{C}$. These coefficients are polynomials in $\Bbbk[\mathbf{g}]$. By requiring that $\det(\mathbf{g}) \neq 0$ and by imposing the conditions "$= 0$" and "$\neq 0$" on these coefficient polynomials in all possible ways, we can read off all possible initial ideals $\text{in}(g \cdot I)$. Since $\mathcal{C}$ is finite, there are only finitely many possibilities, and hence the number of distinct ideals in$(g \cdot I)$ as g runs over $GL_n(\Bbbk)$ is finite. The unique Zariski open equivalence class U can be specified by imposing the condition "$\neq 0$" on all the leading coefficients of the polynomials in the comprehensive Gröbner basis $\mathcal{C}$. $\square$

The previous lemma tells us that the next definition makes sense.

Definition 2.7 Fix a term order $<$ on S. The initial ideal in$_<(g \cdot I)$ that, as a function of g, is constant on a Zariski open subset U of GL_n is called the **generic initial ideal** of I for the term order $<$. It is denoted by

$$\text{gin}_<(I) \;=\; \text{in}_<(g \cdot I).$$

Example 2.8 Let $n = 2$ and consider the ideal $I = \langle x_1^2, x_2^2 \rangle$, where $<$ is the lexicographic order with $x_1 > x_2$. For this term order, the ideal J defined in the proof of Lemma 2.6 has the comprehensive Gröbner basis

$$\mathcal{C} \;=\; \{g_{11}^2 x_1^2 + 2g_{11}g_{12}x_1 x_2 + g_{12}^2 x_2^2,\; g_{21}^2 x_1^2 + 2g_{21}g_{22}x_1 x_2 + g_{22}^2 x_2^2,$$
$$2g_{21}g_{11}(g_{22}g_{11} - g_{21}g_{12})x_1 x_2 \;+\; (g_{22}g_{11} - g_{21}g_{12})(g_{21}g_{12} + g_{22}g_{11})x_2^2,$$
$$(g_{22}g_{11} - g_{21}g_{12})^3 x_2^3\}.$$

The group $GL_2(\Bbbk)$ decomposes into only two equivalence classes in this case:

- in$_<(g \cdot I) = \langle x_1^2, x_2^2 \rangle$ if $g_{11}g_{21} = 0$
- in$_<(g \cdot I) = \langle x_1^2, x_1 x_2, x_2^3 \rangle$ if $g_{11}g_{21} \neq 0$

The second ideal is the generic initial ideal: $\text{gin}(I) = \langle x_1^2, x_1 x_2, x_2^3 \rangle$. $\diamond$

The punch line is the result of Galligo, Bayer, and Stillman describing a general procedure to turn arbitrary ideals into Borel-fixed ideals.

Theorem 2.9 *The generic initial ideal* gin$_<(I)$ *is Borel-fixed.*

Proof. We refer to Eisenbud's commutative algebra textbook, where this result appears as [Eis95, Theorem 15.20]. A complete proof is given there. $\square$

It is important to note that the generic initial ideal gin$_<(I)$ depends heavily on the choice of the term order $<$. Two extreme examples of term orders are the *purely lexicographic term order*, denoted $<_{\text{lex}}$, and the *reverse lexicographic term order*, denoted $<_{\text{revlex}}$. For two monomials $\mathbf{x}^{\mathbf{a}}$ and $\mathbf{x}^{\mathbf{b}}$ of the same degree, we have $\mathbf{x}^{\mathbf{a}} >_{\text{lex}} \mathbf{x}^{\mathbf{b}}$ if the leftmost nonzero entry of the vector $\mathbf{a} - \mathbf{b}$ is positive, whereas $\mathbf{x}^{\mathbf{a}} >_{\text{revlex}} \mathbf{x}^{\mathbf{b}}$ if the rightmost nonzero entry of the vector $\mathbf{a} - \mathbf{b}$ is negative.

Example 2.10 Let $f, g \in \Bbbk[x_1, x_2, x_3, x_4]$ be generic forms of degrees d and e, respectively. Considering the three smallest nontrivial cases, we list the generic initial ideal of $I = \langle f, g \rangle$ for both the lexicographic order and the reverse lexicographic order. The ideals $J = \mathrm{gin}_{\mathrm{lex}}(I)$ are:

$(d, e) = (2, 2)$ $J = \langle x_2^4, x_1 x_3^2, x_1 x_2, x_1^2 \rangle$
$\qquad\qquad\quad = \langle x_1, x_2^4 \rangle \cap \langle x_1^2, x_2, x_3^2 \rangle,$

$(d, e) = (2, 3)$ $J = \langle x_2^6, x_1 x_3^6, x_1 x_2 x_4^4, x_1 x_2 x_3 x_4^2, x_1 x_2 x_3^2, x_1 x_2^2, x_1^2 \rangle$
$\qquad\qquad\quad = \langle x_1, x_2^6 \rangle \cap \langle x_1^2, x_2, x_3^6 \rangle \cap \langle x_1^2, x_2^2, x_3, x_4^4 \rangle \cap \langle x_1^2, x_2^2, x_3^2, x_4^2 \rangle,$

$(d, e) = (3, 3)$ $J = \langle x_2^9, x_1 x_3^{18}, x_1 x_2 x_4^{16}, x_1 x_2 x_3 x_4^{14}, \ldots, x_1^3 \rangle$ (26 generators).

On the other hand, the ideals $J = \mathrm{gin}_{\mathrm{revlex}}(I)$ are:

$(d, e) = (2, 2)$ $\qquad\qquad J = \langle x_2^3, x_1 x_2, x_1^2 \rangle$
$\qquad\qquad\qquad\qquad\quad = \langle x_1, x_2^3 \rangle \cap \langle x_1^2, x_2 \rangle,$

$(d, e) = (2, 3)$ $\qquad\qquad J = \langle x_2^4, x_1 x_2^2, x_1^2 \rangle$
$\qquad\qquad\qquad\qquad\quad = \langle x_1, x_2^4 \rangle \cap \langle x_1^2, x_2^2 \rangle,$

$(d, e) = (3, 3)$ $\qquad\qquad J = \langle x_2^5, x_1 x_2^3, x_1^2 x_2, x_1^3 \rangle$
$\qquad\qquad\qquad\qquad\quad = \langle x_1, x_2^5 \rangle \cap \langle x_1^2, x_2^3 \rangle \cap \langle x_1^3, x_2 \rangle,$

The reverse lex **gin** is much nicer than the lex **gin**, mostly because there are fewer generators, but also because they have lower degrees. All six ideals J above are Borel-fixed. $\qquad\qquad\qquad\qquad\qquad\qquad\qquad\qquad \diamond$

Let us conclude this section with one more generality on Gröbner bases: they work for submodules of free S-modules. Suppose that $\mathcal{F} = S^\beta$ is a free module of rank β, with basis $\mathbf{e}_1, \ldots, \mathbf{e}_\beta$. There is a general definition of term order for $\mathcal{F}$, which is a total order on elements of the form $m\mathbf{e}_i$, for monomials $m \in S$, satisfying appropriate analogues of the multiplicative and artinian properties of term orders for S. Initial modules are defined just as they were for ideals (which constitute the case $\beta = 1$). For our purposes, we need only consider term orders on $\mathcal{F}$ obtained from a term order on S by ordering the basis vectors $\mathbf{e}_1 > \cdots > \mathbf{e}_\beta$. To get such a term order, we have to pick which takes precedence, the term order on S or the ordering on the basis vectors. In the former case, we get the TOP order, which stands for *term-over-position*; in the latter case, we get the POT order, for *position-over-term*. In the POT order, for example, $m\mathbf{e}_i > m'\mathbf{e}_j$ if either $i < j$, or else $i = j$ and $m > m'$. If $M \subseteq \mathcal{F}$ is a submodule, then $\{f_1, \ldots, f_r\} \subset M$ is a *Gröbner basis* if $\mathrm{in}(f_1), \ldots, \mathrm{in}(f_r)$ generate $\mathrm{in}(M)$. The notion of reduced Gröbner basis for modules requires only that if $\mathrm{in}(f_k) = m\mathbf{e}_i$, then m does not divide m' for any other term $m'\mathbf{e}_i$ with the same $\mathbf{e}_i$ appearing in any f_j.

2.3 The Eliahou–Kervaire resolution

Next we describe the minimal free resolution, Betti numbers and Hilbert series of a Borel-fixed ideal I. The same construction works also for the larger class of so-called "stable ideals", but we restrict ourselves to the Borel-fixed case here. Throughout this section, the monomials $m_1, \ldots, m_r$ *minimally* generate the Borel-fixed ideal I, and for every monomial m, we write $\max(m)$ for the largest index of a variable dividing m. For instance, $\max(x_1^7 x_2^3 x_4^5) = 4$ and $\max(x_2 x_3^7) = 3$. Similarly, let $\min(m)$ denote the smallest index of a variable dividing m.

Lemma 2.11 *Each monomial m in the Borel-fixed ideal $I = \langle m_1, \ldots, m_r \rangle$ can be written uniquely as a product $m = m_i m'$ with $\max(m_i) \leq \min(m')$.*

In what follows, we abbreviate $u_i = \max(m_i)$ for $i = 1, \ldots, r$.

Proof. Uniqueness: Suppose $m = m_i m_i' = m_j m_j'$ both satisfy the condition, with $u_i \leq u_j$. Then m_i and m_j agree in every variable with index $< u_i$. If x_{u_i} divides m_j', then $u_i = u_j$ by the assumed condition, whence one of m_i and m_j divides the other, so $i = j$. Otherwise, x_{u_i} does not divide m_j'. In this case the degree of x_{u_i} in m_i is at most the degree of x_{u_i} in m_j, which equals the degree of x_{u_i} in m, so that again m_i divides m_j and $i = j$.

Existence: Suppose that $m = m_j m'$ for some j, but that $u_j > u := \min(m')$. Proposition 2.3 says that we can replace m_j by any minimal generator m_i dividing $m_j x_u / x_{u_j}$. By construction, $u_i \leq u_j$, so either $u_i < u_j$, or $u_i = u_j$ and the degree of x_{u_i} in m_i is $<$ the degree of x_{u_i} in m_j. This shows that we cannot keep going on making such replacements forever. $\square$

Recall that a quotient of S by a monomial ideal I has a *K-polynomial* if the $\mathbb{N}^n$-graded Hilbert series of S/I agrees with a rational function having denominator $(1 - x_1) \cdots (1 - x_n)$, in which case $\mathcal{K}(S/I; \mathbf{x})$ is the numerator.

Proposition 2.12 *For the Borel-fixed ideal $I = \langle m_1, \ldots, m_r \rangle$, the quotient S/I has K-polynomial*

$$\mathcal{K}(S/I; \mathbf{x}) \;=\; 1 - \sum_{i=1}^{r} m_i \prod_{j=1}^{u_i - 1} (1 - x_j).$$

Proof. By Lemma 2.11, the set of monomials in I is the disjoint union over $i = 1, \ldots, r$ of the monomials in $m_i \cdot \Bbbk[x_{u_i}, \ldots, x_n]$. The sum of all monomials in such a translated subalgebra of S equals the series

$$\frac{m_i}{\prod_{l=1}^{n}(1 - x_l)} \prod_{j=1}^{u_i - 1} (1 - x_j)$$

by Example 1.11. Summing this expression from $i = 1$ to r yields the Hilbert series of I, and subtracting this from the Hilbert series of S yields the Hilbert series of S/I. Clear denominators to get the K-polynomial. $\square$

Example 2.13 Let I be the ideal in Example 2.4. Its K-polynomial is

$$\begin{aligned}
\mathcal{K}(S/I; \mathbf{x}) &= 1 - x_1^2 - x_1 x_2 (1 - x_1) - x_2^3 (1 - x_1) - x_1 x_3^3 (1 - x_1)(1 - x_2) \\
&= 1 - x_1^2 - x_1 x_2 - x_2^3 - x_1 x_3^3 \\
&\quad + x_1^2 x_3^3 + x_1 x_2 x_3^3 + x_1 x_2^3 + x_1^2 x_2 \\
&\quad - x_1^2 x_2 x_3^3.
\end{aligned}$$

This expansion suggests that the minimal resolution of S/I has the form

$$0 \leftarrow S \longleftarrow S^4 \longleftarrow S^4 \longleftarrow S \leftarrow 0,$$

and this is indeed the case, by the formula in Theorem 2.18. ◇

The simplicial complexes that arise in connection with Borel-fixed ideals have rather simple geometry. Since we will need this geometry in the proof of Theorem 2.18, via Lemma 2.15, let us make a formal definition.

Definition 2.14 A simplicial complex Δ on the vertices $1, \ldots, k$ is **shifted** if $(\tau \smallsetminus \alpha) \cup \beta$ is a face of Δ whenever τ is a face of Δ and $1 \leq \alpha < \beta \leq k$.

The distinction between faces and facets will be crucial in what follows.

Lemma 2.15 *Fix a shifted simplicial complex Γ on $1, \ldots, k$, and let $\Delta \subseteq \Gamma$ consist of the faces of Γ not having k as a vertex. Then $\dim_{\Bbbk} \widetilde{H}_i(\Gamma; \Bbbk)$ equals the number of dimension i facets τ of Δ such that $\tau \cup k$ is not a face of Γ.*

Proof. Γ is a subcomplex of the cone $k * \Delta$ from the vertex k over Δ. By Definition 2.14, if $\tau \in \Delta$ is a face, then Γ contains every proper face of the simplex $\tau \cup k$. In other words, Γ is a *near-cone* over Δ, which is by definition obtained from $k * \Delta$ by removing the interior of the simplex $\tau \cup k$ for some of the facets τ of Δ.

The only i-faces of Γ are (i) the i-faces of Δ, (ii) the cones $\sigma \cup k$ over some subset of the $(i-1)$-facets $\sigma \in \Delta$, and (iii) the cones from k over all non-facet $(i-1)$-faces of Δ. If σ is an $(i-1)$-facet of Δ, then $\sigma \cup k \in \Gamma$ cannot have nonzero coefficient $c \in \Bbbk$ in any i-cycle of Γ, because σ would have coefficient $\pm c$ in its boundary.

For each $j \geq 0$, let $\Delta_j \subseteq \Delta$ be the subcomplex that is the union of all (closed) j-faces of Δ. For the purpose of computing $\widetilde{H}_i(\Gamma; \Bbbk)$, we assume using the previous paragraph that Δ has no facets of dimension less than i, by replacing Δ with $\Delta_{\geq i} = \bigcup_{j \geq i} \Delta_j$ and taking only those faces of Γ contained in $k * \Delta_{\geq i}$. Thus every i-face of $k * \Delta$ lies in Γ. Since we are interested in the ith homology of Γ, we also assume that $\dim(\Delta) \leq i + 1$.

There can be $(i+1)$-faces of the cone $k * \Delta$ that do not lie in Γ, but these missing $(i+1)$-faces all have the form $\tau \cup k$ for a facet τ of dimension i in Δ. Now consider the long exact homology sequence arising from the inclusion $\Gamma \to k * \Delta$. It contains the sequence $\widetilde{H}_{i+1}(k * \Delta) \to \widetilde{H}_{i+1}(k * \Delta, \Gamma) \to \widetilde{H}_i(\Gamma) \to \widetilde{H}_i(k * \Delta)$. The outer terms are zero because $k * \Delta$ is a cone.

When Γ is the *minimal* near-cone over Δ, the dimension of the relative homology $\widetilde{H}_{i+1}(k * \Delta, \Gamma)$ is the number of i-facets of Δ, because the only faces of $k*\Delta$ contributing to the relative chain complex are $\tau \cup k$ for i-facets τ of Δ. Hence the isomorphism $\widetilde{H}_{i+1}(k * \Delta, \Gamma) \to \widetilde{H}_i(\Gamma)$ proves the lemma in this case. For general Γ, adding a face $\tau \cup k$ can only cancel at most one i^{th} homology class of Γ, so it must cancel exactly one, because adding all of the faces $\tau \cup k$ for i-facets of Δ yields $k * \Delta$, which has no homology. $\square$

The main theorem of this section refers to an important notion that will resurface again in Chapter 5. For any vector $\mathbf{b} = (b_1, \ldots, b_n) \in \mathbb{N}^n$, let $|\mathbf{b}| = b_1 + \cdots + b_n$.

Definition 2.16 An $\mathbb{N}^n$-graded free resolution $\mathcal{F}_{\bullet}$ is **linear** if there is a choice of monomial matrices for the differentials of $\mathcal{F}_{\bullet}$ such that in each matrix, $|\mathbf{a}_p - \mathbf{a}_q| = 1$ whenever the scalar entry λ_{qp} is nonzero. A module M has **linear free resolution** if its minimal free resolution is linear.

Using the ungraded notation for maps between free S-modules, a $\mathbb{Z}$-graded free resolution is *linear* if the nonzero entries in some choice of matrices for all of its differentials are linear forms. When the resolution is $\mathbb{N}^n$-graded, the linear forms can be taken to be scalar multiples of variables.

Example 2.17 Let M be an $\mathbb{N}^n$-graded module whose generators all lie in degrees $\mathbf{b} \in \mathbb{N}^n$ satisfying $|\mathbf{b}| = d$ for some fixed integer $d \in \mathbb{N}$. Then M has linear resolution if and only if for all $i \geq 0$, the minimal i^{th} syzygies of M lie in degrees $\mathbf{b} \in \mathbb{N}^n$ satisfying $|\mathbf{b}| = d + i$. $\diamond$

Theorem 2.18 *Let M be the module of first syzygies on the Borel-fixed ideal $I = \langle m_1, \ldots, m_r \rangle$. Then M has a Gröbner basis such that its initial module $\mathrm{in}(M)$ has linear free resolution. Moreover, $S^r / \mathrm{in}(M)$ has the same number of minimal i^{th} syzygies as $I \cong S^r / M$, namely $\sum_{j=1}^{r} \binom{\max(m_j)-1}{i}$.*

Proof. The idea of the proof is to compare the minimal free resolution of M to a direct sum of Koszul complexes. We make the following crucial labeling assumption, in which $\deg_u(m)$ is the degree of x_u in each monomial m, and again $u_i = \max(m_i)$ for $i = 1, \ldots, r$:

$$i > j \quad \Rightarrow \quad u_i \leq u_j \quad \text{and} \quad \deg_{u_j}(m_i) \leq \deg_{u_j}(m_j).$$

Let us begin by constructing some special elements in the syzygy module M. Consider any product $m = x_u m_j$ in which $u < u_j$. By Lemma 2.11, this monomial can be rewritten uniquely as

$$m = x_u \cdot m_j = m' \cdot m_i \quad \text{with} \quad u_i \leq \min(m').$$

Since $u < u_j$, we must have $\min(m') \leq u_j$. Moreover, if $\min(m') = u_j$, then $\deg_{u_j}(m_i) < \deg_{u_j}(m_j)$. Therefore $i > j$ with our labeling assumption. This means that the following vector is a nonzero first syzygy on I:

$$x_u \cdot \mathbf{e}_j - m' \cdot \mathbf{e}_i \in M. \tag{2.1}$$

Fix any term order on S^r that picks the underlined term as the leading term for every $j = 1, \ldots, r$ and $u = 1, \ldots, u_j$; the POT order induced by $\mathbf{e}_1 > \mathbf{e}_2 > \cdots > \mathbf{e}_r$ will do, for instance. We claim that the set of syzygies (2.1), as u and j run over all pairs satisfying $u < u_j$, equals the reduced Gröbner basis of M, and in particular, generates M.

If the Gröbner basis property does not hold, then some nonzero syzygy

$$m'' \cdot \mathbf{e}_j - m' \cdot \mathbf{e}_i \in M$$

has the property that neither $m'' \cdot \mathbf{e}_j$ nor $m' \cdot \mathbf{e}_i$ lies in the submodule of S^r generated by the underlined leading terms in (2.1). This means that

$$\min(m') \geq \max(m_i) \quad \text{and} \quad \min(m'') \geq \max(m_j).$$

The identity $m' \cdot m_i = m'' \cdot m_j$ contradicts the uniqueness statement in Lemma 2.11. This contradiction proves that the relations (2.1) constitute a Gröbner basis for the submodule $M \subset S^r$. This Gröbner basis is reduced because no leading term $x_u \mathbf{e}_j$ divides either term of another syzygy (2.1).

We have shown that the initial module $\mathrm{in}(M)$ under the given term order is minimally generated by the monomials $x_u \cdot \mathbf{e}_j$ for which $u < u_j$. Hence this initial module decomposes as the direct sum

$$\mathrm{in}(M) \quad = \quad \bigoplus_{j=1}^{r} \langle x_1, x_2, \ldots, x_{u_j-1} \rangle \cdot \mathbf{e}_j. \tag{2.2}$$

The minimal free resolution of $\mathrm{in}(M)$ is the direct sum of the minimal free resolutions of the r summands in (2.2). The minimal free resolution of the ideal $\langle x_1, x_2, \ldots, x_{u_j-1} \rangle$ is a Koszul complex, which is itself a linear resolution. Moreover, the number of ith syzygies in this Koszul complex equals $\binom{u_j-1}{i}$. We conclude that $\mathrm{in}(M)$ has linear resolution and that its number of minimal i^{th} syzygies equals the desired number, namely $\sum_{j=1}^{r} \binom{u_j-1}{i}$.

We have reduced Theorem 2.18 to the claim that the Betti numbers of M equal those of its initial module $\mathrm{in}(M)$ in every degree $\mathbf{b} \in \mathbb{N}^n$. In fact, we only need to show that $\beta_{i,\mathbf{b}}(M) \geq \beta_{i,\mathbf{b}}(\mathrm{in}(M))$, because it is always the case that $\beta_{i,\mathbf{b}}(M) \leq \beta_{i,\mathbf{b}}(\mathrm{in}(M))$ for all $\mathbf{b} \in \mathbb{N}^n$ (we shall prove this in a general context in Theorem 8.29). Fix $\mathbf{b} = (b_1, \ldots, b_n)$ with $\beta_{i,\mathbf{b}}(\mathrm{in}(M)) \neq 0$, and let k be the largest index with $b_k > 0$.

By (2.2), the Betti number $\beta_{i,\mathbf{b}}(\mathrm{in}(M))$ equals the number of indices $j \in \{1, \ldots, r\}$ such that $\mathbf{x}^{\mathbf{b}}/m_j$ is a squarefree monomial $\mathbf{x}^{\tau} \in S$ for some subset $\tau \subseteq \{1, \ldots, u_j - 1\}$ of size $i + 1$. All of these indices j share the property that $\deg_{x_k}(m_j) = b_k$. Each index j arising here leads to a different $(i+1)$-subset τ of $\{1, \ldots, k-1\}$.

The Betti number $\beta_{i,\mathbf{b}}(M) = \beta_{i+1,\mathbf{b}}(I)$ can be computed, by Theorem 1.34, as the dimension of the ith homology group of the upper Koszul simplicial complex $K^{\mathbf{b}}(I)$ in degree $\mathbf{b}$. Applying Proposition 2.3 to monomials $m = \mathbf{x}^{\mathbf{b}-\tau}$ for squarefree vectors τ, we find that $K^{\mathbf{b}}(I)$ is shifted. Hence we deduce from Lemma 2.15 that $\dim_{\Bbbk} \widetilde{H}_i(K^{\mathbf{b}}(I); \Bbbk)$ equals the number of dimension i facets $\tau \in \Delta$ such that $\tau \cup k$ is not a face of $K^{\mathbf{b}}(I)$. But every size $i+1$ subset τ from the previous paragraph is a facet of Δ, and $\tau \cup k$ is not in $K^{\mathbf{b}}(I)$, both because $\mathbf{x}^{\mathbf{b}-\tau} = m_i$ is a minimal generator of I. Therefore $\beta_{i,\mathbf{b}}(M) \geq \beta_{i,\mathbf{b}}(\mathrm{in}(M))$, and the proof is complete. $\square$

We illustrate Theorem 2.18 and its proof with two nontrivial examples.

Example 2.19 Let $n = 4$ and $r = 7$, and consider the following ideal:

$$\langle x_1 x_2 x_4^4, \quad x_1 x_2 x_3 x_4^2, \quad x_1 x_3^6, \quad x_1 x_2 x_3^2, \quad x_2^6, \quad x_1 x_2^2, \quad x_1^2 \rangle.$$

$$
\begin{array}{lllllll}
x_3\,\mathbf{e}_1 & -x_4^2\,\mathbf{e}_2 & & & & & \\
x_2\,\mathbf{e}_1 & & & & & -x_4^4\,\mathbf{e}_6 & \\
x_1\,\mathbf{e}_1 & & & & & & -x_2 x_4^4\,\mathbf{e}_7 \\
 & x_3\,\mathbf{e}_2 & -x_4^2\,\mathbf{e}_4 & & & & \\
 & x_2\,\mathbf{e}_2 & & & & -x_3 x_4^2\,\mathbf{e}_6 & \\
 & x_1\,\mathbf{e}_2 & & & & & -x_2 x_3 x_4^2\,\mathbf{e}_7 \\
 & & x_2\,\mathbf{e}_3 & -x_3^4\,\mathbf{e}_4 & & & \\
 & & x_1\,\mathbf{e}_3 & & & & -x_3^6\,\mathbf{e}_7 \\
 & & & x_2\,\mathbf{e}_4 & & -x_3^2\,\mathbf{e}_6 & \\
 & & & x_1\,\mathbf{e}_4 & & & -x_2 x_3^2\,\mathbf{e}_7 \\
 & & & & x_1\,\mathbf{e}_5 & -x_2^4\,\mathbf{e}_6 & \\
 & & & & & x_1\,\mathbf{e}_6 & -x_2^2\,\mathbf{e}_7
\end{array}
$$

This monomial ideal is Borel-fixed. Beneath the seven generators, we wrote in 12 rows the 12 minimal first syzygies (2.1) on the generators. These form a Gröbner basis for the syzygy module M, and the initial module is

$$
\begin{aligned}
\mathrm{in}(M) \;=\; \langle\, & x_1\,\mathbf{e}_1, \quad x_2\,\mathbf{e}_1, \quad x_3\,\mathbf{e}_1, \\
& x_1\,\mathbf{e}_2, \quad x_2\,\mathbf{e}_2, \quad x_3\,\mathbf{e}_2, \\
& x_1\,\mathbf{e}_3, \quad x_2\,\mathbf{e}_3, \\
& x_1\,\mathbf{e}_4, \quad x_2\,\mathbf{e}_4, \\
& x_1\,\mathbf{e}_5, \\
& x_1\,\mathbf{e}_6 \,\rangle \\
\subset \quad & S^7 = \Bbbk[x_1, x_2, x_3, x_4]^7.
\end{aligned}
$$

Its minimal free resolution is a direct sum of six Koszul complexes:

$$
\begin{aligned}
& (S\,\mathbf{e}_1 \longleftarrow S^3 \longleftarrow S^3 \longleftarrow S \longleftarrow 0) \\
\oplus\; & (S\,\mathbf{e}_2 \longleftarrow S^3 \longleftarrow S^3 \longleftarrow S \longleftarrow 0) \\
\oplus\; & (S\,\mathbf{e}_3 \longleftarrow S^2 \longleftarrow S \longleftarrow 0) \\
\oplus\; & (S\,\mathbf{e}_4 \longleftarrow S^2 \longleftarrow S \longleftarrow 0) \\
\oplus\; & (S\,\mathbf{e}_5 \longleftarrow S \longleftarrow 0) \\
\oplus\; & (S\,\mathbf{e}_6 \longleftarrow S \longleftarrow 0)
\end{aligned}
$$

$$0 \longleftarrow \mathrm{in}(M) \longleftarrow S^{12} \longleftarrow S^8 \longleftarrow S^2 \longleftarrow 0.$$

The resolution of $\mathrm{in}(M)$ is linear and lifts (by adding trailing terms as in Schreyer's algorithm [Eis95, Theorem 15.10]) to the minimal free resolution of M. The resulting resolution of the Borel-fixed ideal S^7/M is called the *Eliahou–Kervaire resolution*:

$$0 \leftarrow S \xleftarrow{\quad (x_1x_2x_4^4 \quad x_1x_2x_3x_4^2 \quad \cdots \quad x_1^2) \quad} S^7 \leftarrow S^{12} \leftarrow S^8 \leftarrow S^2 \leftarrow 0.$$

The reader is encouraged to compute the matrices representing the differentials in a computer algebra system. ◇

Our results on the Betti numbers of Borel-fixed ideals apply in particular to the $GL_n(\Bbbk)$-fixed ideals. By Corollary 2.2, these are the powers $\mathfrak{m}^d$ of the maximal homogeneous ideal $\mathfrak{m} = \langle x_1, \ldots, x_n \rangle$, as follows when $n = d = 3$.

Example 2.20 Let $n = d = 3$, and use the variable set $\{x, y, z\}$. The Betti numbers and Eliahou–Kervaire resolution of the Borel-fixed ideal $I = \langle x, y, z \rangle^3$ can be visualized as follows:

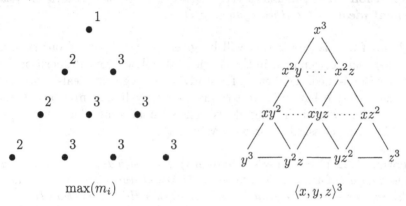

The importance of the dotted lines in the right-hand diagram will be explained in Example 4.22. The numbers in the left-hand diagram determine the binomial coefficients $\binom{\max(m_j)-1}{i}$ from Theorem 2.18, which are given in the triangles below. By adding these triangles, we get the Betti numbers of the minimal free resolution

$$S \xleftarrow{\quad} S^{10} \xleftarrow{\quad} S^{15} \xleftarrow{\quad} S^6 \xleftarrow{\quad} 0$$

$$\begin{array}{ccc}
1 & 0 & 0 \\
1\ 1 & 1\ 2 & 0\ 1 \\
1\ 1\ 1 & 1\ 2\ 2 & 0\ 1\ 1 \\
1\ 1\ 1\ 1 & 1\ 2\ 2\ 2 & 0\ 1\ 1\ 1
\end{array}$$

The triangles show how the resolution of the initial module $\mathrm{in}(M)$ decomposes as a direct sum of 10 Koszul complexes, one for each generator of I. ◇

2.4 Lex-segment ideals

In this section, fix the lexicographic term order $< \,=\, <_{\mathrm{lex}}$ on the polynomial ring $S = \Bbbk[x_1, \ldots, x_n]$. The dth graded component S_d will be identified with the set of all monomials in S of degree d. Fix a function $H : \mathbb{N} \to \mathbb{N}$ that equals the $\mathbb{N}$-graded *Hilbert function* of some homogeneous ideal I in S, meaning that $H(d)$ is the number of $\Bbbk$-linearly independent homogeneous polynomials of degree d lying in the ideal I. There are many choices for I, given our fixed H, and this section is about a certain extreme choice.

Let L_d be the vector space over $\Bbbk$ spanned by the $H(d)$ largest monomials in the lexicographic order on S_d. Define a subspace of S by taking the direct sum of these finite-dimensional spaces of homogeneous polynomials:

$$ L \;=\; \bigoplus_{d=0}^{\infty} L_d. $$

The following result is due to Macaulay [Mac27].

Proposition 2.21 *The graded vector space L is an ideal, called the **lex-segment ideal** for the Hilbert function H.*

A proof of this proposition will be given later, as part of our general combinatorial development in this section. It follows from Proposition 2.3 that L is Borel-fixed. The reason for studying lex-segment ideals is because their numerical behavior is so extreme that they bound from above the numerical behavior of all other ideals. The seminal result along these lines is the following classical theorem of Macaulay.

Theorem 2.22 (Macaulay's Theorem) *For every degree $d \geq 0$, the lex-segment ideal L for the Hilbert function H has at least as many generators in degree d as every other (monomial) ideal with Hilbert function H.*

Example 2.23 Let $n = 4$ and let H be the Hilbert function of the ideal generated by two generic forms of degrees d and e. The lex-segment ideal L for this Hilbert function has more generators than the lexicographic initial ideal in Example 2.10. The first two ideals in this family are

$$(d, e) = (2, 2): \quad L \;=\; \langle x_2^4 x_3^2, x_2^5, x_1 x_4^4, x_1 x_3 x_4^2, x_1 x_3^2, x_1^2, x_1 x_2 \rangle,$$
$$(d, e) = (2, 3): \quad L \;=\; \langle x_2^6 x_3^6, x_2^7 x_4^4, x_2^7 x_3 x_4^2, x_2^9, x_2^8 x_3, x_2^7 x_3^2, x_2^8 x_4, x_1 x_3^2 x_4^5, $$
$$x_1 x_3 x_4^6, x_1 x_4^7, x_1 x_3^4 x_4^2, x_1 x_3^3 x_4^3, x_1 x_3^5, x_1 x_2 x_4^4, $$
$$x_1 x_2 x_3 x_4^2, x_1 x_2 x_3^2, x_1 x_2^2, x_1^2 \rangle.$$

How many monomial generators does L have for $(d, e) = (3, 3)$?   ◇

In Theorem 2.22, it is enough to restrict our attention to monomial ideals, since any initial ideal of an $\mathbb{N}$-graded ideal I has a least as many generators in each degree d as I does. In fact, in view of Theorem 2.9 on generic initial ideals, it suffices to consider only Borel-fixed monomial ideals, as $g \cdot I$ has the same number of generators in each degree as I does.

The degrees of the generators of an ideal measure its zeroth Betti numbers. One can also ask which ideals have the worst behavior with respect to the degrees of the higher Betti numbers. The ultimate statement is that lex-segment ideals take the cake simultaneously for all Betti numbers.

Theorem 2.24 (Bigatti–Hulett Theorem) *For every $i \in \{0, 1, \ldots, n\}$ and $d \geq 0$, the lex-segment ideal L has the most degree d minimal i^{th} syzygies among all (monomial) ideals I with the same fixed Hilbert function H.*

In this section we present proofs for Theorems 2.22 and 2.24 and, of course, also for Proposition 2.21. For the Bigatti–Hulett Theorem, it also suffices to consider only Borel-fixed monomial ideals I. The reason is that Betti numbers can only increase when we pass to an initial ideal (we will prove this in Theorem 8.29), and generic initial ideals are Borel-fixed. To begin with, we need to introduce some combinatorial definitions.

Let W be any finite set of monomials in the polynomial ring S, and write $W_d = W \cap S_d$ for the subset of monomials in W of degree d. For $i \in \{1, \ldots, n\}$, set

$$\mu_i(W) = \big|\{m \in W \mid \max(m) = i\}\big|,$$
$$\mu_{\leq i}(W) = \big|\{m \in W \mid \max(m) \leq i\}\big|.$$

Call W a *Borel set* of monomials if $mx_i/x_j \in W$ whenever x_j divides $m \in W$ and $i < j$. We call W a *lex segment* if $m \in W$ and $m' >_{\text{lex}} m$ implies $m' \in W$. If W is a Borel set then, by Lemma 2.11, every monomial m in $\{x_1, \ldots, x_n\} \cdot W$ factors uniquely as $m = x_i \cdot \tilde{m}$ for some $\tilde{m} \in W$ with $\max(\tilde{m}) \leq i$. This implies the following identity, which holds for all Borel sets W and all $i \in \{1, \ldots, n\}$:

$$\mu_i(\{x_1, \ldots, x_n\} \cdot W) = \mu_{\leq i}(W). \tag{2.3}$$

In the next lemma, we consider sets of monomials all having equal degree d.

Lemma 2.25 *Let L be a lex segment in S_d and B a Borel set in S_d. If $|L| \leq |B|$ then $\mu_{\leq i}(L) \leq \mu_{\leq i}(B)$ for all i.*

Proof. The prove is by induction on n. We distinguish three cases according to the value of i. If $i = n$ then the asserted inequality is obvious:

$$\mu_{\leq n}(L) = |L| \leq |B| = \mu_{\leq n}(B).$$

Suppose now that $i = n - 1$. Partition the Borel set B by powers of x_n:

$$B = B[0] \cup (x_n \cdot B[1]) \cup (x_n^2 \cdot B[2]) \cup \cdots \cup (x_n^d \cdot B[d]).$$

Then $B[i]$ is a Borel set in $\Bbbk[x_1, \ldots, x_{n-1}]_{d-i}$. Similarly, decompose the lex segment L, so $L[i]$ is a lex segment in $\Bbbk[x_1, \ldots, x_{n-1}]_{d-i}$. Let $C[i]$ denote the lex segment in $\Bbbk[x_1, \ldots, x_{n-1}]_{d-i}$ of the same cardinality as $B[i]$. Set

$$C = C[0] \cup (x_n \cdot C[1]) \cup (x_n^2 \cdot C[2]) \cup \cdots \cup (x_n^d \cdot C[d]).$$

By induction, Lemma 2.25 is true in $n-1$ variables, so we have inequalities

$$\mu_{\leq j}(C[i]) \;\leq\; \mu_{\leq j}(B[i]) \quad \text{for all } i, j. \tag{2.4}$$

We claim that C is a Borel set. Since B is a Borel set, $\{x_1, \ldots, x_{n-1}\}B[i]$ is a subset of $B[i-1]$. The inductive hypothesis (2.4) together with (2.3) implies

$$
\begin{aligned}
|\{x_1, \ldots, x_{n-1}\} \cdot C[i]| \;=\; \sum_{j=1}^{n-1} \mu_j(\{x_1, \ldots, x_{n-1}\} \cdot C[i]) \;&=\; \sum_{j=1}^{n-1} \mu_{\leq j}(C[i]) \\
&\leq\; \sum_{j=1}^{n-1} \mu_{\leq j}(B[i]) \\
&=\; \sum_{j=1}^{n-1} \mu_j(\{x_1, \ldots, x_{n-1}\} \cdot B[i]) \\
&=\; |\{x_1, \ldots, x_{n-1}\} \cdot B[i]| \\
&\leq\; |B[i-1]| \;=\; |C[i-1]|.
\end{aligned}
$$

Since $\{x_1, \ldots, x_{n-1}\} \cdot C[i]$ and $C[i-1]$ are lex segments, we deduce that

$$\{x_1, \ldots, x_{n-1}\} \cdot C[i] \;\subseteq\; C[i-1],$$

which means that C is a Borel set in S_d.

Since L is a lex segment and since $|L| \leq |B| = |C|$, the lexicographically minimal monomials in C and L respectively satisfy

$$\min_{\text{lex}}(C) \;\leq_{\text{lex}}\; \min_{\text{lex}}(L).$$

Since both C and L are Borel-fixed, this implies that

$$\min_{\text{lex}}(C[0]) \;\leq_{\text{lex}}\; \min_{\text{lex}}(L[0]).$$

Thus $L[0] \subseteq C[0]$ since both are lex segments in $\Bbbk[x_1, \ldots, x_{n-1}]_d$. Hence

$$\mu_{\leq n-1}(L) \;=\; |L[0]| \;\leq\; |C[0]| \;=\; |B[0]| \;=\; \mu_{\leq n-1}(B), \tag{2.5}$$

which completes the proof for $i = n-1$.

Finally, consider the case $i \leq n-2$. From (2.5) we have $|L[0]| \leq |B[0]|$, so Lemma 2.25 can be applied inductively to the sets $B[0]$ and $L[0]$ to get

$$\mu_{\leq i}(L) \;=\; \mu_{\leq i}(L[0]) \;\leq\; \mu_{\leq i}(B[0]) \;=\; \mu_{\leq i}(B) \quad \text{for } 1 \leq i \leq n-2.$$

Here, the middle inequality is the one from the inductive hypothesis. □

For any finite set W of monomials, define

$$\beta_i(W) \;=\; \sum_{m \in W} \binom{\max(m) - 1}{i}. \tag{2.6}$$

If W minimally generates a Borel-fixed ideal I, then according to Theorem 2.18, $\beta_i(W)$ is the number of minimal i^{th} syzygies of I. But certainly we can consider the combinatorial number $\beta_i(W)$ for any set of monomials.

Lemma 2.26 *If B is a Borel set in S_d then*

$$\beta_i(B) = \binom{n-1}{i} \cdot |B| - \sum_{j=1}^{n-1} \mu_{\leq j}(B) \binom{j-1}{i-1}.$$

Proof. Rewrite (2.6) for $W = B$ as follows:

$$\beta_i(B) = \sum_{j=1}^{n} \mu_j(B) \binom{j-1}{i}$$

$$= \sum_{j=1}^{n} (\mu_{\leq j}(B) - \mu_{\leq j-1}(B)) \binom{j-1}{i}$$

$$= \mu_{\leq n}(B) \binom{n-1}{i} + \sum_{j=1}^{n-1} \mu_{\leq j}(B) \binom{j-1}{i} - \sum_{j=2}^{n} \mu_{\leq j-1}(B) \binom{j-1}{i}$$

$$= |B| \binom{n-1}{i} + \sum_{j=1}^{n-1} \mu_{\leq j}(B) \left(\binom{j-1}{i} - \binom{j}{i} \right).$$

The binomial identity $\binom{j-1}{i} - \binom{j}{i} = -\binom{j-1}{i-1}$ completes the proof. □

Lemma 2.27 *Let L be a lex segment in S_d and B a Borel set in S_d with $|L| = |B|$. Then the following inequalities hold:*

1. $\beta_i(L) \geq \beta_i(B)$.
2. $\beta_i(\{x_1,\ldots,x_n\} \cdot L) \leq \beta_i(\{x_1,\ldots,x_n\} \cdot B)$.

Proof. The proof of part 1 is immediate from Lemmas 2.25 and 2.26:

$$\beta_i(L) = \binom{n-1}{i} \cdot |L| - \sum_{j=1}^{n-1} \mu_{\leq j}(L) \binom{j-1}{i-1}$$

$$\geq \binom{n-1}{i} \cdot |B| - \sum_{j=1}^{n-1} \mu_{\leq j}(B) \binom{j-1}{i-1}$$

$$= \beta_i(B).$$

For part 2, apply the identity (2.3) for both B and L to get

$$\beta_i(\{x_1,\ldots,x_n\} \cdot L) = \sum_{j=1}^{n} \mu_j(\{x_1,\ldots,x_n\} \cdot L) \cdot \binom{j-1}{i}$$

$$= \sum_{j=1}^{n} \mu_{\leq j}(L) \binom{j-1}{i}$$

$$\leq \sum_{j=1}^{n} \mu_{\leq j}(B) \binom{j-1}{i}$$

$$= \sum_{j=1}^{n} \mu_j(\{x_1,\ldots,x_n\} \cdot B) \cdot \binom{j-1}{i}.$$

This quantity equals $\beta_i(\{x_1, \ldots, x_n\} \cdot B)$, and the proof is complete. □

We are now ready to tie up all loose ends and prove the three assertions.

Proof of Proposition 2.21. The function H is the Hilbert function of some ideal B, which we may assume to be Borel-fixed by Theorem 2.9, because Hilbert series are preserved under the operations $I \rightsquigarrow g \cdot I$ and $I \rightsquigarrow \text{in}(I)$ (the latter uses that the standard monomials constitute a vector space basis modulo each of I and $\text{in}(I)$). For any degree d, we have $|L_d| = |B_d|$. Using Lemma 2.25 and (2.3), we find that

$$
\begin{aligned}
|\{x_1, \ldots, x_n\} \cdot L_d| &= \sum_{j=1}^{n} \mu_j(\{x_1, \ldots, x_n\} \cdot L_d) \\
&= \sum_{j=1}^{n} \mu_{\leq j}(L_d) \\
&\leq \sum_{j=1}^{n} \mu_{\leq j}(B_d) \\
&= |\{x_1, \ldots, x_n\} \cdot B_d| \\
&\leq |B_{d+1}| \\
&= |L_{d+1}|.
\end{aligned}
$$

Both $\{x_1, \ldots, x_n\} \cdot L_d$ and L_{d+1} are lex segments in S_{d+1}. The inequality between their cardinalities implies the inclusion

$$
\{x_1, \ldots, x_n\} \cdot L_d \subseteq L_{d+1}.
$$

Since this holds for all d, we conclude that L is an ideal. □

Proof of Theorem 2.22. For any graded ideal I, any term order, and any $d \geq 0$, the number of minimal generators of $\text{in}(I)$ in degree d cannot be smaller than the number of minimal generators of I in degree d, because every Gröbner basis for I contains a minimal generating set. Therefore, replacing I with $\text{gin}(I)$, we need only compare L to Borel-fixed ideals B.

In the previous proof, we derived the inequalities

$$
|\{x_1, \ldots, x_n\} \cdot L_d| \leq |\{x_1, \ldots, x_n\} \cdot B_d| \leq |B_{d+1}| = |L_{d+1}|.
$$

The number of minimal generators of L in degree $d+1$ is the difference $|L_{d+1}| - |\{x_1, \ldots, x_n\} \cdot L_d|$ between the outer two terms. The corresponding number for B is the difference $|B_{d+1}| - |\{x_1, \ldots, x_n\} \cdot B_d|$ between the middle two terms, which can only be smaller. This proves Macaulay's Theorem. □

Next we rewrite the Eliahou–Kervaire formula for the Betti numbers of a Borel-fixed ideal I. If $\text{gens}(I)$ is the set of minimal generators of I, then

$$
\beta_i(\text{gens}(I)) = \sum_{d>0} \left(\beta_i(I_d) - \beta_i(\{x_1, \ldots, x_n\} \cdot I_{d-1}) \right). \qquad (2.7)
$$

Since I is finitely generated, all but finitely many terms in this sum cancel. Thus the right side of (2.7) reduces to the finite sum (2.6) for $W = \text{gens}(I)$.

Proof of Theorem 2.24. Let B be a Borel-fixed ideal and L the lex-segment ideal with the same Hilbert function as B. Our claim is the inequality

$$\beta_i(\text{gens}(B)) \leq \beta_i(\text{gens}(L)) \quad \text{for } i = 0, 1, \ldots, n.$$

Expanding both sides using (2.7), we find that the desired inequality follows immediately from parts 1 and 2 of Lemma 2.27.   □

Exercises

2.1 Give necessary and sufficient conditions, in terms of $i_1, \ldots, i_r$ and $a_1, \ldots, a_r$, for an irreducible monomial ideal $I = \langle x_{i_1}^{a_1}, \ldots, x_{i_r}^{a_r} \rangle$ to be Borel-fixed.

2.2 Can you find a general formula for the number $\mathcal{B}(r, d)$ of Borel-fixed ideals generated by r monomials of degree d in three unknowns $\{x_1, x_2, x_3\}$?

2.3 Show that all associated primes of a Borel-fixed ideal are also Borel-fixed.

2.4 Is the class of Borel-fixed ideals closed under the ideal-theoretic operations of taking intersections, sums, and products?

2.5 Fix a term order on $\Bbbk[x_1, \ldots, x_n]$. Use the artinian property of term orders to show that every ideal has a unique reduced Gröbner basis. Do the same for submodules of free S-modules under any TOP or POT order.

2.6 Find a Borel-fixed ideal that is not the initial monomial ideal of any homogeneous prime ideal in $\Bbbk[x_1, \ldots, x_n]$. Are such examples rare or abundant?

2.7 Prove that if I is Borel fixed and $<$ is any term order, then $\text{gin}_<(I) = I$.

2.8 Let $I = \langle x_1 x_2, x_1 x_3 \rangle$ and fix the lexicographic term order on $S = \Bbbk[x_1, x_2, x_3]$. List all distinct monomial ideals $\text{in}_<(g \cdot I)$ as g runs over $GL_3(\Bbbk)$. Find a comprehensive Gröbner basis $\mathcal{C}$ as in the proof of Lemma 2.6.

2.9 Let P be the *parabolic subgroup* of $GL_4(\Bbbk)$ corresponding to the partition $4 = 2 + 2$, so P consists of all matrices of the form

$$\begin{bmatrix} * & * & * & * \\ * & * & * & * \\ 0 & 0 & * & * \\ 0 & 0 & * & * \end{bmatrix}.$$

Derive a combinatorial condition characterizing P-fixed ideals in $\Bbbk[x_1, x_2, x_3, x_4]$.

2.10 Let I be the ideal generated by two general homogeneous polynomials of degree 3 and 4 in $\Bbbk[x_1, x_2, x_3, x_4]$. Compute the generic initial ideal $\text{gin}_<(I)$ for the lexicographic term order and for the reverse lexicographic term order. Also compute the lex-segment ideal with the same Hilbert function.

2.11 Let $I = \langle x_1 x_2 x_3, x_1 x_2 x_4, x_1 x_3 x_4, x_2 x_3 x_4 \rangle$. Compute the generic initial ideal $\text{gin}_<(I)$ for the lexicographic and reverse lexicographic term orders. Also compute the lex-segment ideal with the same Hilbert function.

2.12 Compute the Betti numbers and Hilbert series of the ideal

$$I = \langle x_1, x_2, x_3, x_4, x_5 \rangle^5.$$

2.13 If $\mathcal{F}.$ is a linear free resolution, must *every* choice of matrices for its differentials have only linear forms for nonzero entries? Must $\mathcal{F}.$ be minimal?

2.14 Given a Borel-fixed ideal I, compute $K^{\mathbf{b}}(I)$ in any degree $\mathbf{b} \in \mathbb{N}^n$.

2.15 Let M be the first syzygy module of any Borel-fixed ideal. Give an example to show that even though $\text{in}(M)$ has linear resolution, M itself need not. More generally, write down explicitly all of the boundary maps in the Eliahou–Kervaire resolution. Hint: Feel free to consult [EK90].

2.16 Is lexicographic order the only one for which Proposition 2.21 holds?

2.17 Can you find a monomial ideal that is not lex-segment but has the same graded Betti numbers as the lex-segment ideal with the same Hilbert function?

Notes

The original motivation for generic initial ideals, and hence Borel-fixed ideals, came from Hartshorne's proof of the connectedness of the Hilbert scheme of subschemes of projective space [Har66a]. Galligo proved Theorem 2.9 in characteristic zero [Gal74], and then Bayer and Stillman worked out the case of arbitrary characteristic [BS87]. It is worth noting that some of the other results in this chapter do not hold verbatim in positive characteristic, partially because the notion of "Borel-fixed" has a different combinatorial characterization due to Pardue [Par94]. See Eisenbud's textbook [Eis95, Section 15.9] for an exposition of Borel-fixed and generic initial ideals, including the finite characteristic case as well as more history and references.

The Eliahou–Kervaire resolution first appeared in [EK90], where it was derived for the class of *stable ideals*, which is slightly more general than Borel-fixed ideals. The passage from a monomial ideal to its generic initial ideals with respect to various term orders is called *algebraic shifting* in the combinatorics literature. This is an active area of research at the interface of combinatorics and commutative algebra; see the articles by Aramova–Herzog–Hibi [AHH00] and Babson–Novik–Thomas [BNT02] as well as the references given there. The explicit identification of cycles representing homology classes in shifted complexes, such as the boundaries of the missing faces $\tau \cup k$ in Lemma 2.15, is typical; in fact, it is a motivating aspect of their combinatorics (see [BK88, BK89], for example).

Theorem 2.22 is one of Macaulay's fundamental contributions to the theory of Hilbert functions [Mac27]. Theorem 2.24 is due independently to Bigatti [Big93] and Hulett [Hul93]; the proof given here is Bigatti's. The geometry of lexicographic generic initial ideals is a promising direction of future research, toward which first steps have been taken in recent work of Conca and Sidman [CS04].

Chapter 3

Three-dimensional staircases

Squarefree and Borel-fixed ideals each have their own advantages, the former yielding insight into the combinatorics of simplicial complexes and the latter into extremal numerical behavior in algebraic geometry. In both cases we can express relevant data in terms of the defining properties of these special classes of monomial ideals, and in the Borel-fixed case, we can actually write down an explicit minimal free resolution.

However, such explicit minimal resolutions are not available for general monomial ideals, at least not without making choices that are arbitrary. Even in the Borel-fixed case, the choices have already been made for us—in the order of the variables, for instance—and it may well be that an ideal is Borel-fixed with respect to more than one such order. This occurs for powers of the maximal ideal $\mathfrak{m} = \langle x_1, \ldots, x_n \rangle$. Our inability to write down explicit canonical minimal (or at least "small") resolutions leads us to examine intrinsic geometric properties of monomial ideals resulting from the inclusion of the lattice $\mathbb{Z}^n$ into the vector space $\mathbb{R}^n$.

The coming chapters use convex geometric techniques, along with the combinatorial and algebraic topological methods surrounding them, to express data associated to arbitrary monomial ideals (and even some binomial ideals as well, in Chapter 9). The details of the multiple facets of this theory in higher dimensions are the subjects of later chapters in Part I. Here, we start out by letting the *staircases* speak for themselves in the case of two and three variables. The main result, Theorem 3.17, describes how planar graphs arise as minimal free resolutions of monomial ideals over polynomial rings $\Bbbk[x, y, z]$ in three variables.

3.1   Monomial ideals in two variables

Consider an arbitrary monomial ideal I in the bivariate polynomial ring $S = \Bbbk[x, y]$. It can be written in terms of minimal monomial generators as

$$I \;=\; \langle m_1, \ldots, m_r \rangle \;=\; \langle x^{a_1}y^{b_1}, x^{a_2}y^{b_2}, \ldots, x^{a_r}y^{b_r} \rangle,$$

where $a_1 > a_2 > \cdots > a_r \geq 0$ and $0 \leq b_1 < b_2 < \cdots < b_r$. The *staircase diagram* for the ideal I shows the interface between regions of the plane containing (exponent vectors of) monomials in I and those not in I:

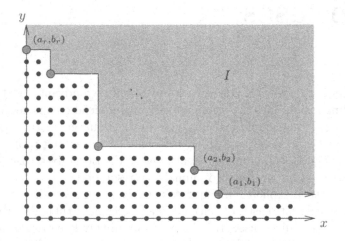

The black lattice points, contained completely within the unshaded region, form a $\Bbbk$-basis for S/I. The Hilbert series $H(S/I; x, y)$ is the formal sum of all monomials not in I. This generating function is a rational function with denominator $(1 - x)(1 - y)$. One way to see this is by inclusion–exclusion: start with all of the monomials in S; then, for each minimal generator m_i, subtract off the monomials in the principal ideal $\langle m_i \rangle$, which looks like a shifted positive orthant; of course, now we have subtracted the monomials in the principal ideal $\langle m_i \rangle \cap \langle m_j \rangle = \langle \mathrm{lcm}(m_i, m_j) \rangle$ generated by the least common multiple of m_i and m_j too many times, so we have to add those monomials back in. Continuing in this way, we eventually (after at most r steps) count each monomial the right number of times. But this procedure produces 2^r terms, which is many more terms than are necessary. Almost all terms in the naïve inclusion–exclusion formulas cancel in the end.

There is a more efficient way to do the inclusion–exclusion: after we have subtracted off the principal ideals $\langle m_i \rangle$, we add back in not *all* of the principal ideals $\langle \mathrm{lcm}(m_i, m_j) \rangle$, but only those which come from monomials m_i and m_j that are *adjacent pairs*—that is, where $j = i + 1$. This yields the Hilbert series after just two steps. The numerator of the Hilbert series therefore simplifies to

$$
\begin{aligned}
\mathcal{K}(S/I; x, y) &= (1-x)(1-y)\, H(S/I; x, y) \\
&= (1-x)(1-y) \sum_{x^i y^j \notin I} x^i y^j
\end{aligned}
$$

$$
\text{(naïve inclusion–exclusion)} \quad = \sum_{\sigma \subseteq \{1,\ldots,r\}} (-1)^{|\sigma|}\, \mathrm{lcm}(x^{a_i} y^{b_i} \mid i \in \sigma)
$$

$$
\text{(efficient inclusion–exclusion)} \quad = 1 - \sum_{i=1}^{r} x^{a_i} y^{b_i} + \sum_{j=1}^{r-1} x^{a_j} y^{b_{j+1}}
$$

$$
= 1 - \text{inner corners} + \text{outer corners}.
$$

The naïve inclusion–exclusion process reflects a highly nonminimal free resolution of S/I called the *Taylor resolution*, to be introduced later. Our more efficient way of doing things yields a *minimal* free resolution of S/I.

Proposition 3.1 *The minimal free resolution of an ideal generated by r monomials in $S = \Bbbk[x, y]$ has the format*

$$
0 \longleftarrow S \longleftarrow S^r \longleftarrow S^{r-1} \longleftarrow 0.
$$

The minimal first syzygies are the vectors $y^{b_{i+1}-b_i} \mathbf{e}_i - x^{a_i - a_{i+1}} \mathbf{e}_{i+1}$ corresponding to adjacent pairs $\{x^{a_i} y^{b_i},\ x^{a_{i+1}} y^{b_{i+1}}\}$ of minimal generators of I.

Proof. The kernel of the map $S \leftarrow S^r$ requires at least $r - 1$ generators, as can be seen by passing to the field $\Bbbk(x, y)$ of fractions of S. The adjacent syzygies $y^{b_{i+1}-b_i} \mathbf{e}_i - x^{a_i - a_{i+1}} \mathbf{e}_{i+1}$ not only span this kernel, but they in fact constitute a Gröbner basis in the position-over-term (POT) order. Indeed, it is easy to see that every syzygy on I can be reduced to zero by successively replacing occurrences of $y^{b_{i+1}-b_i} \mathbf{e}_i$ by $x^{a_i - a_{i+1}} \mathbf{e}_{i+1}$ for $i = 1, \ldots, r-1$. $\square$

The natural adjacency relation among minimal generators of a bivariate monomial ideal I also determines an *irredundant irreducible decomposition* of I. By definition, such a decomposition expresses I as an intersection of monomial ideals generated by powers of the variables (*irreducible monomial ideals*), in such a way that no intersectands can be omitted.

Proposition 3.2 *$I \subset \Bbbk[x, y]$ has the irredundant irreducible decomposition*

$$
I = \langle y^{b_1} \rangle \cap \langle x^{a_1}, y^{b_2} \rangle \cap \langle x^{a_2}, y^{b_3} \rangle \cap \cdots \cap \langle x^{a_{r-1}}, y^{b_r} \rangle \cap \langle x^{a_r} \rangle,
$$

where the first or last components are to be deleted if $b_1 = 0$ or $a_r = 0$.

Proof. After removing common factors from the generators, we may assume that $b_1 = 0$ and $a_r = 0$, so that I is artinian. The given ideals $\langle x^{a_i}, y^{b_i+1} \rangle$ are irreducible and clearly contain I. Inspection of the staircase diagram shows that each monomial in their intersection must also lie in I. $\square$

In view of the previous two propositions concerning $\Bbbk[x, y]$, it is natural to wonder how the notion of adjacent monomials can be generalized to ideals in three or more variables. An answer will be offered in Section 3.3.

3.2   An example with six monomials

A standard method in commutative algebra for treating homological and enumerative questions about arbitrary monomial ideals is to reduce to the squarefree or Borel-fixed case. This allows us to apply specific techniques suited to these classes of ideals. This section describes these two approaches for a particular monomial ideal in three variables, along with their advantages and drawbacks, and compares them with resolution by a planar graph.

We will study the following artinian monomial ideal:

$$J = \langle x^4, y^4, z^4, x^3y^2z, xy^3z^2, x^2yz^3 \rangle \subset \Bbbk[x,y,z] = S.$$

Method 1: Reduction to the squarefree case. The homological behavior of any monomial ideal is preserved under passing to a certain related squarefree monomial ideal, called its *polarization*. In the polarization process, each power of a variable, say the power x^d of the variable x, is replaced by a product of d new variables, say $x_1x_2\cdots x_d$. Thus the polarization of our ideal J is

$$\begin{aligned} I_\Delta &= \langle x_1x_2x_3x_4,\ y_1y_2y_3y_4,\ z_1z_2z_3z_4,\ x_1x_2x_3y_1y_2z_1, \\ &\quad x_1y_1y_2y_3z_1z_2,\ x_1x_2y_1z_1z_2z_3 \rangle. \end{aligned}$$

This is now a monomial ideal in the polynomial ring in 12 variables,

$$\widetilde{S} = \Bbbk[x_1, x_2, x_3, x_4, y_1, y_2, y_3, y_4, z_1, z_2, z_3, z_4].$$

The ideal I_Δ still has codimension 3. The key feature of polarization is that

$$x_1 - x_2,\ x_2 - x_3,\ x_3 - x_4,\ y_1 - y_2,\ y_2 - y_3,\ y_3 - y_4,\ z_1 - z_2,\ z_2 - z_3,\ z_3 - z_4$$

is a regular sequence in the ring $\widetilde{S}/I_\Delta$, meaning that each element is a nonzerodivisor modulo the ideal generated by all previous elements. Taking the quotient of $\widetilde{S}/I_\Delta$ modulo the ideal generated by this regular sequence, we obtain precisely the ring S/J we started with (and homological information is preserved; see Exercise 3.15 for details on the transition $\widetilde{S}/I_\Delta \rightsquigarrow S/J$). Therefore, to get information about S/J, we first compute the minimal free resolution and Hilbert series of I_Δ. The resolution looks like

$$0 \longleftarrow \widetilde{S} \longleftarrow \widetilde{S}^6 \longleftarrow \widetilde{S}^{12} \longleftarrow \widetilde{S}^7 \longleftarrow 0. \tag{3.1}$$

A minimal free resolution of J is obtained by erasing the indices from the variables, or equivalently by substituting $x_i \mapsto x, y_i \mapsto y, z_i \mapsto z$ for every variable in each matrix of the resolution (3.1), and likewise for the Hilbert series and K-polynomial. The ideal I_Δ corresponds to a simplicial complex Δ on 12 vertices, and according to Hochster's formula in Corollary 1.40, the multigraded Betti numbers of (3.1) are encoded in this complex.

The drawback of polarization is that Δ is much too large. In our example, Δ is *pure* (the dimensions of its facets are all equal) of dimension 8 and has 51 facets. Its f-vector $(f_{-1}, f_0, f_1, \ldots)$, whose entry f_d for $d \geq -1$ counts the number of faces of dimension d, reads

$$f(\Delta) \quad = \quad (1, 12, 66, 220, 492, 768, 837, 264, 51).$$

Passing from a monomial ideal to its polarization is a nice theoretical tool, but rarely used in practice due to the size of the resulting simplicial complex.

Method 2: Data from the Borel-fixed case. The process of replacing a monomial ideal I by its generic initial ideal is called *(symmetric) algebraic shifting*. This replaces I by the Borel-fixed monomial ideal $\text{gin}(I)$. Shifting our example with the reverse lexicographic term order yields

$$\text{gin}_{\text{revlex}}(J) \quad = \quad \langle x^4, x^3 y, x^2 y^2, xy^4, y^5, x^3 z^3, x^2 yz^3, xy^3 z^2, xy^2 z^3, y^4 z^2,$$
$$x^2 z^5, xyz^5, xz^6, y^3 z^4, y^2 z^5, yz^6, z^7 \rangle.$$

Both ideals have colength 51, the number of cubes in the staircase, but the generic initial ideal is much more complicated than J itself, the grading by $\mathbb{N}^3$ is lost, and the Betti numbers might have increased from those of J to those of $\text{gin}_{\text{revlex}}(J)$ (see Theorem 8.29). But the $\mathbb{N}$-grading is retained, and we can compute the coarse Hilbert series and the K-polynomial using the Eliahou–Kervaire formula from Proposition 2.12. We find that

$$\mathcal{K}(S/J; t, t, t) \quad = \quad 1 - 3t^4 - 3t^6 + 3t^7 + 9t^8 - 7t^9$$
$$= \quad (1 - t)^3 \cdot (1 + 3t + 6t^2 + 10t^3 + 12t^4 + 12t^5 + 7t^6).$$

The last factor of degree 6 is the Hilbert series of S/J. A theorem of Bayer and Stillman [BS87] states that the *Castelnuovo–Mumford regularity* of J can be read off as the largest degree of a minimal generator of $\text{gin}_{\text{revlex}}(J)$. This number is an important invariant, and it equals 7 in our example.

Method 3: Resolution by picture. Our main tools for studying monomial ideals in two variables were *staircase diagrams*. These are also possible to draw for monomial ideals in three variables. For instance, Fig. 3.1 depicts a staircase diagram for our ideal $J = \langle x^4, y^4, z^4, x^3 y^2 z, xy^3 z^2, x^2 yz^3 \rangle$. The surface we see is the interface between being in J and not being in J, with the lattice points strictly behind the interface being those outside of J. Thus any lattice point that is visible in the staircase diagram is the exponent vector on a monomial in our ideal J. Dark dots correspond to the minimal generators of J; note how they sit at the "inner" corners.

Consider the graph in Fig. 3.2, in which we have connected the minimal generators of J according to when they "look adjacent" (we will make this precise soon). Each edge and each triangular face is labeled by the exponent vector of the least common multiple of its vertices. As will be explained in

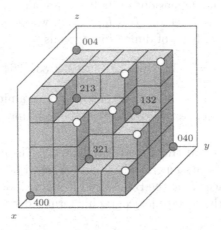

Figure 3.1: A staircase diagram in three variables

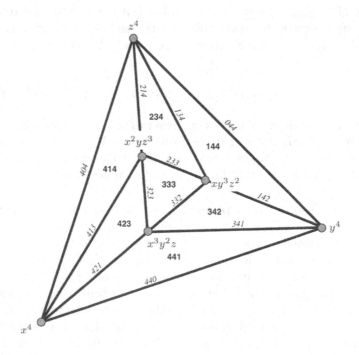

Figure 3.2: Resolution by picture

the following chapters, much of the structure of the monomial ideal can be read off from this figure. For example, vertices correspond to generators, edges to first syzygies, and facets to second syzygies. In the particular case of Fig. 3.1, where the monomial ideal is *artinian*, the facets also reveal the irreducible components, which correspond to the white dots on the "outer" corners of the staircase surface. Overall, the information we can get includes the following:

Irreducible decomposition (labels on triangles in Fig. 3.2):

$$
\begin{aligned}
J \quad &:= \quad \langle x^4, y^4, z^4, x^3y^2z, xy^3z^2, x^2yz^3 \rangle \\
&= \quad \langle x^4, y^4, z \rangle \cap \langle x^4, y, z^4 \rangle \cap \langle x, y^4, z^4 \rangle \cap \langle x^4, y^2, z^3 \rangle \cap \\
&\quad\quad \langle x^3, y^4, z^2 \rangle \cap \langle x^2, y^3, z^4 \rangle \cap \langle x^3, y^3, z^3 \rangle
\end{aligned}
$$

Minimal free resolution (chain complex of the triangulation):

$$
0 \longleftarrow S \longleftarrow S^6 \longleftarrow S^{12} \longleftarrow S^7 \longleftarrow 0
$$

The summands correspond to the 6 vertices, 12 edges, and 7 facets of the triangulated triangle in Fig. 3.2.

Numerator of the Hilbert series (alternating sum of all face labels):

$$
1 - x^4 - \ldots - x^2yz^3 + x^4y^4 + \ldots + xy^3z^4 - x^4y^4z - \ldots - x^3y^3z^3
$$

This is the K-polynomial $\mathcal{K}(S/J; x, y, z)$. Note that by specializing, we get $\mathcal{K}(S/J; t, t, t) = 1 - 3t^4 - 3t^6 + 3t^7 + 9t^8 - 7t^9$, as we did earlier.

3.3 The Buchberger graph

Finding minimal sets of first syzygies for monomial ideals has an impact on algorithmic computation for arbitrary ideals. The connection is through Gröbner bases. We recall Buchberger's Criterion from Gröbner basis theory.

Theorem 3.3 (Buchberger's Criterion) *A set* $\{f_i\}_{i=1}^r$ *of polynomials*

$$
f_i \quad = \quad m_i + \text{trailing terms under the term order} <
$$

is a Gröbner basis under the term order $<$ *if each s-pair*

$$
s(f_i, f_j) \quad := \quad \frac{\text{lcm}(m_i, m_j)}{m_i} f_i - \frac{\text{lcm}(m_i, m_j)}{m_j} f_j
$$

can be reduced to zero by $\{f_1, \ldots, f_r\}$ *using the division algorithm.*

Each s-pair $s(f_i, f_j)$ yields an element σ_{ij} of the free module S^r, namely

$$
\sigma_{ij} \quad = \quad \frac{\text{lcm}(m_i, m_j)}{m_i} \mathbf{e}_i - \frac{\text{lcm}(m_i, m_j)}{m_j} \mathbf{e}_j.
$$

The $\binom{r}{2}$ elements σ_{ij} generate the module of first syzygies

$$\mathrm{syz}(I) \;=\; \ker_S[m_1\, m_2\, \cdots\, m_r]$$

of the monomial ideal $I = \langle m_1, m_2, \ldots, m_r \rangle$, but often they do not generate minimally. In order to make Buchberger's Criterion (and hence Buchberger's algorithm for computing Gröbner bases) more efficient, it is important to take advantage of the structure of the syzygy module $\mathrm{syz}(I)$. *Buchberger's Second Criterion* states that Theorem 3.3 can be strengthened as follows: if G is any subset of the pairs (i,j) with $1 \le i < j \le r$ such that the set $\{\sigma_{ij} \mid (i,j) \in G\}$ generates $\mathrm{syz}(I)$, then it suffices that only the s-pairs $s(f_i, f_j)$ with $(i,j) \in G$ reduce to zero in order to imply the Gröbner basis property for $\{f_1, f_2, \ldots, f_r\}$. This leads us to the following.

Definition 3.4 The **Buchberger graph** $\mathrm{Buch}(I)$ of a monomial ideal $I = \langle m_1, \ldots, m_r \rangle$ has vertices $1, \ldots, r$ and an edge (i,j) whenever there is no monomial m_k such that m_k divides $\mathrm{lcm}(m_i, m_j)$ and the degree of m_k is different from $\mathrm{lcm}(m_i, m_j)$ in every variable that occurs in $\mathrm{lcm}(m_i, m_j)$.

For example, if I is a monomial ideal in two variables, then $\mathrm{Buch}(I)$ consists of the $r - 1$ consecutive pairs of minimal generators.

Proposition 3.5 *The syzygy module* $\mathrm{syz}(I)$ *is generated by syzygies* σ_{ij} *corresponding to edges* (i,j) *in the Buchberger graph* $\mathrm{Buch}(I)$.

Proof. The following identity holds for all $i, j, k \in \{1, \ldots, r\}$:

$$\frac{\mathrm{lcm}(m_i, m_j, m_k)}{\mathrm{lcm}(m_i, m_j)}\,\sigma_{ij} + \frac{\mathrm{lcm}(m_i, m_j, m_k)}{\mathrm{lcm}(m_j, m_k)}\,\sigma_{jk} + \frac{\mathrm{lcm}(m_i, m_j, m_k)}{\mathrm{lcm}(m_k, m_i)}\,\sigma_{ki} \;=\; 0.$$

If (i,j) is not an edge of $\mathrm{Buch}(I)$, then for some k, the coefficient of σ_{ij} is 1 while the coefficients of σ_{jk} and σ_{ki} are nonconstant monomials. Hence σ_{ij} lies in the S-module generated by other first syzygies of strictly smaller degree. This means that we can remove σ_{ij} from the generators of $\mathrm{syz}(I)$ without running into a cycle. $\square$

The Buchberger graph for our example J in the previous section can be embedded nicely into the staircase diagram:

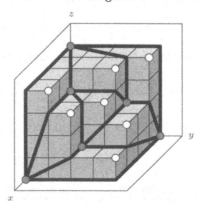

Each edge consists of two straight segments connecting the minimal generators m_1 and m_2 to the exponent vector on $\mathrm{lcm}(m_1, m_2)$. To be precise in what follows, let us make the notion of a staircase formal.

Definition 3.6 The **staircase surface** of a monomial ideal I in $\Bbbk[x, y, z]$ is the topological boundary of the set of vectors $(v_x, v_y, v_z) \in \mathbb{R}^3$ for which there is some monomial $x^{u_x} y^{u_y} z^{u_z} \in I$ satisfying $u_i \leq v_i$ for all $i \in \{x, y, z\}$.

Staircase surfaces are homeomorphic to $\mathbb{R}^2$ by orthogonal projection with kernel $(1, 1, 1)$. In the above illustration of an embedded Buchberger graph, each region contains precisely one white dot situated on an outside corner, each vertex is a dark dot on an inside corner, and each edge passes through one corner that is neither inside nor outside. The label on each vertex, edge, and region is the vector represented by the corresponding corner. In this example, the Buchberger graph $\mathrm{Buch}(J)$ is therefore a planar graph, its edges minimally generate the first syzygies on J, and its canonical planar embedding coincides with the minimal free resolution of S/J, as in Fig. 3.2. Do these properties hold for any trivariate monomial ideal? The following example shows that the answer is "no" in general. Proposition 3.9 in the next section shows that the answer is "yes" for the special class of *generic* monomial ideals. Our previous example J is generic. Generic monomial ideals in any number of variables are the topic of Chapter 6.

Example 3.7 Consider the ideal $I' = \langle x^2 z, xyz, y^2 z, x^3 y^5, x^4 y^4, x^5 y^3 \rangle$, whose staircase is depicted in Fig. 3.3. The drawing of $\mathrm{Buch}(I')$ there contains the nonplanar complete bipartite graph $K_{3,3}$ as a subgraph. ◇

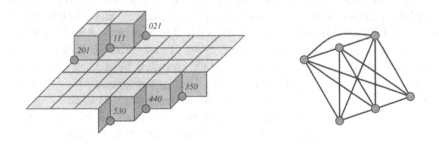

Figure 3.3: $\mathrm{Buch}(I')$ contains $K_{3,3}$

3.4 Genericity and deformations

The nonplanarity encountered at the end of the previous section never occurs for a certain special class of ideals: the strongly generic ideals. By a process called "deformation", it yields tight complexity bounds in Corollary 3.15. The two main results in this section, the planarity in Proposition 3.9 and the free resolution in Theorem 3.11, admit proofs relying only on three-dimensional geometric methods. However, as with many results concerning planar graphs, these intuitive statements require more technicality than one might expect. Since we will in any case prove these results in more generality later, we only sketch their proofs here.

Definition 3.8 A monomial ideal I in $\Bbbk[x, y, z]$ is **strongly generic** if every pair of minimal generators $x^i y^j z^k$ and $x^{i'} y^{j'} z^{k'}$ of I satisfies

$$(i \neq i' \text{ or } i = i' = 0) \text{ and } (j \neq j' \text{ or } j = j' = 0) \text{ and } (k \neq k' \text{ or } k = k' = 0).$$

In other words, no two generators agree in the exponent on any variable that actually appears in both of them.

The following result is a special case of Theorem 6.13, because for generic monomial ideals, the edges of the Scarf complex coincide with the Buchberger graph (Lemma 6.10), and the hull complex of an ideal in $\Bbbk[x, y, z]$ is planar and connected (Theorem 4.31 and Proposition 4.5).

Proposition 3.9 *If I is a strongly generic monomial ideal in $\Bbbk[x, y, z]$, then the Buchberger graph* Buch(I) *is planar and connected. If, in addition, I is artinian, then* Buch(I) *consists of the edges in a triangulated triangle.*

Sketch of proof. First observe that it suffices to consider artinian monomial ideals I, meaning that the minimal generators of I include pure powers in each of the three variables, say x^a, y^b, and z^c. Indeed, erasing all edges and regions incident to one or more of $\{x^a, y^b, z^c\}$ yields the Buchberger graph for the ideal without the corresponding generator, and what results is connected because planar triangulations are 3-connected (Definition 3.16).

The idea now is that the bounded faces in the staircase surface of the monomial ideal I form a topological disk bounded by a piecewise linear triangle with vertices x^a, y^b, and z^c, the pure power generators of I. Each edge $\{m, m'\}$ of Buch(I) is drawn in the staircase surface as the union of the two line segments from m to lcm(m, m') and from m' to lcm(m, m'). The fact that lcm(m, m') lies in the staircase surface is a consequence of genericity, which also implies that lcm(m, m') has no other edges passing through it. We thus obtain an embedding of Buch(I) in the staircase surface. What remains to be shown is that each region of that subdivision is a triangle (that is, bounded by exactly three Buchberger edges.) This is proved by showing that each of the two regions containing any interior Buchberger edge $\{m, m'\}$ is a triangle. This triangle is produced by finding a uniquely determined third generator m'' such that the least common multiple of $\{m, m', m''\}$ lies in the staircase surface; the region is then bounded by the Buchberger edges $\{m, m'\}$, $\{m, m''\}$, and $\{m', m''\}$.   □

Planar graphs G can usually be embedded in $\mathbb{R}^2$ in many ways, making the notion of "the regions of G" ambiguous. It is customary to distinguish a planar graph from a particular embedding of that graph.

Definition 3.10 A **planar map** is a graph G together with an embedding of G into a surface homeomorphic to the plane $\mathbb{R}^2$.

That being said, we refer to the planar map simply as G if its embedding is given. We require the surface to be homeomorphic rather than equal to $\mathbb{R}^2$ to encourage the drawing of planar maps in staircase surfaces. Indeed, the proof of Proposition 3.9 endows the Buchberger graph of a generic monomial ideal with a canonical embedding in its staircase surface.

Theorem 3.11 says that planar maps encode minimal free resolutions insofar as they organize into single diagrams the syzygies and their interrelations. The free resolution given by a planar map G with v vertices, e edges, and f faces, all labeled by monomials, has the form

$$\mathcal{F}_G: \quad 0 \leftarrow S \longleftarrow S^v \xleftarrow{\partial_E} S^e \xleftarrow{\partial_F} S^f \leftarrow 0. \tag{3.2}$$

If we express the differentials by monomial matrices as in Chapter 1, then the scalar entries are precisely those coming from the usual differentials on a planar map (after choosing orientations on the edges), but with monomial row and column labels. For instance, the matrix for ∂_F has the edge monomials for row labels and the face monomials for column labels, while its scalar entries take each face to the signed sum of the oriented edges on its boundary. To express the differentials in ungraded notation, we write $m_{ij} = \mathrm{lcm}(m_i, m_j)$ for each edge $\{i, j\}$ of G, and m_R for the least common multiple of the monomial labels on the edges in each region R. Then

$$\partial_E(\mathbf{e}_{ij}) \quad = \quad \frac{m_{ij}}{m_j} \cdot \mathbf{e}_j - \frac{m_{ij}}{m_i} \cdot \mathbf{e}_i$$

if an edge oriented toward m_j joins the vertices labeled m_i and m_j, whereas

$$\partial_F(\mathbf{e}_R) \quad = \quad \sum_{\substack{\text{edges} \\ \{i,j\} \subset R}} \pm \frac{m_R}{m_{ij}} \cdot \mathbf{e}_{ij}$$

for each region R, where the sign is positive precisely when the edge $\{i, j\}$ is oriented counterclockwise around R. The construction of these differentials in arbitrary dimensions is the subject of Chapter 4. The rigorous n-dimensional proof of the next result will follow even later, in Theorem 6.13.

Theorem 3.11 *Given a strongly generic monomial ideal I in $\Bbbk[x, y, z]$, the planar map $\mathrm{Buch}(I)$ provides a minimal free resolution of I.*

Sketch of proof. Begin by throwing high powers x^a, y^b, and z^c into I. What results is still strongly generic, but now artinian. If we are given a

minimal free resolution of this new ideal by a planar map, then deleting all
edges and regions incident to one or more of $\{x^a, y^b, z^c\}$ leaves a minimal
free resolution of I. Indeed, these deletions have no effect on the $\mathbb{N}^3$-graded
components of degree $\preceq (a-1, b-1, c-1)$, which remain exact, and I has
no syzygies in any other degree. Therefore we assume that I is artinian.

Each triangle in $\text{Buch}(I)$ contains a unique "mountain peak" in the
surface of the staircase, located at the outside corner $\text{lcm}(m, m', m'')$. That
peak is surrounded by three "mountain passes" $\text{lcm}(m, m')$, $\text{lcm}(m, m'')$,
and $\text{lcm}(m', m'')$, each of which represents a minimal first syzygy of I by
Theorem 1.34 (check that the simplicial complex $K^{\mathbf{b}}(I)$ from Definition 1.33
is disconnected precisely when a mountain pass sits in degree $\mathbf{b}$). The
mountain peak represents a second syzygy relating these three first syzygies
by the identity in the proof of Proposition 3.5, and all minimal second
syzygies arise this way by Theorem 1.34. $\square$

Next we show how to approximate arbitrary monomial ideals by strongly
generic ones. The idea is to add small rational numbers to the exponents
on the generators of I without reversing any strict inequalities between the
degrees in x, y, or z of any two generators. This process occurs inside a
polynomial ring $S_\epsilon = \Bbbk[x^\epsilon, y^\epsilon, z^\epsilon]$, where $\epsilon = 1/N$ for some large positive integer N, which contains $S = \Bbbk[x, y, z]$ as a subring. Equalities among x-, y-,
and z-degrees can turn into strict inequalities potentially going either way.

Definition 3.12 Let $I = \langle m_1, \dots, m_r \rangle$ and $I_\epsilon = \langle m_{\epsilon,1}, \dots, m_{\epsilon,r} \rangle$ be monomial ideals in S and S_ϵ, respectively. Call I_ϵ a **strong deformation** of I
if the partial order on $\{1, \dots, r\}$ by x-degree of the $m_{\epsilon,i}$ refines the partial
order by x-degree of the m_i, and the same holds for y and z. We also say
that I is a **specialization** of I_ϵ.

Constructing a strong deformation I_ϵ of any given monomial ideal I is
easy: simply replace each generator m_i by a nearby generator $m_{\epsilon,i}$ in such
a way that $\lim_{\epsilon \to 0} m_{\epsilon,i} = m_i$. The ideal I_ϵ need not be strongly generic;
however, it will be if the strong deformation is chosen randomly.

Example 3.13 The ideal in S_ϵ given by

$$\langle x^3, x^{2+\epsilon} y^{1+\epsilon}, x^2 z^1, x^{1+2\epsilon} y^2, x^{1+\epsilon} y^1 z^{1+\epsilon}, x^1 z^{2+\epsilon}, y^3, y^{2-\epsilon} z^{1+2\epsilon}, y^{1+2\epsilon} z^2, z^3 \rangle$$

is one possible strongly generic deformation of the ideal $\langle x, y, z \rangle^3$ in S. $\diamond$

Proposition 3.14 *Suppose I is a monomial ideal in $\Bbbk[x, y, z]$ and I_ϵ is a*
strong deformation resolved by a planar map G_ϵ. Specializing the vertices
(hence also the edges and regions) of G_ϵ yields a planar map resolution of I.

Proof. Consider the minimal free resolution $\mathcal{F}_{G_\epsilon}$ determined by the triangulation G_ϵ as in (3.2). The specialization G of the labeled planar map G_ϵ
still gives a complex $\mathcal{F}_G$ of free modules over $\Bbbk[x, y, z]$, and we need to

demonstrate its exactness. Considering any fixed $\mathbb{N}^3$-degree $\omega = (a, b, c)$, we must demonstrate exactness of the complex of vector spaces over $\Bbbk$ in the degree ω part of $\mathcal{F}_G$. Define ω_ϵ as the exponent vector on

$$\operatorname{lcm}(m_{\epsilon,i} \mid m_i \text{ divides } x^a y^b z^c).$$

The summands contributing to the degree ω part of $\mathcal{F}_G$ are exactly those summands of $\mathcal{F}_{G_\epsilon}$ contributing to its degree ω_ϵ part, which is exact. $\square$

In the next section we will demonstrate how any planar map resolution can be made minimal by successively removing edges and joining adjacent regions. For now, we derive a sharp complexity bound from Proposition 3.14 using *Euler's formula*, which states that $v - e + f = 1$ for any connected planar map with v vertices, e edges, and f bounded faces [Wes01, Theorem 6.1.21], plus its consequences for simple planar graphs with at least three vertices: $e \leq 3v - 6$ [Wes01, Theorem 6.1.23] and $f \leq 2v - 5$.

Corollary 3.15 *An ideal I generated by $r \geq 3$ monomials in $\Bbbk[x, y, z]$ has at most $3r - 6$ minimal first syzygies and $2r - 5$ minimal second syzygies. These Betti number bounds are attained if I is artinian, strongly generic, and xyz divides all but three minimal generators.*

Proof. Choose a strong deformation I_ϵ of I that is strongly generic. Proposition 3.14 implies that I has Betti numbers no larger than those of I_ϵ, so we need only prove the first sentence of the theorem for I_ϵ. Theorem 3.11 implies that I_ϵ is resolved by a planar map, so Euler's formula and its consequences give the desired result.

For the second statement, let x^a, y^b, and z^c be the three special generators of I. Every other minimal generator $x^i y^j z^k$ satisfies $i \geq 1$, $j \geq 1$, and $k \geq 1$, so that $\{x^a, y^b\}$, $\{x^a, z^c\}$, and $\{y^b, z^c\}$ are edges in $\operatorname{Buch}(I)$. By Proposition 3.9, $\operatorname{Buch}(I)$ is a triangulation of a triangle with r vertices such that $r - 3$ vertices lie in the interior. It follows from Euler's formula and the easy equality $2e = 3(f + 1)$ for any such triangulation that the number of edges is $3r - 6$ and the number of triangles is $2r - 5$. The desired result is now immediate from the minimality in Theorem 3.11. $\square$

3.5 The planar resolution algorithm

Our goal for the rest of this chapter is to demonstrate how the nonplanarity obstacles at the end of Section 3.3 can be overcome.

Definition 3.16 A graph G with at least three vertices is **3-connected** if deleting any pair of vertices along with all edges incident to them leaves a connected graph. Given a set $\mathcal{V}$ of vertices in G, define the **suspension** of G over $\mathcal{V}$ by adding a new vertex to G and connecting it by edges to all vertices in $\mathcal{V}$. The graph G is **almost 3-connected** if it comes with a set $\mathcal{V}$ of three distinguished vertices such that the suspension of G over $\mathcal{V}$ is 3-connected.

The vertex sets of our graphs will be monomials minimally generating some ideal I inside $\Bbbk[x, y, z]$. Note that when I is artinian, such a vertex set contains a distinguished set $\mathcal{V}$ of three vertices: the pure-power generators x^a, y^b, and z^c. Now we come to the main result in this chapter.

Theorem 3.17 *Every monomial ideal I in $\Bbbk[x, y, z]$ has a minimal free resolution by some planar map. If I is artinian then the graph G underlying any such planar map is almost 3-connected.*

The vertices, edges, and bounded regions of this planar map are labeled by their associated "staircase corners" as in the examples above. This determines a complex of free modules over $S = \Bbbk[x, y, z]$ as in (3.2). Let us begin by presenting an algorithm for finding a planar map resolution as in Theorem 3.17 for artinian ideals.

Given a deformation I_ϵ of a monomial ideal $I = \langle m_1, \ldots, m_r \rangle$ with $m_i = x^{a_i} y^{b_i} z^{c_i}$, write the ith deformed generator as $m_{\epsilon,i} = x^{a_{\epsilon,i}} y^{b_{\epsilon,i}} z^{c_{\epsilon,i}}$. Algorithm 3.18 requires a generic deformation satisfying the condition

$$\text{if} \quad a_i = a_j \text{ and } c_i < c_j \quad \text{then} \quad a_{\epsilon,i} < a_{\epsilon,j} \tag{3.3}$$

as well as its analogues via cyclic permutation of (a, b, c). Observe that $c_i < c_j$ is equivalent to $b_i > b_j$ when the condition $a_i = a_j$ is assumed; in other words, if two generators lie at the same distance in front of the yz-plane, then the lower one lies farther to the right (as seen from far out on the x-axis). Condition (3.3) says that among generators that start at the same distance from the yz-plane, the deformation pulls increasingly farther from the yz-plane as the generators move up and to the left.

Keep in mind while reading the algorithm that its geometric content will be explained in the course of its proof of correctness.

Algorithm 3.18 Fix an artinian monomial ideal I inside $\Bbbk[x, y, z]$.

- **initialize** I_ϵ = the strongly generic deformation of I in (3.3), and $G = \text{Buch}(I_\epsilon)$.
- **while** $I_\epsilon \neq I$ **do**
    - **choose** $u \in \{a, b, c\}$ and an index i such that $u_{\epsilon,i}$ is minimal among the deformed u-coordinates satisfying $u_{\epsilon,i} \neq u_i$. Assume (for the sake of notation) that $u = a$, by cyclic symmetry of (a, b, c).
    - **find** the region of G whose monomial label $x^\alpha y^\beta z^\gamma$ has $\alpha = a_{\epsilon,i}$ and γ minimal.
    - **find** the generator $m_{\epsilon,j}$ with the least x-degree among those with y-degree β and z-degree strictly less than γ.
    - **redefine** I_ϵ and G by setting $a_{\epsilon,i} = a_i$ and leaving all other generators alone.
    - **if** $a_j = a_i$ **then** delete from G the edge labeled $x^{a_i} y^\beta z^\gamma$, **else** leave G unchanged
- **output** G

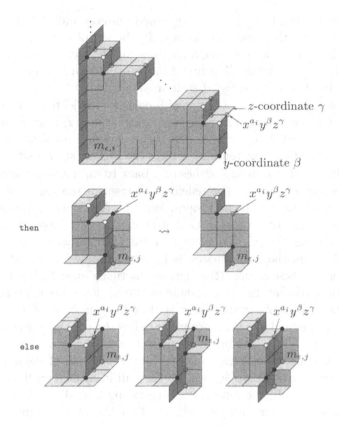

Figure 3.4: The geometry of Algorithm 3.18

The reason for choosing such a specific strongly generic deformation and then being so careful about how the specialization proceeds is that we need control over which syzygy degrees collide at any given stage. In particular, at most one edge should disappear at a time.

Proof of correctness. If I is generic then the algorithm terminates immediately and correctly by Theorem 3.11. By induction on the number of passes through the `while-do` loop, assume that I_ϵ at the beginning of the loop is minimally resolved by the regions, edges, and vertices of G. Once the staircase is rotated so that $u = a$ in the first stage of the loop, it looks near $m_{\epsilon,i}$ like the top image in Fig. 3.4 (this will become clearer as the proof progresses). Gray dots represent minimal generators of I_ϵ, white dots represent regions of G ($=$ second syzygies of I_ϵ), and black dots represent first syzygies.

Even though I_ϵ need not be generic (if the loop has run a few times), $m_{\epsilon,i}$ is still the only generator of I_ϵ lying on the plane $x = a_{\epsilon,i}$, by genericity. Looking from far down the x-axis, it follows that the monomial $m_{\epsilon,i}$ has a vertical plateau behind it (the large medium-gray wall depicted parallel to the yz-plane) that does not continue to the left of $y = b_{\epsilon,i}$. It also follows that there must be an outside corner sharing the same x-coordinate as $m_{\epsilon,i}$, because I_ϵ is artinian. The first find routine simply captures the lowest (and therefore farthest right) such outside corner $x^\alpha y^\beta z^\gamma$.

The right-hand wall of this outside corner, parallel to the xz-plane, must have an inside corner in its relative interior because I_ϵ is artinian (i.e., some generator must divide $x^\alpha y^\beta z^\gamma$ strictly in x and z but share the exponent on y). The other find routine captures the highest such inside corner $m_{\epsilon,j}$.

The redefine routine pushes $m_{\epsilon,i}$ back to m_i, moving the vertical wall back a small amount. The redefined G resolves the redefined I_ϵ, though perhaps not minimally, by Proposition 3.14. The only monomial labels on regions, edges, or vertices of G that change at this stage are those whose x-coordinates change. Therefore, if a relabeled corner of any type (inside, outside, or neither) now shares its label with some other relabeled corner, then one of these corners (the first, say) actually moved while the other did not. In particular, the x-coordinate of the unmoved second corner is a_i.

The crucial observation now is that no generator of I_ϵ can have x-coordinate a_i and also have y-coordinate less than β, because condition (3.3) prevents it. The y-coordinate of any unmoved corner with x-coordinate a_i must therefore be at least β. On the other hand, all of the moved corners of I_ϵ have y-coordinate at most β (they all in fact lie on the boundary of the vertical wall), because these corners must be divisible by $m_{\epsilon,i}$. The only moved corners with y-coordinate β are the outside corner $x^\alpha y^\beta z^\gamma$ and the first syzygy beneath it.

The first syzygy would have to collide with an outside corner in order to become nonminimal, and this is prohibited because that outer corner would divide $x^\alpha y^\beta z^\gamma$. But $x^\alpha y^\beta z^\gamma$ becomes nonminimal if and only if it collides with a first syzygy at $x^{a_i} y^\beta z^\gamma$. This explains the if-then-else routine, keeping in mind that $a_{\epsilon,j} = a_j$, and completes the proof. □

Example 3.19 If $I = \langle x^2, xy, xz, y^2, yz, z^2 \rangle$ is the square of the maximal ideal $\langle x, y, z \rangle$, then $I_\epsilon = \langle x^2, xy^{1.1}, x^{1.1}z, y^2, yz^{1.1}, x^2 \rangle$ is a strongly generic deformation satisfying the condition of Algorithm 3.18. Furthermore, the Buchberger planar map of I_ϵ is the triangle with its edge midpoints connected, as in the left-hand side of Fig. 3.5. If Algorithm 3.18 is run on this I_ϵ, then one of the three nonminimal edges is removed on the first iteration of the while-do loop. Precisely which of the nonminimal edges is removed depends on which $u \in \{x, y, z\}$ is chosen first; any u will work, not just $u = x$ (we have drawn the case $u = x$ in Fig. 3.5). In the remaining two iterations of the while-do loop, no further edges are removed. It is instructive to work out this example by hand, drawing the staircases as well. ◇

Proof of Theorem 3.17. The argument beginning the proof of Theorem 3.11 also works here, reducing everything to the artinian case. Algorithm 3.18

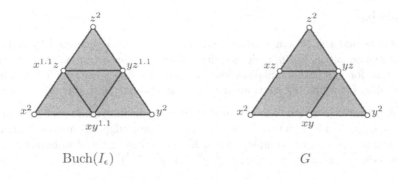

Buch(I_ϵ) $\qquad\qquad\qquad\qquad$ G

Figure 3.5: Algorithmic specialization from Example 3.19

produces a minimal planar map resolution. What remains is to show that the underlying graph G is almost 3-connected.

It is enough to produce three independent paths, one to each of the pure powers x^a, y^b, and z^c, from each generator m_i of I (*independent* means that the paths intersect only at m_i). Leaving the inside corner m_i parallel to the x-axis eventually hits a first syzygy degree. That first syzygy corresponds to an edge e of G. The other endpoint of e is a monomial m_j whose y and z-coordinates are at most those of m_i. Continuing in this manner creates a sequence of edges in G whose vertices have strictly increasing x-coordinates but weakly decreasing y- and z-coordinates. Repeating the procedure for the cyclic permutations of (x, y, z) yields the three desired paths. They intersect only at m_i because of their monotonicity. $\qquad\qquad\square$

Remark 3.20 The independent paths produced in the previous proof constitute an instance of *Menger's Theorem* in the suspended graph. In general, Menger's Theorem says that any two distinct vertices v and w in a k-connected graph have at least k independent paths between them (so the paths pairwise intersect in $\{v, w\}$). An example of three paths produced as in the proof of Theorem 3.17 is illustrated in Fig. 3.6.

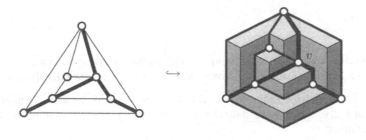

Figure 3.6: Menger's theorem illustrated geometrically

Exercises

3.1 Prove that the $\mathbb{N}^3$-graded Betti numbers of the ideal generated by a fixed set of monomials in $\Bbbk[x, y, z]$ do not depend on the characteristic of $\Bbbk$. Is the same true for sets of monomials in four variables? What is the smallest number of variables for which the Betti numbers can depend on the characteristic of $\Bbbk$?

3.2 A **minor** of a graph G is obtained from G by deleting some vertices (along with all edges incident to them) and contracting some edges. Draw the staircase diagram, and exhibit the complete graph K_5 as a minor in the Buchberger graph of the ideal $\langle x^5, y^5, z^5, x^2yz, xy^2z, x^3z^2, y^3z^2, x^4y^3, x^3y^4 \rangle$.

3.3 Find a family of monomial ideals in $\Bbbk[x, y, z]$ whose Buchberger graphs equal the complete graphs K_n for $n \in \mathbb{N}$.

3.4 Exhibit minimal planar map resolutions of $\langle x, y, z \rangle^r$ for $r \equiv 0, 1 \pmod 3$ that are symmetric under the action of S_3 permuting the variables.

3.5 Fix an integer $r \equiv 2 \pmod 3$. Prove that no minimal planar map resolution of $\langle x, y, z \rangle^r$ can be symmetric under the action of S_3 permuting the variables.

3.6 Let I be the monomial ideal in $\Bbbk[x, y, z]$ whose staircase diagram is presented below. Is I strongly generic? Draw the Buchberger graph of I. Turn the picture upside down and do the same thing. What would you call the first of these two graphs? (It comes up in the context of simplicial topology.)

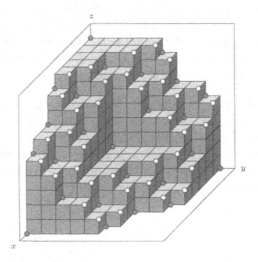

3.7 Describe how to find the uniquely determined third monomial m'' in the proof sketch of Proposition 3.9, given the interior Buchberger edge $\{m, m'\}$.

3.8 Show that an irredundant irreducible decomposition of any artinian monomial ideal I in $\Bbbk[x, y, z]$ can be read off the labels on the regions in any minimal planar map resolution.

3.9 Prove that the K-polynomial of $I \subseteq \Bbbk[x, y, z]$ is the alternating sum of the vertex, edge, and face labels on any planar map G resolving I. Interpret Euler's formula for $v - e + f$ in this context as a statement about the ranks of the free modules occurring in any such resolution of I. Show that no cancellation occurs in the $\mathbb{N}^3$-graded alternating sum if and only if G is minimal.

3.10 Call a monomial ideal I in $\Bbbk[x, y, z]$ **rigid** if its Buchberger graph is naturally *embedded* inside its staircase surface. What conditions guarantee that I is rigid?

3.11 For a rigid monomial ideal, the Buchberger graph comes with a canonical embedding into the staircase surface, so the Buchberger map is well-defined. Prove that the Buchberger map is the *only* planar map resolution of a rigid ideal.

3.12 Exhibit an ideal in $\Bbbk[x, y, z]$ having two distinct minimal planar map resolutions, neither of which is obtained from the other by permuting the variables.

3.13 Exhibit a sequence of monomial ideals in $\Bbbk[x, y, z]$ showing that the number of distinct minimal planar map resolutions of an ideal can be arbitrarily large.

3.14 Prove that the Buchberger graph of any artinian monomial ideal in $\Bbbk[x, y, z]$ is almost 3-connected. More generally, a graph G with at least n vertices *n-connected* if deleting any $n - 1$ vertices along with all their incident edges leaves a connected graph. Call G **almost n-connected** if its suspension over a set $\mathcal{V}$ of n distinguished vertices is n-connected. Prove that the Buchberger graph of any artinian monomial ideal in the polynomial ring $\Bbbk[x_1, \ldots, x_n]$ is almost n-connected.

3.15 Fix a non-squarefree monomial ideal $J = \langle m_1, \ldots, m_r \rangle$ with r minimal generators in $S = \Bbbk[x_1, \ldots, x_n]$, and let $I_\Delta \subseteq \widetilde{S}$ be the polarization of J, as in Section 3.2. The goal of this exercise is to prove that a minimal $\widetilde{S}$-free resolution of $\widetilde{S}/I_\Delta$ descends to a minimal S-free resolution of S/J, in the precise sense of (f) and (g), below. The argument starts with two general lemmas, in (a) and (b).

(a) Let R be an $\mathbb{N}$-graded ring, M an $\mathbb{N}$-graded R-module, and $\theta \in R$ a homogeneous element of degree k. Show that θ is not a zerodivisor on M if and only if the unvariate Hilbert series of M and $M/\theta M$ satisfy

$$H(M, t) = \frac{H(M/\theta M, t)}{1 - t^k}.$$

Hint: See Claim 13.38 in Chapter 13.

(b) Let R be a polynomial ring and $\mathcal{F}_\bullet$ a free resolution of an R-module M. If $\theta \in R$ is not a zerodivisor on M, prove that $\mathcal{F}_\bullet/\theta\mathcal{F}_\bullet$ is a free resolution of $M/\theta M$ over the quotient ring $R/\theta R$. Hint: See Lemma 8.27 in Chapter 8.

The idea will be to apply (b) repeatedly, as one undoes the polarization one step at a time, using (a) at each stage to verify the nonzerodivisor hypothesis. Assume that the highest power of x_j dividing any of the monomials m_i is x_j^a for some $a > 1$. Define the *partial polarization* $J' = \langle m_1', \ldots, m_r' \rangle$ in the polynomial ring $S' = \Bbbk[x_1, \ldots, x_n, y]$ by setting

$$m_i' = \begin{cases} \frac{y}{x_j} m_i & \text{if } x_j^a \text{ divides } m_i \\ m_i & \text{if } x_j^a \text{ does not divide } m_i. \end{cases}$$

(c) Prove that the map $\{\text{monomials } m \in J'\} \rightarrow \mathbb{N} \times \{\text{monomials in } S\}$ sending

$$y^b m \mapsto \begin{cases} (b - 1, x_j \cdot m) & \text{if } m \notin J \\ (b, m) & \text{if } m \in J \end{cases}$$

induces a bijection $\{\text{monomials in } J'\} \rightarrow \mathbb{N} \times \{\text{monomials in } J\}$.

(d) Deduce from (a) and (c) that $\theta = y - x_j$ is not a zerodivisor on S'/J'.

(e) Construct a sequence of partial polarizations starting at J and ending at I_Δ.

(f) Show that the kernel of the map $\widetilde{S} \to S$ is generated by a regular sequence $\Theta = (\theta_1, \theta_2, \ldots)$ in $\widetilde{S}$ such that each θ is a difference of two variables.

(g) Conclude that if $\mathcal{F}_\bullet$ is a minimal free resolution of $\widetilde{S}/I_\Delta$ over $\widetilde{S}$, then $\mathcal{F}_\bullet/\Theta\mathcal{F}_\bullet = \mathcal{F}_\bullet \otimes_{\widetilde{S}} \widetilde{S}/\Theta$ is a minimal free resolution of S/J over S.

Notes

The term "Buchberger graph" appears explicitly here for the first time. These graphs were always lurking as one of the motivations for the concept of genericity in exponents (as opposed to coefficients). Various developers of Gröbner basis software, including Gebauer and Möller [GM88], used versions of the Buchberger graph to avoid unnecessary reductions of s-pairs.

Even in dimension 3, the notion of strong genericity is strictly stronger (meaning "less inclusive") than the genericity that is the subject of Chapter 6. Similarly, strong deformations are particular cases of the deformations in Chapter 6. See the Notes to Chapter 6 for more on the development of these ideas.

The converse to Theorem 3.17 holds as well: every planar graph G that is almost 3-connected appears as the minimal free resolution of some monomial ideal. In fact, there exists a monomial ideal whose staircase surface contains G embedded as its Buchberger graph [Mil02b] via the procedure in the proof of Proposition 3.9. Such *rigid embeddings* connect the algebra and geometry of monomial ideals to order dimension theory for planar maps [Fel01, Fel03, Mil02b].

Solutions to Exercises 3.4 and 3.5 can be found in [MS99].

Solutions to Exercises 3.10 and 3.11 can be found in [Mil02b].

Chapter 4

Cellular resolutions

For monomial ideals in three variables, we found that free resolutions can be described in terms of planar graphs. In this chapter we study the higher-dimensional geometric objects involved in doing similar things for monomial ideals in four and more variables. These geometric objects are derived from the combinatorial data hidden in the generators and their least common multiples. Our aim in this chapter is to show how all monomial ideals "resolve themselves" via geometric resolutions, as suggested by the following picture. Here, the 12 vertices, 18 edges, and 8 two-dimensional faces of the polytope correspond to the Betti numbers 12, 18, and 8.

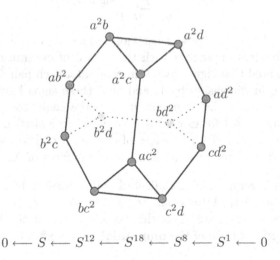

$$0 \longleftarrow S \longleftarrow S^{12} \longleftarrow S^{18} \longleftarrow S^{8} \longleftarrow S^{1} \longleftarrow 0$$

This is the minimal free resolution of $\Bbbk[a, b, c, d]/I$, where I is the ideal generated by the 12 monomials that label the vertices of this polytope.

4.1   Construction and exactness

Definition 4.1 A **polyhedral cell complex** X is a finite collection of convex polytopes (in a real vector space $\mathbb{R}^m$), called **faces** of X, satisfying two properties:

- If $\mathcal{P}$ is a polytope in X and F is a face of $\mathcal{P}$, then F is in X.
- If $\mathcal{P}$ and $\mathcal{Q}$ are in X, then $\mathcal{P} \cap \mathcal{Q}$ is a face of both $\mathcal{P}$ and $\mathcal{Q}$. ·

Here are some examples. The set of all faces of a fixed polytope is a polyhedral cell complex X. For instance, we have just seen such a complex consisting of one 3-polytope, 8 polygons, 18 edges, and 12 vertices. Any simplicial complex on m vertices can be realized as a polyhedral cell complex in $\mathbb{R}^m$. Any planar graph together with its bounded regions can be realized as a polyhedral cell complex in $\mathbb{R}^3$ (this is a consequence of the *Steinitz Theorem* on three-dimensional polytopes [Zie95, Theorem 4.1]).

The polyhedral cell complex X comes equipped with a *reduced chain complex*, which specializes to the usual reduced chain complex for simplicial complexes X. All of the notation and conventions in Chapter 1 regarding reduced chain complexes of simplicial complexes works just as well for polyhedral cell complexes, except that the signs are specified by (arbitrarily) *orienting* the faces of X. (For simplicial complexes, the orientations came implicitly from the ordering on the vertices.) Thus the *boundary chain* of a given face F in X is the signed sum of its facets:

$$\partial(F) \;\; = \sum_{\text{facets } G \subset F} \text{sign}(G, F) \cdot G,$$

where $\text{sign}(G, F)$ is $+1$ if F's orientation induces G's orientation, and -1 otherwise. Readers unfamiliar with the notion of orientation can simply take it for granted that signs have been chosen for each pair $G \subset F$ of faces in X differing in dimension by 1, and that these signs have been chosen consistently, to make the boundary map in the chain complex square to zero. See Example 4.4 for examples of (induced) orientations.

Just as in Chapter 3, the vertices of our cell complexes will come with labels from $\mathbb{N}^n$, and then we can label all of the faces of X.

Definition 4.2 Suppose X is a **labeled cell complex**, by which we mean that its r vertices have **labels** that are vectors $\mathbf{a}_1, \ldots, \mathbf{a}_r$ in $\mathbb{N}^n$. The **label** on an arbitrary face F of X is the exponent $\mathbf{a}_F$ on the least common multiple $\text{lcm}(\mathbf{x}^{\mathbf{a}_i} \mid i \in F)$ of the **monomial labels** $\mathbf{x}^{\mathbf{a}_i}$ on vertices in F.

The point of labeling a cell complex X is to get enough data to construct a monomial matrix for a complex of $\mathbb{N}^n$-graded free modules over the polynomial ring $S = \Bbbk[x_1, \ldots, x_n]$.

Definition 4.3 Let X be a labeled cell complex. The **cellular monomial matrix** supported on X uses the reduced chain complex of X for scalar entries, with $\varnothing$ in homological degree 0. Row and column labels are those on the corresponding faces of X. The **cellular free complex** $\mathcal{F}_X$ **supported** on X is the complex of $\mathbb{N}^n$-graded free S-modules (with basis) represented by the cellular monomial matrix supported on X. The free complex $\mathcal{F}_X$ is a **cellular resolution** if it is acyclic (homology only in degree 0).

By convention, the label on the empty face $\varnothing \in X$ is $\mathbf{0} \in \mathbb{N}^n$, which is the exponent on $1 \in S$, the least common multiple of no monomials. It is also possible to write down the differential ∂ of $\mathcal{F}_X$ without using monomial matrices, where it can be written as

$$\mathcal{F}_X = \bigoplus_{F \in X} S(-\mathbf{a}_F), \qquad \partial(F) = \sum_{\text{facets } G \text{ of } F} \text{sign}(G, F) \mathbf{x}^{\mathbf{a}_F - \mathbf{a}_G} G.$$

The symbols F and G here are thought of both as faces of X and as basis vectors in degrees $\mathbf{a}_F$ and $\mathbf{a}_G$. The sign for (G, F) equals ± 1 and is part of the data in the boundary map of the chain complex of X.

Example 4.4 The following labeled hexagon appears as a face of the three-dimensional polytope at the beginning of this chapter:

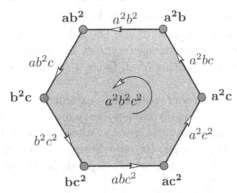

Given the orientations that we have chosen for the faces of X, the cellular free complex $\mathcal{F}_X$ supported by this labeled hexagon is written as follows:

	a^2c^2	a^2bc	a^2b^2	ab^2c	b^2c^2	abc^2			$a^2b^2c^2$
a^2c	1	-1	0	0	0	0		a^2c^2	1
a^2b	0	1	-1	0	0	0		a^2bc	1
ab^2	0	0	1	-1	0	0		a^2b^2	1
b^2c	0	0	0	1	-1	0		ab^2c	1
bc^2	0	0	0	0	1	-1		b^2c^2	1
ac^2	-1	0	0	0	0	1		abc^2	1

$$S^6 \longleftarrow \qquad\qquad\qquad\qquad\qquad S^6 \longleftarrow S \leftarrow 0$$

	a^2c	a^2b	ab^2	b^2c	bc^2	ac^2
1	1	1	1	1	1	1

$$0 \leftarrow S \longleftarrow$$

This is the representation of the resolution in terms of cellular monomial matrices. The arrows drawn in and on the hexagon denote the orientations of its faces, which determine the values of $\text{sign}(G, F)$. For example,

$$\partial\left(\langle\!\!\!\bigcirc\!\!\!\rangle\right) \quad = \qquad b^2\cdots \qquad \Big/ + bc\cdots \qquad \diagdown + c^2\cdots \qquad \overline{}$$

$$+\, ac\cdot \diagup \qquad \cdot + a^2\cdot \diagdown \qquad \cdot + ab\cdots \qquad \underline{}$$

in the non-monomial matrix way of writing cellular free complexes.   ◇

Given two vectors $\mathbf{a}, \mathbf{b} \in \mathbb{N}^n$, we write $\mathbf{a} \preceq \mathbf{b}$ and say that $\mathbf{a}$ *precedes* $\mathbf{b}$, if $\mathbf{b} - \mathbf{a} \in \mathbb{N}^n$. A subset $Q \subseteq \mathbb{N}^n$ is an *order ideal* if $\mathbf{a} \in Q$ whenever $\mathbf{b} \in Q$ and $\mathbf{a} \preceq \mathbf{b}$. Loosely, Q is "closed under going down" in the partial order on $\mathbb{N}^n$. For an order ideal Q, define the labeled subcomplex

$$X_Q \;=\; \{F \in X \mid \mathbf{a}_F \in Q\}$$

of a labeled cell complex X. For each $\mathbf{b} \in \mathbb{N}^n$ there are two important such subcomplexes. By $X_{\preceq \mathbf{b}}$ we mean the subcomplex of X consisting of all faces with labels coordinatewise at most $\mathbf{b}$. Similarly, denote by $X_{\prec \mathbf{b}}$ the subcomplex of X consisting of all faces with labels $\prec \mathbf{b}$, where $\mathbf{b}' \prec \mathbf{b}$ if $\mathbf{b}' \preceq \mathbf{b}$ and $\mathbf{b}' \neq \mathbf{b}$.

A fundamental property of cellular free complexes is that their acyclicity can be determined using merely the geometry of polyhedral cell complexes. Let us call a cell complex *acyclic* if it is either empty or has zero reduced homology. In the empty case, its only homology lies in homological degree -1. The property of being acylic depends on the underlying field $\Bbbk$, as we shall see in Section 4.3.5.

Proposition 4.5 *The cellular free complex $\mathcal{F}_X$ supported on X is a cellular resolution if and only if $X_{\preceq \mathbf{b}}$ is acyclic over $\Bbbk$ for all $\mathbf{b} \in \mathbb{N}^n$. When $\mathcal{F}_X$ is acyclic, it is a free resolution of S/I, where $I = \langle \mathbf{x}^{\mathbf{a}_v} \mid v \in X \text{ is a vertex}\rangle$ is generated by the monomial labels on vertices.*

Proof. The free modules contributing to the part of $\mathcal{F}_X$ in degree $\mathbf{b} \in \mathbb{N}^n$ are precisely those generated in degrees $\preceq \mathbf{b}$. This proves the criterion for acyclicity, noting that if this degree $\mathbf{b}$ complex is acyclic, then its homology contributes to the homology of $\mathcal{F}_X$ in homological degree 0. If $\mathcal{F}_X$ is acyclic, then it resolves S/I because the image of its last map equals $I \subseteq S$.   □

Example 4.6 Let I be the ideal whose generating exponents are the vertex labels on the right-hand cell complex in Fig. 4.1. The label '215' in the diagrams is short for $(2, 1, 5)$. The labeled complex X on the left supports a cellular minimal free resolution of $S/(I + \langle x^5, y^6, z^6\rangle)$, so Proposition 4.5 implies that the subcomplex $\mathcal{F}_{X_{\preceq 455}}$ resolves S/I.   ◇

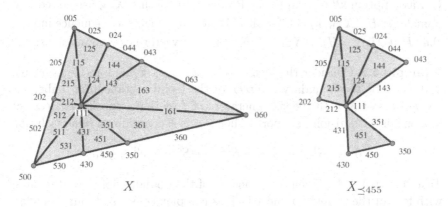

Figure 4.1: The cell complexes from Example 4.6

4.2 Betti numbers and K-polynomials

Given a monomial ideal I with a cellular resolution $\mathcal{F}_X$, we next see how the Betti numbers and the K-polynomial of the monomial ideal I can be computed from the labeled cell complex X. The key is that X satisfies the acylicity criterion of Proposition 4.5. In the forthcoming statement and its proof, we use freely the fact that $\beta_{i,\mathbf{b}}(I) = \beta_{i+1,\mathbf{b}}(S/I)$. As in Chapter 1 for the simplicial case, if X is a polyhedral cell complex and $\Bbbk$ is a field then $\widetilde{H}_\bullet(X;\Bbbk)$ denotes the homology of the reduced chain complex $\widetilde{C}_\bullet(X;\Bbbk)$.

Theorem 4.7 *If $\mathcal{F}_X$ is a cellular resolution of the monomial quotient S/I, then the Betti numbers of I can be calculated for $i \geq 1$ as*

$$\beta_{i,\mathbf{b}}(I) \;=\; \dim_\Bbbk \widetilde{H}_{i-1}(X_{\prec\mathbf{b}};\Bbbk).$$

Proof. When $\mathbf{x^b}$ does not lie in I, the complex $X_{\prec\mathbf{b}}$ consists at most of the empty face $\varnothing \in X$, which has no homology in homological degrees ≥ 0. This is good, because $\beta_{i,\mathbf{b}}(I)$ is zero unless $\mathbf{x^b} \in I$, as $K^{\mathbf{b}}(I)$ is void if $\mathbf{x^b} \notin I$. Now assume $\mathbf{x^b} \in I$, and calculate Betti numbers as in Lemma 1.32 by tensoring $\mathcal{F}_X$ with $\Bbbk$. The resulting complex in degree $\mathbf{b}$ is the complex of vector spaces over $\Bbbk$ obtained by taking the quotient of the reduced chain complex $\widetilde{C}_\bullet(X_{\preceq\mathbf{b}};\Bbbk)$ modulo its subcomplex $\widetilde{C}_\bullet(X_{\prec\mathbf{b}};\Bbbk)$. In other words, the desired Betti number $\beta_{i,\mathbf{b}}(I)$ is the dimension over $\Bbbk$ of the ith homology of the rightmost complex in the following exact sequence of complexes:

$$0 \longrightarrow \widetilde{C}_\bullet(X_{\prec\mathbf{b}};\Bbbk) \longrightarrow \widetilde{C}_\bullet(X_{\preceq\mathbf{b}};\Bbbk) \longrightarrow \widetilde{C}_\bullet(\mathbf{b}) \longrightarrow 0.$$

The long exact sequence for homology reads

$$\cdots \to \widetilde{H}_i(X_{\preceq\mathbf{b}};\Bbbk) \to \widetilde{H}_i(\widetilde{C}_\bullet(\mathbf{b})) \to \widetilde{H}_{i-1}(X_{\prec\mathbf{b}};\Bbbk) \to \widetilde{H}_{i-1}(X_{\preceq\mathbf{b}};\Bbbk) \to \cdots$$

Our assumption $\mathbf{x^b} \in I$ implies by Proposition 4.5 that $X_{\preceq \mathbf{b}}$ has no reduced homology: $\widetilde{H}_j(X_{\preceq \mathbf{b}}; \Bbbk) = 0$ for all j. Hence the long exact sequence implies that $\widetilde{H}_i(\widetilde{\mathcal{C}}_{\bullet}(\mathbf{b})) \cong \widetilde{H}_{i-1}(X_{\prec \mathbf{b}}; \Bbbk)$. Now take $\Bbbk$-vector space dimensions.   □

Example 4.8 Consider the ideal $I = \langle x_1 x_2, x_1 x_3, x_1 x_4, x_2 x_3, x_2 x_4, x_3 x_4 \rangle$, and let X be the boundary complex of the (solid) octahedron. Label the six vertices of X with the six generators of I so that opposite vertices get monomials with disjoint support. Then $\mathcal{F}_X$ is a nonminimal free resolution

$$0 \longleftarrow S^1 \longleftarrow S^6 \longleftarrow S^{12} \longleftarrow S^8 \longleftarrow S^1 \longleftarrow 0.$$

Take $\mathbf{b} = (1,1,1,1)$. Then $X_{\prec \mathbf{b}}$ consists of the boundary of the octahedron with four of the triangles removed. This complex consists of four triangles. Since its reduced homology in homological degree 1 has dimension 3, Theorem 4.7 implies that $\beta_{2,\mathbf{b}}(I) = 3$. If we take $\mathbf{b} = (1,1,1,0)$, then $X_{\prec \mathbf{b}}$ consists of three isolated points, so $\beta_{1,\mathbf{b}}(I) = 2$. Applying these considerations to all squarefree degrees, we conclude that the minimal free resolution of the monomial quotient S/I looks like $0 \leftarrow S^6 \leftarrow S^8 \leftarrow S^3 \leftarrow 0$.   ◇

After labeling the faces of a cell complex X with vectors in $\mathbb{N}^n$, we were able to get homological information about its vertex labels from various subcomplexes of X defined via its face labels. Now let us "$\mathbb{N}^n$-grade" another invariant of X.

Definition 4.9 The **Euler characteristic** of a cell complex X is the alternating sum $\sum_{d \geq -1} (-1)^d f_d(X)$ of the numbers of faces of varying dimensions. The $\mathbb{N}^n$-**graded Euler characteristic** of a labeled cell complex X is the alternating sum of its monomial face labels:

$$\chi(X; x_1, \ldots, x_n) = \sum_{F \in X} (-1)^{1 + \dim F} \mathbf{x}^{\mathbf{a}_F}.$$

The difference in sign from $(-1)^d$ in the ungraded case to $(-1)^{1+\dim F}$ in the $\mathbb{N}^n$-graded case is because cellular free complexes place the empty face $\varnothing \in X$ in homological degree 0 instead of -1.

Lemma 4.10 *The Euler characteristic of a nonempty acyclic cell complex is zero. The Euler characteristic of the irrelevant cell complex $\{\varnothing\}$ is -1.*

Proof. In the irrelevant case, there is only one nonzero chain group; it has rank 1 and homological degree -1. In the nonempty case, the reduced chain complex has zero homology. Therefore the result is precisely the rank-nullity theorem from linear algebra: the alternating sum of the dimensions of vector spaces in an exact sequence of any finite length is zero.   □

If we take Euler characteristics while keeping track of the monomial labels on faces, then we end up with K-polynomials.

Theorem 4.11 *If a labeled cell complex X supports a cellular free resolution of a monomial quotient S/I, then the K-polynomial of S/I equals the $\mathbb{N}^n$-graded Euler characteristic of X:*

$$\mathcal{K}(S/I; x_1, \ldots, x_n) = \chi(X; x_1, \ldots, x_n).$$

Proof. Dividing $\chi(X; \mathbf{x})$ by $(1 - x_1) \cdots (1 - x_n)$ yields an alternating sum of power series that we wish to show is the Hilbert series of S/I. However, the number of times a monomial $\mathbf{x}^\mathbf{b}$ appears in this alternating sum is simply the negative of the ordinary Euler characteristic of the ungraded cell complex underlying $X_{\preceq \mathbf{b}}$. Now apply Lemma 4.10 to Proposition 4.5. (Section 4.3.2 or Corollary 4.20 will show that S/I has a K-polynomial.) $\square$

4.3 Examples of cellular resolutions

In this section we present numerous examples of cellular resolutions. The important case of generic monomial ideals, which are resolved by their *Scarf complexes*, will not be treated here but will be deferred to Chapter 6.

4.3.1 Planar maps

Having now introduced cellular free resolutions, we finally know precisely how planar maps resolve trivariate monomial ideals. The reader might wish to look back at Theorem 3.17 and Algorithm 3.18 to see how they interact with the acyclicity criterion (Proposition 4.5) and the calculation of Betti numbers (Theorem 4.7).

4.3.2 Taylor resolution

The most basic example in arbitrary dimensions is the *Taylor resolution*, where X is the full $(r-1)$-dimensional simplex whose r vertices are labeled by given monomials $\mathbf{x}^{\mathbf{a}_1}, \ldots, \mathbf{x}^{\mathbf{a}_r}$. For any vector $\mathbf{b} \in \mathbb{N}^n$, the subcomplex $X_{\preceq \mathbf{b}}$ is a face of X; namely, it is the full simplex on all monomials $\mathbf{x}^{\mathbf{a}_i}$ dividing $\mathbf{x}^\mathbf{b}$. In particular, $X_{\preceq \mathbf{b}}$ is contractible, and hence the resulting cellular free complex is a cellular resolution by Proposition 4.5.

Note that $\mathcal{F}_X$ is the Taylor resolution of S/I, where $I = \langle \mathbf{x}^{\mathbf{a}_1}, \ldots, \mathbf{x}^{\mathbf{a}_r} \rangle$ is the ideal generated by all vertex labels of X. The Betti numbers of S/I are given by the homology of the simplicial complexes $X_{\preceq \mathbf{b}}$. Therefore, since the faces of X are labeled by least common multiples of the generators of I, the Betti numbers can occur only in such degrees.

Of course, the Taylor resolution tends not to be minimal: its length is r and its rank is 2^r. Combinatorics underlying the Taylor resolution generalize to arbitrary dimension the naïve inclusion–exclusion in Section 3.1. In Chapter 6, we will demonstrate that the Taylor resolution always contains a much smaller resolution of length at most n, namely the Scarf complex of any "generic deformation".

4.3.3 Permutohedron ideals

Let $\mathbf{u} = (u_1, u_2, \ldots, u_n) \in \mathbb{N}^n$ with $u_1 < u_2 < \cdots < u_n$. By permuting the coordinates of $\mathbf{u}$, we obtain $n!$ points in $\mathbb{N}^n \subset \mathbb{R}^n$ constituting the vertices of an $(n-1)$-dimensional polytope called a *permutohedron* $\mathcal{P}(\mathbf{u})$. The *permutohedron ideal* is the ideal $I(\mathbf{u})$ whose (minimal) generators are those monomials obtained by permuting the exponents of the monomial $\mathbf{x}^{\mathbf{u}} = x_1^{u_1} x_2^{u_2} \cdots x_n^{u_n}$. Labeling the vertices of the permutohedron with the generators of the permutohedron ideal in the natural way, we get a cellular resolution minimally resolving $I(\mathbf{u})$.

We now describe the degrees associated to each face of $\mathcal{P}(\mathbf{u})$. Set $[n] = \{1, \ldots, n\}$ and let $\mathbf{v} \in \mathbb{R}^n$. For each subset $\sigma \subseteq [n]$, define $v_\sigma = \sum_{i \in \sigma} v_i$ and $\alpha_\sigma = \sum_{i=1}^{|\sigma|} u_i$. The permutohedron has the inequality description

$$\mathcal{P}(\mathbf{u}) \quad = \quad \{\mathbf{v} \in \mathbb{R}^n \mid v_{[n]} = \alpha_{[n]} \text{ and } v_\sigma \geq \alpha_\sigma \text{ for all } \sigma \subset [n]\}.$$

Each i-dimensional face is determined by a chain of distinct proper subsets

$$\sigma_1 \subset \sigma_2 \subset \cdots \subset \sigma_{n-i-1}$$

of $[n]$ by setting $v_{\sigma_i} = \alpha_{\sigma_i}$ in the inequality description for $\mathcal{P}(\mathbf{u})$. Given any such chain, define $\sigma_0 = \varnothing$ and $\sigma_{n-i} = [n]$. For the corresponding face F,

$$\mathbf{x}^{\mathbf{a}_F} \quad = \quad \prod_{j=1}^{n-i} \prod_{\ell \in \sigma_j \setminus \sigma_{j-1}} x_\ell^{\max\{\sigma_j \setminus \sigma_{j-1}\}}$$

is its monomial label. The hexagon in Example 4.4 is the minimal resolution of a permutohedron ideal $I(0, 1, 2)$. The staircase surface for the *standard* $n = 3$ permutohedron ideal is on the left-hand side of Fig. 4.2, and its minimal cellular resolution is the permutohedron at right.

The ideal $I(\mathbf{u})$ and the polytope $\mathcal{P}(\mathbf{u})$ make perfect sense even if some of the coordinates u_i are equal. In that case, $I(\mathbf{u})$ has fewer than $n!$ generators and $\mathcal{P}(\mathbf{u})$ has fewer than $n!$ vertices. The boundary of the "generalized permutohedron" $\mathcal{P}(\mathbf{u})$ is still a cellular resolution of $I(\mathbf{u})$, but it is not always minimal. For instance, the octahedron $\mathcal{P}(1, 1, 0, 0)$ gives a nonminimal resolution of the ideal $I(1, 1, 0, 0)$, whereas the truncated octahedron $\mathcal{P}(2, 1, 0, 0)$ does give a minimal resolution of $I(2, 1, 0, 0)$. The latter is the example depicted at the beginning of this chapter.

4.3.4 Tree ideals

The *tree ideal* in n variables is defined as

$$I^\star \quad = \quad \left\langle \left(\prod_{s \in \sigma} x_s\right)^{n - |\sigma| + 1} \Bigm| \varnothing \neq \sigma \subseteq \{1, \ldots, n\} \right\rangle.$$

The tree ideal for $n = 3$ has staircase surface at left in Fig. 4.3.

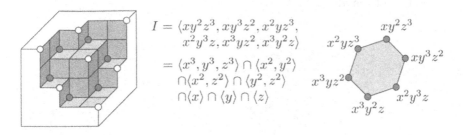

$$I = \langle xy^2z^3, xy^3z^2, x^2yz^3,$$
$$x^2y^3z, x^3yz^2, x^3y^2z \rangle$$
$$= \langle x^3, y^3, z^3 \rangle \cap \langle x^2, y^2 \rangle$$
$$\cap \langle x^2, z^2 \rangle \cap \langle y^2, z^2 \rangle$$
$$\cap \langle x \rangle \cap \langle y \rangle \cap \langle z \rangle$$

Figure 4.2: Permutohedron ideal for $n = 3$

The name "tree ideal" comes from the fact that $I^\star$ has the same number $(n+1)^{n-1}$ of standard monomials as there are labeled trees on $n+1$ vertices. The minimal resolution of $S/I^\star$ is cellular, supported on the barycentric subdivision of an $(n-1)$-simplex. These ideals will be investigated in Chapter 6, where it is shown that tree ideals are generic and also Alexander dual (Chapter 5) to permutohedron ideals. At this point, let us simply note that the cellular resolution of the permutohedron ideal is simple, while the resolution of the tree ideal is simplicial.

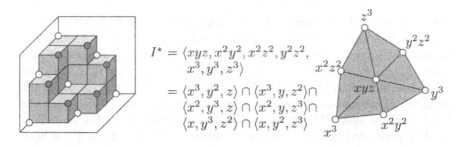

$$I^\star = \langle xyz, x^2y^2, x^2z^2, y^2z^2,$$
$$x^3, y^3, z^3 \rangle$$
$$= \langle x^3, y^2, z \rangle \cap \langle x^3, y, z^2 \rangle \cap$$
$$\langle x^2, y^3, z \rangle \cap \langle x^2, y, z^3 \rangle \cap$$
$$\langle x, y^3, z^2 \rangle \cap \langle x, y^2, z^3 \rangle$$

Figure 4.3: Tree ideal for $n = 3$

To convince yourself of the duality between tree ideals and permutohedron ideals, compare the staircase diagram in Fig. 4.3 to the one in Fig. 4.2. Note how dots of the same color correspond in the two staircases. The minimal resolution is drawn on the right in Fig. 4.3.

4.3.5 The minimal triangulation of $\mathbb{RP}^2$

Consider the Stanley–Reisner ideal of the minimal triangulation of the real projective plane. The cellular dual to the triangulation is a cell complex X consisting of six pentagons, where opposite edges are to be identified in the antiparallel orientations. Label the 10 vertices of X with the minimal generators of the ideal:

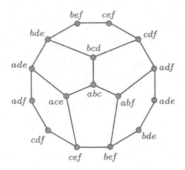

If the field $\Bbbk$ has characteristic not equal to 2, then X is acyclic, and the cellular complex $\mathcal{F}_X$ is the minimal free resolution

$$0 \longleftarrow S \longleftarrow S^{10} \longleftarrow S^{15} \longleftarrow S^6 \longleftarrow 0$$

of the Stanley–Reisner ring of the minimal triangulation of $\mathbb{RP}^2$. On the other hand, if $\Bbbk$ has characteristic 2, then X is not acyclic.

4.3.6   Simple polytopes

A convex polytope of dimension d is *simple* if every vertex meets d edges. Every simple polytope $\mathcal{P}$ gives a minimal cellular resolution of a squarefree monomial ideal in S naturally associated to $\mathcal{P}$, as follows. Suppose $\mathcal{P}$ has facets $F_1, \ldots, F_n$ and vertices $v_1, \ldots, v_r$. Label each vertex v_i of $\mathcal{P}$ by the squarefree monomial $\prod_{v_i \notin F_j} x_j$. Each face is labeled by the product of the variables x_i corresponding to the facets not containing that face. Then the labeled cell complex $\mathcal{P}$ supports a cellular resolution $\mathcal{F}_{\mathcal{P}}$ of the monomial ideal $I_{\mathcal{P}}$ generated by the labels on its vertices. The resolution $\mathcal{F}_{\mathcal{P}}$ is both minimal and linear (Definition 2.16). These properties rely on the fact that $\mathcal{P}$ is a simple polytope; the reader is asked to supply a proof in Exercise 4.5.

Example 4.12 When $\mathcal{P}$ is two-dimensional, so $\mathcal{P}$ is a polygon, these resolutions follow the pattern

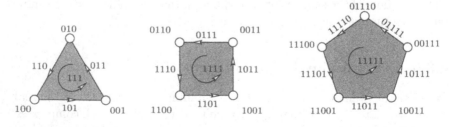

The number of variables equals the number of facets of $\mathcal{P}$.   ◇

The ideal $I_{\mathcal{P}}$ plays an important role in the study of toric varieties (cf. Chapter 10). Briefly, each smooth (or just simplicial) projective toric variety is specified by a simple polytope $\mathcal{P}$ called its *moment polytope*. The facets of $\mathcal{P}$ correspond to the torus invariant divisors on the toric variety. The ideal $I_{\mathcal{P}}$ is the *irrelevant ideal* of the toric variety inside its *homogeneous coordinate ring* $S = \Bbbk[x_1, \dots, x_n]$, which means that sheaves on the toric variety are represented by suitably graded S-modules that are saturated with respect to $I_{\mathcal{P}}$. Hence the cellular resolution $\mathcal{F}_{\mathcal{P}}$ is closely related to computation of sheaf cohomology on toric varieties.

4.3.7 Squarefree monomial ideals revisited

In this subsection we generalize the octahedron in Example 4.8. Suppose that I is an ideal generated by squarefree monomials $\mathbf{x}^\sigma$ of degree d, so each one satisfies $\sigma \in \{0,1\}^n$ and $d = |\sigma|$. The *Newton polytope* of I is the convex hull of the exponent vectors σ of the generators of I. This is a polytope of dimension $\leq n-1$ because it lies inside the $(n-1)$-simplex Δ consisting of all nonnegative vectors in $\mathbb{R}^n$ with coordinate sum d.

Let X be the labeled boundary complex of the Newton polytope of I. It will follow from Theorem 4.17 that $\mathcal{F}_X$ is a cellular free resolution of S/I. Using Theorem 4.7, we can determine the Betti numbers of I as follows. For $\tau \in \{0,1\}^n$, let Δ^τ be the relatively open face of Δ consisting of all points whose support equals the support of τ. In particular, $\Delta_{(1,\dots,1)}$ denotes the interior of Δ. Let $\partial\Delta^\tau$ denote the boundary of the simplex Δ^τ. Then

$$X_{\prec\tau} = X \cap \partial\Delta^\tau \quad \text{and} \quad \beta_{i,\tau}(I) = \dim_{\Bbbk} \widetilde{H}_{i-1}(X \cap \partial\Delta^\tau; \Bbbk).$$

The minimal free resolution of I measures homologically how the inclusion of polytopes $X \subset \Delta$ restricts to the boundary of each face of the simplex.

4.4 The hull resolution

In Chapter 3 we exploited the geometry of staircases—that is, the manner in which exponent vectors in $\mathbb{N}^n$ also sit in $\mathbb{R}^n$—to produce free resolutions via planar graphs. Now, with the machinery of cellular resolutions, we construct canonical free resolutions of monomial ideals in arbitrary dimension from this geometry. These resolutions will generally be nonminimal, but their length is always bounded above by n. Given a real number $t \in \mathbb{R}$ and a vector $\mathbf{a} \in \mathbb{N}^n$, set

$$t^{\mathbf{a}} = (t^{a_1}, \dots, t^{a_n}) \in \mathbb{R}^n.$$

Fix a monomial ideal I and $t \in \mathbb{R}$. Consider the closed convex set

$$\mathcal{P}_t = \text{conv}\{t^{\mathbf{a}} \mid \mathbf{x}^{\mathbf{a}} \in I\} \subset \mathbb{R}^n$$

and assume that $t \geq 1$. We show that the finite set $\min(I)$ of minimal generators of I contains all the extreme points of the convex set $\mathcal{P}_t$. The reverse inclusion is also true, but we defer this to Corollary 4.19.

Lemma 4.13 *The set $\mathcal{P}_t$ is a polyhedron in $\mathbb{R}^n$. More precisely, we have*

$$\mathcal{P}_t \;=\; \mathbb{R}^n_{\geq 0} + \operatorname{conv}\{t^{\mathbf{a}} \mid \mathbf{x}^{\mathbf{a}} \in \min(I)\}.$$

Here, $\mathbb{R}^n_{\geq 0}$ denotes the orthant consisting of all nonnegative real vectors.

Proof. First we prove the inclusion $\subseteq$. Let $\mathbf{x}^{\mathbf{b}}$ be any monomial in I. Then there is a minimal generator $\mathbf{x}^{\mathbf{a}} \in \min(I)$ dividing $\mathbf{x}^{\mathbf{b}}$. This implies $t^{a_i} \leq t^{b_i}$ for all i, and hence $t^{\mathbf{b}} - t^{\mathbf{a}}$ lies in $\mathbb{R}^n_{\geq 0}$. Thus $t^{\mathbf{b}}$ lies in $t^{\mathbf{a}} + \mathbb{R}^n_{\geq 0}$, which is contained in $\mathbb{R}^n_{\geq 0} + \operatorname{conv}\{t^{\mathbf{a}} \mid \mathbf{x}^{\mathbf{a}} \in \min(I)\}$. Since this latter set is convex, it must contain the convex hull $\mathcal{P}_t$ of all points $t^{\mathbf{b}}$ with $\mathbf{x}^{\mathbf{b}} \in I$.

For the other inclusion, we prove that $t^{\mathbf{a}} + \mathbb{R}^n_{\geq 0} \subseteq \mathcal{P}_t$ if $\mathbf{x}^{\mathbf{a}} \in \min(I)$. Fix $t^{\mathbf{a}} + \mathbf{u} \in t^{\mathbf{a}} + \mathbb{R}^n_{\geq 0}$ for a nonnegative real vector $\mathbf{u} = (u_1, \ldots, u_n)$. Choose a positive integer r such that $0 \leq u_j \leq t^{a_j + r} - t^{a_j}$ for $j = 1, \ldots, n$. Let C be the convex hull of the 2^n points $t^{\mathbf{a}} + \sum_{j \in J}(t^{a_j + r} - t^{a_j}) \cdot \mathbf{e}_j$ where J runs over all subsets of $\{1, \ldots, n\}$. These points represent the monomials $\mathbf{x}^{\mathbf{a}} \cdot \prod_{j \in J} x_j^r$. The cube C is contained in $\mathcal{P}_t$ and contains $t^{\mathbf{a}} + \mathbf{u}$. $\square$

Proposition 4.14 *The **face poset** (i.e., the set of faces partially ordered by inclusion) of the polyhedron $\mathcal{P}_t$ is independent of $t \in \mathbb{R}$ for $t > (n+1)!$. The same holds for the subposet consisting of all bounded faces of $\mathcal{P}_t$.*

Proof. The face poset of $\mathcal{P}_t$ can be computed as follows. Let $C_t \subset \mathbb{R}^{n+1}$ be the cone spanned by the vectors $(t^{\mathbf{a}}, 1)$ for all minimal generators $\mathbf{x}^{\mathbf{a}}$ of I together with the unit vectors $(\mathbf{e}_i, 0)$ for $i = 1, \ldots, n$. The faces of $\mathcal{P}_t$ are in order-preserving bijection with the faces of C_t that do not lie in the hyperplane $x_{n+1} = 0$ "at infinity". A face of $\mathcal{P}_t$ is bounded if and only if the corresponding face of C_t contains none of the vectors $(\mathbf{e}_i, 0)$. It suffices to prove that the face poset of C_t is independent of t.

Consider any $(n+1)$-tuple of generators of the cone C_t, written as the columns of a square matrix, and compute the sign of its determinant:

$$\operatorname{sign} \det \begin{bmatrix} \mathbf{e}_{i_0} & \cdots & \mathbf{e}_{i_r} & t^{\mathbf{a}_{j_1}} & \cdots & t^{\mathbf{a}_{j_{n-r}}} \\ 0 & \cdots & 0 & 1 & \cdots & 1 \end{bmatrix} \in \{-1, 0, +1\}. \tag{4.1}$$

The list of these signs forms the *oriented matroid* of the cone C_t. It is known that the face poset of a polyhedral cone is determined by its oriented matroid. For details see [BLSWZ99, Chapter 9]. It therefore suffices to show that the sign of the determinant in (4.1) is independent of t as long as $t > (n+1)!$. This follows from the next lemma. $\square$

Lemma 4.15 *Let a_{ij} be integers for $1 \leq i, j \leq r$. Then the Laurent polynomial $f(t) = \det([t^{a_{ij}}]_{1 \leq i, j \leq r})$ either vanishes identically or has no real roots for $t > r!$.*

Proof. Suppose that f is not the zero polynomial and write $f(t) = c_\alpha t^\alpha + \sum_\beta c_\beta t^\beta$, where the first term has the highest degree in t. For $t > r!$ we have the chain of inequalities

$$\left| \sum_\beta c_\beta \cdot t^\beta \right| \leq \sum_\beta |c_\beta| \cdot t^\beta \leq \left(\sum_\beta |c_\beta| \right) \cdot t^{\alpha-1} < r! \cdot t^{\alpha-1} < t^\alpha \leq |c_\alpha \cdot t^\alpha|.$$

Therefore $f(t)$ is nonzero, and $\mathrm{sign}(f(t)) = \mathrm{sign}(c_\alpha)$. $\qquad\square$

Definition 4.16 The **hull complex** $\mathrm{hull}(I)$ of a monomial ideal I is the polyhedral cell complex of all bounded faces of $\mathcal{P}_t$ for $t \gg 0$. This complex is naturally labeled, with each vertex corresponding to a minimal generator of I. The cellular free complex $\mathcal{F}_{\mathrm{hull}(I)}$ is called the **hull resolution** of I.

Our terminology for $\mathcal{F}_{\mathrm{hull}(I)}$ is justified by the next theorem.

Theorem 4.17 *The cellular free complex $\mathcal{F}_{\mathrm{hull}(I)}$ is a resolution of S/I.*

For the proof, we make use of the following general result from topological combinatorics.

Lemma 4.18 *Let F be a face of a polytope Q. If K is the subcomplex of ∂Q consisting of all faces of Q that are disjoint from F, then K is contractible.*

Proof. Consider the barycentric subdivision $B(\partial Q)$ of the boundary of Q. This is a triangulation of ∂Q whose simplices are in bijection with chains (*flags*) of faces of Q. A geometric realization of $B(\partial Q)$ is determined by selecting one point in the relative interior of each face of Q (the faces are then convex hulls of the vertices corresponding to a flag of faces). We construct a particular realization of $B(\partial Q)$ by selecting a hyperplane H that separates the vertices of F from all other vertices of Q. For each face of Q that meets F but is not contained in F, select the point in the relative interior of that face to lie in the hyperplane H.

Let $H_{\geq 0}$ denote the closed half-space of H containing F and let $H_{<0}$ denote the complementary open half-space. Then $\partial Q \cap H_{<0}$ is an open ball that comes with a distinguished triangulation by $B(\partial Q)$, using a *Schlegel diagram* [Zie95, Definition 5.5]. We construct a deformation retraction from $\partial Q \cap H_{<0}$ to its subcomplex $B(K)$ as follows. Each point p in the difference $(\partial Q \cap H_{<0}) \setminus B(K)$ lies interior to a unique simplex of $B(\partial Q)$. That simplex has a nonempty face in $B(K)$, and it contains a unique point P such that p is a convex combination of P and the vertices of the complementary face. The deformation retraction is obtained by linearly connecting each point p in $(\partial Q \cap H_{<0}) \setminus B(K)$ to the corresponding point P in $B(K)$.

This shows that the simplicial complex $B(K)$ is contractible, and hence so is the polyhedral cell complex K. $\qquad\square$

Proof of Theorem 4.17. Let $X = (\mathrm{hull}(I))_{\preceq \mathbf{b}}$ for some degree $\mathbf{b}$. By Proposition 4.5, we need to show that the cell complex X is acyclic over $\Bbbk$. We will even show that X is contractible. If X is empty or a single vertex then this is immediate. Otherwise, choose a real number $t > (n+1)!$, and let $\mathbf{v} = t^{-\mathbf{b}}$. If $t^{\mathbf{a}}$ is a vertex of X then $\mathbf{a} \preceq \mathbf{b}$, so

$$t^{\mathbf{a}} \cdot \mathbf{v} \;=\; t^{-\mathbf{b}} \cdot t^{\mathbf{a}} \;\leq\; t^{-\mathbf{b}} \cdot t^{\mathbf{b}} \;=\; n,$$

whereas for any other monomial $\mathbf{x}^{\mathbf{c}} \in I$ we have $c_i \geq b_i + 1$ for some i, so

$$t^{\mathbf{a}} \cdot \mathbf{v} \;=\; t^{-\mathbf{b}} \cdot t^{\mathbf{c}} \;\geq\; t^{c_i - b_i} \;\geq\; t \;>\; n.$$

Thus the hyperplane H defined by $\mathbf{x} \cdot \mathbf{v} = n$ separates the vertices of X from all other vertices of $\mathcal{P}_t$. Let $H_{\geq 0}$ be the half-space containing the vertices of X. Then $F = \mathcal{P}_t \cap H$ is a face of the polytope $\mathcal{Q} = \mathcal{P}_t \cap H_{\geq 0}$ and X is precisely the complex of faces of $\mathcal{Q}$ which are disjoint from F. Lemma 4.18 implies that X is contractible and hence acyclic.   □

Corollary 4.19 *The vertices of the polyhedron $\mathcal{P}_t$ are exactly the points $t^{\mathbf{a}}$ for which $\mathbf{x}^{\mathbf{a}}$ is a minimal generator of I.*

The hull resolution is a canonical (though usually nonminimal) cellular resolution of S/I for any monomial ideal I. Used with Theorem 4.11, this yields an explicit formula for the numerator of the $\mathbb{N}^n$-graded Hilbert series.

Corollary 4.20 *Every monomial quotient S/I has a K-polynomial; it is given by the $\mathbb{N}^n$-graded Euler characteristic $\chi(\mathrm{hull}(I); \mathbf{x})$ of its hull complex.*

Of course, we could have stated the same corollary using the full Taylor resolution (Section 4.3.2) instead of the hull complex; but the hull complex has length at most n (the number of variables), and it has far fewer free summands. All $\mathbb{N}^n$-graded degrees of cells in $\mathrm{hull}(I)$ are exponents of least common multiples of some generators of I. The hull complex efficiently organizes inclusion–exclusion on the poset of least common multiples (the *lcm-lattice*). Compare our discussion of inclusion–exclusion in Section 3.1.

Example 4.21 If $I = \mathfrak{m}$ and $t > 1$, then $\mathcal{P}_t$ has n vertices $(t, 1, 1, \ldots, 1)$, $(1, t, 1, \ldots, 1), \ldots, (1, 1, 1, \ldots, t)$. Their convex hull is an $(n-1)$-simplex, and this is the only bounded facet of $\mathcal{P}_t$. Hence $\mathrm{hull}(I)$ is an $(n-1)$-simplex and the cellular complex $\mathcal{F}_{\mathrm{hull}(I)}$ is the Koszul complex, as in Section 1.4. ◇

Example 4.22 The staircase diagram of $I = \langle x, y, z \rangle^5$ is at left below. For the hull resolution, consider the convex hull of $\{(t^i, t^j, t^k) \mid i + j + k = 5\}$, for $t > 1$, and look at this convex polyhedron from the point $(1, 1, 1)$:

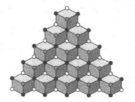 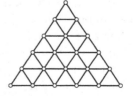

The hull resolution of $\langle x, y, z \rangle^5$, depicted at right above, respects the S_3-symmetry. In general, the hull resolution of $\langle x, y, z \rangle^d$ has two classes of second syzygies: the "up" triangles and the "down" triangles. The three edges of any "down" triangle have the same label (the coordinates of the black dots in the staircase), and they are the reason for nonminimality: there should be two edges in each such degree.

It is always possible using Algorithm 3.18 to remove edges from a cellular resolution of an ideal in three variables to get a minimal cellular resolution. In the present case, we have to remove one edge from each "down" triangle. When the power d is congruent to 0 or 1 (mod 3), it is even possible to retain the S_3-symmetry. (The reader is invited to produce symmetric resolutions and to prove that none exist for 2 (mod 3); check [MS99] if necessary.) ◇

Example 4.23 Consider the ideal $I = \langle x^2 z, xyz, y^2 z, x^3 y^5, x^4 y^4, x^5 y^3 \rangle$. In Chapter 3 we saw that the Buchberger graph of I contains the nonplanar graph $K_{3,3}$. The minimal free resolution of S/I has the format

$$0 \longleftarrow S^1 \longleftarrow S^6 \longleftarrow S^7 \longleftarrow S^2 \longleftarrow 0,$$

where $S = \Bbbk[x, y, z]$. The computer algebra system Macaulay 2 [GS04] produces the minimal cellular resolution on the left:

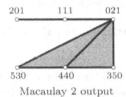

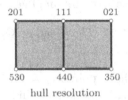

Macaulay 2 output   hull resolution

The hull resolution for this example is depicted on the right. It is also minimal, but it is more symmetric than the minimal resolution on the left. ◇

Example 4.24 The minimal resolution of the permutohedron ideal in Section 4.3 is its hull resolution. ◇

Example 4.25 Not every minimal cellular resolution is a hull resolution. We will see systematic failures in Chapter 6, provided by cogeneric monomial ideals. For a specific counterexample, the cellular resolution of the real projective plane $\mathbb{RP}^2$ in Section 4.3.5 is not the hull resolution. It is a good exercise to check this directly by writing down the hull complex explicitly, but there is a much easier reason: the hull complex is independent of characteristic, whereas the minimal resolution of $\mathbb{RP}^2$ is not. ◇

Example 4.26 In fact, the cellular resolutions associated to simple polytopes in Section 4.3.6 are usually not hull resolutions, either. This is essentially because not every simple polytope has the same combinatorial type as the convex hull of a collection of squarefree vectors (i.e., in $\{0, 1\}^n$). ◇

4.5   Subdividing the simplex

For artinian ideals, there are some useful geometric properties of hull resolutions that makes them a little more tangible. Apart from the intrinsic interest in the main result of this section, Theorem 4.31, it has applications to the duality for Betti numbers (Theorem 5.48) and the characterization of generic monomial ideals (Theorem 6.26).

Suppose that J is an *artinian* monomial ideal, meaning that $x_1^{d_1}, \ldots, x_n^{d_n}$ are among its minimal generators for some strictly positive $d_1, \ldots, d_n$. Choose $t > (n+1)!$, and let $v_1, \ldots, v_n$ be the vertices of the polyhedron $\mathcal{P}_t$ from Definition 4.16 determined by these minimal generators. The convex hull of the points $v_1, \ldots, v_n$ is an $(n-1)$-dimensional simplex that we denote by $\Delta(J)$. Join this simplex with the special point $\mathbf{1} = (1, \ldots, 1) = t^{\mathbf{0}}$ to form an n-dimensional simplex, and intersect this n-simplex with the polyhedron $\mathcal{P}_t$ to get the following new convex polytope:

$$\mathcal{Q}_t \;=\; \operatorname{conv}\big(\{\mathbf{1}\} \cup \Delta(J)\big) \cap \mathcal{P}_t.$$

If J has no minimal generators other than $x_1^{d_1}, \ldots, x_n^{d_n}$, then $\mathcal{Q}_t = \Delta(J)$. In this case, the hull complex equals the Koszul complex on $x_1^{d_1}, \ldots, x_n^{d_n}$ and is a minimal free resolution of J. In what follows, we assume that J has at least one more minimal generator $x_1^{i_1} \cdots x_n^{i_n}$ with $i_1 < d_1, \ldots, i_n < d_n$.

Lemma 4.27 *The polytope $\mathcal{Q}_t$ is n-dimensional, and it has the simplex $\Delta(J)$ as a facet.*

Proof. The n points $v_i = (1, \ldots, 1, t^{d_i}, 1, \ldots, 1)$ and the additional point $\mathbf{1} = (1, \ldots, 1)$ are affinely independent. Their convex hull is the translate by $\mathbf{1}$ of an n-simplex given by the origin and n points on the n positive coordinate rays. Let w be the vector in $\mathbb{R}^n$ whose i^{th} coordinate is

$$w_i \;=\; \frac{1}{t^{d_i} - 1} \;>\; 0.$$

This vector has the same inner product with each of the vertices $v_1, \ldots, v_n$ of the $(n-1)$-simplex $\Delta(J)$. The value of this inner product tends to n for $t \to \infty$. The inner product of w with $\mathbf{1}$ tends to 0 for $t \to \infty$. Hence w is an outer normal vector for the facet $\Delta(J)$ of the n-simplex $\operatorname{conv}(\{\mathbf{1}\} \cup \Delta(J))$.

Now consider any other vertex of $\mathcal{P}_t$. It has the form

$$(t^{i_1}, t^{i_2}, \ldots, t^{i_n}), \quad \text{where} \quad i_1 < d_1, \ldots, i_n < d_n.$$

The inner product of w with this vector tends to 0 for $t \to \infty$. This implies that this vector lies in the simplex $\operatorname{conv}(\{\mathbf{1}\} \cup \Delta(J))$ but not on the facet $\Delta(J)$. Therefore $\mathcal{Q}_t$ is an n-dimensional polytope, and the face of $\mathcal{Q}_t$ with outer normal vector w is the $(n-1)$-simplex $\Delta(J)$. $\square$

Lemma 4.28 *Every bounded face of $\mathcal{P}_t$ is a face of $\mathcal{Q}_t$.*

Proof. If F is a face of $\mathcal{P}_t$ then $F \cap \mathcal{Q}_t$ is a face of $\mathcal{Q}_t$ because $\mathcal{Q}_t \subseteq \mathcal{P}_t$. Suppose that F is bounded. Then F is the convex hull of a subset of the vertices of $\mathcal{P}_t$. But since all vertices of $\mathcal{P}_t$ lie in $\mathcal{Q}_t$, it follows that $F = F \cap \mathcal{Q}_t$. □

Lemma 4.29 *A face F of $\mathcal{Q}_t$ is a face of $\mathcal{P}_t$ if and only if F has a strictly positive inner normal vector (all coordinates positive). The collection of such faces F is the hull complex* hull(J).

Proof. Suppose F is the face of $\mathcal{Q}_t$ at which a strictly positive vector w attains its minimum. Then the face of $\mathcal{P}_t$ at which w attains its minimum is bounded, and it contains F, so by the previous lemma it must equal F. Hence $F \in$ hull(J). For the converse, suppose that F is a face of $\mathcal{Q}_t$ at which a vector w with $w_i = 0$ for some i attains its minimum. Let F' be the face of $\mathcal{P}_t$ at which w attains its minimum. Then $F' + \mathbb{R}_{\geq 0} e_i \subseteq F'$, which means that F' is unbounded. Hence $F' \cap \mathcal{Q}_t = F$ but $F' \neq F$, which means that F is not a face of $\mathcal{P}_t$. In particular, $F \notin$ hull(J). □

Definition 4.30 A **polyhedral subdivision** of a polytope $\mathcal{P}$ in $\mathbb{R}^n$ is a polyhedral cell complex X whose underlying space $|X|$ equals $\mathcal{P}$. This means that $\mathcal{P}$ is the union of the polytopes in X. If all polytopes in X are simplices, then X is a **triangulation** of $\mathcal{P}$. More generally, X is a **polyhedral subdivision** of a polyhedral cell complex Y if $|X| = |Y|$ and every face F in Y is polyhedrally subdivided by the cells of X contained in F.

There is a general construction [Zie95, Definition 5.5] in the theory of convex polytopes that uses a polytope $\mathcal{Q}$ to induce a polyhedral subdivision of a chosen facet Δ. This subdivision is called the *Schlegel diagram* of the polytope $\mathcal{Q}$ on the facet Δ. It is a technique for visualizing n-dimensional polytopes in dimension $n-1$, in particular for $n = 3$ and $n = 4$. If all extra generators $x_1^{i_1} \cdots x_n^{i_n}$ have full support, then hull(J) is precisely the Schlegel diagram for $\mathcal{Q}_t$ on the facet $\Delta(J)$. If some extra generators $x_1^{i_1} \cdots x_n^{i_n}$ do not have full support, then what we get is not the Schlegel diagram, but it is almost as good. Here is our main result in this section.

Theorem 4.31 *The hull complex* hull(J) *of an artinian monomial ideal J in n variables is a polyhedral subdivision of the $(n-1)$-simplex $\Delta(J)$. A face G lies in the boundary of* hull(J) *if and only if $\mathbf{a}_G$ fails to have full support.*

Proof. Pick any point p in the simplex $\Delta(J)$ and imagine walking from p toward the point $\mathbf{1}$ along a straight line segment ℓ. Since $\mathcal{Q}_t$ is a closed subset of $\mathbb{R}^n$, there is a unique last point $\ell(p)$ along ℓ that is still in $\mathcal{Q}_t$. Let $L(p)$ denote the unique face of $\mathcal{Q}_t$ that contains $\ell(p)$ in its relative interior. Consider an inner normal ν to $\mathcal{Q}_t$ along $L(p)$. The set $\mathcal{I}(p)$ of coordinates i such that $p_i > 1$ coincides with the set of coordinates i such that $\ell(p)_i > 1$, because $\mathbf{1} \notin \mathcal{Q}_t$. The vector ν is strictly positive in each coordinate $i \in \mathcal{I}(p)$, since otherwise ν would be smaller at some vertex of $\Delta(J)$.

On the other hand, ν_i for $i \notin \mathcal{I}(p)$ can be made as large as desired without changing the fact that ν is an inner normal to $\mathcal{Q}_t$ along $L(p)$. Lemma 4.29 implies that $L(p)$ is a face of $\mathcal{P}_t$, and hence that $L(p)$ is in hull(J).

Let us call two points p and p' *equivalent* if $L(p) = L(p')$. The set of equivalence classes defines a subdivision of $\Delta(J)$, with the cell containing p in this subdivision being affinely isomorphic to the polytope $L(p)$ by projection from $\mathbf{1}$. Hence we get a polyhedral subdivision of $\Delta(J)$ that is isomorphic to a subcomplex of hull(J). This subcomplex is all of hull(J) because the ray from $\mathbf{1}$ to any point of $\mathcal{P}_t$ eventually pierces the simplex $\Delta(J)$.

The second claim follows because the support of $\mathbf{a}_{L(p)}$ is $\mathcal{I}(p)$.   □

Subdivisions of convex polytopes that arise from a polytope in one higher dimension in the manner described earlier are called *regular subdivisions*. We can therefore summarize our discussion as follows.

Corollary 4.32 *The hull complex of an artinian monomial ideal is a regular subdivision of the simplex.*

For details on the construction and algorithmic aspects of regular subdivisions we refer to the book of De Loera, Rambau, and Santos [DRS04]. We close with an instructive example in three dimensions.

Example 4.33 Consider the following subideal of the one in Example 4.22:

$$J \;=\; \langle x^5,\, y^5,\, z^5,\, x^3 y^2,\, x^2 y^3,\, x^3 yz,\, x^2 yz^2,\, xy^3 z,\, xy^2 z^2 \rangle.$$

The three-dimensional polytope $\mathcal{Q}_t$ has two distinguished facets, namely the triangle $\Delta(J)$ with vertex set $\{(t^5, 1, 1), (1, t^5, 1), (1, 1, t^5)\}$ and the hexagon with vertex set $\{(t^3, t^2, 1), (t^2, t^3, 1), (t^3, t, t), (t^2, t, t^2), (t, t^3, t), (t, t^2, t^2)\}$. These two facets are joined by a band of six additional facets, namely three triangles and three quadrangles. In total, $\mathcal{Q}_t$ has 9 vertices, 15 edges, and 8 facets. The construction in the previous proof amounts to looking at the polytope $\mathcal{Q}_t$ from the eye point $(1, 1, 1)$. The hull complex hull(J) is the subcomplex of the boundary of $\mathcal{Q}_t$ that is visible from $(1, 1, 1)$. Of course, the triangle $\Delta(J)$ is not visible. Also not visible are the edge connecting $(t^5, 1, 1)$ and $(1, t^5, 1)$ and the facet formed by this edge with the edge connecting $(t^3, t^2, 1)$ and $(t^2, t^3, 1)$. Thus hull(J) is a subdivision of a triangle with one hexagon, two quadrangles, and three triangles. In total, the hull complex hull(J) has 9 vertices, 14 edges, and 6 facets. The algebraic hull complex $\mathcal{F}_{\text{hull}(J)}$ is a minimal free resolution of $\Bbbk[x, y, z]/J$.   ◇

Exercises

4.1 Draw pictures illustrating Example 4.33, or get a computer to do it for you.

4.2 For an arbitrary monomial ideal I and an arbitrary positive integer t, prove that the K-polynomial of the Frobenius power $I^{[t]}$ of I (Exercise 1.6) satisfies $\mathcal{K}(I^{[t]}; \mathbf{x}) = \mathcal{K}(I; \mathbf{x}^{[t]})$, where $\mathbf{x}^{[t]} = (x_1^t, \ldots, x_n^t)$.

4.3 A **weakly labeled cell complex** X has labels $\mathbf{a}_G \in \mathbb{N}^n$ attached to its faces $G \in X$ in such a way that $\mathbf{a}_G \preceq \mathbf{a}_{G'}$ when $G \subseteq G'$. A free complex or resolution supported on a weakly labeled cell complex is **weakly cellular**. Show that if a weakly cellular resolution $\mathcal{F}_X$ resolves the quotient S/I, then I equals the ideal $\langle \mathbf{x}^{\mathbf{a}_v} \mid v$ is a vertex of $X \rangle$; in other words, $\mathbf{x}^{\mathbf{a}_v} \in I$ for all vertices $v \in X$.

4.4 Find an artinian monomial ideal I inside $\Bbbk[x, y, z]$ and a cell complex that supports a minimal free resolution of I such that the edge graph of the cell complex is not planar. Hint: Exercise 3.2.

4.5 Give a full proof of the claim in Section 4.3.6 to the effect that every simple polytope $\mathcal{P}$ supports a minimal linear free resolution.

4.6 Extend the construction of Section 4.3.6 from simple polytopes to (possibly unbounded) simple polyhedra. In this general case, the cellular resolution is supported on the complex of bounded faces of the simple polyhedron.

4.7 Prove that the union of all (closed) bounded faces of a convex polyhedron in $\mathbb{R}^n$ is always a contractible topological space.

4.8 Show that if the face F in Lemma 4.18 is a vertex, then K is a union of facets.

4.9 Draw the polyhedron $\mathcal{P}_t$ corresponding to the monomial ideal in Example 4.23 and verify that the hull resolution is indeed minimal.

4.10 Consider an arrangement of n hyperplanes in a real affine space. Label each cell of the arrangement with the squarefree monomial $x_{i_1} x_{i_2} \cdots x_{i_r}$ such that $i_1, i_2, \ldots, i_r$ are the indices of the hyperplanes not containing this cell. Prove that the complex of bounded cells is a minimal cellular resolution.

4.11 The *lcm-lattice* of a monomial ideal I is the set of all least common multiples of subsets of the minimal generators of I, ordered by divisibility. Show that the Betti numbers of I are determined by the poset homology of intervals in the lcm-lattice of I.

4.12 Fix a cellular resolution $\mathcal{F}_X$ of S/I, and let J be another monomial ideal. In the spirit of Theorem 4.7, write down a cellular description of $\mathrm{Tor}_S^i(S/I, S/J)_{\mathbf{a}}$ in terms of the topology of the labeled cell complex X and the combinatorics of J.

4.13 Describe the hull resolution of the ideal $\langle x_1, x_2, x_3, x_4 \rangle^m$ and compare it with the Eliahou–Kervaire resolution.

4.14 Explain how the face poset of a polyhedral cone can be read off from the oriented matroid of its generators. (This is the construction used in the proof of Proposition 4.14.)

4.15 Does the converse to Corollary 4.32 hold? That is, does every regular subdivision of a simplex arise as the hull complex of some artinian monomial ideal?

4.16 Describe an algorithm for computing the hull complex of a given monomial ideal. Analyze the running time of your algorithm.

Notes

Cellular resolutions and the hull complex were introduced by Bayer and Sturmfels in [BS98], as an extension of the simplicial construction of Bayer, Peeva, and

Sturmfels [BPS98]. Most of the chapter is based on [BS98].

Example 4.6 is taken from [Mil98, Example 5.4]. The Taylor resolution in Section 4.3.2 is due to D. Taylor [Tay60]. A more efficient version of the Taylor resolution, which takes advantage of the ordering of the monomial generators, was given by Lyubeznik [Lyu88]. For more on the structure of permutohedra see [BiS96], or check [Mil98, Section 5] for extra details on the connections with cellular resolutions. Tree ideals from Section 4.3.4 arise in connection with the algebra generated by the Chern 2-forms of the tautological line bundles on the flag variety [PSS99]. The minimal cellular resolutions of tree ideals appeared in [MSY00]. The real projective plane comes up often as an example in combinatorial commutative algebra because of its sensitivity to the characteristic of the field $\Bbbk$; see, for example, [BH98, Chapter 5].

Lemma 4.18 is known, but it seems difficult to locate a reference for the proof.

One possible solution to Exercise 4.4 appears in [Mil02b, Example 9.2]. The construction of Section 4.3.6 and Exercises 4.5 and 4.6 is essentially equivalent to that of Exercise 4.10, which appeared in [NPS02]. The lcm-lattice of a monomial ideal in Exercise 4.11 was introduced by Gasharov, Peeva, and Welker [GPW99] as an analogue for monomial ideals of the intersection lattice of a hyperplane arrangement, including all of the encoded homological information.

Chapter 5

Alexander duality

Duality gives rise to fundamental notions in many parts of algebra, combinatorics, topology, and geometry. In our context, the intersection of these notions is *Alexander duality*. Its essence for arbitrary monomial ideals is the familiar optical illusion in which isometric drawings of cubes look alternately like they are pointing "in" or "out" (see Fig. 5.1). Alexander duality extends the combinatorial notion for simplicial complexes by exchanging generators of ideals for irreducible components. More generally, this exchange works on cellular resolutions of monomial ideals, where it is manifested as topological duality. Roughly speaking, data contained in the least common multiples of minimal generators are equivalent (but dual) to data contained in the greatest common divisors of irreducible components.

5.1 Simplicial Alexander duality

Combinatorial duality on simplicial complexes is imposed by switching the roles of minimal generators and prime components: a minimal generator of the form $\mathbf{x}^\sigma = \prod_{i \in \sigma} x_i$ becomes a prime component $\mathfrak{m}^\sigma = \langle x_i \mid i \in \sigma \rangle$, as in Definition 1.35. Our first observation here is that Alexander duality really is a duality, in the sense that repeating it yields back the original.

Proposition 5.1 *If I is a squarefree monomial ideal, then $(I^\star)^\star = I$. Equivalently, $(\Delta^\star)^\star = \Delta$ for any simplicial complex Δ.*

Proof. View Alexander duality as poset duality in the Boolean lattice $2^{[n]}$ of subsets of $[n] := \{1, \ldots, n\}$, as follows. Proposition 1.37 says that removing Δ from $2^{[n]}$ leaves a poset isomorphic to $\Delta^\star$, but with containments reversed under the operation $\tau \mapsto \overline{\tau}$. Removing $\Delta^\star$ from $2^{[n]}$ therefore leaves Δ, but with containments reversed under the operation $\overline{\tau} \mapsto \tau$. $\square$

Example 5.2 There are self-dual simplicial complexes, such as the two-dimensional simplicial complex consisting of an empty triangle and a single fourth vertex. There are also complexes that are isomorphic to their duals (after relabeling the vertices), but not equal. For example, the *stick twisted cubic* with ideal $I = \langle ab, bc, cd \rangle = \langle a, c \rangle \cap \langle b, c \rangle \cap \langle b, d \rangle$ has this property. ◇

Example 5.3 Fix a simple polytope $\mathcal{P}$ with n facets $F_1, \ldots, F_n$ and r vertices $v_1, \ldots, v_r$. If Δ is the boundary of the simplicial d-polytope polar to $\mathcal{P}$, so that the n vertices of Δ are in bijection with the n facets of $\mathcal{P}$, then I_Δ is Alexander dual to the ideal $I_\mathcal{P}$ introduced in Section 4.3.6.

For example, let Δ be the octahedron

$$
\begin{aligned}
I_\Delta = {} & \langle x_0, y_0, z_0 \rangle \cap \langle x_0, y_0, z_1 \rangle \cap \langle x_0, y_1, z_0 \rangle \\
& \cap \langle x_0, y_1, z_1 \rangle \cap \langle x_1, y_0, z_0 \rangle \cap \langle x_1, y_0, z_1 \rangle \\
& \cap \langle x_1, y_1, z_0 \rangle \cap \langle x_1, y_1, z_1 \rangle \\
= {} & \langle x_0 x_1, y_0 y_1, z_0 z_1 \rangle
\end{aligned}
$$

whose vertices are labeled by variables x_i, y_i, or z_i depending on which axis they lie. The Alexander dual ideal is

$$
\begin{aligned}
I_\Delta^\star = {} & \langle x_0 y_0 z_0, x_0 y_0 z_1, x_0 y_1 z_0, \\
& x_0 y_1 z_1, x_1 y_0 z_0, x_1 y_0 z_1, \\
& x_1 y_1 z_0, x_1 y_1 z_1 \rangle \\
= {} & \langle x_0, x_1 \rangle \cap \langle y_0, y_1 \rangle \cap \langle z_0, z_1 \rangle
\end{aligned}
$$

with the labeling described in Section 4.3.6 on the cube $\mathcal{P}$ polar to Δ. ◇

One theme that we will develop in this chapter is that Alexander duality extends from Stanley–Reisner ideals and simplicial complexes to free resolutions. In the squarefree context, this effect can be seen most simply on Koszul complexes. The idea is that instead of using the reduced chain complex of the simplex $\{1, \ldots, n\}$, we can use its reduced *cochain* complex. This change produces another version of the Koszul complex.

Definition 5.4 Monomial matrices for the **coKoszul complex** $\mathbb{K}^\bullet$ have scalar entries given by the reduced cochain complex of the full simplex on

$\{1, \ldots, n\}$, with the label $\mathbf{x}^\tau$ on the column and row corresponding to $e^*_{\overline{\tau}}$, where $\overline{\tau} = \{1, \ldots, n\} \smallsetminus \tau$. The homological degrees are shifted so that $e^*_{\overline{\tau}}$ sits in homological degree $n - |\overline{\tau}| = |\tau|$.

For example, $e^*_{\varnothing}$ sits in homological degree n, while $e^*_{\{1,\ldots,n\}}$ sits in homological degree 0. The following is dual to Proposition 1.28.

Proposition 5.5 *The coKoszul complex* $\mathbb{K}^\bullet$ *minimally resolves* $\Bbbk = S/\mathfrak{m}$.

Proof. Suppose $\mathbf{b} \in \mathbb{N}^n$ has support σ. The degree $\mathbf{b}$ part $(\mathbb{K}^\bullet)_{\mathbf{b}}$ of the complex $\mathbb{K}^\bullet$ comes from those rows and columns labeled by faces $\tau \subseteq \sigma$. These rows and columns correspond to the basis vectors $e^*_{\overline{\tau}}$ for $\tau \subseteq \sigma$. Therefore, as a complex of $\Bbbk$-vector spaces, $(\mathbb{K}^\bullet)_{\mathbf{b}}$ is the subcomplex of the cochain complex of the entire simplex on n vertices spanned as a vector space by the basis elements $\{e^*_{\overline{\tau}} \mid \tau \subseteq \sigma\}$. (The reader should verify that this subvector space is closed under the coboundary maps.) Replacing each τ by $\sigma - \tau$, this set can also be written as $\{e^*_{\tau \cup \overline{\sigma}} \mid \tau \subseteq \sigma\}$. With this indexing, $(\mathbb{K}^\bullet)_\sigma$ is more clearly isomorphic (up to homological shift) to the cochain complex of the simplicial complex consisting of all faces of σ:

$$\begin{aligned}
(\mathbb{K}^\bullet)_\sigma &\cong \widetilde{\mathcal{C}}^\bullet(\sigma; \Bbbk) \\
e^*_{\tau \cup \overline{\sigma}} &\mapsto \mathrm{sign}(\tau, \overline{\sigma}) e^*_\tau,
\end{aligned} \tag{5.1}$$

where $\mathrm{sign}(\tau, \overline{\sigma})$ is the sign of the permutation that puts the list $(\tau, \overline{\sigma})$ into increasing order. Now use the fact that nonempty simplices have zero cohomology, while the irrelevant complex $\{\varnothing\}$ has cohomology in degree -1. $\square$

The Koszul and coKoszul complexes are abstractly isomorphic as $\mathbb{N}^n$-graded complexes. Combinatorially, however, their $\mathbb{N}^n$-graded degrees have different interpretations, and such variations can be important in applications. In particular, comparing the Koszul and coKoszul points of view will result in our next theorem.

Many readers who have previously encountered Alexander duality will have done so in a topological context, where it manifests itself as an isomorphism between the reduced homology of a closed topological subspace of a sphere and the reduced cohomology of the complement. In combinatorial language, this isomorphism reads as follows.

Theorem 5.6 (Alexander duality) $\widetilde{H}_{i-1}(\Delta^\star; \Bbbk) \cong \widetilde{H}^{n-2-i}(\Delta; \Bbbk)$.

Proof. We have already calculated the left-hand side to be $\mathrm{Tor}^S_{i+1}(\Bbbk, S/I_\Delta)_{\mathbf{1}}$ for $\mathbf{1} = (1, \ldots, 1)$ in the proof of Theorem 1.34, by tensoring the Koszul complex $\mathbb{K}_\bullet$ with I_Δ and taking i^{th} homology. Now let us instead calculate this Tor module by tensoring the coKoszul complex $\mathbb{K}^\bullet$ with S/I_Δ.

The $\mathbb{N}^n$-graded degree $\mathbf{1}$ part $(\mathbb{K}^\bullet \otimes S/I_\Delta)_{\mathbf{1}}$ is a quotient of the cochain complex $(\mathbb{K}^\bullet)_{\mathbf{1}} = \widetilde{\mathcal{C}}^\bullet(2^{[n]}; \Bbbk)$ of the full simplex $2^{[n]}$, namely

$$(\mathbb{K}^\bullet \otimes S/I_\Delta)_{\mathbf{1}} = (\mathbb{K}^\bullet)_{\mathbf{1}} / (I_\Delta \cdot \mathbb{K}^\bullet)_{\mathbf{1}}.$$

Arguing as in the proof of Proposition 5.5, this quotient complex is naturally the reduced cochain complex $\mathcal{C}^\bullet(\Gamma; \Bbbk)$ for some simplicial complex Γ. Writing $1_{\overline{\tau}}$ for the basis vector of $\Bbbk^\bullet$ in $\mathbb{N}^n$-graded degree $\overline{\tau}$, and noting that $\mathbf{x}^\tau \cdot 1_{\overline{\tau}} \in (\Bbbk^\bullet)_1$ corresponds to $e_\tau^* \in \mathcal{C}^\bullet(2^{[n]}; \Bbbk)$, we find that $\Gamma = \Delta$ because

$$\tau \in \Gamma \quad \Leftrightarrow \quad \mathbf{x}^\tau \cdot 1_{\overline{\tau}} \notin (I_\Delta \cdot \Bbbk^\bullet)_1 \quad \Leftrightarrow \quad \mathbf{x}^\tau \notin I_\Delta \quad \Leftrightarrow \quad \tau \in \Delta.$$

Since $e_\varnothing^*$ sits in homological degree n instead of cohomological degree -1, $(\Bbbk^\bullet \otimes S/I_\Delta)_1$ has i^{th} homology $\widetilde{H}^{n-1-i}(\Delta; \Bbbk)$. Taking $(i+1)^{\text{st}}$ homology, we conclude that $\widetilde{H}_{i-1}(\Delta^\star; \Bbbk) \cong \operatorname{Tor}_{i+1}^S(\Bbbk, S/I_\Delta)_1 \cong \widetilde{H}^{n-2-i}(\Delta; \Bbbk)$. $\square$

Remark 5.7 The direct connection between combinatorial Alexander duality and the usual topological notion uses the fact that a simplicial complex Δ is a closed subcomplex of the $(n-2)$-sphere constituting the boundary of the simplex $2^{[n]}$, as long as Δ is not the whole simplex $2^{[n]}$. The complement of Δ in this sphere retracts onto the simplicial complex $\Delta^\star$. Therefore, Theorem 5.6 expresses the topological Alexander duality relation inside the $(n-2)$-sphere.

Note that our proof does not use any properties of $\Bbbk$ and can be applied over the integers $\mathbb{Z}$ or any other ring R, since the Koszul complex still resolves R as a module over $R[x_1, \ldots, x_n]$. This naturality explains why, despite the fact that dual vector spaces are isomorphic over fields, one side of the isomorphism in Theorem 5.6 uses cohomology and the other uses homology: the extension to arbitrary rings (and not just fields) would fail over general rings if both sides used homology, or if both used cohomology.

Example 5.8 As we mentioned in Example 1.36, the simplicial complexes Δ and Γ from Examples 1.8 and 1.14 are Alexander dual, so $\Gamma = \Delta^\star$. The fact that Δ has two components means that $\widetilde{H}_0(\Delta; \Bbbk) = \Bbbk$ for any field $\Bbbk$. On the dual side, this homology corresponds to the fact that $\widetilde{H}_0(\Delta; \Bbbk) = \widetilde{H}_{1-1}(\Delta; \Bbbk) \cong \widetilde{H}^{5-2-1}(\Gamma; \Bbbk) = \widetilde{H}^2(\Gamma; \Bbbk) = \Bbbk$. The simplicial complexes Δ and Γ each have one remaining nonzero reduced (co)homology group. In what (co)homological degrees do they lie? $\diamond$

The proof of Theorem 5.6 required only the degree $\mathbf{1}$ part of $\Bbbk^\bullet \otimes S/I_\Delta$, but we can similarly calculate $\mathbb{N}^n$-graded Betti numbers for arbitrary monomial ideals, using the Alexander dual to Definition 1.33.

Definition 5.9 For each vector $\mathbf{b} \in \mathbb{N}^n$, define $\mathbf{b}'$ by subtracting 1 from each nonzero coordinate of $\mathbf{b}$. Given a monomial ideal I and a degree $\mathbf{b} \in \mathbb{N}^n$, the **(lower) Koszul simplicial complex** of S/I in degree $\mathbf{b}$ is

$$K_{\mathbf{b}}(I) \;=\; \{\text{squarefree vectors } \tau \preceq \mathbf{b} \mid \mathbf{x}^{\mathbf{b}'+\tau} \notin I\}.$$

The reason for our terminology "upper" and "lower" for the Koszul simplicial complexes $K^{\mathbf{b}}$ and $K_{\mathbf{b}}$ can now be made explicit. The following is immediate from the definitions.

Lemma 5.10 *For any monomial ideal I and degree $\mathbf{b} \in \mathbb{N}^n$, the upper and lower Koszul simplicial compelexes $K^{\mathbf{b}}$ and $K_{\mathbf{b}}$ are Alexander dual inside the full simplex whose vertices are $\text{supp}(\mathbf{b}) = \{i \mid b_i \neq 0\}$.*

Theorem 5.11 *Given a vector $\mathbf{b} \in \mathbb{N}^n$ with support $\sigma = \{i \mid b_i \neq 0\}$, the Betti numbers of I and S/I in degree $\mathbf{b}$ can be expressed as*

$$\beta_{i-1,\mathbf{b}}(I) = \beta_{i,\mathbf{b}}(S/I) = \dim_{\Bbbk} \widetilde{H}^{|\sigma|-i-1}(K_{\mathbf{b}}(I); \Bbbk).$$

Proof. Apply Theorem 5.6 to Theorem 1.34, using Lemma 5.10. □

As a consequence, we derive Hochster's original formulation of the result whose "dual form" appeared in Corollary 1.40. For each $\sigma \subseteq \{1, \ldots, n\}$, define the *restriction* of Δ to σ by

$$\Delta|_{\sigma} = \{\tau \in \Delta \mid \tau \subseteq \sigma\}.$$

Corollary 5.12 (Hochster's formula) *The nonzero Betti numbers of I_{Δ} and S/I_{Δ} lie only in squarefree degrees σ, and we have*

$$\beta_{i-1,\sigma}(I_{\Delta}) = \beta_{i,\sigma}(S/I_{\Delta}) = \dim_{\Bbbk} \widetilde{H}^{|\sigma|-i-1}(\Delta|_{\sigma}; \Bbbk).$$

Proof. The nonzero Betti numbers lie in squarefree degrees by Corollary 1.40. Hence the result is obtained by applying Theorem 5.6 to Theorem 5.11, once we show that $K_{\sigma}(I_{\Delta})$ is the restriction $\Delta|_{\sigma}$. This follows directly from the definitions of I_{Δ}, $K_{\sigma}(I_{\Delta})$, and $\Delta|_{\sigma}$. □

Example 5.13 Let Γ be as in Example 1.14. Taking the subset $\sigma = \{a, b, c, d, e\}$, corresponding to the monomial $abcde$, we have $\Gamma|_{\sigma} = \Gamma$. From the labels on the monomial matrices from Example 1.25, we see that $\beta_{3,\sigma}(S/I_{\Gamma}) = \beta_{2,\sigma}(S/I_{\Gamma}) = 1$, while the other Betti numbers in this degree are zero. Hochster's formula computes the dimensions of the cohomology groups of Γ: we find that $\widetilde{H}^1(\Gamma; \Bbbk) \cong \widetilde{H}^2(\Gamma; \Bbbk) \cong \Bbbk$, whereas the other reduced cohomology groups of Γ are 0. The nonzero cohomology comes from the "empty" circle $\{a, b, e\}$ and the "empty" sphere $\{a, b, c, d\}$.

For another example, take $\sigma = \{a, b, c, e\}$, corresponding to the monomial $abce$. The restriction $\Gamma|_{\sigma}$ is the simplicial complex

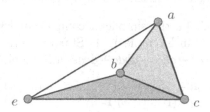

for which Hochster's formula gives $\widetilde{H}^{|\sigma|-1-2}(\Gamma|_{\sigma}; \Bbbk) = \widetilde{H}^1(\Gamma|_{\sigma}; \Bbbk) \cong \Bbbk$. The other cohomology groups of $\Gamma|_{\sigma}$ are trivial. ◇

Comparing the two versions of Hochster's formula, Corollary 1.40 and Corollary 5.12, we see that the links of faces in a simplicial complex carry the same homological (and in fact combinatorial) information as the restrictions of its Alexander dual to subsets of its vertices. Although restrictions may seem easier to visualize, it is the links of faces that more often carry geometric significance. For example, if Δ is a simplicial manifold, then all links of nonempty faces of Δ are spheres. That being said, when working with Koszul simplicial complexes that have nonzero homology—that is, at the "corners" of the staircase diagram of an arbitrary monomial ideal I—it is best to heed the advice of Dave Bayer [Bay96]:

> In choosing how to view a corner of I, one is deciding which of two dual simplicial complexes to favor. Often, the relationship between a corner and properties of I is inscrutable viewed one way, but obvious viewed the other way. One wants to develop the reflex of always looking at corners both ways, rather than assuming that one's initial vantage point is preferable.

In particular, should one wish to study a simplicial complex Δ via Stanley–Reisner theory, one should consider perhaps to use not I_Δ, but $I_{\Delta^\star}$ instead!

As an example, here is a device to recover the Hilbert series of I_Δ and $I_{\Delta^\star}$ from one another. In applications, the forthcoming inversion formula allows one to follow Bayer's advice, by using one or the other of Corollary 1.40 and Corollary 5.12. Denote by $\mathcal{K}(\mathbf{1} - \mathbf{x})$ the polynomial that results after substituting $(1 - x_1, \ldots, 1 - x_n)$ for $(x_1, \ldots, x_n)$ in a polynomial $\mathcal{K}(\mathbf{x})$.

Theorem 5.14 (Alexander inversion formula) *If Δ is any simplicial complex, then the K-polynomial of its Stanley-Reisner ring satisfies*

$$\mathcal{K}(S/I_\Delta; \mathbf{x}) \;=\; \mathcal{K}(I_{\Delta^\star}; \mathbf{1} - \mathbf{x}).$$

Proof. By Proposition 1.37, the Hilbert series of $I_{\Delta^\star}$ is the sum of all monomials $\mathbf{x}^\mathbf{b}$ divisible by $\prod_{j \notin \sigma} x_j$ for some face $\sigma \in \Delta$:

$$H(I_{\Delta^\star}; \mathbf{x}) \;=\; \sum_{\sigma \in \Delta} \prod_{j \notin \sigma} \frac{x_j}{1 - x_j} \;=\; \sum_{\sigma \in \Delta} \frac{\prod_{i \in \sigma}(1 - x_i) \cdot \prod_{j \notin \sigma} x_j}{\prod_{i=1}^{n}(1 - x_i)}.$$

Now compare the numerator in the above expression with Theorem 1.13. $\square$

Example 5.15 Let Δ be the simplicial complex in Example 1.5, which is Alexander dual to Γ in Example 1.14. Starting with the K-polynomial $1 - abcd - abe - ace - de + abce + abde + acde$ in Example 1.14, we calculate

$$
\begin{aligned}
&1 - (1-a)(1-b)(1-c)(1-d) - (1-a)(1-b)(1-e) - (1-a)(1-c)(1-e) - (1-d)(1-e) \\
&\quad + (1-a)(1-b)(1-c)(1-e) + (1-a)(1-b)(1-d)(1-e) + (1-a)(1-c)(1-d)(1-e) \\
&= ad + ae + be + ce + de + bcd - abe - ace - bce - 2ade - bde - cde + abce + abde + acde
\end{aligned}
$$

to be the K-polynomial of the Stanley–Reisner ideal I_Δ. $\diamond$

5.2 Generators versus irreducible components

In this section we prove uniqueness of irredundant decompositions of monomial ideals as intersection of irreducible monomial ideals from the (seemingly easier) uniqueness of minimal monomial generating sets. The tool that interpolates between these two is Alexander duality, suitably generalized. First, what are irreducible monomial ideals and decompositions?

Definition 5.16 A monomial ideal in $S = \mathbb{k}[x_1, \ldots, x_n]$ is **irreducible** if it is generated by powers of variables. Such an ideal can be expressed as

$$\mathfrak{m}^{\mathbf{b}} = \langle x_i^{b_i} \mid b_i \geq 1 \rangle$$

for some vector $\mathbf{b} \in \mathbb{N}^n$. An **irreducible decomposition** of a monomial ideal I is an expression as follows, for vectors $\mathbf{b}_1, \ldots, \mathbf{b}_r \in \mathbb{N}^n$:

$$I = \mathfrak{m}^{\mathbf{b}_1} \cap \cdots \cap \mathfrak{m}^{\mathbf{b}_r}.$$

This decomposition is called **irredundant** (and the ideals $\mathfrak{m}^{\mathbf{b}_1}, \ldots, \mathfrak{m}^{\mathbf{b}_r}$ are called **irreducible components** of I) if no intersectands can be omitted.

Thus $\mathfrak{m}^{(1,0,5)}$ is the ideal $\langle x, z^5 \rangle$ when $S = \mathbb{k}[x, y, z]$. In examples, we might write an expression such as $\mathfrak{m}^{105}$ instead of $\mathfrak{m}^{(1,0,5)}$ when all the integers involved have just one digit. In general, the notation $\mathfrak{m}^{\mathbf{b}}$ takes the monomial $\mathbf{x}^{\mathbf{b}}$ and inserts commas between the variables, ignoring those variables with exponent 0. We use the symbol $\mathfrak{m}$ because it commonly denotes the maximal monomial ideal $\langle x_1, \ldots, x_n \rangle$.

Remark 5.17 In the context of general commutative algebra, an arbitrary (not necessarily monomial) ideal in $S = \mathbb{k}[x_1, \ldots, x_n]$ is called *irreducible* if it is not the intersection of two strictly larger ideals. For our purposes in this chapter, we will not need that irreducible monomial ideals are irreducible in this usual commutative algebra sense (Exercise 5.7). However, we prove it more generally for semigroup rings in Chapter 11 (Proposition 11.41).

Before defining Alexander duality, let us describe a fun algorithm to produce irreducible decompositions of a given monomial ideal.

Lemma 5.18 *Every monomial ideal has an irreducible decomposition.*

Proof. If m is a minimal generator of I and $m = m'm''$ is a product of relatively prime monomials m' and m'', then $I = (I + \langle m' \rangle) \cap (I + \langle m'' \rangle)$. Iterating this process eventually writes the monomial ideal I as an intersection of ideals generated by powers of some of the variables. □

Example 5.19 $I = \langle xy^2, z \rangle = (I + \langle x \rangle) \cap (I + \langle y^2 \rangle) = \langle x, z \rangle \cap \langle y^2, z \rangle.$ ◇

For squarefree monomial ideals, Alexander duality can be confusing, with too many $\{0,1\}$ vectors and subsets of $[n] = \{1, \ldots, n\}$ creeping around along with their complements. When the duality is generalized to arbitrary monomial ideals, the confusion subsides a little, as the various squarefree vectors begin to take different roles: we are forced to forgo our conventions of automatically identifying any two objects representing a subset of $[n]$.

Of course, the definition of Alexander dual must necessarily become more complicated. Nonetheless, the basic idea remains the same: make the irreducible components into generators.

Definition 5.20 Given two vectors $\mathbf{a}, \mathbf{b} \in \mathbb{N}^n$ with $\mathbf{b} \preceq \mathbf{a}$ (that is, $b_i \leq a_i$ for $i = 1, \ldots, n$), let $\mathbf{a} \smallsetminus \mathbf{b}$ denote the vector whose i^{th} coordinate is

$$
a_i \smallsetminus b_i \;\; = \;\; \begin{cases} a_i + 1 - b_i & \text{if } b_i \geq 1 \\ 0 & \text{if } b_i = 0. \end{cases}
$$

If I is a monomial ideal whose minimal generators all divide $\mathbf{x}^{\mathbf{a}}$, then the **Alexander dual of I with respect to $\mathbf{a}$** is

$$
I^{[\mathbf{a}]} \;\; = \;\; \bigcap \{ \mathfrak{m}^{\mathbf{a} \smallsetminus \mathbf{b}} \mid \mathbf{x}^{\mathbf{b}} \text{ is a minimal generator of } I \}.
$$

For an example of complementation, $(7,6,5) \smallsetminus (2,0,3) = (6,0,3)$.

Example 5.21 Let $\mathbf{a} = (4,4,4)$. Then

$$
\begin{aligned}
I &= \langle x^3, xy, yz^2 \rangle & I^{[\mathbf{a}]} &= \langle x^2 \rangle \cap \langle x^4, y^4 \rangle \cap \langle y^4, z^3 \rangle \\
&= \langle x^3, y \rangle \cap \langle x, z^2 \rangle & \Longrightarrow \qquad &= \langle x^2 y^4, x^4 z^3 \rangle.
\end{aligned}
$$

Note that $(I^{[\mathbf{a}]})^{[\mathbf{a}]} = I$. We will see that this holds in general. $\diamond$

Example 5.22 Let $n = 3$, so that $S = \Bbbk[x, y, z]$. Fig. 5.1 lists the minimal generators and irreducible components of an ideal $I \subseteq S$ and its dual $I^{[455]}$ with respect to $\mathbf{a} = (4,5,5)$. The (truncated) staircase diagrams representing the monomials not in these ideals are also rendered in Fig. 5.1, where the black lattice points are generators and the white lattice points indicate irreducible components. The numbers are to be interpreted as vectors, so $205 = (2,0,5)$, for example. The arrows attached to a white lattice point indicate the directions in which the component continues to infinity; it should be noted that a white point has a zero in some coordinate precisely when it has an arrow pointing in the corresponding direction.

Alexander duality in three dimensions comes down to the familiar optical illusion in which isometrically rendered cubes appear alternately to point "in" or "out". In fact, the staircase diagram for $I^{[455]}$ in Fig. 5.1 is obtained by literally turning the staircase diagram for I upside down (the reader is encouraged to try this). Notice that each minimal generator of I has the same support as the corresponding irreducible component of $I^{[455]}$. $\diamond$

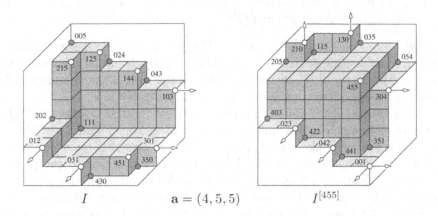

$$I \qquad\qquad \mathbf{a} = (4,5,5) \qquad\qquad I^{[455]}$$

$$
\begin{aligned}
I &= \langle z^5, x^2 z^2, x^4 y^3, x^3 y^5, y^4 z^3, y^2 z^4, xyz \rangle \\
&= \langle x^2, y, z^5 \rangle \cap \langle y, z^2 \rangle \cap \langle y^3, z \rangle \cap \langle x^4, y^5, z \rangle \cap \langle x^3, z \rangle \cap \langle x, z^3 \rangle \cap \langle x, y^4, z^4 \rangle \cap \langle x, y^2, z^5 \rangle
\end{aligned}
$$

$$
\begin{aligned}
I^{[455]} &= \langle z \rangle \cap \langle x^3, z^4 \rangle \cap \langle x, y^3 \rangle \cap \langle x^2, y \rangle \cap \langle y^2, z^3 \rangle \cap \langle y^4, z^2 \rangle \cap \langle x^4, y^5, z^5 \rangle \\
&= \langle x^3 y^5 z, y^5 z^4, y^3 z^5, xyz^5, x^2 z^5, x^4 z^3, x^4 y^2 z^2, x^4 y^4 z \rangle.
\end{aligned}
$$

Figure 5.1: Truncated staircase diagrams of I and $I^{[455]}$ from Example 5.22

The definition of Alexander duality is consistent with our earlier definition in the squarefree case: if $I = I_\Delta$ is a squarefree monomial ideal, then $I_{\Delta^\star} = (I_\Delta)^{\mathbf{1}}$ is the Alexander dual of I_Δ with respect to $\mathbf{1} = (1, \ldots, 1)$. Further statements beyond the definition of Alexander duality also have analogues for arbitrary ideals. Next we generalize Proposition 1.37.

Proposition 5.23 *Suppose that all minimal generators of the ideal I divide $\mathbf{x^a}$. If $\mathbf{b} \preceq \mathbf{a}$, then $\mathbf{x^b}$ lies outside I if and only if $\mathbf{x^{a-b}}$ lies inside $I^{[\mathbf{a}]}$.*

Proof. Suppose $I = \langle \mathbf{x^c} \mid \mathbf{c} \in C \rangle$. Then $\mathbf{x^b} \notin I$ if and only if we have $\mathbf{b} \not\succeq \mathbf{c}$, or equivalently, $\mathbf{a} - \mathbf{b} \not\preceq \mathbf{a} - \mathbf{c}$, for all $\mathbf{c} \in C$. This means that for each $\mathbf{c} \in C$, some coordinate of $\mathbf{a} - \mathbf{b}$ equals at least the corresponding coordinate of $\mathbf{a} + \mathbf{1} - \mathbf{c}$; that is, $\mathbf{x^{a-b}} \in \mathfrak{m}^{\mathbf{a+1-c}}$ for all $\mathbf{c} \in C$. Equivalently, $\mathbf{x^{a-b}}$ lies in the intersection $\bigcap_{\mathbf{c} \in C} \mathfrak{m}^{\mathbf{a+1-c}}$, which equals $I^{[\mathbf{a}]} + \mathfrak{m}^{\mathbf{a+1}}$ by definition. But $\mathbf{x^{a-b}} \in I^{[\mathbf{a}]} + \mathfrak{m}^{\mathbf{a+1}}$ exactly when $\mathbf{x^{a-b}} \in I^{[\mathbf{a}]}$, because $\mathbf{a} - \mathbf{b} \preceq \mathbf{a}$. $\qquad\square$

The complementation identity for vectors $\mathbf{b} \preceq \mathbf{a}$ in $\mathbb{N}^n$ reads

$$\mathbf{a} \smallsetminus (\mathbf{a} \smallsetminus \mathbf{b}) = \mathbf{b} \tag{5.2}$$

and generalizes the squarefree relation $\overline{\overline{\sigma}} = \sigma$; it follows from the obvious complementation $a \smallsetminus (a \smallsetminus b) = b$ for natural numbers $b \leq a$. Moreover, the next theorem generalizes the squarefree result in Proposition 5.1.

Theorem 5.24 *If all minimal generators of I divide $\mathbf{x}^{\mathbf{a}}$, then all minimal generators of $I^{[\mathbf{a}]}$ divide $\mathbf{x}^{\mathbf{a}}$, and $(I^{[\mathbf{a}]})^{[\mathbf{a}]} = I$.*

Proof. Suppose $I = \langle \mathbf{x}^{\mathbf{b}_1}, \ldots, \mathbf{x}^{\mathbf{b}_r} \rangle$. The powers of variables generating the irreducible components of $I^{[\mathbf{a}]}$ all divide $\mathbf{x}^{\mathbf{a}}$ by definition. Since every minimal generator of $I^{[\mathbf{a}]}$ can be expressed as the least common multiple of some of these powers of variables, these generators divide $\mathbf{x}^{\mathbf{a}}$.

Now generalize the proof of Proposition 5.1 as follows. Consider the set $[\mathbf{0}, \mathbf{a}]$ of vectors in $\mathbb{N}^n$ preceding $\mathbf{a}$ as a poset (think geometrically: a product of intervals, shaped like a box). Proposition 5.23 says that removing from $[\mathbf{0}, \mathbf{a}]$ all monomials outside of I leaves a poset isomorphic to the poset of monomials in $[\mathbf{0}, \mathbf{a}]$ inside $I^{[\mathbf{a}]}$, but with the order reversed under the operation $\mathbf{b} \mapsto \mathbf{a} - \mathbf{b}$. It follows that removing from $[\mathbf{0}, \mathbf{a}]$ all monomials outside of $I^{[\mathbf{a}]}$ leaves a poset isomorphic to the poset of monomials in $[\mathbf{0}, \mathbf{a}]$ inside I, but with the order reversed under the operation $\mathbf{a} - \mathbf{b} \mapsto \mathbf{b}$. This argument shows that for $\mathbf{b} \preceq \mathbf{a}$, we have $\mathbf{x}^{\mathbf{b}} \in I$ if and only if $\mathbf{x}^{\mathbf{b}} \in (I^{[\mathbf{a}]})^{[\mathbf{a}]}$. The result follows because the previous paragraph implies as well that all minimal generators of $(I^{[\mathbf{a}]})^{[\mathbf{a}]}$ divide $\mathbf{x}^{\mathbf{a}}$. $\square$

Referring to Fig. 5.1 might help the reader understand the above proof, which explains how to generalize the optical illusion to higher dimensions.

Proposition 5.23 and Theorem 5.24 together imply an algebraic statement of Alexander duality in the language of colon ideals.

Corollary 5.25 *If all generators of I divide $\mathbf{x}^{\mathbf{a}}$, then $I^{[\mathbf{a}]}$ is the unique ideal with generators dividing $\mathbf{x}^{\mathbf{a}}$ that satisfies $(\mathfrak{m}^{\mathbf{a}+\mathbf{1}} : I) = I^{[\mathbf{a}]} + \mathfrak{m}^{\mathbf{a}+\mathbf{1}}$.*

Proof. Observe that $\mathbf{x}^{\mathbf{b}} \notin I$ if and only if all monomials dividing $\mathbf{x}^{\mathbf{b}}$ lie outside of I. If $\mathbf{b} \preceq \mathbf{a}$, then this occurs precisely when all monomials dividing $\mathbf{x}^{\mathbf{a}}$ lie outside of $\mathbf{x}^{\mathbf{a}-\mathbf{b}} \cdot I$, which is equivalent to $\mathbf{x}^{\mathbf{a}-\mathbf{b}} \cdot I \subseteq \mathfrak{m}^{\mathbf{a}+\mathbf{1}}$. $\square$

The next lemma is for the proof of uniqueness of irredundant irreducible decompositions in Theorem 5.27. It explains the odd definition of $\mathbf{a} \smallsetminus \mathbf{b}$.

Lemma 5.26 *Suppose that $\mathbf{b} \preceq \mathbf{a}$ and $\mathbf{c} \preceq \mathbf{a}$ in $\mathbb{N}^n$. Then $\mathbf{x}^{\mathbf{a} \smallsetminus \mathbf{b}}$ divides $\mathbf{x}^{\mathbf{a} \smallsetminus \mathbf{c}}$ if and only if $\mathfrak{m}^{\mathbf{b}} \subseteq \mathfrak{m}^{\mathbf{c}}$.*

Proof. We have $\mathfrak{m}^{\mathbf{b}} \subseteq \mathfrak{m}^{\mathbf{c}}$ if and only if $b_i \geq c_i$ whenever $c_i \geq 1$ and also $b_i = 0$ whenever $c_i = 0$. This occurs if and only if $a_i - b_i \leq a_i - c_i$ whenever $c_i \geq 1$ and also $b_i = 0$ whenever $c_i = 0$; that is, $a_i \smallsetminus b_i \leq a_i \smallsetminus c_i$ for all i. $\square$

Theorem 5.27 *Assume that all minimal generators of I divide $\mathbf{x}^{\mathbf{a}}$. Then I has a unique irredundant irreducible decomposition, and it is given by*

$$I \;=\; \bigcap \{ \mathfrak{m}^{\mathbf{a} \smallsetminus \mathbf{b}} \mid \mathbf{x}^{\mathbf{b}} \text{ is a minimal generator of } I^{[\mathbf{a}]} \}.$$

Equivalently, the Alexander dual of I is given by minimal generators as

$$I^{[\mathbf{a}]} \;=\; \langle \mathbf{x}^{\mathbf{a} \smallsetminus \mathbf{b}} \mid \mathfrak{m}^{\mathbf{b}} \text{ is an irreducible component of } I \rangle.$$

Proof. The given intersection is equal to I by Theorem 5.24. It is irredundant by Lemma 5.26 because the intersection is taken over *minimal* generators of $I^{[\mathbf{a}]}$. Now suppose that we are given any irredundant irreducible decomposition $I = \bigcap_{\mathbf{b} \in B} \mathfrak{m}^{\mathbf{b}}$, and choose $\mathbf{a}$ so that $\mathbf{b} \preceq \mathbf{a}$ for all $\mathbf{b} \in B$. The ideals $\{\mathfrak{m}^{\mathbf{b}} \mid \mathbf{b} \in B\}$ are pairwise incomparable by irredundancy, so the set $\{\mathbf{x}^{\mathbf{a} \smallsetminus \mathbf{b}} \mid \mathbf{b} \in B\}$ minimally generates some ideal J by Lemma 5.26. Furthermore, the Alexander dual of J is $J^{[\mathbf{a}]} = I$ by definition, whence $J = I^{[\mathbf{a}]}$ by Theorem 5.24. It follows that $B = \{\mathbf{a} \smallsetminus \mathbf{c} \mid \mathbf{x}^{\mathbf{c}}$ is a minimal generator of $I^{[\mathbf{a}]}\}$. Therefore, the decomposition is unique, and in particular it is independent of the choice of $\mathbf{a}$. Apply (5.2) for the "Equivalently" statement. $\square$

Remark 5.28 Theorem 5.27 along with Corollary 5.25 provides a useful way to compute the irreducible components of I given its minimal generators: simply take those generators $\mathbf{x}^{\mathbf{b}}$ of $(\mathfrak{m}^{\mathbf{a}+\mathbf{1}} : I)$ dividing $\mathbf{x}^{\mathbf{a}}$, and replace each one by $\mathfrak{m}^{\mathbf{a} \smallsetminus \mathbf{b}}$. It turns out that computing colon ideals is fast on many symbolic algebra systems. Of course, we can also compute the generators of I from its irreducible components this way, by turning each component $\mathfrak{m}^{\mathbf{b}}$ into a generator $\mathbf{x}^{\mathbf{a} \smallsetminus \mathbf{b}}$ for $I^{[\mathbf{a}]}$ and computing I using Corollary 5.25.

Remark 5.29 By a Noetherian induction argument, every (not necessarily monomial) ideal I can be written as an intersection $Q_1 \cap \cdots \cap Q_r$ of irreducible ideals, as defined in Remark 5.17. Such intersections are not unique—it might be that intersecting all but one of the Q_i still yields I. But even assuming this is not so (i.e., that the intersection is irredundant), the irreducible decomposition still need not be unique. Theorem 5.27 says that the situation changes dramatically when the ideal I and all of the intersectands Q_i are required to be monomial ideals.

5.3 Duality for resolutions

We have already seen that Alexander duality produces fun optical illusions on staircases in three dimensions and provides a useful way to think about irreducible decompositions, by relating them to minimal generators of the dual ideal. Moreover, we have seen connections to topological duality when dealing with squarefree ideals. In this section we explore a deeper connection: applying Alexander duality to a cellular resolution supported on a cell complex X corresponds to topological duality on X itself, rather than duality on Koszul simplicial complexes $K^{\mathbf{b}}$ of the monomial ideal it resolves. In this way, "global" topological duality on free resolutions induces "local" topological dualities at every $\mathbb{N}^n$-graded degree.

Let us start by reviewing a little relative cellular topology. If X is a cell complex, then its *cochain complex* $C^{\bullet}(X; \Bbbk)$ is the $\Bbbk$-vector space dual of the chain complex $C_{\bullet}(X; \Bbbk)$; its differential, called the *coboundary map*, is transpose to the boundary map. We saw this notion for simplicial complexes in Chapter 1. If $X' \subset X$ is a subcomplex, then of course X'

also has chain and cochain complexes. The inclusion $C_{\bullet}(X';\Bbbk) \subset C_{\bullet}(X;\Bbbk)$ is naturally dual to a surjection of cochain complexes the other way, and its kernel is an object that is central to duality for resolutions.

Definition 5.30 The **cochain complex** $C^{\bullet}(X, X';\Bbbk)$ of the **pair** $X' \subseteq X$ of cell complexes is defined by the exact sequence

$$0 \longrightarrow C^{\bullet}(X, X';\Bbbk) \longrightarrow C^{\bullet}(X;\Bbbk) \longrightarrow C^{\bullet}(X';\Bbbk) \longrightarrow 0.$$

The ith **relative cohomology** of the pair is $H^i(X, X';\Bbbk) = H^i C^{\bullet}(X, X';\Bbbk)$.

When we use language such as "Y is a pair of cell complexes", we think of $Y = (X' \subset X)$ as the set of faces in X that lie outside X'. Thus, for instance, we will use the term *facet* of Y to mean a facet of X that happens not to lie in X', noting that every maximal face of Y is also maximal in X, because X' is a subcomplex of X.

It will be convenient to use the language of distributive lattices instead of referring to greatest common divisors and least common multiples of monomials. Thus, for two vectors $\mathbf{a}$ and $\mathbf{b}$ in $\mathbb{N}^n$, we write $\mathbf{a} \wedge \mathbf{b}$ and $\mathbf{a} \vee \mathbf{b}$ for the *meet* and *join*, respectively. These vectors satisfy $\mathbf{x}^{\mathbf{a} \wedge \mathbf{b}} = \gcd(\mathbf{x}^{\mathbf{a}}, \mathbf{x}^{\mathbf{b}})$ and $\mathbf{x}^{\mathbf{a} \vee \mathbf{b}} = \text{lcm}(\mathbf{x}^{\mathbf{a}}, \mathbf{x}^{\mathbf{b}})$, so their i^{th} coordinates are

$$\begin{aligned}
(\mathbf{a} \wedge \mathbf{b})_i &= \min(a_i, b_i), \\
(\mathbf{a} \vee \mathbf{b})_i &= \max(a_i, b_i).
\end{aligned}$$

Definition 5.31 Let Y be a cell complex or a cellular pair. Then Y is **weakly colabeled** if the labels on faces $G \subseteq F$ satisfy $\mathbf{a}_G \succeq \mathbf{a}_F$, and Y is **colabeled** if, in addition, every face label $\mathbf{a}_G$ equals the join $\bigvee \mathbf{a}_F$ of all the labels on facets $F \supseteq G$.

The point of a colabeling is that it is dual to a labeling, with the roles of vertices and facets being switched: subtracting all labels on a labeled complex from a fixed vector yields a weakly colabeled complex (this is the next lemma; its proof is immediate from the definitions). When speaking of cell complexes endowed with multiple labelings, it is helpful to have a notation $\underline{X}$ for the *underlying unlabeled cell complex* X.

Lemma 5.32 *If a cell complex X is labeled, and the label on every face $G \in X$ satisfies $\mathbf{a}_G \preceq \mathbf{c}$, then relabeling each face $G \in \underline{X}$ by $\mathbf{c} - \mathbf{a}_G$ yields a weakly colabeled complex $\mathbf{c} - X$.*

Weakly colabeled cell complexes give rise to monomial matrices, just as labeled cell complexes do, but using the coboundary map instead of the boundary map of the underlying cell complex.

Definition 5.33 Let Y be a cell complex or a cellular pair $X' \subset X$, (weakly) colabeled. The (weakly) **cocelluar monomial matrix** supported on Y has the cochain complex $C^{\bullet}(Y;\Bbbk)$ for scalar entries, with faces of dimension $n-1$ in homological degree 0; its row and column labels are the face labels on Y. The (weakly) **cocellular free complex** $\mathcal{F}^Y$ **supported** on Y is the complex of $\mathbb{N}^n$-graded free S-modules (with basis) represented by the cocellular monomial matrix supported on Y. If $\mathcal{F}^Y$ is acyclic (so its homology lies only in degree 0), then $\mathcal{F}^Y$ is a (weakly) **cocellular resolution**.

Example 5.34 Starting with the labeled complex X in Fig. 4.1, form the

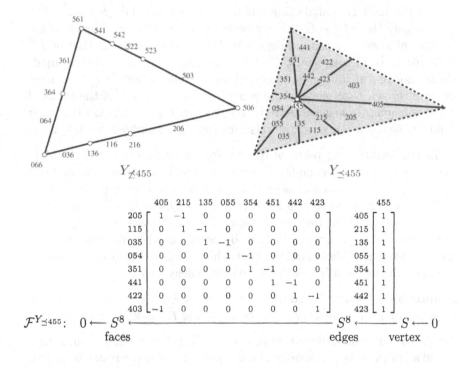

Figure 5.2: Colabeled relative complex from Example 5.34; compare Fig. 4.1

weakly colabeled cell complex $Y = (5,6,6) - X$ as in Lemma 5.32. The set $Y_{\npreceq 455}$ of faces sharing at least one coordinate with $(5,6,6)$ constitutes a weakly colabeled subcomplex of Y, depicted on the left in Fig. 5.2. The cellular pair $Y_{\preceq 455}$ of complexes $Y_{\npreceq 455} \subset Y$ is depicted on the right in Fig 5.2. Observe that $Y_{\preceq 455}$ is colabeled, not just weakly colabeled; it supports a cocellular free complex $\mathcal{F}^{Y_{\preceq 455}}$ written down in full detail in Fig. 5.2. To write the scalar matrices, orient all edges toward the center and all faces counterclockwise. The left copy of S^8 represents the 2-cells in clockwise order starting from 361, the right copy of S^8 represents the edges clockwise starting from 161, and the copy of S represents the lone vertex. The other vertices and edges are missing because they lie in the subcomplex $Y_{\npreceq 455}$. As it turns out, $\mathcal{F}^{Y_{\preceq 455}}$ resolves the ideal $I^{[455]}$ from Example 5.22; this will be a consequence of Theorem 5.37, given Example 4.6. ◇

In the next lemma, we write $M_{\preceq \mathbf{a}} = \bigoplus_{\mathbf{b} \preceq \mathbf{a}} M_{\mathbf{b}}$ for the quotient of an $\mathbb{N}^n$-graded module M modulo its elements of degree not preceding $\mathbf{a}$. Note that when I is an ideal, $I_{\preceq \mathbf{a}}$ is not an ideal of S, but rather the S-module $I/(I \cap \mathfrak{m}^{\mathbf{a}+1})$.

Lemma 5.35 *Fix an ideal I generated in degrees preceding $\mathbf{a}$. If $\mathcal{F}^Y$ is a cocellular resolution of $I_{\preceq \mathbf{a}}$, and $Y_{\prec \mathbf{a}}$ is the set of faces of Y whose labels precede $\mathbf{a}$, then $\mathcal{F}^{Y_{\preceq \mathbf{a}}}$ is a cocellular resolution of I.*

Proof. The faces G contributing a nonzero monomial to degree $\mathbf{b}$ of $\mathcal{F}^{Y_{\preceq \mathbf{a}}}$ are precisely those faces $G \in Y$ whose labels $\mathbf{a}_G$ precede $\mathbf{a}$. Therefore the complex of $\Bbbk$-vector spaces in degree $\mathbf{b}$ of $\mathcal{F}^{Y_{\preceq \mathbf{a}}}$ is the same as that of $\mathcal{F}^Y$ in degree $\mathbf{a} \wedge \mathbf{b}$. Consequently, $\mathcal{F}^{Y_{\preceq \mathbf{a}}}$ is a cocellular resolution of some module M. Looking at the generators and relations tells us that $M = I$. Indeed, we have thrown away none of the generators of I, nor any of the minimal syzygies among these generators, by results in Section 4.3.2. On the other hand, we have thrown away the relations saying that $\mathbf{x}^{\mathbf{b}} = 0$ for $\mathbf{b} \succeq \mathbf{a}$. $\square$

In the forthcoming proof of duality for resolutions, we will calculate the homology of a cell complex Y using an *acyclic cover* by subcomplexes $U_1, \ldots, U_n$ of Y. This means that we will be given an expression $Y = U_1 \cup \cdots \cup U_n$, referred to as a *cover* $\mathcal{U}$ of Y, in which $U_\sigma = \bigcap_{i \in \sigma} U_i$ either has zero reduced homology or is empty for each subset $\sigma \subseteq \{1, \ldots, n\}$. The *nerve* of this (or any) cover $\mathcal{U}$ is, by definition, the simplicial complex $\mathcal{N}(\mathcal{U})$ consisting of those subsets σ for which U_σ is nonempty. For acyclic covers $\mathcal{U}$, the nerve of $\mathcal{U}$ has the same cohomology as Y.

Lemma 5.36 (Nerve lemma) *If $\mathcal{U}$ is an acyclic cover of a polyhedral cell complex Y by polyhedral subcomplexes, then $\widetilde{H}^i(Y; \Bbbk) \cong \widetilde{H}^i(\mathcal{N}(\mathcal{U}); \Bbbk)$.*

Proof. By barycentrically subdividing every face of Y, we may assume that Y and all of the subcomplexes in $\mathcal{U}$ are simplicial. Now the result is [Rot88, Theorem 7.26], but for cohomology instead of homology. (The argument in [Rot88] works just as well for cohomology; alternatively, use that we are working over a field $\Bbbk$, so homology and cohomology are isomorphic.) $\square$

Now we come to the main general theorem concerning duality for resolutions. In the course of its proof, we apply Proposition 4.5 so many times that we will not explicitly mention it.

Theorem 5.37 *Fix a monomial ideal I generated in degrees preceding $\mathbf{a}$ and a length n cellular resolution $\mathcal{F}_X$ of $S/(I + \mathfrak{m}^{\mathbf{a}+1})$ such that all face labels on X precede $\mathbf{a} + 1$. If $Y = \mathbf{a} + 1 - X$, then $\mathcal{F}^Y$ is a weakly cocellular resolution of $(I^{[\mathbf{a}]})_{\preceq \mathbf{a}}$, and $\mathcal{F}^{Y_{\preceq \mathbf{a}}}$ is a weakly cocellular resolution of $I^{[\mathbf{a}]}$. Both Y and $Y_{\preceq \mathbf{a}}$ support minimal cocellular resolutions if $\mathcal{F}_X$ is minimal.*

The assumption $X_{\preceq \mathbf{a}+1} = X$, which prevents generators of $\mathcal{F}^{\mathbf{a}+1-X}$ from occurring in degrees outside $\mathbb{N}^n$, is there to simplify the proof, but the theorem is true without it. The assumption is pretty harmless: in all naturally occurring cellular resolutions, every vertex label precedes $\mathbf{a} + 1$; consequently $X_{\preceq \mathbf{a}+1} = X$, since face labels on X are joins of vertex labels.

Proof. By Lemma 5.35, it is enough to show that $\mathcal{F}^{\mathbf{a}+\mathbf{1}-X}$ is a weakly cocellular resolution of $(I^{[\mathbf{a}]})_{\preceq \mathbf{a}}$. The faces of $\underline{X}$ contributing monomials to degree $\mathbf{b}$ in $\mathcal{F}^{\mathbf{a}+\mathbf{1}-X}$ are precisely those whose labels in $\mathbf{a}+\mathbf{1}-X$ precede $\mathbf{b}$. These are the faces in $X^{\succeq \mathbf{a}+\mathbf{1}-\mathbf{b}}$. Therefore $(\mathcal{F}^{\mathbf{a}+\mathbf{1}-X})_{\mathbf{b}} = \widetilde{C}^{\bullet}(X^{\succeq \mathbf{a}+\mathbf{1}-\mathbf{b}}; \Bbbk)$, up to a homological shift that we will identify precisely later.

Let $X_{\not\succeq \mathbf{a}+\mathbf{1}-\mathbf{b}}$ consist of those faces of X whose labels are not preceded by $\mathbf{a}+\mathbf{1}-\mathbf{b}$. Then we have an exact sequence

$$0 \longrightarrow \widetilde{C}^{\bullet}(X_{\not\succeq \mathbf{a}+\mathbf{1}-\mathbf{b}}; \Bbbk) \longrightarrow \widetilde{C}^{\bullet}(X; \Bbbk) \longrightarrow \widetilde{C}^{\bullet}(X^{\succeq \mathbf{a}+\mathbf{1}-\mathbf{b}}; \Bbbk) \longrightarrow 0$$

of complexes of vector spaces over $\Bbbk$. Since $\underline{X}$ is acyclic, the long exact cohomology sequence implies that $\widetilde{H}^{i-1}(X^{\succeq \mathbf{a}+\mathbf{1}-\mathbf{b}}; \Bbbk) \cong \widetilde{H}^{i-2}(X_{\not\succeq \mathbf{a}+\mathbf{1}-\mathbf{b}}; \Bbbk)$.

The cell complex $X_{\not\succeq \mathbf{a}+\mathbf{1}-\mathbf{b}}$ is covered by its subcomplexes $U_1, \ldots, U_n$, where U_i consists of those faces $G \in X$ whose labels $\mathbf{a}_G$ have i^{th} coordinate at most $a_i - b_i$. Setting $\mathbf{c}_{\sigma} = (\mathbf{a}-\mathbf{b}) + d \cdot \overline{\sigma}$ for each subset $\sigma \subseteq \{1, \ldots, n\}$ and some fixed $d \gg 0$, we find that $U_{\sigma} = \bigcap_{i \in \sigma} U_i$ in fact equals $X_{\preceq \mathbf{c}_{\sigma}}$, which has zero homology when it is nonempty. By the nerve lemma, $\widetilde{H}^i(X_{\not\succeq \mathbf{a}+\mathbf{1}-\mathbf{b}}; \Bbbk)$ can be calculated as the cohomology $\widetilde{H}^i(\mathcal{N}(\mathcal{U}); \Bbbk)$ of the nerve of $\mathcal{U}$.

The key point will be that $(I^{[\mathbf{a}]})_{\preceq \mathbf{a}}$ has an "artinian" relation along the ith axis, with degree preceding $(a_i - b_i + d)\mathbf{e}_i$, for each $i = 1, \ldots, n$.

First assume the set τ of indices i such that $b_i \geq a_i + 1$ is nonempty. Then U_{σ} is empty unless $\sigma \subseteq \overline{\tau}$. On the other hand, when $\sigma \subseteq \overline{\tau}$, the faces corresponding to artinian relations in degrees preceding $d\mathbf{e}_i$ for each $i \in \tau$ all lie in U_{σ}. Therefore $\mathcal{N}(\mathcal{U})$ is a simplex, which has zero cohomology.

Now assume $\mathbf{b} \preceq \mathbf{a}$. Then U_{σ} is nonempty, except perhaps when $\sigma = \{1, \ldots, n\}$, because of artinian relations. Therefore the nerve $\mathcal{N}(\mathcal{U})$ is either a full $(n-1)$-simplex or it is an $(n-2)$-sphere. The latter case occurs exactly when $X_{\preceq \mathbf{a}-\mathbf{b}}$ is empty, or equivalently when $\mathbf{x}^{\mathbf{a}-\mathbf{b}}$ does not lie in I. Using Proposition 5.23, we find that $\mathcal{N}(\mathcal{U})$ is an $(n-2)$-sphere when $\mathbf{x}^{\mathbf{b}} \in I^{[\mathbf{a}]}$.

The isomorphism of $\widetilde{C}^{\bullet}(X^{\succeq \mathbf{a}+\mathbf{1}-\mathbf{b}}; \Bbbk)$ with $(\mathcal{F}^{\mathbf{a}+\mathbf{1}-X})_{\mathbf{b}}$ reindexes the former to be a chain complex (differentials decrease indices), with its faces of dimension $n-1$ in homological degree 0. This makes $H_i(\mathcal{F}^{\mathbf{a}+\mathbf{1}-X})_{\mathbf{b}}$ equal to $\widetilde{H}^{n-i+2}(X_{\not\succeq \mathbf{a}+\mathbf{1}-\mathbf{b}}; \Bbbk)$ and results in the only homology of the spheres in the previous paragraph being placed in homological degree 0. $\square$

Remark 5.38 The most natural setting in which to carry out Alexander duality is that of injective resolutions. These explain, for instance, why the boundary of the triangle had to be removed in Example 5.34. Injective resolutions are main characters in Chapter 11, and some of their connections to Alexander duality are treated in the exercises there.

5.4 Cohull resolutions and other applications

As a first indication of the usefulness of cocellular resolutions, let us derive some important properties of minimal cellular resolutions. (We challenge the reader to prove them without Theorem 5.37; we do not know how.)

Corollary 5.39 *If the labeled cell complex X supports a minimal resolution of an artinian monomial quotient of S, then X is pure of dimension $n - 1$.*

Proof. $\mathcal{F}_X$ resolves S/I for an ideal I containing $\mathfrak{m}^{\mathbf{a}+1}$ for some $\mathbf{a}$. If G is a facet of X, then the differential of $\mathcal{F}^{\mathbf{a}+1-X}$ is zero on G. Minimality of $\mathcal{F}^{\mathbf{a}+1-X}$ implies that G represents a nonzero homology class. Hence G, which must sit in homological degree 0 of $\mathcal{F}^{\mathbf{a}+1-X}$, has dimension $n - 1$. $\square$

Proposition 5.40 *Suppose G is a face of a labeled cell complex X supporting a minimal cellular resolution of an artinian quotient of S. If the ith coordinate $(\mathbf{a}_G)_i$ of the face label $\mathbf{a}_G$ is nonzero, then $(\mathbf{a}_G)_i = (\mathbf{a}_{G'})_i$ for some face $G' \in X$ containing G that is maximal among faces whose labels have the same support as $\mathbf{a}_G$. Any such face G' satisfies $\dim(G') = |\mathrm{supp}(G)| - 1$.*

Proof. Let X_G be the subcomplex of X on faces whose labels have support contained in $\mathrm{supp}(\mathbf{a}_G)$. Since X_G equals $X_{\preceq d \cdot \mathrm{supp}(\mathbf{a}_G)}$ for $d \gg 0$, it supports a minimal cellular resolution of an artinian quotient of the polynomial ring $\Bbbk[x_i \mid i \in \mathrm{supp}(\mathbf{a}_G)]$ by Proposition 4.5. Restricting to X_G reduces us to the case where $X = X_G$, so $\mathbf{a}_G$ has full support $\mathrm{supp}(\mathbf{a}_G) = \{1, \ldots, n\}$.

It is enough to show that if G has dimension $d < n - 1$, then G is strictly contained inside a face whose label shares its ith coordinate with $\mathbf{a}_G$. Supposing that this is not the case, we show that $\mathcal{F}_X$ is not minimal. This assumption means that x_i divides the coefficient of G on $\partial(G')$ for all faces G' under the differential ∂ of $\mathcal{F}_X$ (where by convention, x_i divides 0).

The differential δ on $\mathcal{F}^{\mathbf{a}+1-X}$ (given by the transpose of ∂) is nonzero on G by Corollary 5.39. Hence δ must take G to an element $x_i y$ for some $y \in \mathcal{F}^{\mathbf{a}+1-X}$ by the previous paragraph. However, $\delta(y) = 0$ because $\mathcal{F}^{\mathbf{a}+1-X}$ is a torsion-free S-module and $x_i \delta(y) = \delta(x_i y) = \delta^2(G)$ is zero. Therefore G does not map under δ to a minimal generator of $\ker(\delta)$, because $x_i y$ lies in $x_i \ker(\delta) \subseteq \mathfrak{m} \ker(\delta)$. It follows that $\mathcal{F}^{\mathbf{a}+1-X}$ is not minimal. $\square$

Corollary 5.41 *If the cellular resolution $\mathcal{F}_X$ in Theorem 5.37 is minimal, then the adjective "weakly" may be dropped from that theorem's conclusion.*

Proof. For faces G whose labels $\mathbf{a}_G$ have full support, Proposition 5.40 says that $\mathbf{a}_G$ equals the meet of the labels on all facets of X containing G. $\square$

Alexander duality is based on the principle that irreducible decompositions are dual to generating sets. Duality in polyhedral geometry is based on the principle that vertices are dual to facets. Our next application unifies these two principles: irreducible decompositions of I can be read off the facet labels on minimal cellular resolutions of "artinianizations" of S/I.

Theorem 5.42 *Fix a monomial ideal I generated in degrees preceding $\mathbf{a}$, and let $\mathcal{F}_X$ be a minimal cellular resolution of $S/(I + \mathfrak{m}^{\mathbf{a}+1})$. Writing $\hat{\mathbf{b}} = \sum_{b_i \leq a_i} b_i \mathbf{e}_i$ for the vector obtained from $\mathbf{b} \in \mathbb{N}^n$ by setting each ith coordinate greater than a_i to zero, the intersection $\bigcap_G \mathfrak{m}^{\widehat{\mathbf{a}}_G}$ over facets G of X is an irredundant irreducible decomposition of I.*

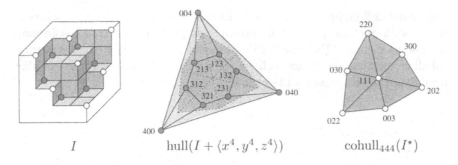

I $\mathrm{hull}(I + \langle x^4, y^4, z^4 \rangle)$ $\mathrm{cohull}_{444}(I^*)$

Figure 5.3: I and I^* are the permutohedron and tree ideals when $n = 3$

Proof. Corollary 5.39 says that every facet G has dimension $n - 1$, so Proposition 5.40 implies that $\mathbf{a}_G$ has full support. Therefore we find that $\mathbf{a} + \mathbf{1} - \mathbf{a}_G \preceq \mathbf{a}$ for all facets $G \in X$. But then $\mathbf{x}^{\mathbf{a}+\mathbf{1}-\mathbf{a}_G}$ is a minimal generator of $I^{[\mathbf{a}]}$ by Theorem 5.37, and these are in bijection with irreducible components of $I^{[\mathbf{a}]}$ by Theorem 5.27. Now note that $\mathbf{a}_G$ has i^{th} coordinate $a_i + 1$ if and only if $\mathbf{a} + \mathbf{1} - \mathbf{a}_G$ has i^{th} coordinate zero, which occurs if and only if $\mathbf{a} \smallsetminus (\mathbf{a} + \mathbf{1} - \mathbf{a}_G)$ has i^{th} coordinate zero. $\qquad\square$

Example 5.43 Theorem 5.42 is evident for the cellular resolution illustrated in Section 4.3.4, as well as for the one in Example 4.6, which resolves the ideal whose staircase is on the left-hand side of Fig. 5.1. $\qquad\diamond$

Example 5.44 The $n = 3$ example I^* in Section 4.3.4 is Alexander dual to the ideal I in Section 4.3.3 with respect to $\mathbf{a} = (3, 3, 3)$. It so happens that the hull resolution of $\Bbbk[x, y, z]/(I + \mathfrak{m}^{(4,4,4)})$ is minimal; see the middle of Fig. 5.3. Therefore Theorem 5.37 produces a minimal cocellular resolution of I^*, supported on the interior faces of the center diagram in Fig. 5.3, but with the labels subtracted from $(4, 4, 4)$. $\qquad\diamond$

Definition 5.45 Given an ideal I generated in degrees preceding $\mathbf{a}$, the **cohull complex** of I with respect to $\mathbf{a}$ is the weakly colabeled complex

$$\mathrm{cohull}_{\mathbf{a}}(I) \;=\; (\mathbf{a} + \mathbf{1} - X)_{\preceq \mathbf{a}} \quad \text{for} \quad X = \mathrm{hull}(I^{[\mathbf{a}]} + \mathfrak{m}^{\mathbf{a}+\mathbf{1}}),$$

and $\mathcal{F}^{\mathrm{cohull}_{\mathbf{a}}(I)}$ is the **cohull resolution** of I with respect to $\mathbf{a}$.

Theorem 5.37 justifies our terminology.

Corollary 5.46 $\mathcal{F}^{\mathrm{cohull}_{\mathbf{a}}(I)}$ *is a weakly cocellular free resolution of I.*

Proof. The complex $\mathcal{F}^{\mathrm{cohull}_{\mathbf{a}}(I)}$ is Alexander dual to the hull resolution of $S/(I^{[\mathbf{a}]} + \mathfrak{m}^{\mathbf{a}+\mathbf{1}})$, which satisfies the hypotheses of Theorem 5.37. $\qquad\square$

The center diagram in Fig. 5.3 betrays the fact that the cohull resolution of I^* can also be construed as a cellular resolution supported on the

right-hand cell complex of Fig. 5.3. In fact, this is the cellular resolution
we drew in Section 4.3.4. This example suggests that cohull resolutions
are always cellular (Exercise 5.16). It is not hard to show that arbitrary
cohull resolutions are *weakly cellular* (Exercise 4.3), and therefore cellular
if minimal; see Exercises 5.13–5.15.

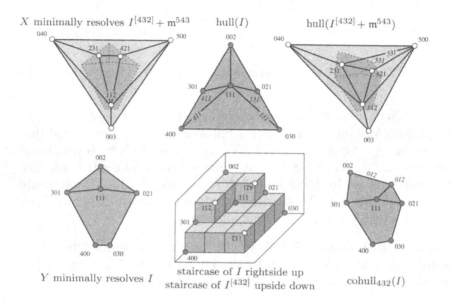

Figure 5.4: The cellular resolutions of Example 5.47

Example 5.47 Not all cellular resolutions come directly from hull and
cohull resolutions. All resolutions in this example can be construed as being cellular, supported on labeled cell complexes depicted in Fig. 5.4. Set
$I = \langle z^2, x^3z, x^4, y^3, y^2z, xyz \rangle$ so that $I^{[432]} = \langle xyz^2, x^2y^3z, x^4y^2z \rangle$. Then
hull(I) and cohull$_{432}$(I) are not minimal (the offending cells have italic la-
bels); moreover, cohull$_{\mathbf{a}}$(I) = cohull$_{432}$(I) for all $\mathbf{a} \succeq (4, 3, 2)$. Nonetheless,
$I^{[432]} + \mathbf{m}^{432}$ has a minimal cellular resolution $\mathcal{F}_X$, so Theorem 5.37 yields
a minimal cocellular resolution for I. In fact, this cocellular resolution is
cellular, supported on the labeled cell complex Y.   ◇

   The next theorem can be thought of as the reflection for arbitrary monomial ideals of the fact that Hochster's formula has two equivalent and dual
statements. In the case where $I = I_\Delta$ and $\mathbf{a} = (1, \dots, 1)$, it reduces to
simplicial Alexander duality, Theorem 5.6.

Theorem 5.48 (Duality for Betti numbers) *If I is generated in de-
grees preceding $\mathbf{a}$ and $\mathbf{1} \preceq \mathbf{b} \preceq \mathbf{a}$, then* $\beta_{n-i,\mathbf{b}}(S/I) = \beta_{i,\mathbf{a}+1-\mathbf{b}}(I^{[\mathbf{a}]})$.

Proof. Let $X = \text{hull}(I + \mathfrak{m}^{\mathbf{a}+1})$ and $Y = \text{cohull}_{\mathbf{a}}(I^{[\mathbf{a}]})$. By Theorem 4.7 applied to X, we get the equality $\beta_{i,\mathbf{b}}(S/I) = \beta_{i,\mathbf{b}}(S/(I + \mathfrak{m}^{\mathbf{a}+1}))$ when $\mathbf{b} \preceq \mathbf{a}$. Now calculate the Betti numbers of S/I and $I^{[\mathbf{a}]}$ as in Lemma 1.32 by tensoring $\mathcal{F}_X$ and $\mathcal{F}^Y$ with $\Bbbk$. By Theorem 4.31 and Theorem 5.37, the resulting complexes $\Bbbk \otimes_S \mathcal{F}_X$ and $\Bbbk \otimes_S \mathcal{F}^Y$ in degrees $\mathbf{b}$ and $\mathbf{a} + 1 - \mathbf{b}$ are vector space duals over $\Bbbk$, and their homological indexing has been reversed (subtracted from n). Therefore the $(n - i)^{\text{th}}$ homology of $\Bbbk \otimes_S \mathcal{F}_X$ has the same vector space dimension as the ith homology of $\Bbbk \otimes_S \mathcal{F}^Y$ over $\Bbbk$. $\quad\square$

When $S/(I+\mathfrak{m}^{\mathbf{a}+1})$ has a minimal cellular resolution $\mathcal{F}_X$, the equality of Betti numbers in Theorem 5.48 comes from a geometric bijection of syzygies rather than an equality of vector space dimensions: the $(n - i - 1)$-faces labeled by $\mathbf{b}$ in X are the *same* faces of $\underline{X}$ labeled by $\mathbf{a} + 1 - \mathbf{b}$ in Y. It is just that $G \in X$ represents a minimal $(n - i)^{\text{th}}$ syzygy of S/I, whereas $G \in Y$ represents a minimal i^{th} syzygy of $I^{[\mathbf{a}]}$.

Example 5.49 The following table lists some instances where the Betti numbers are 1 for the permutohedron and tree ideals I and $I^{\star} = I^{[333]}$ of Sections 4.3.3 and 4.3.4:

$3 - i$	$\mathbf{b}$	i	$\mathbf{a} + 1 - \mathbf{b}$
0	$(1, 2, 3)$	2	$(3, 2, 1)$
1	$(1, 3, 3)$	1	$(3, 1, 1)$
2	$(3, 3, 3)$	0	$(1, 1, 1)$

$$\beta_{3-i,\mathbf{b}}(I) = \beta_{i,444-\mathbf{b}}(I^{[333]}) = 1$$

Look at the figures in Sections 4.3.3 and 4.3.4 to verify these equalities, noting both the positions of these degrees in the staircase diagrams and which faces correspond in the cellular resolutions. Fig. 5.3 may also be helpful. $\diamond$

Alexander duality for resolutions in three variables has a striking interpretation for planar graphs. To state it, let us call *axial* an almost 3-connected planar map that minimally resolves an artinian ideal in $\Bbbk[x, y, z]$. This term refers to the three *axial vertices* each labeled by a power of a variable and lying on the corresponding axis in the staircase surface. An axial planar map has a well-defined outer cycle. The *planar dual* of a given map G is the planar map $\hat{G}$ obtained by placing a vertex in each region of G and connecting pairs of vertices if they are in adjacent regions. For axial planar maps, we omit the vertex of $\hat{G}$ in the unique unbounded region of G, and we instead draw infinite arcs emanating from vertices of $\hat{G}$ in bounded regions of G adjacent to the unbounded region. The resulting dual of an axial planar map is called its *dual radial map*.

Theorem 5.50 *Let $I \supseteq \mathfrak{m}^{\mathbf{a}}$, where $\mathfrak{m} = \langle x, y, z \rangle$. An axial planar map G supports a minimal cellular resolution of $\Bbbk[x, y, z]/I$ if and only if its dual radial map $\hat{G}$ supports a minimal cellular resolution of $\Bbbk[x, y, z]/I^{[\mathbf{a}]}$.*

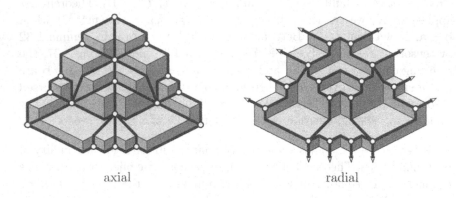

axial   radial

Figure 5.5: Duality for planar graphs as Alexander duality

Example 5.51 In nice cases, the dual axial and radial graphs can both be embedded in their staircase surfaces. We shall not make this precise here, but we instead present an example in Fig. 5.5 that we hope is convincing. Note that both surfaces are the same; this makes it easier to compare the planar maps drawn on them. Turning the picture upside down yields two pictures of the Alexander dual staircase surface, with the radial embedding appearing the right way out and the axial embedding backward. Note how the irreducible components form natural spots to place the dual vertices and how the "outer" ridges naturally carry edges of the planar dual.   ◇

The reader is invited to produce their own proof for Theorem 5.50 (the key being duality for resolutions) or to see the Notes for references. It is an open question how to generalize the embeddings of planar maps in 3-dimensional staircases to get embeddings of cellular resolutions inside staircases—canonically or otherwise—in higher dimensions.

5.5   Projective dimension and regularity

The interaction of Alexander duality with the commutative algebra of arbitrary monomial ideals, as developed in this chapter, was sparked in large part by a fundamental observation relating free resolutions of Alexander dual squarefree ideals. Specifically, duality interchanges two standard types of homological invariants, which we introduce in Definitions 5.52 and 5.54.

Definition 5.52 The length of a minimal resolution of a module M is the **projective dimension** $pd(M)$. The module M is **Cohen–Macaulay** if $pd(M)$ equals the codimension of M.

The Auslander–Buchsbaum formula [BH98, Theorem 1.3.3] implies that the projective dimension of M is at least its codimension, which—if M is a monomial quotient S/I—equals the smallest number of generators of any irreducible component of I. Hence the Cohen–Macaulay condition is a certain kind of desirable minimality: the free resolution is as short as possible.

There are many useful criteria for determining when a Stanley–Reisner ring is Cohen–Macaulay; we shall see some in Chapter 13, including general criteria such as those in Theorem 13.37 and a specific combinatorial condition (shellability) in Theorem 13.45. The most widely used criterion, and the most useful here, is the one due to Reisner, which says that links have only top homology. It is a consequence of the general Cohen–Macaulay characterization in Chapter 13, specifically part 9 of Theorem 13.37, based on local cohomology. Therefore, although we present Reisner's criterion here for use in the Eagon–Reiner Theorem, we postpone its proof until Section 13.5. (No results between here and Section 13.5 depend logically on the Eagon–Reiner Theorem or on Reisner's criterion.)

Theorem 5.53 (Reisner's criterion) *The Stanley–Reisner ring S/I_Δ is Cohen–Macaulay if and only if, for every face $\sigma \in \Delta$, the link satisfies*

$$\widetilde{H}^i(\mathrm{link}_\Delta(\sigma); \Bbbk) = 0 \quad for \quad i \neq \dim(\Delta) - |\sigma|.$$

*(We say that Δ is a **Cohen–Macaulay** simplicial complex.)*

Cohen–Macaulayness is a length condition on free resolutions. On the other hand, here is a measure of how "wide" a free resolution is.

Definition 5.54 The **regularity** of a $\mathbb{N}^n$-graded module M is

$$\mathrm{reg}(M) \;\;=\;\; \max\{|\mathbf{b}| - i \mid \beta_{i,\mathbf{b}}(M) \neq 0\}, \quad \text{where } |\mathbf{b}| = \sum_{j=1}^n b_j.$$

The next lemma follows immediately from the definitions. The converse to the second sentence holds when M is a monomial ideal (Exercise 5.19).

Lemma 5.55 *The regularity of M is at least the smallest total degree of a generator of M. If all of the minimal generators of M lie in the same degree, then M has linear free resolution precisely when that degree equals $\mathrm{reg}(M)$.*

The duality theorem of Eagon and Reiner says that the conditions of minimality in the regularity and projective dimension are Alexander dual: for free resolutions, minimal length is dual to minimal width.

Theorem 5.56 (Eagon–Reiner Theorem) *S/I_Δ is Cohen–Macaulay if and only if $I_\Delta^\star$ has linear free resolution.*

Proof. Suppose that the ideal $I_\Delta^\star$ is generated in degree d. Then $I_\Delta^\star$ has linear free resolution if and only if $\beta_{i,\sigma}(I_\Delta^\star)$ is zero whenever $|\sigma| \neq d + i$. The dual version of Hochster's formula, Corollary 1.40, says that the ideal $I_\Delta^\star$ has linear free resolution if and only if for every face $\overline{\sigma} \in \Delta$,

$$\widetilde{H}_{i-1}(\text{link}_\Delta(\overline{\sigma}); \Bbbk) = 0 \quad \text{for} \quad i \neq |\sigma| - d. \tag{5.3}$$

The ideal $I_\Delta^\star$ being generated in degree d is equivalent to Δ having dimension $n - d - 1$, so $\dim(\Delta) - |\overline{\sigma}|$ equals $n - d - 1 - (n - |\sigma|) = |\sigma| - d - 1$. Hence (5.3) is Reisner's criterion for Δ to be Cohen–Macaulay. $\quad\square$

Example 5.57 The face ideal of a simplicial sphere Δ is Cohen–Macaulay. In particular, if Δ is the boundary of a simplicial polytope as in Example 5.3, then I_Δ is Cohen–Macaulay. By Theorem 5.56, $I_\Delta^\star$ has a linear resolution. Of course, we already know from Section 4.3 (and Exercise 4.5) that this linear resolution is cellular, supported on the polar polytope $\mathcal{P}$. See Example 5.3 for an illustration of this linear resolution. $\quad\diamond$

Example 5.58 The stick twisted cubic (Example 5.2) is Cohen–Macaulay because the simplicial complex is 1-dimensional and connected. On the other hand, we found that the Alexander dual of the stick twisted cubic is just another stick twisted cubic, and therefore also Cohen–Macaulay. Thus Theorem 5.56 implies that its face ideal has a linear resolution, as well. $\diamond$

The rest of this chapter, which contains no proofs, surveys some generalizations of Theorem 5.56; references can be found in the end-of-chapter Notes. The first generalization, still in the context of squarefree ideals, says that in addition to transposing the properties of length-minimality and width-minimality for free resolutions, Alexander duality in fact transposes the *deviation* from minimality: for free resolutions, length is dual to width.

Theorem 5.59 *The projective dimension and regularity of Alexander dual squarefree ideals satisfy* $\text{pd}(S/I_\Delta) = \text{reg}(I_\Delta^\star)$.

Note that Theorem 5.56 follows immediately from Theorem 5.59, because the codimension of I_Δ equals the smallest degree of a generator of $I_\Delta^\star$ by the very definition of Alexander dual ideal ($\mathfrak{m}^\sigma \leftrightarrow \mathbf{x}^\sigma$). Theorem 5.59 has an elementary proof relying only on Hochster's formulas.

The relation between projective dimension and regularity can be viewed as the boundary case of a duality that preserves a family of homological invariants interpolating between them.

Definition 5.60 An ith Betti number $\beta_{i,j}(M) \neq 0$ of an $\mathbb{N}$-graded module M in degree j is **extremal** if $\beta_{p,q}(M) = 0$ for all p and q satisfying the following three conditions: (i) $p \geq i$, (ii) $p - q \geq i - j$, and (iii) $q \geq j + 1$.

In the *Macaulay* betti *diagram* of M, the Betti number $\beta_{i,j}(M)$ is plotted in column i and row $j - i$. Using this notation, condition (i) says that $\beta_{p,q}(M)$ lies in a column weakly east of $\beta_{i,j}(M)$, condition (ii) says that

$\beta_{p,q}(M)$ lies in a row weakly south of $\beta_{i,j}(M)$, and imposing condition (iii) is equivalent to the additional requirement that $(p,q) \neq (i,j)$. Thus a nonzero Betti number $\beta_{i,j}(M)$ is extremal if it is the only nonzero *Macaulay* betti entry in the quadrant of which it is the northwest corner.

Projective dimension measures the column index of the easternmost extremal Betti number, whereas regularity measures the row index of the southernmost extremal Betti number. The following theorem implies, in particular, that these roles are switched under Alexander duality.

Theorem 5.61 *The Betti number* $\beta_{i,j}(S/I_\Delta)$ *is extremal if and only if* $\beta_{j-i-1,j}(S/I_\Delta^\star)$ *is extremal, and in this case* $\beta_{i,j}(S/I_\Delta) = \beta_{j-i-1,j}(S/I_\Delta^\star)$.

Theorem 5.59 is refined by Theorem 5.61 for squarefree monomial ideals, in the sense that the former is an immediate consequence of the latter. For arbitrary monomial ideals, even Theorem 5.59 cannot hold verbatim, since one side of the equality (projective dimension) is bounded while the other (regularity) is not. On the other hand, regularity is not a particularly $\mathbb{N}^n$-graded thing to measure—the definition requires us to sum the coordinates of the degree $\mathbf{b}$, which is more of a $\mathbb{Z}$-graded procedure. The generalization to arbitrary monomial ideals of Theorems 5.56 and 5.59 needs an $\mathbb{N}^n$-graded analogue of regularity.

Definition 5.62 The **support-regularity** of a monomial ideal I is

$$\text{supp.reg}(I) = \max\left\{|\text{supp}(\mathbf{b})| - i \mid \beta_{i,\mathbf{b}}(I) \neq 0\right\},$$

and I is said to have a **support-linear free resolution** if there is a $d \in \mathbb{N}$ such that $|\text{supp}(m)| = d = \text{supp.reg}(I)$ for all minimal generators m of I.

For squarefree ideals the notions of regularity and support-regularity coincide, because the only degrees we ever care about are squarefree. In particular, the two sentences in the following result specialize to the Eagon–Reiner Theorem and Theorem 5.59 when $\mathbf{a} = (1, \ldots, 1)$.

Theorem 5.63 *If a monomial ideal I is generated in degrees preceding* $\mathbf{a}$, *then S/I is Cohen–Macaulay if and only if the Alexander dual ideal $I^{[\mathbf{a}]}$ has support-linear free resolution. More generally,* $\text{pd}(S/I) = \text{supp.reg}(I^{[\mathbf{a}]})$.

The optimal insight provided by Theorem 5.63 comes in a context combining monomial matrices for free and *injective* resolutions, the latter of which we will introduce in Chapter 11. For a glimpse of this context, see Exercise 11.2. Essentially, decreases in the dimensions of the indecomposable injective summands in a minimal injective resolution of S/I correspond precisely to increases in the supports of the degrees in a minimal free resolution of $I^{[\mathbf{a}]}$. The former detect the projective dimension of S/I by the Auslander–Buchsbaum formula. Thus, when the supports of syzygy degrees of $I^{[a]}$ increase as slowly as possible, so that $I^{[a]}$ has support-linear free resolution, the dimensions of indecomposable summands in a minimal injective resolution of S/I decrease as slowly as possible. This slowest possible decrease in dimension postpones the occurrence of summands isomorphic to injective hulls of $\Bbbk$ as long as possible, making the depth of S/I as large as possible. As a result, S/I must be Cohen–Macaulay (see Theorem 13.37.7).

At the beginning of this section, we noted that Alexander duality interchanges two types of homological invariants, by which we meant projective dimension and regularity. Theorem 5.61 extends this interchange to a flip on a family of refinements of this pair of invariants. In contrast, the crux of Theorem 5.63 is that we could have meant a different interchange: namely the switch of Betti numbers for *Bass numbers* (Definition 11.37): whereas Betti numbers determine the regularity, the projective dimension can be reinterpreted in terms of depth—and hence in terms of Bass numbers—via the Auslander–Buchsbaum formula.

Exercises

5.1 Prove Theorem 5.11 directly, by tensoring the coKoszul complex $\mathbb{K}^{\bullet}$ with S/I.

5.2 Prove Corollary 5.12 by applying Theorem 5.6 to Corollary 1.40.

5.3 Compute the Alexander dual of $\langle x^4, y^4, x^3 z, y^3 z, x^2 z^2, y^2 z^2, xz^3, yz^3 \rangle$ with respect to $\mathbf{a} = (5, 6, 8)$.

5.4 Resume the notation from Exercise 3.6.

(a) Turning the picture there upside down yields the staircase diagram for an Alexander dual ideal $I^{[a]}$. What is $\mathbf{a}$?

(b) On a photocopy of the upside down staircase diagram, draw the Buchberger graph of $I^{[a]}$. Compare it to the graph $\mathrm{Buch}(I)$ that you drew in Exercise 3.6.

(c) Use the labels on the planar map determined by $\mathrm{Buch}(I^{[a]})$ to relabel the vertices, edges, and regions in the planar map determined by $\mathrm{Buch}(I)$.

(d) Show that this relabeled planar map is colabeled and determines the resolution Alexander dual to the usual one from $\mathrm{Buch}(I)$, as in Theorem 5.37.

5.5 For any monomial ideal I, let $\mathbf{a}_I$ be the exponent on the least common multiple of all minimal generators of I, and define the *tight Alexander dual* $I^{\star} = I^{[\mathbf{a}_I]}$. Find a monomial ideal I such that $(I^{\star})^{\star} \neq I$. Characterize such ideals I.

5.6 Show that tight Alexander duality commutes with radicals: $\mathrm{rad}(I)^{\star} = \mathrm{rad}(I^{\star})$.

5.7 Prove from first principles that a monomial ideal is irreducible as in Definition 5.16 if and only if it cannot be expressed as an intersection of two (perhaps ungraded) ideals strictly containing it.

5.8 The **socle** of a module M is the set $\mathrm{soc}(M) = (0 :_M \mathfrak{m})$ of elements in M annihilated by every variable. If $M = S/I$ is artinian, prove that $\mathbf{x}^{\mathbf{b}} \in \mathrm{soc}(M)$ if and only if $\mathfrak{m}^{\mathbf{b}+1}$ is an irreducible component of I. Use Corollary 5.39 and Hochster's formula to construct another proof of Theorem 5.42.

5.9 The **monomial localization** of a monomial ideal $I \subseteq \Bbbk[\mathbf{x}]$ at x_i is the ideal $I|_{x_i=1} \in \Bbbk[\mathbf{x} \smallsetminus x_i]$ that results after setting $x_i = 1$ in all generators of I. Suppose that a labeled cell complex X supports a minimal cellular resolution of $S/(I + \mathfrak{m}^{\mathbf{a}+1})$. Explain how to recover a minimal cellular resolution of $I|_{x_i=1}$ from the faces of X containing the vertex $v \in X$ labeled by $\mathbf{a}_v = x_i^{a_i+1}$. This set of faces is called the *star* of v, and the minimal cellular resolution will be supported on the *link* of v (also known as the *vertex figure* of X in a neighborhood of v).

5.10 Suppose that a colabeled cell complex Y supports a minimal cocellular resolution of $S/(I + \mathfrak{m}^{\mathbf{a}+1})$. Explain why the set of faces of Y whose labels have i^{th} coordinate $a_i + 1$ is another colabeled complex. Show that it supports a minimal cocellular resolution of the monomial localization $I|_{x_i=1}$ (Exercise 5.9).

5.11 Exhibit an example demonstrating that if the condition of minimality in Theorem 5.42 is omitted, then the intersection given there can fail to be an irreducible decomposition—even a redundant one. Nonetheless, prove that if the intersection is taken over a suitable subset of facets, then the conclusion still holds.

5.12 If $\mathcal{F}_X$ is a minimal cellular resolution of an artinian quotient, then a face $G \in X$ is in the boundary of X if and only if its label $\mathbf{a}_G$ fails to have full support.

5.13 Prove that weakly cellular resolutions (Exercise 4.3) of artinian quotients are cellular if they are minimal.

5.14 Prove that the cohull resolution $\mathcal{F}^{\mathrm{cohull}_{\mathbf{a}}(I)}$ of I with respect to $\mathbf{a}$ can be viewed as a weakly cellular free resolution $\mathcal{F}_{\mathrm{cohull}_{\mathbf{a}}(I)}$. Hint: Consider the polyhedron dual to $\mathcal{P}_t$ from Definition 4.16, and use Theorem 4.31.

5.15 Prove that if $\mathrm{hull}(I^{[\mathbf{a}]} + \mathfrak{m}^{\mathbf{a}+1})$ is minimal, then $\mathcal{F}_{\mathrm{cohull}_{\mathbf{a}}(I)}$ is a minimal cellular (not weakly cellular) resolution.

5.16* *Open problem:* Prove that all cohull resolutions are cellular.

5.17 Replace "$\mathcal{F}_X$ a minimal cellular resolution" in Theorem 5.42 by "$\mathcal{F}_X$ the (possibly nonminimal) hull resolution", and conclude with these hypotheses that the intersection $\bigcap_G \mathfrak{m}^{\widehat{\mathbf{a}}_G}$ over facets $G \in X$ is a (possibly redundant) irreducible decomposition of I. Hint: Use Exercises 4.3 and 5.14.

5.18 Define a vector $\mathbf{b} \in \mathbb{N}^n$ to lie on the **staircase surface** of a monomial ideal I if $\mathbf{x}^{\mathbf{b}} \in I$ but $\mathbf{x}^{\mathbf{b}-\mathrm{supp}(\mathbf{b})} \notin I$. Prove that every face label on the hull complex $\mathrm{hull}(I)$ lies on the staircase surface of I. Hint: This can be done directly, using the convex geometry of hull complexes, or with Exercises 4.3 and 5.14.

5.19 Prove that if a monomial ideal I is not generated in a single $\mathbb{N}$-graded degree, then I has a minimal first syzygy between two generators of different $\mathbb{N}$-degrees. Conclude that if the module M in Lemma 5.55 is a monomial ideal, then M can only have linear free resolution if its generators all have the same total degree.

Notes

In one form or another, Alexander duality has been appearing in the context of commutative algebra for decades. A seminal such use of it came in Hochster's paper [Hoc77]; our proof of Theorem 5.6 more or less constitutes his proof of Corollary 5.12. Sharper focus has been given to the notion of Alexander dual simplicial complex, as a combinatorial object, ever since its appearance in the work of Eagon and Reiner [ER98]. The Eagon–Reiner Theorem initiated the subsequent active research on interactions of Alexander duality with commutative algebra, including all of the results after Section 5.1 in this chapter.

The Alexander inversion formula seems to have been noticed first in [Mil00b, Theorem 4.36], where it is proved for the *squarefree modules* of Yanagawa [Yan00]. It was motivated by connections to equivariant K-theory of vector spaces with algebraic group actions, but in applications it is used as a tool to help calculate the K-polynomial of an ideal through its dual, in keeping with Dave Bayer's advice. As an example, see [KnM04a], where the for subword complexes (generalizing the ones to be introduced in Chapter 16) are computed this way.

Our presentation of irreducible decomposition is adapted from [Mil00b, Section 1.1]. The algorithm in Remark 5.28 for computing irreducible decompositions has been implemented in Macaulay 2 by G. Smith [GS04, HoS02]. The special case of Alexander duality in the context of planar graphs was originally stated in [Mil02b, Theorem 15.1].

Background on relative (co)homology can be found in a number of good textbooks such as [Hat02, Mun84, Rot88] on Algebraic Topology.

Duality for resolutions in the form of Theorem 5.37 is a special case of the Grothendieck–Serre *local duality theorem* [BH98, Section 3.6]. The proof here using cellular resolutions to avoid the technology of general homological algebra is new. There is a generalization of Grothendieck–Serre duality, due to Greenlees and May [GM92]; correspondingly, there is strengthening of Alexander duality, in the context of free and injective resolutions [Mil02a].

Theorem 5.59 is due to Terai [Ter99a]. It inspired Bayer, Charalambous, and Popescu to introduce extremal Betti numbers and prove Theorem 5.61 [BCP99]. The robustness of these $\mathbb{N}$-graded homological invariants is supported by their stability under taking reverse-lexicographic generic initial ideals [BCP99]. The natural $\mathbb{N}^n$-graded refinements of extremal Betti numbers for squarefree monomial ideals are also preserved numerically while their locations are flipped by Alexander duality [BCP99]. Extremal Betti numbers can be defined for graded modules over exterior algebras; Aramova and Herzog proved that taking generic initial ideals preserves extremal Betti numbers in that setting [AH00], just as it does over polynomial rings, and they consequently gave new proofs of Kalai's theorems on algebraic shifting (see the Notes to Chapter 2). In general, reworking many of the results in this book for exterior algebras should be a fruitful line of future research.

Theorem 5.63 is a consequence of a general result for arbitrary $\mathbb{N}^n$-graded modules [Mil00a, Theorem 4.25] that describes how Alexander duality extends to a functor interchanging free and injective resolutions. This functorial Alexander duality for resolutions implies Theorem 5.48 and generalizes it to $\mathbb{N}^n$-graded degrees without full support, where Bass numbers are more natural invariants to use. Solutions to Exercises 5.9, 5.10, 5.12, and 5.13 can be extracted from [Mil00a].

Reisner's criterion (Theorem 5.53) is one of the fundamental results that connects simplicial topology to commutative algebra and algebraic geometry. It originated in the thesis of Gerald Reisner [Rei76], who (according to his advisor, Mel Hochster) pronounces his last name "reess'- nər".

Chapter 6

Generic monomial ideals

We have already seen in Chapter 2 that monomial ideals derived from certain kinds of randomness have more concrete homological algebra. In our discussion of three-dimensional staircases, we saw that randomness of the exponent vectors on the minimal generators has similar consequences. In this chapter we study generic monomial ideals in any number of variables. Their minimal free resolutions are cellular. The underlying complex is simplicial and is known as the *Scarf complex*. Certain questions about arbitrary monomial ideals can be reduced to questions about generic ideals by a process called *deformation of exponents*. It is in this context that the naturality of genericity is borne out. We close with a discussion of cogeneric monomial ideals, which are Alexander dual to generic monomial ideals.

6.1 Taylor complexes and genericity

Consider an arbitrary monomial ideal $I = \langle m_1, \ldots, m_r \rangle$ in the polynomial ring $S = \Bbbk[x_1, \ldots, x_n]$. For any subset σ of $\{1, \ldots, r\}$, we write m_σ for the least common multiple of $\{m_i \mid i \in \sigma\}$ and set $\mathbf{a}_\sigma = \deg(m_\sigma) \in \mathbb{N}^n$.

Definition 6.1 Let Δ be a labeled simplicial complex on $\{1, \ldots, r\}$. The **Taylor complex** $\mathcal{F}_\Delta$ is defined by putting the reduced chain complex of Δ into a sequence of monomial matrices with the face label $m_\sigma = \mathbf{x}^{\mathbf{a}_\sigma}$ on the row and column corresponding to the (unlabeled) face $\sigma \in \Delta$.

The Taylor complex $\mathcal{F}_\Delta$ is a cellular free complex supported on Δ. It is therefore an $\mathbb{N}^n$-graded complex of free S-modules, and assuming that each singleton $\{i\}$ is a face of Δ, its zeroth homology module equals S/I.

Let us also describe $\mathcal{F}_\Delta$ without referring to monomial matrices. Introduce a basis vector $\mathbf{e}_\sigma$ in $\mathbb{N}^n$-graded degree $\deg(m_\sigma)$ and homological degree $|\sigma|$ for each face σ of Δ. The free S-module

$$\mathcal{F}_\Delta \;=\; \bigoplus_{\sigma \in \Delta} S \cdot \mathbf{e}_\sigma$$

with differential

$$\partial(\mathbf{e}_\sigma) \;=\; \sum_{i \in \sigma} \operatorname{sign}(i,\sigma) \frac{m_\sigma}{m_{\sigma \smallsetminus i}} \; \mathbf{e}_{\sigma \smallsetminus i}$$

is the Taylor complex. Here, $\operatorname{sign}(i,\sigma) = (-1)^{j-1}$ if i is the jth element of σ when the elements of the set σ are listed in increasing order. In the literature, the term "Taylor complex" has almost always referred to the Taylor resolution of Section 4.3.2, which is the special case when Δ is the full $(r-1)$-simplex consisting of *all* subsets of $\{1, \dots, r\}$; but Definition 6.1 should raise no confusion.

Example 6.2 Taking $I = \langle x^2, xy, y^2z, z^2 \rangle$, let Δ be the simplicial complex consisting of the two triples $\{1,2,4\}$ and $\{2,3,4\}$ and their subsets. Here is a picture of Δ, with each face accompanied by its monomial label.

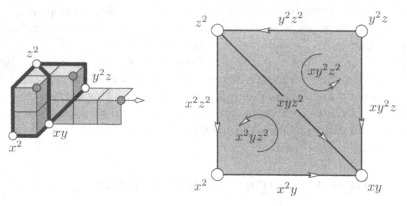

The Taylor complex $\mathcal{F}_\Delta$ is given by the following monomial matrices:

$$0 \leftarrow S \xleftarrow{\ \begin{array}{cccc} xy & x^2 & z^2 & y^2z \\ 1 \begin{bmatrix} 1 & 1 & 1 & 1 \end{bmatrix} \end{array}\ } S^4 \xleftarrow{\ \begin{array}{c} \begin{array}{ccccc} xy^2z & y^2z^2 & x^2z^2 & x^2y & xyz^2 \end{array} \\ \begin{array}{c} xy \\ x^2 \\ z^2 \\ y^2z \end{array} \begin{bmatrix} 1 & 0 & 0 & 1 & 1 \\ 0 & 0 & 1 & -1 & 0 \\ 0 & 1 & -1 & 0 & -1 \\ -1 & -1 & 0 & 0 & 0 \end{bmatrix} \end{array}\ } S^5 \xleftarrow{\ \begin{array}{c} \begin{array}{cc} xy^2z^2 & x^2yz^2 \end{array} \\ \begin{array}{c} xy^2z \\ y^2z^2 \\ x^2z^2 \\ x^2y \\ xyz^2 \end{array} \begin{bmatrix} -1 & 0 \\ 1 & 0 \\ 0 & 1 \\ 0 & 1 \\ 1 & -1 \end{bmatrix} \end{array}\ } S^2 \leftarrow 0$$

For an example of the non-monomial matrix way to write this complex, note that the left column in the rightmost map corresponds to

$$\partial(\mathbf{e}_{234}) \;=\; z\mathbf{e}_{23} + x\mathbf{e}_{34} - y\mathbf{e}_{24},$$

where $\mathbf{e}_{234}$ is the basis vector of $\mathcal{F}_\Delta$ in degree $\mathbf{a}_{\{2,3,4\}} = (1,2,2)$.　　　　◇

The Taylor complex $\mathcal{F}_\Delta$ in the above example is both exact and minimal, so it is a minimal free resolution of $I = \langle x^2, xy, y^2z, z^2 \rangle$. However, if we were to flip the diagonal and redefine Δ as the simplicial complex with facets $\{1, 2, 3\}$ and $\{1, 3, 4\}$, then $\mathcal{F}_\Delta$ would not be exact. (Check this.) This raises the question of under what conditions $\mathcal{F}_\Delta$ is exact or minimal.

Lemma 6.3 *The Taylor complex $\mathcal{F}_\Delta$ is acyclic if and only if for every monomial m, the simplicial subcomplex $\Delta_{\preceq m} = \{\sigma \in \Delta \mid m_\sigma \text{ divides } m\}$ is acyclic over $\Bbbk$ (homology only in degree 0).*

Proof. This is a special case of Proposition 4.5. $\square$

Lemma 6.4 *The Taylor complex $\mathcal{F}_\Delta$ is minimal if and only if for all faces $\sigma \in \Delta$ and all indices $i \in \sigma$, the monomials m_σ and $m_{\sigma \smallsetminus i}$ are different.*

Proof. A complex of $\mathbb{N}^n$-graded free S-modules is minimal if in its representation by monomial matrices, every nonzero matrix entry has its column label different from its row label. Here, these labels are m_σ and $m_{\sigma \smallsetminus i}$. $\square$

In Chapter 4 we constructed the hull resolution, which is a cellular free resolution of length $\leq n$ for an arbitrary monomial ideal I in $\Bbbk[x_1, \ldots, x_n]$. In this chapter we will see that I also has a simplicial free resolution of length $\leq n$; that is, there exists a simplicial complex Δ of dimension $\leq n-1$ on the generators of I whose Taylor complex $\mathcal{F}_\Delta$ is acyclic. The basic idea in constructing such resolutions is to wiggle the exponents and to consider generic monomial ideals first. In the next section we show that for generic ideals, the hull resolution is both minimal and simplicial, and in Theorem 6.24 we show how to "unwiggle" the exponents.

Let us close this section with the definition of "generic".

Definition 6.5 A monomial m' **strictly divides** another monomial m if m' divides m/x_i for all variables x_i dividing m. A monomial ideal $\langle m_1, \ldots, m_r \rangle$ is **generic** if whenever two distinct minimal generators m_i and m_j have the same positive (nonzero) degree in some variable, a third generator m_k strictly divides their least common multiple $\text{lcm}(m_i, m_j)$.

Equivalently, a monomial ideal $I = \langle m_1, \ldots, m_r \rangle$ is generic if the two monomials in any edge $\{m_i, m_j\}$ of the Buchberger graph $\text{Buch}(I)$ do not have the same positive degree in any variable. This definition is more inclusive than the notion of strongly generic in Chapter 3. For instance, the ideal $\langle x^2, xy, y^2z, z^2 \rangle$ in Example 6.2 is strongly generic and hence also generic. The ideal $\langle x^2z, xy, y^2z, z^2 \rangle$ is generic but not strongly generic. The ideal $\langle x^2, xy, yz, z^2 \rangle$ is neither strongly generic nor even generic.

Example 6.6 The tree ideal I^* in Section 4.3.4 is generated by the monomials $\omega_\sigma = \prod_{s \in \sigma} x_s^{n-|\sigma|+1}$ for the nonempty subsets $\sigma \subseteq \{1, \ldots, n\}$. If σ and σ' are distinct subsets, then $\omega_{\sigma \cup \sigma'}$ strictly divides the least common multiple of ω_σ and $\omega_{\sigma'}$. This shows that $I^\star$ is generic. Let Δ be the first barycentric subdivision of the $(n-1)$-simplex. The vertices of Δ are labeled by nonempty subsets of $\{1, \ldots, n\}$ and hence by the generators of $I^\star$. For $n = 3$ this is depicted in Section 4.3.4. Using Lemma 6.3 we can see that the Taylor complex $\mathcal{F}_\Delta$ is a minimal free resolution of $I^\star$. This is an instance of the Scarf complex construction in the next section. $\diamond$

6.2   The Scarf complex

To every monomial ideal we can associate a simplicial complex as follows.

Definition 6.7 Let I be a monomial ideal with minimal generating set $\{m_1, \ldots, m_r\}$. The **Scarf complex** Δ_I is the collection of all subsets of $\{m_1, \ldots, m_r\}$ whose least common multiple is unique:

$$\Delta_I \;=\; \{\sigma \subseteq \{1, \ldots, r\} \mid m_\sigma = m_\tau \Rightarrow \sigma = \tau\}.$$

We will now show that a subset of a set in Δ_I is again a set in Δ_I.

Lemma 6.8 *The Scarf complex Δ_I is a simplicial complex. Its dimension is at most $n - 1$.*

Proof. If σ is a face of the Scarf complex and i is an element of σ, let $\tau = \sigma \setminus i$. Suppose that $m_\tau = m_\rho$ for some index set ρ. Then $m_\sigma = m_{\rho \cup i}$ and consequently $\rho \cup i = \sigma$, because σ lies in the Scarf complex. It follows that either $\rho = \tau$ or $\rho = \sigma$. However, the latter is impossible, since that would mean $m_\tau = m_\sigma$. Hence $\tau = \rho$ and we conclude that τ is a face of Δ_I.

For the dimension count, a facet σ of Δ_I has cardinality at most n because for each index $i \in \sigma$, the generator m_i contributes at least one coordinate to m_σ—that is, there is some variable x_k such that m_i is the only generator dividing m_σ and having the same degree in x_k as m_σ. $\square$

If $n = 2$ then the Scarf complex is one-dimensional, and its facets are the adjacent pairs of generators in the staircase. For an example with $n = 3$, the complex Δ of two triangles in Example 6.2 is the Scarf complex of the given monomial ideal. Note that the Scarf complex may be disconnected.

Example 6.9 When $I = \langle xy, xz, yz \rangle$, the Scarf complex Δ_I consists of three isolated points, its 1-skeleton edges(Δ_I) is the empty graph on three nodes, and Buch(I) is the triangle. The minimal free resolution is given by any two of the three edges. $\diamond$

In all dimensions, every edge of the Scarf complex of a monomial ideal is an edge of the Buchberger graph:

$$\text{edges}(\Delta_I) \;\subseteq\; \text{Buch}(I),$$

but the converse is usually not true unless I is generic; this is the content of the next lemma, whose proof we leave to the reader.

Lemma 6.10 *For I a generic monomial ideal,* $\mathrm{edges}(\Delta_I) = \mathrm{Buch}(I)$.

We now consider the cellular free complex defined by the Scarf complex.

Definition 6.11 *The Taylor complex $\mathcal{F}_{\Delta_I}$ supported on the Scarf complex Δ_I is called the* **algebraic Scarf complex** *of the monomial ideal I.*

Whether or not I is generic, its Scarf complex always shows up.

Proposition 6.12 *If I is a monomial ideal in S, then every free resolution of S/I contains the algebraic Scarf complex $\mathcal{F}_{\Delta_I}$ as a subcomplex.*

Proof. Every free resolution contains a minimal free resolution (Exercise 1.11), so it is enough to show that $\mathcal{F}_{\Delta_I}$ is contained in some minimal free resolution $\mathcal{F}$ of S/I. In particular, we may choose $\mathcal{F}$ to be a subcomplex of the *full* Taylor resolution, which is supported on the entire simplex whose vertices are the minimal generators of I. Every basis vector $\mathbf{e}_\sigma$ for $\sigma \in \Delta_I$ must lie in $\mathcal{F}$ by Theorem 4.7 and the uniqueness of $\mathbf{a}_\sigma$ as a face label. $\square$

The algebraic Scarf complex solves the problem of finding the best possible cellular (in fact, simplicial) resolutions for generic monomial ideals.

Theorem 6.13 *If I is a monomial ideal, then its Scarf complex Δ_I is a subcomplex of the hull complex $\mathrm{hull}(I)$. If I is generic then $\Delta_I = \mathrm{hull}(I)$, so its algebraic Scarf complex $\mathcal{F}_{\Delta_I}$ minimally resolves the quotient S/I.*

Proof. Let $F = \{\mathbf{x}^{\mathbf{a}_1}, \ldots, \mathbf{x}^{\mathbf{a}_p}\}$ be a face of the Scarf complex Δ_I with $m_F = \mathbf{x}^{\mathbf{u}}$. For any index $i \in \{1, \ldots, p\}$, the least common multiple $m_{F \smallsetminus i}$ of $F \smallsetminus \{\mathbf{x}^{\mathbf{a}_i}\}$ strictly divides m_F in at least one variable. After relabeling, we may assume that this variable is x_i. Hence the x_i-degree of $\mathbf{x}^{\mathbf{a}_i}$ is strictly larger than the x_i-degree of $m_{F \smallsetminus i}$. We conclude that $a_{ki} < a_{ii}$ for any two distinct indices i and k in $\{1, \ldots, p\}$. This condition ensures that the determinant of the $p \times p$ matrix (t^{ki}) is nonzero, so the points $t^{\mathbf{a}_1}, \ldots, t^{\mathbf{a}_p}$ are affinely independent in $\mathbb{R}^n$, and their convex hull is a simplex.

The points $t^{\mathbf{a}_1}, \ldots, t^{\mathbf{a}_p}$ constitute the vertex set of the restricted hull complex $\mathrm{hull}(I)_{\preceq \mathbf{u}}$. It follows that every face of $\mathrm{hull}(I)$ labeled by $\mathbf{x}^{\mathbf{u}}$ has vertices with labels from among $\{\mathbf{x}^{\mathbf{a}_1}, \ldots, \mathbf{x}^{\mathbf{a}_p}\}$. There can be at most one such face of $\mathrm{hull}(I)$, since F is a Scarf face, and there must be at least one by Proposition 6.12. We conclude that the simplex F is a face of the polyhedral cell complex $\mathrm{hull}(I)_{\preceq \mathbf{u}}$. This completes the proof of the first assertion in Theorem 6.13.

For the second assertion in Theorem 6.13, we need the following lemma.

Lemma 6.14 *Let I be a monomial ideal and F a face of $\mathrm{hull}(I)$. For each monomial $m \in I$ there is a variable x_j such that $\deg_{x_j}(m) \geq \deg_{x_j}(m_F)$.*

Proof. Suppose that $m = \mathbf{x}^{\mathbf{u}}$ strictly divides m_F in each coordinate. Let $t^{\mathbf{a}_1}, \ldots, t^{\mathbf{a}_p}$ be the vertices of the face F and consider their barycenter

$$\mathbf{v}(t) \;=\; \frac{1}{p} \cdot (t^{\mathbf{a}_1} + \cdots + t^{\mathbf{a}_p}) \;\in\; F.$$

The jth coordinate of $\mathbf{v}(t)$ is a polynomial in t of degree equal to $\deg_{x_j}(m_F)$. The jth coordinate of $t^{\mathbf{u}}$ is a monomial of strictly lower degree. Hence $t^{\mathbf{u}} < \mathbf{v}(t)$ coordinatewise for $t \gg 0$. Let $\mathbf{w}$ be a nonzero linear functional that is nonnegative on $\mathbb{R}^n_+$ and whose minimum over $\mathcal{P}_t$ is attained at the face F. Then $\mathbf{w} \cdot \mathbf{v}(t) = \mathbf{w} \cdot \mathbf{a}_1 = \cdots = \mathbf{w} \cdot \mathbf{a}_p$, but our discussion implies $\mathbf{w} \cdot t^{\mathbf{u}} < \mathbf{w} \cdot \mathbf{v}(t)$, a contradiction. $\square$

Continuing with the proof of Theorem 6.13, let F be any face of hull(I) and let $\mathbf{x}^{\mathbf{a}_1}, \ldots, \mathbf{x}^{\mathbf{a}_p}$ be the monomial generators of I corresponding to the vertices of F. We may assume that all n variables x_j appear in the monomial $m_F = \operatorname{lcm}(\mathbf{x}^{\mathbf{a}_1}, \ldots, \mathbf{x}^{\mathbf{a}_p})$. Suppose that F is not a face of the Scarf complex Δ_I. Then either

(i) $\operatorname{lcm}(\mathbf{x}^{\mathbf{a}_1}, \ldots, \mathbf{x}^{\mathbf{a}_{i-1}}, \mathbf{x}^{\mathbf{a}_{i+1}}, \ldots, \mathbf{x}^{\mathbf{a}_p}) = m_F$ for some $i \in \{1, \ldots, p\}$, or
(ii) there exists another generator $\mathbf{x}^{\mathbf{u}}$ of I such that $t^{\mathbf{u}} \notin F$ and $\mathbf{x}^{\mathbf{u}}$ divides m_F.

Consider first case (i). By Lemma 6.14 applied to $m = \mathbf{x}^{\mathbf{a}_i}$, there exists a variable x_j such that $\deg_{x_j}(\mathbf{x}^{\mathbf{a}_i}) = \deg_{x_j}(m_F)$, and hence $\deg_{x_j}(\mathbf{x}^{\mathbf{a}_i}) = \deg_{x_j}(\mathbf{x}^{\mathbf{a}_k})$ for some $k \neq i$. Since I is generic, there exists another generator m of I strictly dividing $\operatorname{lcm}(\mathbf{x}^{\mathbf{a}_i}, \mathbf{x}^{\mathbf{a}_k})$ in all of its positive coordinates. Since $\operatorname{lcm}(\mathbf{x}^{\mathbf{a}_i}, \mathbf{x}^{\mathbf{a}_k})$ divides m_F, it follows that m divides m_F in all n coordinates. This is a contradiction to Lemma 6.14.

Consider now case (ii), and suppose that we are not in case (i). For any variable x_j there exists $i \in \{1, \ldots, p\}$ such that $\deg_{x_j}(\mathbf{x}^{\mathbf{a}_i}) = \deg_{x_j}(m_F) \geq \deg_{x_j}(\mathbf{x}^{\mathbf{u}})$. If the inequality "$\geq$" is an equality "$=$", then there exists a new monomial generator m strictly dividing m_F in all of its positive coordinates, a contradiction to Lemma 6.14, as before. Therefore "$\geq$" is a strict inequality "$>$" for all variables x_j. This means that $\mathbf{x}^{\mathbf{u}}$ strictly divides m_F in all coordinates, again a contradiction to Lemma 6.14.

Hence both cases (i) and (ii) lead to a contradiction, and we conclude that every face of the hull complex hull(I) is a face of the Scarf complex Δ_I. This implies that hull$(I) = \Delta_I$, by the first part of Theorem 6.13. The algebraic Scarf complex $\mathcal{F}_{\Delta_I}$ is minimal because no two faces in Δ_I have the same degree. $\square$

In what follows we draw some algebraic conclusions from Theorem 6.13.

Corollary 6.15 *The minimal free resolution of a generic monomial ideal I is independent of the characteristic of the field $\Bbbk$. The total Betti number $\beta_i(I) = \sum_{\mathbf{a} \in \mathbb{N}^n} \beta_{i,\mathbf{a}}(I)$ equals the number $f_i(\Delta_I)$ of i-dimensional faces of its Scarf complex Δ_I.*

Corollary 6.16 *The K-polynomial of S/I for a generic monomial ideal I equals the $\mathbb{N}^n$-graded Euler characteristic of the Scarf complex Δ_I:*

$$\mathcal{K}(S/I; x_1, \ldots, x_n) \;=\; \sum_{\sigma \in \Delta_I} (-1)^{|\sigma|} m_\sigma.$$

Moreover, there is no cancellation of terms in this formula.

Proof. The Euler characteristic statement follows from Theorems 6.13 and 4.11. There can be no cancellation by definition of Δ_I. $\qquad\square$

Example 6.17 If $I = \langle x^2, xy, y^2z, z^2 \rangle$ as in Example 6.2, then

$$1 - x^2 - xy - y^2z - z^2 + x^2z^2 + x^2y + xy^2z + y^2z^2 + xyz^2 - x^2yz^2 - xy^2z^2$$

is the K-polynomial of S/I. $\qquad\diamond$

We close with another trivariate example to show that Scarf complexes of generic monomial ideals need not be pure.

Example 6.18 The generic ideal $I = \langle x^2z^2, xyz, y^2z^4, y^4z^3, x^3y^5, x^4y^3 \rangle$ has staircase diagram and Scarf complex as follows:

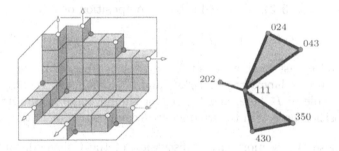

Observe that this Scarf complex is not pure, though it is still contractible. $\diamond$

The Scarf complex best reflects all the properties of a generic ideal I when S/I is artinian, so that I contains a power of each variable.

Corollary 6.19 *If $I = \langle m_1, \ldots, m_r \rangle$ is generic and S/I is artinian, with $m_i = x_i^{d_i}$ for $i = 1, \ldots, n$, then the Scarf complex Δ_I is a regular triangulation (usually with additional vertices, some of which may lie on the boundary) of the $(n-1)$-simplex with vertex set $\{1, \ldots, n\}$.*

Proof. This follows from Theorem 4.31 and Theorem 6.13. $\qquad\square$

The Scarf complex Δ_I in Corollary 6.19 has an additional vertex on the boundary of the $(n-1)$-simplex if and only if the squarefree monomial $x_1 \cdots x_n$ fails to divide some generator of I that is not a power of a variable.

It is not true that every triangulation is the Scarf complex of a generic artinian monomial ideal. A first condition is that the triangulation be regular, but even being regular is not enough: for $n \geq 4$, there are many regular triangulations of the $(n-1)$-simplex that cannot be realized as the Scarf complex of a monomial ideal. One example is the Schlegel diagram of the cyclic 4-polytope with 13 vertices. That this triangulation of the tetrahedron is not a Scarf complex will follow from the results in Section 6.4.

The next result gives a formula for the irreducible decomposition of a generic monomial ideal. It generalizes the irreducible decompositions for monomial ideals in $n \leq 3$ variables that we saw in Chapter 3. We use the same notation as in Chapter 5; for instance, if $\mathbf{c} \in \mathbb{N}^n$ then $\mathfrak{m}^{\mathbf{c}}$ denotes the ideal generated by the powers $x_i^{c_i}$ where i ranges over all indices with $c_i > 0$.

Corollary 6.20 *Let I be a generic monomial ideal, and fix $\mathbf{u} \in \mathbb{N}^n$ such that each minimial generator of I divides $\mathbf{x}^{\mathbf{u}}$. Set $I^* = I + \mathfrak{m}^{\mathbf{u}+\mathbf{1}}$, and for any $\mathbf{b} \in \mathbb{N}^n$, abbreviate $\hat{\mathbf{b}} = \sum_{b_i \leq u_i} b_i \mathbf{e}_i$. Then the intersection $\bigcap_G \mathfrak{m}^{\hat{\mathbf{a}}_G}$ over all facets $G \in \Delta_{I^*}$ is the irredundant irreducible decomposition of I.*

Proof. Use Theorems 5.42 and 6.13, since I^* is still generic (check this!). □

Example 6.21 Let $I = \langle x^3 y^2 z, x^2 y z^3, x y^3 z^2 \rangle$ be the ideal J from Section 3.2, but without any of the artinian generators $\{x^4, y^4, z^4\}$. Here, we can take $\mathbf{u} = (3, 3, 3)$. The irreducible decomposition of I is

$$I = \langle z \rangle \cap \langle y \rangle \cap \langle x \rangle \cap \langle y^2, z^3 \rangle \cap \langle x^3, z^2 \rangle \cap \langle x^2, y^3 \rangle \cap \langle x^3, y^3, z^3 \rangle.$$

The second-to-last component is $\mathfrak{m}^{\hat{\mathbf{a}}_G} = \langle x^2, y^3 \rangle$, where G is the triangle in Δ_{I^*} with vertex labels $x^2 y z^3$, $x y^3 z^2$, and z^4. The ideal J in Section 3.2 plays the role of I^* here, and the reader should compare the irreducible decomposition here with the irreducible decomposition of I^* there.   ◇

We close this section with a discussion of the Cohen–Macaulay condition (Definition 5.52) for a monomial quotient. A necessary condition for S/I to be Cohen–Macaulay is that all associated primes of I have the same dimension, but this condition is generally not sufficient. For instance, the Stanley–Reisner ring of the projective plane in Section 4.3.5 is a counterexample when $\text{char}(\Bbbk) = 2$. It turns out, however, that the necessary condition is also sufficient when the monomial ideal I is generic.

Theorem 6.22 *Let I be generic. The quotient S/I is Cohen–Macaulay if and only if all irreducible components of I have the same dimension. More generally, the projective dimension of S/I equals the maximum number of generators of an irreducible component of I.*

Proof. By Theorem 6.13, S/I has projective dimension equal to the maximum cardinality $|\sigma|$ of a facet $\sigma \in \Delta_I$. Suppose every generator of I divides $\mathbf{x}^{\mathbf{a}}$, and set $I^* = I + \mathfrak{m}^{\mathbf{a}+\mathbf{1}}$. By Corollary 6.19, every facet of Δ_I

extends to a facet of Δ_{I^*} by adding vertices of the form $x_i^{a_i+1}$. For a given facet $\sigma \in \Delta_I$, the number of such vertices added to get a facet of Δ_{I^*} equals $n - |\sigma|$. On the other hand, Corollary 6.20 implies that the minimum of these numbers $n - |\sigma|$ over the facets $\sigma \in \Delta_I$ equals n minus the maximum number of generators of an irreducible component of I. This proves the second statement of the theorem. To get the first, note that I has codimension equal to the minimum number of generators of an irreducible component. $\square$

6.3 Genericity by deformation

As we have used it in Definition 6.5, the word "generic" is basically taken to mean "random", as applied to the exponent vectors on monomial generators of ideals. However, there is another mathematical meaning of the word "generic", namely "invariant under deformation", that also reflects the nature of generic monomial ideals. This meaning of "generic" can be made precise using Definition 6.23, allowing us to *characterize* generic monomial ideals in terms of it. As a result, in Theorem 6.26, we get a host of equivalent algebraic, combinatorial, and geometric conditions equivalent to genericity. Let us begin with the definition of "deformation".

Definition 6.23 A **deformation** ϵ of a monomial ideal $I = \langle m_1, \dots, m_r \rangle$ is a choice of vectors $\epsilon_i = (\epsilon_{i1}, \dots, \epsilon_{in}) \in \mathbb{R}^n$ for $i \in \{1, \dots, r\}$ satisfying

$$a_{is} < a_{js} \;\Rightarrow\; a_{is} + \epsilon_{is} < a_{js} + \epsilon_{js} \qquad \text{and} \qquad a_{is} = 0 \;\Rightarrow\; \epsilon_{is} = 0,$$

where $\mathbf{a}_i = (a_{i1}, \dots, a_{in})$ is the exponent vector of m_i. Formally introduce the monomial ideal (in a polynomial ring with real exponents):

$$I_\epsilon \;=\; \langle m_1 \cdot \mathbf{x}^{\epsilon_1}, m_2 \cdot \mathbf{x}^{\epsilon_2}, \dots, m_r \cdot \mathbf{x}^{\epsilon_r} \rangle \;=\; \langle \mathbf{x}^{\mathbf{a}_1+\epsilon_1}, \mathbf{x}^{\mathbf{a}_2+\epsilon_2}, \dots, \mathbf{x}^{\mathbf{a}_r+\epsilon_r} \rangle.$$

A deformation ϵ is called **generic** if I_ϵ is a generic monomial ideal.

The Scarf complex Δ_{I_ϵ} of the deformed ideal I_ϵ still makes sense, as a combinatorial object, and has the same vertex set $\{1, \dots, r\}$ as Δ_I. The reader uncomfortable with real exponents can safely ignore them and view them simply as symbols to break ties between equal coordinates of generating exponents. Indeed, the combinatorics of a deformation depends only on the coordinatewise order that results on generating exponents, and there is always a choice of deformation that results in integer exponents inducing the same coordinatewise order.

For generic deformations ϵ, the Scarf complex Δ_{I_ϵ} of the deformed ideal gives an easy simplicial (but typically nonminimal) free resolution of I.

Theorem 6.24 *Fix a monomial ideal I and a generic deformation ϵ. Define Δ_I^ϵ by relabeling each face σ in the Scarf complex Δ_{I_ϵ} by m_σ instead of $\mathrm{lcm}(m_i \mathbf{x}^{\epsilon_i} \mid i \in \sigma)$. The resulting Taylor complex $\mathcal{F}_{\Delta_I^\epsilon}$ resolves S/I.*

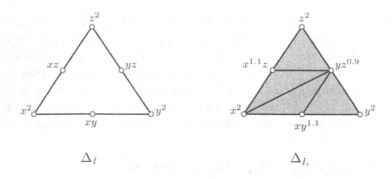

Figure 6.1: Generic deformation of $\langle x, y, z \rangle^2$

Proof. Given any vector $\mathbf{b} \in \mathbb{N}^n$, the (unlabeled) simplicial subcomplex $(\Delta_I^\epsilon)_{\preceq \mathbf{b}}$ can also be expressed as $(\Delta_{I_\epsilon})_{\preceq \mathbf{b}'}$ for the least common multiple

$$\mathbf{x}^{\mathbf{b}'} = \operatorname{lcm}(m_i \mathbf{x}^{\epsilon_i} \mid m_i \text{ divides } \mathbf{x}^{\mathbf{b}})$$

of the deformations of all generators dividing $\mathbf{x}^{\mathbf{b}}$. Now use Proposition 4.5 along with the acyclicity of $\mathcal{F}_{\Delta_{I_\epsilon}}$ that results from Theorem 6.13. $\qquad\square$

The resolution $\mathcal{F}_{\Delta_I^\epsilon}$ in Theorem 6.24 has length less than or equal to the bound n provided by the Hilbert Syzygy Theorem, by Lemma 6.8, but it is generally not minimal. Note that unlike the reductions to squarefree or Borel-fixed ideals, this reduction to the generic situation actually produces a free resolution of S/I for any I. (Sticklers may argue that depolarization of a minimal free resolution of the polarization yields a resolution of the depolarization, but that is reducing the problem to one we also cannot solve: finding the minimal free resolution of a squarefree monomial ideal.)

Example 6.25 The square $\mathfrak{m}^2$ of the maximal ideal $\mathfrak{m} = \langle x, y, z \rangle$ is not generic, and indeed, its Scarf complex is 1-dimensional and not contractible. However, we can find a generic deformation as depicted in Fig. 6.1. The resolution of $\mathfrak{m}^2$ afforded by the right-hand diagram but with labels as in the left-hand diagram is not minimal. Compare Example 3.19. $\qquad\Diamond$

We are now prepared to state the main theorem of this chapter. It provides appropriate converses to Theorem 6.13 and Corollary 6.20. The result is independent of the particular choice of the vector $\mathbf{u} \in \mathbb{N}^n$ used to define I^*, as long as all generators of I divide $\mathbf{x}^{\mathbf{u}}$. As before, $\mathfrak{m}^{\mathbf{u}+1}$ denotes the irreducible artinian ideal $\langle x_1^{u_1+1}, \ldots, x_n^{u_n+1} \rangle$.

Theorem 6.26 *Fix an ideal I generated by monomials dividing $\mathbf{x}^{\mathbf{u}}$, and set $I^* = I + \mathfrak{m}^{\mathbf{u}+1}$. The following are equivalent.*

(a) I is generic.

(b) $\mathcal{F}_{\Delta_{I^*}}$ is a minimal free resolution of S/I^*.

(c) $\Delta_{I^*} = \mathrm{hull}(I^*)$.

(d) $I = \bigcap \{ \mathfrak{m}^{\hat{\mathbf{a}}_\sigma} \mid \sigma \in \Delta_{I^*} \text{ and } |\sigma| = n \}$ is the irredundant irreducible decomposition of I, where $\hat{\mathbf{b}} = \sum_{\{i \mid b_i \le a_i\}} b_i \mathbf{e}_i$.

(e) For each irreducible component $\mathfrak{m}^{\mathbf{b}}$ of I^*, there is a face $\sigma \in \Delta_{I^*}$ labeled by $\mathbf{a}_\sigma = \mathbf{b}$.

(f) $\mathcal{F}_{\Delta_I}$ is a free resolution of S/I, and no variable x_k appears with the same nonzero exponent in m_i and m_j for any edge $\{i,j\}$ of Δ_I.

(g) If $\sigma \notin \Delta_{I^*}$, then some monomial $m \in I$ strictly divides m_σ.

(h) The Scarf complex Δ_{I^*} is unchanged by arbitrary deformations of I^*.

Proof. The scheme of the proof is

$$(b) \Rightarrow (c) \Rightarrow (d) \Rightarrow (e) \Rightarrow (b) \quad \text{and} \quad (c) \Rightarrow (f) \Rightarrow (a) \Rightarrow (g) \Rightarrow (h) \Rightarrow (b).$$

(b) $\Rightarrow$ (c): Use induction on n. If $n = 2$, this is obvious, so suppose (b) $\Rightarrow$ (c) for $\le n-1$ variables. The fact that S/I^* is artinian implies that Δ_{I^*} is pure of dimension $n-1$ by Corollary 5.39. The restriction of Δ_{I^*} to those vertices whose monomial labels are not divisible by x_k is, by definition, the Scarf complex of the ideal $I_k^* = (I^* + \langle x_k \rangle)/\langle x_k \rangle$ in $\Bbbk[x_1, \ldots, x_n]/\langle x_k \rangle$ generated by those monomials in I^* not divisible by x_k. On the other hand, this restriction $\Delta_{I_k^*}$ also equals $(\Delta_{I^*})_{\preceq \mathbf{b}}$ for $\mathbf{b} = \mathbf{u} + 1 - (u_k + 1)\mathbf{e}_k$. By induction, we therefore find that $\Delta_{I_k^*} = \mathrm{hull}(I_k^*)$, because $\mathcal{F}_{\Delta_{I_k^*}}$ is acyclic by Proposition 4.5.

The topological boundary of $\mathrm{hull}(I^*)$ is by Theorem 4.31 the union over $k \in \{1, \ldots, n\}$ of the complexes $\mathrm{hull}(I_k^*) = \Delta_{I_k^*}$. On the other hand, by Theorem 6.13, we know that the acyclic simplicial complex Δ_{I^*} is a subcomplex of the polyhedral cell complex $\mathrm{hull}(I^*)$. The latter being a polyhedral subdivision of the $(n-1)$-simplex, and both complexes containing the boundary of $\mathrm{hull}(I^*)$, we can conclude that $\Delta_{I^*} = \mathrm{hull}(I^*)$.

(c) $\Rightarrow$ (d): Holds for any minimal cellular resolution by Theorem 5.42.

(d) $\Rightarrow$ (e): Trivial, given that $b_i > a_i$ implies $b_i = a_i + 1$ for $\mathbf{b} = \mathbf{a}_\sigma$.

Lemma 6.27 *If $\mathbf{b} \in \mathbb{N}^n$ and $\beta_{i,\mathbf{b}}(S/I^*) \ne 0$ for some i, then there is an irreducible component $\mathfrak{m}^{\mathbf{c}}$ of S/I^* such that $\mathbf{b} \preceq \mathbf{c}$.*

Proof. If $\beta_{i,\mathbf{b}}(S/I^*)$ is nonzero, then the upper Koszul simplicial complex $K^{\mathbf{b}}(I)$ is not the whole simplex $2^{[n]}$, so $\mathbf{x}^{\mathbf{b}-1}$ lies outside of I^*. Since S/I^* is artinian, some monomial multiple $\mathbf{x}^{\mathbf{c}-\mathbf{b}} \cdot \mathbf{x}^{\mathbf{b}-1} = \mathbf{x}^{\mathbf{c}-1}$ lies in $(I^* : \mathfrak{m})$. This means precisely that $K^{\mathbf{c}}(I)$ is the $(n-2)$-sphere consisting of all proper faces of $2^{[n]}$. It follows that $\mathbf{c}$ has full support, and that $\beta_{n,\mathbf{c}}(S/I^*) = 1$ by Theorem 1.34. Using Theorem 5.48 we find that $\mathbf{x}^{\mathbf{u}+1+1-\mathbf{c}}$ is a minimal generator of $(I^*)^{[\mathbf{u}+1]}$, and we conclude using Theorem 5.27 that $\mathfrak{m}^{\mathbf{c}}$ is an irreducible component of I^*. $\square$

(e) $\Rightarrow$ (b): The full Taylor resolution supported on the entire simplex whose vertices are the generators of I^* contains a minimal free resolution $\mathcal{F}$ of S/I^* as an algebraic subcomplex (Exercise 1.11). But $\beta_{i,\mathbf{c}}(S/I^*)$ is nonzero only when $\mathbf{c} = \mathbf{a}_\sigma$ for some face $\sigma \in \Delta_{I^*}$ by hypothesis and Lemma 6.27, so $\mathcal{F}$ must be contained inside the subcomplex $\mathcal{F}_{\Delta_{I^*}}$ of the full Taylor resolution. Proposition 6.12 implies that $\mathcal{F}_{\Delta_{I^*}} = \mathcal{F}$.

(c) $\Rightarrow$ (f): Acyclicity follows from the criterion of Proposition 4.5, because Δ_I is the subcomplex $(\Delta_{I^*})_{\preceq \mathbf{u}}$ consisting of the faces whose labels divide $\mathbf{x}^{\mathbf{u}}$. It therefore suffices to show the condition on edges when $I = I^*$.

Suppose σ is a face of Δ_{I^*} such that $|\sigma| = |\mathrm{supp}(\mathbf{a}_\sigma)|$. For each index $k \in \mathrm{supp}(\mathbf{a}_\sigma)$, there is, by Definition 6.7, a unique vertex $i \in \sigma$ such that $\mathbf{a}_i$ shares its kth coordinate with $\mathbf{a}_\sigma$. It follows that

$$\text{if } |\sigma| = |\mathrm{supp}(\mathbf{a}_\sigma)| \text{ then no two exponent vectors on distinct} \atop \text{vertices of } \sigma \text{ share the same nonzero coordinate with } \mathbf{a}_\sigma. \qquad (*)$$

Suppose now that two generators m_i and m_j have the same degree in x_k and that $\{i, j\} \in \Delta_{I^*}$ is an edge. Proposition 5.40 implies that some face σ containing $\{i, j\}$ satisfies $|\sigma| = |\mathrm{supp}(\mathbf{a}_\sigma)|$ and shares its kth coordinate with $\mathbf{a}_{\{i,j\}}$, so that $\mathbf{a}_i$ and $\mathbf{a}_j$ contradict $(*)$ in coordinate k.

(f) $\Rightarrow$ (a): For any generator m_i let

$$A_i \;=\; \{m_j \mid m_j \neq m_i \text{ and } \deg_{x_k} m_j = \deg_{x_k} m_i > 0 \text{ for some } k\}.$$

The set A_i can be partially ordered by letting $m_j \preceq m_{j'}$ if $m_{\{i,j\}}$ divides $m_{\{i,j'\}}$. It is enough to produce a monomial m_l that strictly divides $m_{\{i,j\}}$ whenever $m_j \in A_i$ is a minimal element for this partial order. Supposing that m_j is minimal, use acyclicity to write

$$\frac{m_{\{i,j\}}}{m_i} \cdot \mathbf{e}_i - \frac{m_{\{i,j\}}}{m_j} \cdot \mathbf{e}_j \;=\; \sum_{\{u,v\} \in \Delta_I} b_{u,v} \cdot d(\mathbf{e}_{\{u,v\}}),$$

where we may assume (by picking such an expression with a minimal number of nonzero terms) that the monomials $b_{u,v}$ are 0 unless $m_{\{u,v\}}$ divides $m_{\{i,j\}}$. There is at least one monomial m_l such that $b_{l,j} \neq 0$, and we claim $m_l \notin A_i$. Indeed, m_l divides $m_{\{i,j\}}$ because $m_{\{l,j\}}$ does; therefore, if $\deg_{x_t} m_i < \deg_{x_t} m_j$ (which must occur for some t because m_j does not divide m_i), then $\deg_{x_t} m_l \leq \deg_{x_t} m_j$. Applying the second half of (f) to $m_{\{l,j\}}$, we get $\deg_{x_t} m_l < \deg_{x_t} m_j$, and furthermore $\deg_{x_t} m_{\{i,l\}} < \deg_{x_t} m_{\{i,j\}}$, whence $m_l \notin A_i$ by minimality of m_j. Therefore, if $\deg_{x_k} m_{\{i,j\}} > 0$ for some k, then either $\deg_{x_k} m_l < \deg_{x_k} m_j$ by the second half of (f), or $\deg_{x_k} m_l < \deg_{x_k} m_i$ because $m_l \notin A_i$.

(a) $\Rightarrow$ (g): Choose $\sigma \notin \Delta_{I^*}$ maximal among subsets with label m_σ. Then $m_\sigma = m_{\sigma \smallsetminus i}$ for some $i \in \sigma$. If $\mathrm{supp}(m_\sigma/m_i) = \mathrm{supp}(m_\sigma)$, the proof is done. Otherwise, there is some $j \in \sigma \smallsetminus i$ with $\deg_{x_k} m_i = \deg_{x_k} m_j > 0$ for some x_k. Then neither m_i nor m_j is a power of a variable, so $m_i, m_j \in I$.

Since I is generic, some monomial $m \in I$ strictly divides $m_{\{i,j\}}$, which in turn divides m_σ.

(g) $\Rightarrow$ (h): The strict inequalities defining the conditions "m_i does not divide m_σ" and "m_i strictly divides m_σ" persist after deformation. Persistence of the former implies that $\sigma \in \Delta_{I^*}$ remains a face in the deformation, while persistence of the latter implies that $\sigma \notin \Delta_{I^*}$ remains a nonface.

(h) $\Rightarrow$ (b): By Theorem 6.24, there is a deformation ϵ of I^* such that $\Delta_{I_\epsilon^*}$ gives a free resolution of S/I^*. Since $\Delta_{I^*} = \Delta_{I_\epsilon^*}$, the complex $F_{\Delta_{I^*}}$ is a free resolution, which is automatically minimal. $\qquad\square$

Remark 6.28 The equivalence (g) $\Leftrightarrow$ (h) remains true even if every occurrence of I^* is replaced by I, but the resulting conditions are not equivalent to genericity. A counterexample is the non-generic ideal $I = \langle xy, xz, xw \rangle$, whose Scarf complex Δ_I nevertheless does not change under deformation.

6.4 Bounds on Betti numbers

The passage from a monomial ideal to a generic deformation does not change the number of minimal generators, but it generally increases the Betti numbers. In this section we examine the question of how large the Betti numbers can be if the number of variables and the number of generators are fixed. We use the *Upper Bound Theorem* from the theory of convex polytopes to derive a nontrivial bound on Betti numbers of monomial ideals.

According to the Upper Bound Theorem (see [Zie95, Theorem 8.23], for example), there exists a polytope $C_n(r)$, the *cyclic polytope*, that simultaneously attains the maximum possible number $C_{i,n,r}$ of i-faces for each i. For $n < r$, the cyclic polytope $C_n(r)$ can be defined as the convex hull of any r distinct points on the moment curve $t \mapsto (t, t^2, \ldots, t^n)$. The combinatorial type of $C_n(r)$ is independent of the choice of r points, and the r points are precisely the vertices of the convex hull.

The statement and proof of the next result rely only on methods from Chapter 4. We waited until now to present it because the maximal Betti numbers are attained by generic ideals, and because we are prepared at this point to see the dual perspective in Corollary 6.31 using Chapter 5.

Theorem 6.29 *The number $\beta_i(I)$ of minimal i^{th} syzygies of any monomial ideal I with r generators in n variables is bounded above by the number $C_{i,n,r}$ of i-dimensional faces of the cyclic n-polytope with r vertices. If $i = n - 1$ then we even have $\beta_i(I) \leq C_{n-1,n,r} - 1$.*

Proof. The number of i-faces of the hull complex $\mathrm{hull}(I)$ equals $\beta_i(I)$. Consider the polytope $\tilde{\mathcal{Q}}_t = \mathrm{conv}\{t^{\mathbf{a}} \mid \mathbf{x}^{\mathbf{a}} \in \min(I)\}$ that appears as a Minkowski summand in Lemma 4.13. This polytope has dimension $\leq r$ and $\leq n$ vertices. Every face of $\mathrm{hull}(I)$ is a bounded face of $\mathcal{P}_t$ and therefore also a face of $\tilde{\mathcal{Q}}_t$, with the same supporting hyperplane. Hence $\beta_i(I)$ is bounded above by the number of i-dimensional faces of $\tilde{Q}_t$, which is at most $C_{i,n,r}$ by the Upper Bound Theorem. The inequality with $C_{i,n,r}$ is strict for $i = n - 1$ because $\tilde{Q}_t$ must have at least one facet whose inner normal vector has a nonnegative coordinate (or else the recession cone of $\tilde{Q}_t$ would contain $\mathbb{R}_{\geq 0}^n$), and this facet is erased in $\mathcal{P}_t$ by the Minkowski sum. $\square$

In three dimensions, these bounds are the ones given by planar graphs:

$$C_{1,3,r} = 3r - 6 \quad \text{and} \quad C_{2,3,r} = 2r - 4.$$

The first new and interesting case is that of monomial ideals in four unknowns, so $S = \mathbb{k}[a, b, c, d]$. Four-dimensional cyclic polytopes are *neighborly*, which means that every pair of vertices is joined by an edge. Hence $C_{1,4,r} = \binom{r}{2}$. The numbers of edges and vertices, together with Euler's formula "vertices − edges + 2-faces − facets = 0", uniquely determines the number of 2-faces and facets of a simplicial 4-polytope. For neighborly 4-polytopes, such as the cyclic polytope, we find that

$$C_{1,4,r} = \tfrac{1}{2}r(r - 1), \quad C_{2,4,r} = r(r - 3), \quad \text{and} \quad C_{3,4,r} = \tfrac{1}{2}r(r - 3).$$

Here is a concrete example where the bounds of Theorem 6.29 are tight.

Example 6.30 ($n = 4$, $r = 12$) For the generic monomial ideal

$$\begin{aligned} I = \quad &\langle a^9, b^9, c^9, d^9, a^6 b^7 c^4 d, a^2 b^3 c^8 d^5, a^5 b^8 c^3 d^2, \\ &ab^4 c^7 d^6, a^8 b^5 c^2 d^3, a^4 bc^6 d^7, a^7 b^6 cd^4, a^3 b^2 c^5 d^8 \rangle \end{aligned}$$

every pairwise first syzygy is minimal. The minimal free resolution of I is

$$0 \longleftarrow I \longleftarrow S^{12} \longleftarrow S^{66} \longleftarrow S^{108} \longleftarrow S^{53} \longleftarrow 0.$$

Each of the Betti numbers in this resolution is maximal among all monomial ideals generated by 12 monomials in four variables. $\diamond$

From the bound on Betti numbers in Theorem 6.29 we derive the following bound on the number of irreducible components.

Corollary 6.31 *The number of irreducible components of an ideal generated by r monomials in n variables is at most $C_{n-1,n,r+n} - 1$.*

Proof. We assume that I is generic, as the number of irreducible components can only rise under generic deformation (the reader is asked to prove this in Exercise 6.9). Now apply Corollary 6.20: The artinian ideal I^* has at most $n+r$ generators, and its Scarf complex Δ_{I^*} has at most $C_{n-1,n,r+n} - 1$ facets G. These facets index the irreducible components $\mathfrak{m}^{\mathfrak{a}_G}$ of I. $\square$

Example 6.32 $(n = 4, r = 9)$ Consider the generic monomial ideal

$$\langle a^6 b^7 c^4 d, a^2 b^3 c^8 d^5, a^5 b^8 c^3 d^2, ab^4 c^7 d^6, a^8 b^5 c^2 d^3, a^4 bc^6 d^7, a^7 b^6 cd^4, a^3 b^2 c^5 d^8 \rangle,$$

which is obtained from Example 6.30 by removing the artinian generators. This ideal has 53 irreducible components, the maximal number among all ideals generated by nine monomials in four variables. ◇

Generalizing the previous example, we say that a monomial ideal I is *neighborly* if every pair of generators is connected by a minimal first syzygy, or in symbols, $\beta_1(I) = \binom{\beta_0(I)}{2}$. Neighborly monomial ideals are algebraic analogues to neighborly polytopes. The cyclic polytopes show that, in fixed dimension $n \geq 4$, neighborly polytopes can have arbitrarily many vertices. Surprisingly, the analogous statement does not hold for monomial ideals. The following theorem gives a precise bound for neighborly ideals. We refer the reader to the original article [HM99] for the proof.

Theorem 6.33 (Hoşten and Morris [HM99]) *Let HM_n be the number of simplicial complexes on the set $\{1, \ldots, n-1\}$ such that no pair of faces covers all of $\{1, \ldots, n-1\}$. Then the maximum number of generators of a neighborly monomial ideal in n variables equals HM_n.*

The quantity HM_n grows doubly-exponentially in n. The following table contains some small values of the Hoşten–Morris number:

n	3	4	5	6	7	8
HM_n	4	12	81	2,646	1,422,564	229,809,982,112

For example, $HM_4 = 12$ refers to the twelve simplicial complexes on $\{1, 2, 3\}$: the void complex, the irrelevant complex, one point (3), two points (3), a segment (3), and three points. These complexes are in a certain bijection with the minimal generators in Example 6.30. Every monomial ideal in $\Bbbk[a, b, c, d]$ with 13 or more generators has at least one "missing s-pair" (i.e., a pair of generators that does not correspond to a minimal first syzygy). Likewise, every monomial ideal in n variables with more than HM_n generators has at least one missing generator. This implies the following.

Corollary 6.34 *The bounds on Betti numbers in Theorem 6.29 are not tight if $n \geq 4$ and $r \geq HM_{n+1}$.*

We next present the analogue to Example 6.30 for $n = 5$.

Example 6.35 $(n = 5, r = 81)$ What follows is a maximal neighborly monomial ideal in five variables. Each of the following 81 tuples of five positive integers $i_1 i_2 i_3 i_4 i_5$ represents a monomial $x_1^{i_1} x_2^{i_2} x_3^{i_3} x_4^{i_4} x_5^{i_5}$:

81 1 1 1 1	1 81 2 2 2	2 2 81 3 3	3 3 3 81 4	70 56 52 41 5
72 54 50 43 6	71 55 51 42 7	68 58 44 49 8	69 57 45 48 9	64 62 47 46 10
65 61 46 47 11	66 60 49 44 12	67 59 48 45 13	62 36 58 63 14	63 37 57 62 15
58 39 70 52 16	59 38 69 53 17	60 41 68 50 18	61 40 67 51 19	54 44 54 67 20
55 45 53 66 21	56 42 56 65 22	57 43 55 64 23	46 49 65 59 24	47 48 66 58 25
48 47 63 61 26	49 46 64 60 27	50 53 61 55 28	51 52 62 54 29	52 51 59 57 30
53 50 60 56 31	80 32 37 35 32	79 33 36 36 33	78 34 39 33 34	77 35 38 34 35
76 28 41 39 36	75 29 40 40 37	74 30 43 37 38	73 31 42 38 39	41 20 80 26 40
42 19 79 27 41	43 22 78 24 42	44 21 77 25 43	45 18 76 23 44	36 25 75 31 45
37 24 74 32 46	38 27 73 29 47	39 26 72 30 48	40 23 71 28 49	31 10 26 80 50
32 9 27 79 51	33 12 24 78 52	34 11 25 77 53	35 8 23 76 54	26 15 31 75 55
27 14 32 74 56	28 17 29 73 57	29 16 30 72 58	30 13 28 71 59	23 6 35 70 60
24 7 34 69 61	25 5 33 68 62	18 80 13 10 63	19 79 14 9 64	20 78 11 12 65
21 77 12 11 66	22 76 10 8 67	13 75 18 15 68	14 74 19 14 69	15 73 16 17 70
16 72 17 16 71	17 71 15 13 72	10 70 22 6 73	11 69 21 7 74	12 68 20 5 75
7 67 6 22 76	8 66 7 21 77	9 65 5 20 78	5 64 9 19 79	6 63 8 18 80
		4 4 4 4 81		

This example appears in a different notation in [HM99, p. 136]. We invite the computationally minded reader to determine the minimal free resolution and the irreducible decomposition of this neighborly monomial ideal in $\Bbbk[x_1, x_2, x_3, x_4, x_5]$. If you enlarge this ideal by any monomial of your choice, then the new ideal with 82 generators is no longer neighborly. $\diamond$

6.5   Cogeneric monomial ideals

In the paragraph before Theorem 5.42 we remarked on the connection, forged by Alexander duality for resolutions, between dualities on monomial ideals and those on polyhedra. Under this connection, monomial ideals generic with respect to their generating sets correspond more or less to *simplicial* polytopes. Consequently, their duals, which are generic with respect to their irreducible components, correspond to *simple* polytope.

Definition 6.36 A monomial ideal I is **cogeneric** if, whenever distinct irreducible components I_i and I_j of I have a minimal generator in common, there is a third irreducible component $I_\ell \subseteq I_i + I_j$ such that I_ℓ and $I_i + I_j$ do not have a minimal generator in common.

Translating this definition into a statement about the minimal generators of an Alexander dual ideal immediately reveals the duality between genericity and cogenericity.

Lemma 6.37 *A monomial ideal $I = \langle \mathbf{x}^{\mathbf{b}_1}, \dots, \mathbf{x}^{\mathbf{b}_r} \rangle$ is cogeneric if and only if its Alexander dual $I^{[\mathbf{a}]}$ for any (hence every) vector $\mathbf{a} \succeq \bigvee_j \mathbf{b}_j$ is generic.*

Proof. If $I_i = \mathfrak{m}^{\mathbf{a} \smallsetminus \mathbf{b}_i}$, $I_j = \mathfrak{m}^{\mathbf{a} \smallsetminus \mathbf{b}_j}$, and $I_\ell = \mathfrak{m}^{\mathbf{a} \smallsetminus \mathbf{b}_\ell}$, then $I_\ell \subseteq I_i + I_j$ if and only if $\mathbf{x}^{\mathbf{b}_\ell}$ divides $\mathrm{lcm}(\mathbf{x}^{\mathbf{b}_i}, \mathbf{x}^{\mathbf{b}_j})$. Moreover, I_ℓ and $I_i + I_j$ do not have a minimal generator in common if and only if $\mathbf{x}^{\mathbf{b}_\ell}$ strictly divides $\mathrm{lcm}(\mathbf{x}^{\mathbf{b}_i}, \mathbf{x}^{\mathbf{b}_j})$. $\square$

Example 6.38 The permutohedron ideal I in Section 4.3.3 is cogeneric. It is the Alexander dual, with respect to $\mathbf{a} = (n+1, \ldots, n+1)$, of the tree ideal I^* in Section 4.3.4. Hence the permutohedron ideal I is the intersection of the irreducible ideals $\langle x_i^{n-|\sigma|+1} \mid i \in \sigma \rangle$, where σ runs over nonempty subsets of $\{1, \ldots, n\}$. Since the tree ideal I^* is generic, by Example 6.6, its minimal free resolution is the Scarf complex Δ_{I^*}. By Theorem 6.13, the Scarf complex Δ_{I^*} coincides with the hull complex hull(I^*). Applying Alexander duality to this resolution, the results in Section 5.4 show that the cohull resolution of the permutohedron ideal is minimal. Since hull(I^*) is simplicial, its Alexander dual cohull(I) is simple. In fact, cohull(I) is precisely the complex of all faces of the permutohedron. $\diamond$

Example 6.38 is an instance of the following general construction.

Definition 6.39 Fix $\mathbf{a} \in \mathbb{N}^n$ and let I be a cogeneric monomial ideal whose generators all divide $\mathbf{x}^{\mathbf{a}}$. The **coScarf complex** $\Delta^{I,\mathbf{a}}$ is the cohull complex cohull$_{\mathbf{a}}(I)$ as in Definition 5.45. The corresponding cohull resolution is called the **algebraic coScarf complex** and is identified with $\Delta^{I,\mathbf{a}}$.

Theorem 6.40 *For a cogeneric monomial ideal I, the algebraic coScarf complex $\Delta^{I,\mathbf{a}}$ is a minimal cellular free resolution of I.*

Proof. Apply Theorem 5.37 to Theorem 6.13. $\square$

In what follows we give a self-contained description the coScarf complex $\Delta^{I,\mathbf{a}}$ that makes no reference to duality for resolutions (Theorem 5.37). Suppose that we are given a monomial ideal I by its irreducible components but that we do *not* know the minimal generators of I. Suppose further that the given irreducible ideals satisfy the requirements, spelled out in Definition 6.36, for I to be cogeneric. Then the following combinatorial construction yields the minimal free resolution $\Delta^{I,\mathbf{a}}$, and as a byproduct we also obtain the minimal generators of I.

Pick $\mathbf{a} = (a_1, \ldots, a_n) \in \mathbb{N}^n$ such that a_i exceeds the exponent of x_i in any of the given irreducible components. Form the Alexander dual ideal

$$I^{[\mathbf{a}]} \quad = \quad \langle \mathbf{x}^{\mathbf{a} \smallsetminus \mathbf{b}} \mid \mathfrak{m}^{\mathbf{b}} \text{ is an irreducible component of } I \rangle,$$

and make $I^{[\mathbf{a}]}$ artinian by adding powers of the variables: set

$$I^* \quad = \quad I^{[\mathbf{a}]} + \mathfrak{m}^{\mathbf{a}+1} \quad = \quad I^{[\mathbf{a}]} + \langle x_1^{a_1+1}, \ldots, x_n^{a_n+1} \rangle,$$

so the ideal I^* is artinian and generic. Next compute its Scarf complex Δ_{I^*}, which is a regular triangulation of the $(n-1)$-simplex, according to Corollary 6.19. Consider the labels on the Scarf complex as exponent vectors rather than monomials, and subtract each label from $\mathbf{a} + 1$. Now make a complex of free S-modules by using the *cochain* complex of Δ_{I^*} for scalars in monomial matrices, with the new labels from Δ_{I^*} on the rows and columns.

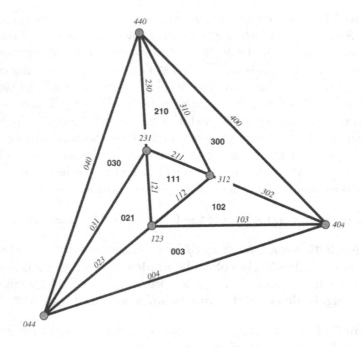

Figure 6.2: The coScarf complex from Example 6.42

Then take the submatrices whose rows and columns are indexed by *interior* faces of Δ_{I^*}. More succinctly, the scalars are the *relative cochain complex* of the pair $(\Delta_{I^*}, \partial\Delta_{I^*})$, where '$\partial$' means "boundary of". At this point, the interior vertices of Δ_{I^*} are labeled by the irreducible components of I, and the facets of Δ_{I^*} are labeled with the minimal generators of I.

Corollary 6.41 *The interior faces of the Scarf complex Δ_{I^*} minimally resolve S/I. This resolution coincides with the coScarf resolution $\Delta^{I,\mathbf{a}}$.*

Proof. The identification between Δ_{I^*}, labeled as described earlier, and the cohull complex $\mathrm{cohull}(I^{[\mathbf{a}]})$ is seen by tracing through the constructions of Section 5.4. Then apply Theorem 6.40. $\square$

Example 6.42 Suppose we are given the task of computing the minimal generators and the free resolution of the trivariate monomial ideal

$$I \;=\; \langle x, y^2, z^3 \rangle \cap \langle x^2, y^3, z \rangle \cap \langle x^3, y, z^2 \rangle.$$

Then what we do is to draw the Scarf complex Δ_{I^*} for $I^* = I^{[\mathbf{a}]} + \langle x^{a_1+1}, y^{a_2+1}, z^{a_3+1} \rangle$. This has been done in Section 3.2, with $\mathbf{a} = (3,3,3)$. Now relabel according to the regimen above, subtracting all of the face labels from $\mathbf{a} + 1 = (4,4,4)$, to get the labeled complex in Fig. 6.2. Reading the facet labels (in nonitalic sans serif font) tells us that

$$I = \langle z^3, y^3, x^3, y^2z, xz^2, x^2y, xyz \rangle.$$

Restricting the cochain complex of the triangulated triangle to the interior faces yields the minimal free resolution

$$0 \longleftarrow S^7 \longleftarrow S^9 \longleftarrow S^3 \longleftarrow 0$$

corresponding to the 3 interior vertices, 9 interior edges, and 7 triangles. $\diamond$

It is instructive to consider the Alexander duals of the various upper bound problems in Section 6.4. This includes the problem of bounding the number of minimal generators in terms of the number of irreducible components. By dualizing Corollary 6.31, we obtain the following.

Corollary 6.43 *The number of minimal generators of an intersection of r irreducible monomial idels in n variables is at most $C_{n-1,n,r+n} - 1$.*

For example, if we intersect 9 irreducible monomial ideals $\langle a^i, b^j, c^k, d^l \rangle$ in $\Bbbk[a, b, c, d]$, then the number of minimal generators is at most 53. That the bound is tight is seen by taking the Alexander dual of Example 6.32.

Exercises

6.1 Prove Lemma 6.10.

6.2 Compute the Scarf complex Δ_I for the generic monomial ideal

$$I = \langle a^5, b^5, c^5, d^5, ab^2c^3d^4, a^2b^3c^4d, a^3b^4cd^2, a^4bc^2d^3 \rangle$$

in $\Bbbk[a, b, c, d]$. This Scarf complex is a triangulation of the tetrahedron; draw it.

6.3 Compute the irreducible decomposition of the ideal I in Exercise 6.2.

6.4 Prove that an edge of $\mathrm{Buch}(I)$ connects two minimal generators of a monomial ideal I if and only if there is a deformation I_ϵ in which the corresponding generators are connected by an edge in the Scarf complex Δ_{I_ϵ}.

6.5 Describe the Stanley–Reisner complex of the radical of I in terms of the Scarf complex Δ_I when I is a generic monomial ideal.

6.6 What is the maximum number of irreducible components of an artinian ideal generated by 10 monomials in 4 variables? Find an example attaining the bound.

6.7 Consider the nongeneric monomial ideal $I = \langle x, y, z \rangle^3$ generated by all monomials of degree 3 in $\{x, y, z\}$. Construct at least three different free resolutions of I by deformation of exponents.

6.8 Express the algebraic coScarf resolution as a cellular free resolution.

6.9 Prove that any generic deformation I_ϵ of a monomial ideal I has at least as many irredundant irreducible components as I does. More precisely, show that every irreducible component of I_ϵ specializes to an irreducible ideal containing I, so the facets of Δ_{I_ϵ} provide a (possibly redundant) irreducible decomposition of I.

6.10 Draw the minimal free resolution of the cogeneric ideal

$$\langle x^1, y^4, z^6 \rangle \cap \langle x^2, y^6, z^1 \rangle \cap \langle x^3, y^3, z^3 \rangle \cap \langle x^4, y^5, z^2 \rangle \cap \langle x^5, y^1, z^5 \rangle \cap \langle x^6, y^2, z^4 \rangle,$$

whose staircase diagram is depicted below:

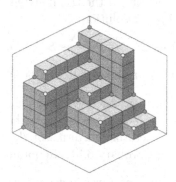

6.11 What is the maximal number of minimal generators of an intersection of 81 irreducible monomial ideals in $\Bbbk[x_1, x_2, x_3, x_4, x_5]$?

6.12 Classify all monomial ideals that are both generic and cogeneric.

6.13 True or false: Every Cohen–Macaulay monomial ideal I possesses a generic deformation that is also Cohen–Macaulay.

6.14 Give a combinatorial characterization, in the spirit of Theorem 6.22, of cogeneric monomial ideals that are Cohen–Macaulay.

6.15 Let P be a finite poset and $\Delta(P)$ the *order complex* of chains in P. Show that there exists a generic monomial ideal whose Scarf complex equals $\Delta(P)$.

Notes

The notions of genericity and deformation were implicit in the work of H. Scarf, who introduced the Scarf complex in the context mathematical economics [Sca86]. The algebraic version is due to Bayer, Peeva, and Sturmfels [BPS98], but was re-worked to its current form by Miller, Sturmfels, and Yanagawa [MSY00] so that genericity can be characterized in terms of invariance under deformation. As a result, monomial ideals called "generic" in the original [BPS98] definition of generic-icity are called "strongly generic" in [MSY00], as we have done in Definition 3.8.

The proof of Theorem 6.26 is one of the main reasons we developed duality for resolutions and its consequences in such detail in Chapter 5. Conditions (b), (d), and (h) in Theorem 6.26 can more naturally be phrased—without referring to I^* and its algebraic properties—in terms of $\mathbb{Z}^n$-graded *injective* resolutions of S/I, which turn out to be equivalent to free resolutions of S/I^* [Mil00a]. See the Exercises and Notes in Chapter 11.

Theorem 6.29 is from [BPS98]; it led to an interplay between commutative algebra and extremal combinatorics, culminating in articles such as [HM99]. The coScarf resolution, introduced in [Stu99] and [Mil98], was one of the points of departure for developing the general theory of Alexander duality in Chapter 5.

Solutions to Exercises 6.5 and 6.14 can be found in [MSY00]. Exercise 6.15 is a result of Postnikov and Shapiro [PS04, Section 6].

Part II

Toric Algebra

Chapter 7

Semigroup rings

The polynomial ring $S = \Bbbk[x_1, \ldots, x_n]$ is the semigroup ring associated with the semigroup $\mathbb{N}^n$. Note that $\mathbb{N}^n$ is the subsemigroup of $\mathbb{Z}^n$ generated by the unit vectors $\mathbf{e}_1, \ldots, \mathbf{e}_n$. In this chapter we replace $\mathbb{Z}^n$ by an arbitrary finitely generated abelian group, and we replace $\mathbb{N}^n$ by a subsemigroup.

The structure of a semigroup ring $\Bbbk[Q]$, including its dimension and whether or not it is an integral domain, is derived from properties of the semigroup Q. When n generators for Q are given, the semigroup ring $\Bbbk[Q]$ is a quotient of the polynomial ring S by a *lattice ideal*, which is characterized in terms of the surjection $\mathbb{N}^n \to Q$. In the case of an *affine semigroup* $Q \subset \mathbb{Z}^d$, the structure of the monomial ideals in $\Bbbk[Q]$ is explicitly described in terms of the polyhedral combinatorics and arithmetic of rational cones in Euclidean space. Our final topic is an introduction to the polyhedral geometry of the initial ideals of lattice ideals under weight orders.

7.1 Semigroups and lattice ideals

Fix an abelian group A together with a distinguished list $\mathbf{a}_1, \ldots, \mathbf{a}_n$ of elements. We write Q for the subsemigroup of A generated by $\mathbf{a}_1, \ldots, \mathbf{a}_n$. By a *semigroup* we will always mean the subsemigroup Q generated by a finite subset of an abelian group A. Thus all our semigroups are finitely generated, cancellative, and come with a zero element (additive identity).

Definition 7.1 The **semigroup ring** $\Bbbk[Q]$ of a semigroup Q is the $\Bbbk$-algebra with $\Bbbk$-basis $\{\mathbf{t}^\mathbf{a} \mid \mathbf{a} \in Q\}$ and multiplication defined by

$$\mathbf{t}^\mathbf{a} \cdot \mathbf{t}^\mathbf{b} \;=\; \mathbf{t}^{\mathbf{a}+\mathbf{b}}.$$

In this chapter we assume that $\Bbbk$ is a field, but the definition makes sense when $\Bbbk$ is any ring. The extra generality will be required in Chapter 8, where we need semigroup rings over the integers.

Let L denote the kernel of the group homomorphism from $\mathbb{Z}^n$ to A that sends $\mathbf{e}_i$ to $\mathbf{a}_i$ for $i = 1, \ldots, n$. Thus L is a *lattice* in $\mathbb{Z}^n$. We have

$$A \supseteq \mathbb{Z}^n/L \quad \text{and} \quad Q \cong \mathbb{N}^n/\sim_L,$$

where $\sim_L$ is the equivalence relation on $\mathbb{N}^n$ given by $\mathbf{u} \sim_L \mathbf{v} \Leftrightarrow \mathbf{u} - \mathbf{v} \in L$. It is useful to translate this relation into multiplicative notation.

Definition 7.2 The **lattice ideal** $I_L \subseteq S$ associated to L is the ideal

$$I_L \;=\; \langle \mathbf{x}^{\mathbf{u}} - \mathbf{x}^{\mathbf{v}} \mid \mathbf{u}, \mathbf{v} \in \mathbb{N}^n \text{ with } \mathbf{u} - \mathbf{v} \in L \rangle.$$

Theorem 7.3 *The semigroup ring $\Bbbk[Q]$ is isomorphic to the quotient S/I_L.*

Proof. Let $\mathbf{t}^{\mathbf{a}_1}, \ldots, \mathbf{t}^{\mathbf{a}_n}$ denote the generators of the semigroup ring $\Bbbk[Q]$ corresponding to the given generators of the semigroup Q. Then $\Bbbk[Q]$ is the free $\Bbbk$-algebra generated by $\mathbf{t}^{\mathbf{a}_1}, \ldots, \mathbf{t}^{\mathbf{a}_n}$ subject to the relations

$$\mathbf{t}^{\mathbf{a}_{i_1}} \cdots \mathbf{t}^{\mathbf{a}_{i_r}} = \mathbf{t}^{\mathbf{a}_{j_1}} \cdots \mathbf{t}^{\mathbf{a}_{j_s}} \quad \text{whenever} \quad \mathbf{a}_{i_1} + \cdots + \mathbf{a}_{i_r} = \mathbf{a}_{j_1} + \cdots + \mathbf{a}_{j_s} \text{ in } A.$$

There is a canonical $\Bbbk$-algebra homomorphism ϕ from S onto $\Bbbk[Q]$ sending x_i to $\mathbf{t}^{\mathbf{a}_i}$. The kernel of ϕ certainly contains the ideal generated by binomials

$$x_{i_1} \cdots x_{i_r} - x_{j_1} \cdots x_{j_s} \quad \text{satisfying} \quad \mathbf{a}_{i_1} + \cdots + \mathbf{a}_{i_r} = \mathbf{a}_{j_1} + \cdots + \mathbf{a}_{j_s} \text{ in } A,$$

which equals the lattice ideal I_L. The question is whether there can be any more relations. But in fact, the kernel of ϕ is spanned as a vector space over $\Bbbk$ by the binomials $\{\mathbf{x}^{\mathbf{u}} - \mathbf{x}^{\mathbf{v}} \mid \mathbf{u} - \mathbf{v} \in L\}$. To see why, consider for each element $\mathbf{a} \in A$ the vector space $S_{\mathbf{a}}$ whose basis consists of the monomials $\mathbf{x}^{\mathbf{u}}$ mapping to $\mathbf{t}^{\mathbf{a}}$. The image of $S_{\mathbf{a}}$ in the quotient by the above binomials has dimension 1 over $\Bbbk$ (assuming that $\mathbf{a} \in Q$, for the dimension of $S_{\mathbf{a}}$ is zero, otherwise), since the images of the basis vectors of $S_{\mathbf{a}}$ are equal in the quotient. The canonical map $S/I_L \to \Bbbk[Q]$ is therefore an isomorphism of vector spaces graded by A and hence an isomorphism of $\Bbbk$-algebras. $\square$

Let us consider some examples of groups generated by three elements. In each case L is a sublattice of $\mathbb{Z}^3$, the abelian group A is $\mathbb{Z}^3/L$, the semigroup Q is $\mathbb{N}^3/\sim_L$, and the lattice ideal I_L lives in $\Bbbk[x, y, z]$.

- $L = \{\mathbf{0}\}$, $I_L = \langle 0 \rangle$, $A = \mathbb{Z}^3$, $Q = \mathbb{N}^3$

- $L = \mathbb{Z}\{(3, 4, 5)\}$, $I_L = \langle x^3 y^4 z^5 - 1 \rangle$, $A = Q = \mathbb{Z}^2$

- $L = \mathbb{Z}\{(3, 4, -5)\}$, $I_L = \langle x^3 y^4 - z^5 \rangle$, $A = \mathbb{Z}^2$, $Q = \mathbb{N}\{(5, 2), (0, 1), (3, 2)\}$

- $L = \{(u, v, w) \in \mathbb{Z}^3 \mid 3u + 4v + 5w = 0\}$, $I_L = \langle x^3 - yz, x^2 y - z^2, xz - y^2 \rangle$, $A = \mathbb{Z}$, $Q = \mathbb{N}\{3, 4, 5\}$

- $L = \{(u, v, w) \in \mathbb{Z}^3 \mid 3u + 4v = 5w\}$, $I_L = \langle x^3 z - y, x^2 yz^2 - 1 \rangle$, $A = Q = \mathbb{Z}$

- $L = \{(u,v,w) \in \mathbb{Z}^3 \mid u+v+w \text{ is even}\}$, $I_L = \langle x^2 - 1, xy - 1, yz - 1 \rangle = \langle x-1, y-1, z-1 \rangle \cap \langle x+1, y+1, z+1 \rangle$, $A = Q = \mathbb{Z}/2\mathbb{Z}$

- $L = \mathbb{Z}^3$, $I_L = \langle x-1, y-1, z-1 \rangle$, $A = Q = \{0\}$

Note that the prime decomposition in the second-to-last example is only valid if $\text{char}(\Bbbk) \neq 2$. If $\text{char}(\Bbbk) = 2$ then $x^2 - 1 = (x-1)^2$ and the corresponding ideal I_L is primary but not prime.

Returning to our general discussion, we have the following result.

Theorem 7.4 *The following are equivalent.*

1. *The lattice ideal I_L is prime.*

2. *The semigroup ring $\Bbbk[Q]$ is an integral domain (has no zerodivisors).*

3. *The group generated by Q inside of A is free abelian.*

4. *The semigroup Q is an **affine semigroup**, meaning that it is isomorphic to a subsemigroup of $\mathbb{Z}^d$ for some d.*

Proof. Replacing A with the subgroup generated by Q if necessary, we may as well assume that Q generates A. The third and fourth conditions are equivalent because every free abelian group is isomorphic to $\mathbb{Z}^d$ for some d. The equivalence of the first two conditions comes from Theorem 7.3. The third condition implies the second because $\Bbbk[A]$ is a Laurent polynomial ring (hence a domain) if A is free abelian, and $\Bbbk[Q]$ is a subalgebra of $\Bbbk[A]$. Finally, suppose the third condition is false. Then A contains a nonzero element $\mathbf{a}$ such that $m \cdot \mathbf{a} = 0$ for some $m > 1$. Write $\mathbf{a} = \mathbf{a}' - \mathbf{a}''$ with $\mathbf{a}'$ and $\mathbf{a}''$ in Q, so $\mathbf{t}^{m\mathbf{a}'} = \mathbf{t}^{m\mathbf{a}''}$ in $\Bbbk[Q]$ but $\mathbf{t}^{\mathbf{a}'} \neq \mathbf{t}^{\mathbf{a}''}$ in $\Bbbk[Q]$. We conclude that $\mathbf{t}^{\mathbf{a}'} - \mathbf{t}^{\mathbf{a}''}$ is a zerodivisor in $\Bbbk[Q]$, and hence $\Bbbk[Q]$ is not a domain. $\square$

Proposition 7.5 *The Krull dimension of $\Bbbk[Q]$ equals $n - \text{rank}(L)$.*

Proof. As the statement does not involve A, we again replace A with its subgroup generated by Q. The inclusion $\Bbbk[Q] \subseteq \Bbbk[A]$ is the localization map inverting the elements $\mathbf{t}^{\mathbf{a}_1}, \ldots, \mathbf{t}^{\mathbf{a}_n}$, and this localization of $\Bbbk[Q]$ equals $\Bbbk[A]$, so the algebras $\Bbbk[A]$ and $\Bbbk[Q]$ have the same Krull dimension. Let A_{tor} be the torsion subgroup of A. Since $A = \mathbb{Z}^n/L$, the group A/A_{tor} is isomorphic to $\mathbb{Z}^{n-\text{rank}(L)}$, and the group algebra $\Bbbk[A/A_{\text{tor}}]$ is a Laurent polynomial ring in $n - \text{rank}(L)$ variables. Choosing a splitting $A/A_{\text{tor}} \hookrightarrow A$, the group algebra $\Bbbk[A]$ becomes a module-finite extension of the Laurent polynomial ring $\Bbbk[A/A_{\text{tor}}]$ and hence also has Krull dimension $n - \text{rank}(L)$. $\square$

In many applications, the semigroup Q will generate the group A, which will be presented to us as the cokernel of an integer matrix $\mathbf{L}$ with n rows. In this case, the lattice L is generated by the columns of $\mathbf{L}$. In order to determine A as an abstract group, we compute the *Smith normal form* of $\mathbf{L}$; that is, we compute invertible integer matrices $\mathbf{U}$ and $\mathbf{V}$ such that

$\mathbf{U} \cdot \mathbf{L} \cdot \mathbf{V}$ is the concatenation of a diagonal matrix and a zero matrix. As an example, let A be the group that is the cokernel of

$$
\mathbf{L} \;=\; \begin{bmatrix} 2 & -4 & 8 \\ 2 & 8 & -4 \\ -4 & 2 & 8 \\ -4 & 8 & 2 \\ 8 & 2 & -4 \\ 8 & -4 & 2 \end{bmatrix} \tag{7.1}
$$

The Smith normal form of the matrix $\mathbf{L}$ equals

$$
\begin{bmatrix} -6 & -2 & 4 & 0 & 0 & 5 \\ -5 & -2 & 3 & 0 & 0 & 4 \\ -12 & -3 & 8 & 0 & 0 & 10 \\ -1 & -1 & 0 & 1 & 0 & 1 \\ -1 & -1 & 1 & 0 & 1 & 0 \\ -3 & -1 & 2 & 0 & 0 & 2 \end{bmatrix} \cdot \begin{bmatrix} 2 & -4 & 8 \\ 2 & 8 & -4 \\ -4 & 2 & 8 \\ -4 & 8 & 2 \\ 8 & 2 & -4 \\ 8 & -4 & 2 \end{bmatrix} \cdot \begin{bmatrix} 0 & 0 & 1 \\ 0 & -1 & 1 \\ 1 & -2 & -2 \end{bmatrix} = \begin{bmatrix} 2 & 0 & 0 \\ 0 & 6 & 0 \\ 0 & 0 & 18 \\ 0 & 0 & 0 \\ 0 & 0 & 0 \\ 0 & 0 & 0 \end{bmatrix}.
$$

We conclude that $A \cong \mathbb{Z}/2\mathbb{Z} \oplus \mathbb{Z}/6\mathbb{Z} \oplus \mathbb{Z}/18\mathbb{Z} \oplus \mathbb{Z}^3$. This computation was done in the computer algebra system Maple using the command `ismith`.

The next question one might ask is how to compute the semigroup Q, or equivalently, its defining lattice ideal I_L, from the matrix $\mathbf{L} = (\ell_{ij})$. This is done as follows. Form the ideal $I_{\mathbf{L}}$ in S that is generated by

$$
\prod_{\substack{i \text{ with} \\ \ell_{ij} > 0}} x_i^{\ell_{ij}} \;-\; \prod_{\substack{i \text{ with} \\ \ell_{ij} < 0}} x_i^{-\ell_{ij}},
$$

where j runs over all column indices of the matrix $\mathbf{L}$.

Lemma 7.6 *The lattice ideal I_L is computed from $I_{\mathbf{L}}$ by taking the saturation with respect to the product of all the variables:*

$$
I_L \;=\; (I_{\mathbf{L}} : \langle x_1 \cdots x_n \rangle^\infty),
$$

which by definition is the ideal $\{ y \in S \mid (x_1 \cdots x_n)^m y \in I_{\mathbf{L}} \text{ for some } m > 0 \}$.

Proof. Clearly $I_{\mathbf{L}}$ is contained in I_L. On the other hand, consider any generator $\mathbf{x}^{\mathbf{u}} - \mathbf{x}^{\mathbf{v}}$ of I_L. We can write $\mathbf{u} - \mathbf{v}$ as a $\mathbb{Z}$-linear combination of the columns of $\mathbf{L}$; hence $\mathbf{x}^{\mathbf{u} - \mathbf{v}}$ is an alternating product of the Laurent monomials $\prod_{\ell_{ij} > 0} x_j^{\ell_{ij}} / \prod_{\ell_{ij} < 0} x_j^{-\ell_{ij}}$. Subtracting 1 from both sides and clearing denominators, we find that a monomial multiple of $\mathbf{x}^{\mathbf{u}} - \mathbf{x}^{\mathbf{v}}$ lies in $I_{\mathbf{L}}$.

The proof is completed by noting that the lattice ideal is saturated: $I_L = (I_L : \langle x_1 \cdots x_n \rangle^\infty)$. This follows from Theorem 7.3 and the observation that none of the monomials $\mathbf{t}^{\mathbf{a}_1}$ are zerodivisors in the semigroup ring $\Bbbk[Q]$. $\square$

For the example in (7.1) we have

$$
I_{\mathbf{L}} \;=\; \langle x_1^2 x_2^2 x_5^8 x_6^8 - x_3^4 x_4^4, \; x_2^8 x_3^2 x_4^8 x_5^2 - x_1^4 x_6^4, \; x_1^8 x_3^8 x_4^2 x_6^2 - x_2^4 x_5^4 \rangle,
$$

but $I_L = (I_{\mathbf{L}} : \langle x_1 x_2 x_3 x_4 x_5 x_6 \rangle^\infty)$ is minimally generated by 28 binomials.

Questions concerning a semigroup Q and the group it generates in A can be answered by examining the lattice ideal I_L. For instance, we may ask whether Q is already a subgroup of A, or equivalently, whether every nonzero element in Q has an additive inverse. It suffices to test whether all of the elements $\mathbf{a}_i$ are invertible.

Remark 7.7 The semigroup Q is a subgroup of A if and only if

$$I_L + \langle x_i \rangle = \langle 1 \rangle \quad \text{for all} \quad i = 1, 2, \ldots, n,$$

so that all of the variables x_i are units modulo I_L.

7.2 Affine semigroups and polyhedral cones

In this section we assume that Q is an affine semigroup; that is, the four equivalent conditions of Theorem 7.4 are satisfied, and the group A is free abelian. The map $\mathbb{Z}^n \to A \subseteq \mathbb{Z}^d$ can therefore be represented by a $d \times n$ integer matrix $\mathbf{A} = (\mathbf{a}_1, \ldots, \mathbf{a}_n)$. This means that Q is the subsemigroup of $\mathbb{Z}^d$ generated by the integer column vectors $\mathbf{a}_1, \ldots, \mathbf{a}_n$. When only $\mathbf{L}$ is given, the matrix $\mathbf{A}$ can be chosen as the last d rows of the invertible $n \times n$ matrix $\mathbf{U}$ in the Smith normal form computation for $\mathbf{L}$.

Definition 7.8 A subset $T \subseteq Q$ is called an **ideal** of Q if $Q + T \subseteq T$. A subsemigroup F of Q is called a **face** if the complement $Q \smallsetminus F$ is an ideal of Q. The affine semigroup Q is **pointed** if its only unit is $\mathbf{0}$, where a **unit** is an element $\mathbf{a} \in Q$ whose additive inverse $-\mathbf{a}$ also lies in Q.

By definition, then, F is a face precisely when each pair of elements $\mathbf{a}, \mathbf{b} \in Q$ satisfies

$$\mathbf{a} + \mathbf{b} \in F \quad \Leftrightarrow \quad \mathbf{a} \in F \text{ and } \mathbf{b} \in F. \tag{7.2}$$

The unique smallest face of Q is its group $Q \cap (-Q)$ of units.

The $\mathbb{N}^n$-graded algebra we did over the polynomial ring $\Bbbk[x_1, \ldots, x_n]$ in Part I generalizes to the $\mathbb{Z}^d$-graded algebra of affine semigroups rings.

Definition 7.9 A **monomial** in the semigroup ring $\Bbbk[Q]$ is an element of the form $\mathbf{t}^\mathbf{a}$ for $\mathbf{a} \in Q$. An ideal $I \subseteq \Bbbk[Q]$ is a **monomial ideal** if it is generated by monomials.

For any subset T of Q, we write $\Bbbk\{T\}$ for the $\Bbbk$-linear span of the monomials $\mathbf{t}^\mathbf{a}$ with $\mathbf{a} \in T$. Thus I is a monomial ideal if and only if $I = \Bbbk\{T\}$ for some ideal T of Q, or equivalently, if I is homogeneous with respect to the tautological A-grading on $\Bbbk[Q]$, which is defined by $\deg(\mathbf{t}^\mathbf{a}) = \mathbf{a}$. For any subset F of Q we abbreviate $P_F = \Bbbk\{Q \smallsetminus F\}$.

All issues concerning primality and primary decomposition are compatible with the A-grading on $\Bbbk[Q]$. For instance, to test whether a homogeneous ideal I is prime or primary, it suffices to check homogeneous polynomials in the definition of "prime" or "primary". Also, the associated primes of an A-homogeneous ideal are automatically A-homogeneous. This follows from [Eis95, Exercise 3.5], because the grading group A can be totally ordered; see Proposition 8.11.

Lemma 7.10 *A subset F of Q is a face if and only if P_F is a prime ideal.*

Proof. The subspace P_F is an ideal if and only if the implication "$\Rightarrow$" holds in (7.2). Assuming that this is the case, the implication "$\Leftarrow$" says that $\mathbf{t}^{\mathbf{a}+\mathbf{b}} \in P_F$ implies $\mathbf{t}^{\mathbf{a}} \in P_F$ or $\mathbf{t}^{\mathbf{b}} \in P_F$. The latter condition is equivalent to P_F being a prime ideal, by the above remark about the A-grading. $\square$

Definition 7.11 If F is a face of Q, then the **localization** of Q along F is the semigroup $Q - F = Q + \mathbb{Z}F$ consisting of all differences $\mathbf{a} - \mathbf{b}$ with $\mathbf{a} \in Q$ and $\mathbf{b} \in F$. The **quotient semigroup** Q/F is the image of Q in the group $\mathbb{Z}^d/\mathbb{Z}F$.

The map $Q \to Q/F$ always factors through the localization $Q \to Q - F$, and the quotient semigroup $Q/F = (Q - F)/\mathbb{Z}F$ is always pointed.

The terms "face" and "pointed" refer to the relationship between affine semigroups and cones, whose polyhedral geometric definitions we recall. A *(polyhedral) cone* in $\mathbb{R}^d$ is the intersection of finitely many closed linear half-spaces in $\mathbb{R}^d$, each of whose bounding hyperplanes contains the origin. We write $\dim(C)$ for the dimension of the linear span of C. Every polyhedral cone C is finitely generated: there exist $\mathbf{c}_1, \ldots, \mathbf{c}_r \in \mathbb{R}^d$ with

$$C \;=\; \{\lambda_1 \mathbf{c}_1 + \cdots + \lambda_r \mathbf{c}_r \mid \lambda_1, \ldots, \lambda_r \in \mathbb{R}_{\geq 0}\}.$$

We call the cone C *rational* if $\mathbf{c}_1, \ldots, \mathbf{c}_r$ can be chosen to have rational coordinates, and we say that C is *simplicial* if $r = \dim(C)$ generators suffice. A *face* of a cone C is a subset of the form $H \cap C$ in which H is the bounding hyperplane of a closed half-space $H_{\geq 0}$ that contains C. The unique smallest face of C is the *lineality space* $C \cap (-C)$. We call the cone C *pointed* if $C \cap (-C) = \{\mathbf{0}\}$.

Lemma 7.12 *The map $F \mapsto \mathbb{R}_{\geq 0}F$ is a bijection from the set of faces of the semigroup Q to the set of faces of the cone $\mathbb{R}_{\geq 0}Q$. In particular, the semigroup Q is pointed if and only if the associated cone $\mathbb{R}_{\geq 0}Q$ is pointed.*

Proof. Let F be a subset of Q and consider the following linear system of equations and inequalities in an indeterminate vector $\mathbf{w} \in \mathbb{R}^d$:

$$\mathbf{w} \cdot \mathbf{a} = 0 \text{ for } \mathbf{a} \in F \quad \text{and} \quad \mathbf{w} \cdot \mathbf{b} > 0 \text{ for } \mathbf{b} \in Q \smallsetminus F.$$

If this system has a solution $\mathbf{w}$, then F is a face of Q by definition. If this system has no solution, then by Farkas' Lemma [Zie95, Proposition 1.7], there exists a linear combination $\mathbf{a}$ of vectors in F that equals a positive linear combination of some vectors $\mathbf{b} \in Q \smallsetminus F$. The vector $\mathbf{a}$ can be moved into F by adding a vector from F, and hence we may assume $\mathbf{a}$ itself lies in F. Since F is a face, some vector $\mathbf{b} \in Q \smallsetminus F$ lies in F as well, a contradiction.

The argument in the previous paragraph shows that a subset F of Q is a face if and only if it has the form $F = H \cap Q$, where H is the bounding

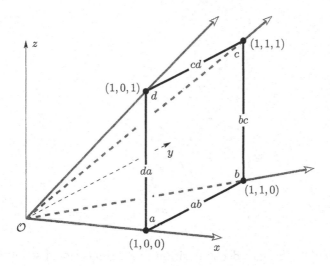

Figure 7.1: The primes in $\Bbbk[Q]$ for Q = the saturated cone over a square

hyperplane of a closed half-space $H_{\geq 0}$ containing Q. If F is a face of Q, then $\mathbb{R}_{\geq 0}F$ is a face of C, and conversely, if F' is a face of C, then $F' \cap Q$ is a face of Q. These two maps are inverses to each other, for if H is a hyperplane satisfying $H \cap Q = F$, then $F \subseteq \mathbb{R}_{\geq 0}F \subseteq H$, whence $Q \cap \mathbb{R}_{\geq 0}F = F$.   □

Lemma 7.12 implies that affine semigroups Q have only finitely many faces F, so affine semigroup rings $\Bbbk[Q]$ have only finitely many homogeneous prime ideals P_F. Computing this list of prime ideals is a valuable preprocessing step in dealing with a semigroup ring. This will be important in our study of injective modules and injective resolutions in Chapter 11.

Example 7.13 Every monomial ideal I in any affine semigroup ring $\Bbbk[Q]$ is an intersection of monomial ideals I_F, at most one for each face F, with I_F primary to P_F. We will prove this in Corollary 11.5, which rests mainly on Proposition 8.11, where we indicate how to derive a more general statement from [Eis95, Exercise 3.5]. For now, we present a 3-dimensional example that also serves to illustrate the other concepts from this section.

Let Q be the subsemigroup of $\mathbb{Z}^3$ generated by $(1,0,0)$, $(1,1,0)$, $(1,1,1)$, $(1,0,1)$. Its semigroup ring equals

$$\Bbbk[Q] \cong \Bbbk[a,b,c,d]/\langle ac - bd \rangle.$$

The cone $\mathbb{R}_{\geq 0}Q$ is the cone over a square and therefore pointed. It has nine faces: one of dimension 0, four of dimension 1, and four of dimension 2. Hence there are precisely nine homogeneous prime ideals in $\Bbbk[Q]$. They are

codim 3 primes: $P_{\mathcal{O}} = \langle a, b, c, d \rangle$

codim 2 primes: $P_a = \langle b, c, d \rangle$, $P_b = \langle a, c, d \rangle$, $P_c = \langle a, b, d \rangle$, $P_d = \langle a, b, c \rangle$

codim 1 primes: $P_{ab} = \langle c, d \rangle$, $P_{bc} = \langle d, a \rangle$, $P_{cd} = \langle a, b \rangle$, $P_{da} = \langle b, c \rangle$.

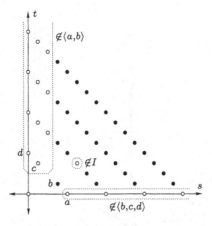

Figure 7.2: Primary decomposition in a 2-dimensional semigroup

The faces of $\mathbb{R}_{\geq 0}Q$ are labeled in Fig. 7.1, where (for example) the ray labeled ab contains all of the monomials outside of P_{ab}.

Computing intersections of monomial ideals in affine semigroup rings is more complicated than in a polynomial ring. Certain bad behavior arises, such as the fact that the intersection of two principal ideals is generally not principal. For instance, in our example, for any $i \in \mathbb{N}$,

$$\langle a^i \rangle \cap \langle d^i \rangle \ = \ \langle a^i d^i,\ a^{i-1}bd^i,\ a^{i-2}b^2d^i,\ a^{i-3}b^3d^i,\ \ldots,\ ab^{i-1}d^i,\ b^id^i \rangle.$$

An arbitrary principal monomial ideal here has

$$\langle a^i b^j c^k d^l \rangle \ = \ \langle a,b \rangle^{i+j} \cap \langle b,c \rangle^{j+k} \cap \langle c,d \rangle^{k+l} \cap \langle a,d \rangle^{i+l} \qquad (7.3)$$

as its primary decomposition. $\diamond$

Example 7.14 Every principal ideal in the ring $\Bbbk[Q]$ from the previous example is pure of codimension 1. Although this may be intuitive from a geometric standpoint, it can fail for arbitrary pointed affine semigroup rings. The simplest example comes from the two-dimensional semigroup Q' generated by $(4,0),(3,1),(1,3),(0,4)$, depicted in Fig. 7.2; note the lack of a dot in the empty space at the midpoint between b and c. (This example gives a reason why we do not assume that our semigroups generate the ambient group A, which in this case is $\mathbb{Z}^2$.) In the semigroup ring

$$\Bbbk[Q'] \ = \ \Bbbk[s^4, s^3t, st^3, t^4] \ = \ \Bbbk[a,b,c,d]/\langle bc - ad, c^3 - bd^2, ac^2 - b^2d, b^3 - a^2c \rangle,$$

the principal ideal generated by b has the minimal primary decomposition

$$\langle b \rangle \ = \ \langle a,b \rangle \cap \langle b,c,d \rangle \cap \langle a^2,b,c^2,d \rangle, \qquad (7.4)$$

so all three monomial prime ideals of $\Bbbk[Q']$ are associated to $\langle b \rangle$. In particular, the maximal ideal $\langle a,b,c,d \rangle$ is an embedded prime of $\langle b \rangle$. In Fig. 7.2, the principal ideal $\langle b \rangle$ consists of the black lattice points, whereas the ideals $\langle a, b \rangle$ and $\langle b, c, d \rangle$ consist of the nonunit lattice points *outside* of the vertical and horizontal strips, respectively. The ideal $I = \langle a^2, b, c^2, d \rangle$ contains all of the nonunit lattice points except for a, c, and ac, but as $a \notin \langle b, c, d \rangle$ and $c \notin \langle a, b \rangle$, we have circled only ac. $\diamond$

All of the primary ideals appearing on the right-hand sides of both (7.3) and (7.4) are actually irreducible ideals. We will treat irreducible decomposition of monomial ideals in semigroup rings from a general perspective in Chapter 11.

The generating sets that we gave for the semigroups in the previous two examples were unique, as we now show more generally.

Proposition 7.15 *Any pointed affine semigroup Q has a unique finite minimal generating set $\mathcal{H}_Q$.*

Proof. Every pointed semigroup Q can be regarded as a partially ordered set (poset) via $\mathbf{a} \preceq \mathbf{b}$ if $\mathbf{b} - \mathbf{a} \in Q$. Moreover, since $\{0\}$ is a face of Q, we can fix a vector $\mathbf{w} \in \mathbb{Z}^d$ such that $\mathbf{w} \cdot \mathbf{a} > 0$ for all $\mathbf{a} \in Q \setminus \{0\}$. We call the positive integer $\mathbf{w} \cdot \mathbf{a}$ the *height* of the element $\mathbf{a} \in Q$.

Let $\mathcal{H}_Q$ be the subset of the generators of Q that are minimal in $Q \setminus \{0\}$ with respect to the partial order on Q. A straightforward argument, by induction on the height, shows that every element $\mathbf{a} \in Q$ is an $\mathbb{N}$-linear combination of elements in $\mathcal{H}_Q$. On the other hand, elements in $\mathcal{H}_Q$ cannot be written in a nontrivial way as $\mathbb{N}$-linear combinations in Q. Hence $\mathcal{H}_Q$ generates Q, and every generating set of Q must contain $\mathcal{H}_Q$. $\square$

Some authors call $\mathcal{H}_Q$ the "Hilbert basis" of Q, but we reserve that term for its use in the next section (Definition 7.17), where Q is saturated.

7.3 Hilbert bases

In the previous section we associated a cone with any affine semigroup. It turns out that we can also go in the opposite direction and associate an affine semigroup to a given cone, provided that the cone is rational.

Theorem 7.16 (Gordan's Lemma) *If C is a rational cone in $\mathbb{R}^d$, then $C \cap A$ is an affine semigroup for any subgroup A of $\mathbb{Z}^d$.*

Proof. Since the intersection of C with the real subspace spanned by A is again a rational cone (with respect to the lattice A), we may as well assume that $A = \mathbb{Z}^d$. What we are claiming is that $C \cap \mathbb{Z}^d$ is finitely generated over $\mathbb{N}$. Since C is rational, there exist integer vectors $\mathbf{b}_1, \ldots, \mathbf{b}_r \in \mathbb{Z}^d$ such that $C = \mathbb{R}_{\geq 0}\{\mathbf{b}_1, \ldots, \mathbf{b}_r\}$. Consider the following compact subset of $\mathbb{R}^d$:

$$K = \left\{ \sum_{i=1}^{r} \lambda_i \mathbf{b}_i \ \middle|\ 0 \leq \lambda_i \leq 1 \text{ for } i = 1, \ldots, r \right\}.$$

Then $K' = K \cap \mathbb{Z}^d$ is a finite subset of our semigroup $C \cap \mathbb{Z}^d$. Every element $\mathbf{a}$ in $C \cap \mathbb{Z}^d$ can be written as $\mathbf{a} = \sum_{i=1}^r \mu_i \mathbf{b}_i$ where μ_i are nonnegative reals. Writing $\lambda_i = \mu_i - \lfloor \mu_i \rfloor$ for the fractional part of μ_i, we find that

$$\mathbf{a} = \sum_{i=1}^r \lfloor \mu_i \rfloor \mathbf{b}_i + \sum_{i=1}^r \lambda_i \mathbf{b}_i.$$

Hence $\mathbf{a}$ is an $\mathbb{N}$-linear combination of elements in K'. We conclude that the finite set K' generates the semigroup $C \cap \mathbb{Z}^d$. $\qquad\square$

Definition 7.17 Let C be a rational pointed cone in $\mathbb{R}^d$. The pointed semi-group $Q = C \cap \mathbb{Z}^d$ has a unique minimal generating set, called the **Hilbert basis** of the cone C and denoted by $\mathcal{H}_C$ or $\mathcal{H}_Q$, afforded by Theorem 7.16 and Proposition 7.15. More generally, a finite subset of $\mathbb{Z}^d$ is a **Hilbert basis** if it coincides with the Hilbert basis of the cone it generates in $\mathbb{R}^d$.

Example 7.18 Let C be the cone in $\mathbb{R}^4$ consisting of all vectors such that the sum of any two distinct coordinates is nonnegative. This is the cone over a 3-dimensional cube. The Hilbert basis $\mathcal{H}_C$ equals $\{(1,0,0,0), (0,1,0,0),$ $(0,0,1,0), (0,0,0,1), (-1,1,1,1), (1,-1,1,1), (1,1,-1,1), (1,1,1,-1)\}$. $\diamond$

It is instructive to examine the Hilbert basis of a cone in the plane.

Example 7.19 (Two-dimensional Hilbert bases) Let C be a rational pointed cone in $\mathbb{R}^2$. The Hilbert basis $\mathcal{H}_C$ is constructed geometrically as follows. Let $\mathcal{P}_C$ denote the unbounded polygon in $\mathbb{R}^2$ obtained by taking the convex hull of all nonzero integer points in C. The polygon $\mathcal{P}_C$ has two unbounded edges and a finite number of bounded edges. The Hilbert basis $\mathcal{H}_C$ is the set of all lattice points that lie on the bounded edges of $\mathcal{P}_C$. We order the elements $\mathbf{a}_1, \mathbf{a}_2, \ldots, \mathbf{a}_n$ of $\mathcal{H}_C$ in counterclockwise order. Then $\mathbf{a}_1$ and $\mathbf{a}_n$ are the primitive lattice points on the boundary of C, and we have

$$\det(\mathbf{a}_i, \mathbf{a}_{i+1}) = 1 \quad \text{for } i = 1, \ldots, n-1$$

because the triangle with vertices $\{\mathbf{0}, \mathbf{a}_i, \mathbf{a}_{i+1}\}$ has no other lattice points in it (Exercise 7.11). It follows that there exists $\lambda_i \in \mathbb{N}$ with

$$\lambda_i \cdot \mathbf{a}_i = \mathbf{a}_{i-1} + \mathbf{a}_{i+1} \quad \text{for } i = 2, 3, \ldots, n-1, \qquad (7.5)$$

which gives rise to the following binomials in the associated lattice ideal:

$$x_{i-1} x_{i+1} - x_i^{\lambda_i} \in I_L \quad \text{for } i = 2, 3, \ldots, n-1. \qquad (7.6)$$

We will return to this ideal in the next section. $\qquad\qquad\qquad\qquad\diamond$

We next describe an algorithm for computing the Hilbert basis of a rational pointed cone C, which we assume has m facets (= maximal faces). As a first step, we embed C as the intersection of a linear subspace V with a positive integer orthant $\mathbb{N}^m$.

Proposition 7.20 *Assume $C \subset \mathbb{R}^d$ is a pointed cone, and let $\nu_1, \ldots, \nu_m$ be the primitive integer inner normals to the facets of C. Define the map $\nu : \mathbb{R}^d \to \mathbb{R}^m$ sending $\mathbf{a} \in \mathbb{R}^d$ to $(\nu_1 \cdot \mathbf{a}, \ldots, \nu_m \cdot \mathbf{a})$, and set $V = \nu(\mathbb{R}^d)$. Then ν is injective, and its restriction to C is an isomorphism to $\nu(C) = \mathbb{R}^m_{\geq 0} \cap V$.*

Proof. The map ν is injective precisely because C is pointed: the intersection of the kernels of $\nu_1, \ldots, \nu_m$ is by definition the lineality space of C, which is zero for pointed cones. Moreover, a point $\mathbf{a} \in \mathbb{R}^d$ lies in C if and only if all $\nu_i \cdot \mathbf{a}$ are nonnegative. $\square$

We wish to compute the Hilbert basis for the pointed semigroup $\mathbb{N}^m \cap V$. Consider the sublattice $\Lambda = \{(\mathbf{v}, -\mathbf{v}) \mid \mathbf{v} \in V \cap \mathbb{Z}^m\}$ of $\mathbb{Z}^{2m}$. The lattice ideal I_Λ is an ideal in $\Bbbk[\mathbf{x}, \mathbf{y}] = \Bbbk[x_1, \ldots, x_m, y_1, \ldots, y_m]$. Such ideals are called *Lawrence ideals*. We can compute a minimal generating set of this ideal using Lemma 7.6. By Theorem 7.21, this solves our problem.

Theorem 7.21 *A vector $\mathbf{a} \in \mathbb{Z}^d$ lies in the Hilbert basis $\mathcal{H}_C$ if and only if the binomial $\mathbf{x}^{\nu \cdot \mathbf{a}} - \mathbf{y}^{\nu \cdot \mathbf{a}}$ appears among the minimal generators of I_Λ.*

Proof. We will equivalently prove that $\mathbf{u} \in \mathcal{H}_{\nu(C)}$ if and only if $\mathbf{x}^{\mathbf{u}} - \mathbf{y}^{\mathbf{u}}$ appears among the minimal generators of I_Λ. Consider a nonzero vector $\mathbf{u}$ in $\mathbb{N}^m \cap V$. If $\mathbf{u}$ is not in $\mathcal{H}_{\nu(C)}$, then we can write $\mathbf{u} = \mathbf{u}_1 + \mathbf{u}_2$ for two nonzero vectors $\mathbf{u}_1$ and $\mathbf{u}_2$ in $\mathbb{N}^m \cap V$. The identity

$$\mathbf{x}^{\mathbf{u}} - \mathbf{y}^{\mathbf{u}} \;=\; \mathbf{x}^{\mathbf{u}_1} \cdot (\mathbf{x}^{\mathbf{u}_2} - \mathbf{y}^{\mathbf{u}_2}) + \mathbf{y}^{\mathbf{u}_2} \cdot (\mathbf{x}^{\mathbf{u}_1} - \mathbf{y}^{\mathbf{u}_1})$$

shows that $\mathbf{x}^{\mathbf{u}} - \mathbf{y}^{\mathbf{u}}$ is not a minimal generator of I_Λ.

For the converse, suppose $\mathbf{x}^{\mathbf{u}} - \mathbf{y}^{\mathbf{u}}$ is not a minimal generator of I_Λ. Then $\mathbf{x}^{\mathbf{u}} - \mathbf{y}^{\mathbf{u}}$ is a (nonconstant) monomial linear combination of some binomials $\mathbf{x}^{\mathbf{v}} \mathbf{y}^{\mathbf{w}} - \mathbf{x}^{\mathbf{w}} \mathbf{y}^{\mathbf{v}}$ with $\mathbf{v}, \mathbf{w} \in \mathbb{N}^m$ and $\mathbf{v} - \mathbf{w} \in V$. We may assume that all terms have the same degree. By setting all x_i equal to zero, we see that at least one appearing binomial satisfies $\mathbf{v} = \mathbf{0}$ or $\mathbf{w} = \mathbf{0}$. Suppose $\mathbf{w} = \mathbf{0}$. Then $\mathbf{x}^{\mathbf{v}}$ properly divides $\mathbf{x}^{\mathbf{u}}$, so $\mathbf{u}$ is not in the Hilbert basis $\mathcal{H}_{\nu(C)}$. $\square$

Example 7.22 Let us find all nonnegative integer solutions to the equation

$$2u_1 + 7u_2 \;=\; 3u_3 + 5u_4. \tag{7.7}$$

The lattice of all integer solutions to this equation has the basis

$$(-1, 0, 1, -1), \; (-1, 1, 0, 1), \; (2, 1, 2, 1).$$

Using this basis we express the corresponding Lawrence ideal I_Λ as follows:

$$(\langle x_1 x_4 y_3 - x_3 y_1 y_4, \; x_2 x_4 y_1 - x_1 y_2 y_4, \; x_1^2 x_2 x_3^2 x_4 - y_1^2 y_2 y_3^2 y_4 \rangle : \langle x_1 x_2 \cdots y_4 \rangle^\infty).$$

This ideal has 30 minimal generators. Eighteen of the generators have the form required in Theorem 7.21 and hence give elements in the Hilbert basis. For example, the generator $x_2^4 x_3 x_4^5 - y_2^4 y_3 y_4^5$ of I_Λ gives $(0, 4, 1, 5) \in \mathcal{H}_C$.

We find that the cone C of nonnegative solutions to (7.7) has Hilbert basis

$$(0,2,3,1),\ (0,3,2,3),\ \underline{(0,3,7,0)},\ (0,4,1,5),\ \underline{(0,5,0,7)},\ (1,1,3,0),$$
$$(1,2,2,2),\ (1,3,1,4),\ (1,4,0,6),\ (2,1,2,1),\ (2,2,1,3),\ (2,3,0,5),$$
$$\underline{(3,0,2,0)},\ (3,1,1,2),\ (3,2,0,4),\ (4,0,1,1),\ (4,1,0,3),\ \underline{(5,0,0,2)}.$$

These 18 vectors minimally generate the semigroup of solutions to (7.7). The underlined vectors will be explained in Example 7.26.   ◇

Proposition 7.20 has another useful consequence.

Corollary 7.23 *Every d-dimensional pointed affine semigroup can be embedded inside $\mathbb{N}^d$.*

Proof. Given a pointed cone C, define $V \subseteq \mathbb{R}^m$ as in Proposition 7.20. Choose $m - d$ standard basis vectors $\mathbf{e}_{i_1}, \ldots, \mathbf{e}_{i_{m-d}}$ so that their images modulo V form a basis for $\mathbb{R}^m/V$. Then the coordinate subspace E spanned by $\mathbf{e}_{i_1}, \ldots, \mathbf{e}_{i_{m-d}}$ intersects V trivially. Under the projection $\mathbb{R}^m \twoheadrightarrow \mathbb{R}^d$ with kernel E, the subspace V maps isomorphically to its image, and $\mathbb{N}^m$ maps to $\mathbb{N}^d$. Therefore, projection modulo E takes any subsemigroup of C isomorphically to its image in $\mathbb{N}^d$.   □

If Q is an arbitrary affine semigroup in $\mathbb{Z}^d$, then $\mathbb{R}_{\geq 0}Q$ is the smallest cone in $\mathbb{R}^d$ containing Q. Similarly, there is a smallest subgroup of $\mathbb{Z}^d$ containing Q. Intersecting these yields an affine semigroup closely related to Q.

Definition 7.24 If A is the subgroup of $\mathbb{Z}^d$ generated by an affine semigroup Q inside of $\mathbb{Z}^d$, then the semigroup $Q_{\mathrm{sat}} = (\mathbb{R}_{\geq 0}Q) \cap A$ is called the **saturation** of the semigroup Q. We call Q **saturated** if $Q = Q_{\mathrm{sat}}$, and we say that its semigroup ring is **normal**.

By the *normalization* of an integral domain R we mean the set of elements in its field of fractions satisfying a monic polynomial in $R[y]$.

Proposition 7.25 *The semigroup ring $\Bbbk[Q_{\mathrm{sat}}]$ of the saturation Q_{sat} is the normalization of the affine semigroup ring $\Bbbk[Q]$.*

Proof. As earlier, we may as well forget the original $\mathbb{Z}^d$ and instead refer to the subgroup A generated by Q as $\mathbb{Z}^d$, after choosing a basis for it. Let $H^1, \ldots, H^r$ be hyperplanes whose associated closed half-spaces $H^i_{\geq 0}$ intersect precisely in $C = \mathbb{R}_{\geq 0}Q$. Then $\Bbbk[Q_{\mathrm{sat}}]$ is the intersection of the rings $\Bbbk[H^i_{\geq 0} \cap \mathbb{Z}^d]$ inside the Laurent polynomial ring $\Bbbk[\mathbb{Z}^d]$. The rings $\Bbbk[H^i_{\geq 0} \cap \mathbb{Z}^d]$ are all normal, being isomorphic to $\Bbbk[\mathbb{N} \times \mathbb{Z}^{d-1}]$, and they all have the same fraction field $\Bbbk(\mathbb{Z}^d)$. Therefore their intersection $\Bbbk[Q_{\mathrm{sat}}]$ is normal, by definition: any element of $\Bbbk(\mathbb{Z}^d)$ satisfying a monic polynomial with coefficients in $\Bbbk[Q_{\mathrm{sat}}]$ lies in each $\Bbbk[H^i_{\geq 0} \cap \mathbb{Z}^d]$. It remains to show that the normalization of $\Bbbk[Q]$ contains $\Bbbk[Q_{\mathrm{sat}}]$. If $\mathbf{t}^{\mathbf{a}} \in \Bbbk[Q_{\mathrm{sat}}]$, a straightforward argument shows that some multiple $m\mathbf{a}$ lies in Q. Thus $\mathbf{t}^{\mathbf{a}} \in \Bbbk[Q_{\mathrm{sat}}]$ satisfies the monic polynomial $f(y) = y^m - \mathbf{t}^{m\mathbf{a}}$ with coefficients in $\Bbbk[Q]$.   □

It is a basic computational task to decide whether a given affine semigroup Q is saturated and, if not, to compute its saturation Q_{sat}. Equivalently, we wish to compute the normalization of a given affine semigroup ring $k[Q]$. Here, the input is any generating set for Q and the desired output is the Hilbert basis for Q_{sat}. For small instances, this task can be accomplished using the algorithm of Theorem 7.21.

Example 7.26 (Computing the saturation of an affine semigroup)
For the semigroup generated by the underlined vectors in Example 7.22,

$$Q = \mathbb{N} \cdot \{(0,3,7,0), (0,5,0,7), (3,0,2,0), (5,0,0,2)\} \subset \mathbb{N}^4,$$

the semigroup ring is not normal:

$$k[Q] \cong k[x_1, x_2, x_3, x_4]/\langle x_1^{10} x_4^{21} - x_2^6 x_3^{35}\rangle.$$

Our four vectors generate the rank 3 lattice defined by (7.7), and the cone $\mathbb{R}_{\geq 0} Q$ is the cone of nonnegative real solutions to (7.7). Therefore the saturation Q_{sat} is precisely the semigroup that was computed in Example 7.22. Its Hilbert basis consists of the 18 listed vectors. ◇

We close this section with an example due to Bruns and Gubeladze showing that Hilbert bases in higher dimensions can be quite complicated and counterintuitive. If Q is a pointed saturated affine semigroup in $\mathbb{Z}^d$, then we say that Q has the *Carathéodory property* if every element in Q is an $\mathbb{N}$-linear combination of a subset of d elements in $\mathcal{H}_C$. Every pointed saturated affine semigroup in dimensions $d \leq 3$ has the Carathéodory property. For example, every nonnegative integer solution to (7.7) can be written as an $\mathbb{N}$-linear combination of only 3 of the 18 listed Hilbert basis elements.

It is unknown whether the Carathéodory property holds for Hilbert bases in dimensions $d = 4$ or $d = 5$. However, it does fail for $d \geq 6$.

Theorem 7.27 (Bruns and Gubeladze, 1999) *There exists a pointed saturated affine semigroup in $\mathbb{Z}^6$ lacking the Carathéodory property.*

Proof. Let C be the semigroup generated by the columns of the matrix

$$\mathbf{A} = \begin{bmatrix} 0 & 0 & 0 & 0 & 0 & 1 & 1 & 1 & 1 & 1 \\ 1 & 0 & 0 & 0 & 0 & 0 & 2 & 1 & 1 & 2 \\ 0 & 1 & 0 & 0 & 0 & 2 & 0 & 2 & 1 & 1 \\ 0 & 0 & 1 & 0 & 0 & 1 & 2 & 0 & 2 & 1 \\ 0 & 0 & 0 & 1 & 0 & 1 & 1 & 2 & 0 & 2 \\ 0 & 0 & 0 & 0 & 1 & 2 & 1 & 1 & 2 & 0 \end{bmatrix}.$$

These 10 vectors coincide with the Hilbert basis $\mathcal{H}_C$. The cone $\mathbb{R}_{\geq 0} C$ is a 6-dimensional pointed cone with 27 facets. The vector

$$\mathbf{a} = (9, 13, 13, 13, 13, 13)^T = \mathbf{A} \cdot (1, 3, 0, 5, 2, 0, 0, 1, 5, 3)^T$$

lies in the semigroup C, but it cannot be written as an $\mathbb{N}$-linear combination of 6 of the 10 columns of $\mathbf{A}$. This can be checked by exhaustively searching over all column bases of $\mathbf{A}$ and solving the associated linear system. □

7.4   Initial ideals of lattice ideals

This section concerns the initial monomial ideals of the lattice ideal I_L. These correspond geometrically to Gröbner degenerations of the variety of I_L and combinatorially to decompositions of the semigroup $Q = \mathbb{N}^n / \sim_L$. We begin by discussing how to get (partial) term orders from weights.

Fix a *positive weight vector* $w \in \mathbb{R}^n_{\geq 0}$. Given any term $c_{\mathbf{u}} \mathbf{x}^{\mathbf{u}}$, where $c_{\mathbf{u}}$ is a nonzero scalar in $\Bbbk$, the *weight* of $c_{\mathbf{u}} \mathbf{x}^{\mathbf{u}}$ is the dot product $w \cdot \mathbf{u} = w_1 u_1 + \cdots + w_n u_n$. For any polynomial $f \in S$, we write $\mathrm{in}_w(f)$ for the *initial form* of f, by which we mean the sum of all terms of f having maximal weight under w. If I is any ideal inside S, then we express the *initial ideal* of I under the *weight order* defined by w as the ideal

$$\mathrm{in}_w(I) \quad = \quad \langle \mathrm{in}_w(f) \mid f \in I \rangle$$

generated by initial forms of all polynomials in I. We say that w is *generic for I* if $\mathrm{in}_w(I)$ is a monomial ideal. In analogy with the Gröbner basics in Chapter 2.2, a finite subset $\mathcal{G}$ of I is a *Gröbner basis* for I with respect to w if $\mathrm{in}_w(I)$ is generated by $\{\mathrm{in}_w(g) \mid g \in \mathcal{G}\}$. If w is generic for I and every element of $\mathcal{G}$ is monic, then $\mathcal{G}$ is a *reduced* Gröbner basis if the initial term of each element in $\mathcal{G}$ does not divide any term of any other element in $\mathcal{G}$. For a fixed weight vector w that is generic for I, the reduced Gröbner basis of I is unique; see Exercise 7.16.

Replacing the initial ideal $\mathrm{in}_w(I)$ by its radical yields a squarefree monomial ideal, whose Stanley–Reisner simplicial complex $\Delta_w(I)$ we call the *initial complex* of I with respect to w. The initial complex can be described as follows: a subset $F \subseteq \{1, \ldots, n\}$ is a face of $\Delta_w(I)$ if there is no polynomial $f \in I$ whose initial monomial $\mathrm{in}_w(f)$ uses only the variables $\{x_i \mid i \in F\}$.

Proposition 7.28 *The Krull dimension of S/I equals $\dim(\Delta_w(I)) + 1$.*

Proof. The three algebras S/I, $S/\mathrm{in}_w(I)$, and $S/\mathrm{rad}(\mathrm{in}_w(I))$ have the same Krull dimension—the first two because their Hilbert series are equal and the latter two because their zero sets are equal. The result follows because $\mathrm{rad}(\mathrm{in}_w(I))$ is the Stanley–Reisner ideal of the simplicial complex $\Delta_w(I)$. □

We now take I to be a fixed lattice ideal I_L and fix a weight vector $w \in \mathbb{R}^n_{>0}$ that is generic for I_L. It is our objective to describe the initial ideal $\mathrm{in}_w(I_L)$ and the initial complex $\Delta_w(I_L)$.

A nonnegative integer vector $\mathbf{u} = (u_1, \ldots, u_n) \in \mathbb{N}^n$ is called *optimal* if $\mathbf{u} - \mathbf{v} \in L$ implies $w \cdot \mathbf{v} > w \cdot \mathbf{u}$ for all $\mathbf{v} \in \mathbb{N}^n \smallsetminus \{\mathbf{0}\}$. Thus $\mathbf{u}$ is not optimal if and only if there exists a vector $\mathbf{v} \in \mathbb{N}^n$ with $w \cdot \mathbf{v} \leq w \cdot \mathbf{u}$ and $\mathbf{x}^{\mathbf{u}} - \mathbf{x}^{\mathbf{v}} \in I_L$.

Corollary 7.29 *The initial monomial ideal $\mathrm{in}_w(I_L)$ equals the vector space spanned over $\Bbbk$ by all monomials $\mathbf{x}^{\mathbf{u}}$ whose exponent vector $\mathbf{u}$ is not optimal.*

Our terminology refers to an optimization problem called "integer programming". Readers familiar with integer programming will note that the initial complex $\Delta_w(I_L)$ is the collection of all optimal bases of the underlying "linear programming relaxation". Here is a tiny optimization example.

Example 7.30 (Making change with the fewest coins) Let L be the kernel of the 1×4 matrix $(1, 5, 10, 25)$ and $w = (1, 1, 1, 1)$. Denoting the four variables in S by the letters p, n, d, and q, we find that

$$\{p^5 - n, \; n^2 - d, \; d^2 n - q, \; d^3 - nq\}$$

is the reduced Gröbner basis for I_L with respect to w. The initial ideal is

$$\mathrm{in}_w(I_L) \;\; = \;\; \langle p^5, n^2, d^2 \rangle \cap \langle p^5, n, d^3 \rangle.$$

A monomial $p^i n^j d^k q^l$ is optimal if and only if $i \leq 4$, $j \leq 1$, and $j + k \leq 2$. Our optimization problem is to replace a collection of i pennies, j nickels, k dimes, and l quarters (U.S. currency) by the fewest coins of equal value. The given collection is optimal if and only if $i \leq 4$, $j \leq 1$, and $j + k \leq 2$. $\diamond$

Let us focus our attention on geometry of the initial complex $\Delta_w(I_L)$. If the semigroup is pointed and the semigroup ring has Krull dimension 1, as in the coin example, then $\Delta_w(I_L)$ is just a point. Therefore let us move on to examples of semigroups in dimension 2.

Example 7.31 Let I_L be the lattice ideal of a two-dimensional Hilbert basis as in Example 7.19. Choose $w \in \mathbb{R}^n$ so that $w_{i-1} + w_{i+1} > \lambda_i \cdot w_i$ for $i = 2, \ldots, n - 1$, where λ_i is the positive integer defined by (7.5). We claim that

$$\mathrm{in}_w(I_L) \;\; = \;\; \langle x_i x_j \mid 1 \leq i \leq j - 2 \leq n - 2 \rangle. \tag{7.8}$$

For all indices i and j with $j - i \geq 2$ there exists a unique relation

$$\mathbf{a}_i + \mathbf{a}_j \;\; = \;\; \mu \mathbf{a}_k + \nu \mathbf{a}_{k+1} \quad \text{with} \quad \mu, \nu \in \mathbb{N} \text{ and } i < k < k + 1 \leq j.$$

The convexity in our construction implies that $x_i x_j$ is the w-initial term in the corresponding binomial $x_i x_j - x_k^\mu x_{k+1}^\nu \in I_L$. Hence the left-hand side contains the right-hand side in (7.8). If containment were strict, then I_L would contain a binomial $x_i^* x_{i+1}^* - x_j^* x_{j+1}^*$ with $i \leq j - 2$. But this is ruled out because the cones $\mathbb{R}_{\geq 0}\{\mathbf{a}_i, \mathbf{a}_{i+1}\}$ and $\mathbb{R}_{\geq 0}\{\mathbf{a}_j, \mathbf{a}_{j+1}\}$ are disjoint.

We conclude that the initial complex consists of $n - 1$ segments:

$$\Delta_w(I_L) \;\; = \;\; \{\{1, 2\}, \{2, 3\}, \ldots, \{n - 1, n\}, \{1\}, \{2\}, \ldots, \{n\}, \emptyset\}.$$

It is customary to identify $\Delta_w(I_L)$ with the triangulation of the cone $C = \mathbb{R}_{\geq 0}\{\mathbf{a}_1, \mathbf{a}_2, \ldots, \mathbf{a}_n\}$ into the subcones $\mathbb{R}_{\geq 0}\{\mathbf{a}_i, \mathbf{a}_{i+1}\}$ for $i = 1, \ldots, n - 1$. $\diamond$

Let $L_{\mathbb{R}} := L \otimes \mathbb{R}$ be the real vector space spanned by L, and let $L_{\mathbb{R}}^{\perp}$ be its orthogonal complement in $\mathbb{R}^n$. Define the closed convex polyhedron

$$\mathcal{P}_w \;=\; \mathbb{R}_{\geq 0}^n \cap \left(w + L_{\mathbb{R}}^{\perp}\right). \tag{7.9}$$

Before proving any results about $\mathcal{P}_w$, let us take a moment to say a word about its geometry.

The vector space $L_{\mathbb{R}}^{\perp}$ is naturally identified with the set of linear functions $A_{\mathbb{R}} \to \mathbb{R}$, where $A_{\mathbb{R}} = \mathbb{R}^n / L_{\mathbb{R}}$. Embed $A_{\mathbb{R}} \cong \mathbb{R}^d$ inside $\mathbb{R}^{d+1}$ with last coordinate zero. Each covector $w + \nu \in w + L_{\mathbb{R}}^{\perp}$ induces a function $\{\mathbf{a}_1, \ldots, \mathbf{a}_n\} \to \mathbb{R}$ that sends $\mathbf{a}_i \mapsto w_i + \nu(\mathbf{a}_i)$. Think of $w + \nu$ as lifting $\mathbf{a}_i$ to the point $\widetilde{\mathbf{a}}_i$ at height $w_i + \nu(\mathbf{a}_i)$ in $\mathbb{R}^{d+1}$. Note that changing ν does not alter the combinatorics of the lower convex hull of $\widetilde{\mathbf{a}}_1, \ldots \widetilde{\mathbf{a}}_n$, as such a change only shears the set $\widetilde{\mathbf{a}}_1, \ldots \widetilde{\mathbf{a}}_n$ by adding the graph of a linear function.

The polyhedron $\mathcal{P}_w$ consists of those lifts $w + \nu$ such that $\widetilde{\mathbf{a}}_1, \ldots, \widetilde{\mathbf{a}}_n$ all lie on or above the hyperplane with last coordinate zero. Among all of these nonnegative lifts $w + \nu$ of $\mathbf{a}_1, \ldots, \mathbf{a}_n$, some leave more of the vectors $\widetilde{\mathbf{a}}_i$ at height zero than any others. The corresponding points $w + \nu \in \mathcal{P}_w$ are vertices. More generally, each face F of $\mathcal{P}_w$ is characterized by the set of indices i such that every covector $w + \nu \in F$ lifts $\mathbf{a}_i$ to height zero in $\mathbb{R}^{d+1}$. This set of indices corresponds to the smallest face of $\mathbb{R}_{\geq 0}^n$ containing F.

Lemma 7.32 $\mathcal{P}_w$ *is a simple polyhedron if* $\mathrm{in}_w(I_L)$ *is a monomial ideal.*

Proof. Let $m = \mathrm{rank}(L)$, so that $\mathcal{P}_w$ has dimension $n - m$. We are claiming that every vertex of $\mathcal{P}_w$ misses exactly m facets. Suppose this is not the case. Then there is a vertex $\mathbf{u} \in \mathcal{P}_w$ such that $|\mathrm{supp}(\mathbf{u})| \leq m - 1$. This implies the existence of a nonzero vector $\mathbf{c} \in L$ with $\mathrm{supp}(\mathbf{c}) \cap \mathrm{supp}(\mathbf{u}) = \varnothing$. Hence $\mathbf{u} \cdot \mathbf{c}_+ = \mathbf{u} \cdot \mathbf{c}_- = 0$, where $\mathbf{c} = \mathbf{c}_+ - \mathbf{c}_-$ is the decomposition into positive and negative parts. Moreover, since $\mathbf{u} - w \in L_{\mathbb{R}}^{\perp}$, we have $w \cdot \mathbf{c} = \mathbf{u} \cdot \mathbf{c} = 0$. Therefore both $\mathbf{x}^{\mathbf{c}_+}$ and $\mathbf{x}^{\mathbf{c}_-}$ lie in the monomial ideal $\mathrm{in}_w(I)$. Hence there exists $\mathbf{b} \in L$ such that $w \cdot \mathbf{b} = \mathbf{u} \cdot \mathbf{b} > 0$ and $\mathbf{x}^{\mathbf{b}_+}$ divides $\mathbf{x}^{\mathbf{c}_+}$. Since $\mathbf{u}$ is nonnegative, we conclude that $0 < \mathbf{u} \cdot \mathbf{b}_+ \leq \mathbf{u} \cdot \mathbf{c}_+$, a contradiction. $\square$

The next theorem is our main result in this section. It tells us that the radical of the initial ideal $\mathrm{in}_w(I_L)$ encodes the faces of the polyhedron $\mathcal{P}_w$. In fact, as the geometry preceding Lemma 7.32 indicates, it says that the initial complex $\Delta_w(I_L)$ is the regular triangulation (see [DRS04]) of the cone $C = \mathbb{R}_{\geq 0} Q$ in $A_{\mathbb{R}}$ corresponding to the lifts $w + \nu \in \mathcal{P}_w$.

Theorem 7.33 *The initial complex* $\Delta_w(I_L)$ *of the lattice ideal* I_L *equals the simplicial complex polar to the boundary of the simple polyhedron* $\mathcal{P}_w$.

Proof. Our assertion is equivalent to the following statement: a subset $F \subseteq \{1, \ldots, n\}$ lies in $\Delta_w(I_L)$ if and only if there exists $\mathbf{u} \in \mathcal{P}_w$ such that $\mathrm{supp}(\mathbf{u}) = \{1, \ldots, n\} \smallsetminus F$. Write $\mathbf{e}_F = \sum_{i \in F} \mathbf{e}_i$ for the incidence (row) vector of a subset F. Let $\mathbf{L}$ be an integer $m \times n$ matrix whose rows form a basis for the lattice L, and let $\mathbf{b} = \mathbf{L} \cdot w$. By linear programming duality,

$$\min\{\mathbf{e}_F \cdot \mathbf{u} \mid \mathbf{u} \in \mathbb{R}^n, \mathbf{u} \geq 0, \mathbf{Lu} = \mathbf{b}\} \quad = \quad \max\{\mathbf{v} \cdot \mathbf{b} \mid \mathbf{v} \in \mathbb{R}^d, \mathbf{vL} \leq \mathbf{e}_F\}.$$

This translates into the equivalent statement

$$\min\{\mathbf{e}_F \cdot \mathbf{u} \mid \mathbf{u} \in \mathcal{P}_w\} \quad = \quad \max\{\mathbf{c} \cdot w \mid \mathbf{c} \in L_{\mathbb{R}}, \mathbf{c} \leq \mathbf{e}_F\}.$$

The left-hand side is nonnegative. It is zero if and only if some point $\mathbf{u} \in \mathcal{P}_w$ has support contained in $\{1, \ldots, n\} \setminus F$. By Lemma 7.32, the family $\{\mathrm{supp}(\mathbf{u}) \mid \mathbf{u} \in \mathcal{P}_w\}$ is closed under taking supersets, so that "is contained in" can be replaced by "equals" in the previous sentence. The maximum on the right-hand side is positive if and only if there exists $\mathbf{c} = \mathbf{c}_+ - \mathbf{c}_- \in L$ with $\mathbf{c}_+ \cdot w > \mathbf{c}_- \cdot w$ and $\mathrm{supp}(\mathbf{c}_+) \subseteq F$. This holds if and only if there exists $f \in I_L$ with $\mathrm{supp}(\mathrm{in}_w(f)) \subseteq F$, which is equivalent to $F \notin \Delta_w(I_L)$. $\square$

The boundary of a polytope is homeomorphic to a sphere, and the boundary of an unbounded polyhedron is homeomorphic to an open ball. If we pass to the normal fan and intersect it with the unit sphere, then we get either a sphere or a closed ball of the same dimension. This implies the following topological result concerning our initial complex.

Corollary 7.34 *The initial complex of a lattice ideal I_L in S is homeomorphic to either a sphere or a ball of dimension $n - \mathrm{rank}(L) - 1$.*

We can use the result of Theorem 7.33 as a method for computing the boundary of a given polyhedron $\mathcal{P}_w$. How this is done in practice depends on how the lattice L and the vector w are given to us. For instance, suppose we are given an $m \times n$ matrix $\mathbf{L} = (\ell_{ij})$ whose rows form a basis for the lattice L and that we are given the vector $\mathbf{b} = \mathbf{L} \cdot w \in \mathbb{R}^m$. Then $\mathcal{P}_w$ is equal to

$$\mathcal{P} = \{\mathbf{u} \in \mathbb{R}^n \mid \mathbf{u} \geq 0, \mathbf{L} \cdot \mathbf{u} = \mathbf{b}\}. \tag{7.10}$$

In combinatorial applications, the matrix $\mathbf{L}$ will have nonnegative entries.

Corollary 7.35 *Let $\mathbf{L}$ be a nonnegative integer $m \times n$ matrix with no zero column, and let $\mathbf{b}$ be a generic point in the cone spanned by the columns of $\mathbf{L}$. Then $\mathcal{P}$ is a simple polytope. Its boundary complex is polar to the initial complex, with respect to any weight vector $w \in \mathcal{P}$, of the ideal*

$$I_{\mathbf{L}} = \langle x_1^{\ell_{i1}} x_2^{\ell_{i2}} \cdots x_n^{\ell_{in}} - 1 \mid i = 1, \ldots, m \rangle.$$

Proof. The assumption that $\mathbf{L}$ is nonnegative and has no zero column implies that $I_{\mathbf{L}} = I_L$. The corollary now follows from Theorem 7.33. $\square$

In order to compute the (supports of the) vertices of the polytope $\mathcal{P}$, it suffices to find the minimal associated primes of the initial ideal $\mathrm{in}_w(I_L)$.

Corollary 7.36 *Using notation as above, we have the prime decomposition*

$$\mathrm{rad}(\mathrm{in}_w(I_L)) \quad = \quad \bigcap_{\substack{vertices \\ \mathbf{v}\ of\ \mathcal{P}}} \langle x_i \mid i \in \mathrm{supp}(\mathbf{v}) \rangle.$$

This algebraic approach to polyhedral computations is surprisingly efficient when the ideal $\mathrm{in}_w(I_L)$ is squarefree, or close to squarefree. In those cases the complexity of $\mathrm{in}_w(I_L)$ is similar to the complexity of $\Delta_w(I_L)$.

Example 7.37 We compute the polytope $\mathcal{P}$ in (7.10) for

$$\mathbf{L} = \begin{bmatrix} 4 & 3 & 0 & 0 & 0 & 1 \\ 0 & 1 & 4 & 3 & 0 & 0 \\ 0 & 0 & 0 & 1 & 4 & 3 \end{bmatrix} \quad and \quad \mathbf{b} = \begin{bmatrix} 7 \\ 8 \\ 5 \end{bmatrix}.$$

For $w = [1, 1, 1, 1, 1, 0] \in \mathcal{P}$, the ideal $I_{\mathbf{L}} = I_L$ has the Gröbner basis

$$\{x_1^4 x_2^3 - x_6^2 x_4 x_5^4,\ x_1^4 x_2^2 x_5^4 x_6^4 - x_3^4 x_4^2,\ x_2 x_3^4 x_4^2 - x_5^4 x_6^3,$$
$$x_3^4 x_4^3 - x_1^4 x_2^2 x_6,\ x_4 x_5^4 x_6^3 - 1,\ x_5^8 x_6^6 - x_2 x_3^6 x_4\}$$

The radical of the initial ideal equals

$$\mathrm{rad}(\mathrm{in}_w(I_L)) \quad = \quad \langle x_1 x_2,\ x_3 x_4,\ x_5 x_6 \rangle.$$

This shows that the initial complex is the boundary of an octahedron,

$$\Delta_w(I_L) \quad = \quad \{135, 136, 145, 146, 235, 236, 245, 246\},$$

and we conclude that $\mathcal{P}$ is combinatorially isomorphic to the 3-cube.   ◇

Example 7.38 Consider finally a case when $A \cong \mathbb{Z}^d$ and the lattice L is given to us as the kernel of the $d \times n$ matrix $\mathbf{A} = [\mathbf{a}_1, \ldots, \mathbf{a}_n]$. Recall that the row space of $\mathbf{A}$ equals $L_{\mathbb{R}}^{\perp}$, and that our simple polyhedron $\mathcal{P}_w$ can be identified with

$$\{\mathbf{u} \in \mathbb{R}^d \mid \mathbf{u} \cdot \mathbf{a}_i \geq -w_i \text{ for } i = 1, 2, \ldots, n\}. \tag{7.11}$$

This polyhedron's normal fan is a regular triangulation of the cone $\mathbb{R}_{\geq 0} Q$ using rays in $\mathbf{a}_1, \ldots, \mathbf{a}_n$, and it coincides with the initial complex of I_L.

Let $\mathbf{a}_1, \ldots, \mathbf{a}_n$ be a Hilbert basis in $\mathbb{Z}^2$ and w as in Example 7.31. Then (7.11) is an unbounded polygon. The normals to its bounded edges are $\mathbf{a}_2, \ldots, \mathbf{a}_{n-1}$. The normals to its unbounded edges are $\mathbf{a}_1$ and $\mathbf{a}_n$.   ◇

Exercises

7.1 Let L be the sublattice of $\mathbb{Z}^d$ that is the integer kernel of a given $d \times n$ matrix $\mathbf{A}$. Show that I_L remains unchanged by elementary row operations on $\mathbf{A}$.

7.2 Suppose that Q and Q' are subsemigroups of two groups A and A', respectively. Prove that Q is isomorphic to Q' if there is an isomorphism $A \xrightarrow{\approx} A'$ taking a generating set of Q to a generating set of Q'.

7.3 Change the matrix $\mathbf{L}$ in (7.1) by replacing each entry -4 by -3. Recompute the Smith normal form. Is the semigroup $Q = \mathbb{N}^n / \sim_L$ affine? Is it pointed? Compute a minimal generating set of binomials for the lattice ideal I_L.

7.4 Let Q be an affine semigroup in $\mathbb{Z}^d$. A subset $T \subseteq \mathbb{Z}^d$ is called a Q-set if $Q + T \subseteq T$. Let us call T **modular** if the complement $(Q + T) \smallsetminus T$ is a Q-set.

  (a) If $T \subseteq \mathbb{Z}^d$ is modular, construct a $\Bbbk[Q]$-module $\Bbbk\{T\}$ with basis $\{\mathbf{t}^{\mathbf{a}} \mid \mathbf{a} \in T\}$.
  (b) Show that T is modular if and only if $-T$ is modular.
  (c) What is the relation between $\Bbbk\{T\}$ and $\Bbbk\{-T\}$?

7.5 Verify the equality (7.3) from Example 7.13.

7.6 Show that the graded maximal ideal is associated to every nonunit principal ideal in the semigroup ring $\Bbbk[Q']$, where $Q' \subset \mathbb{Z}^2$ is generated by the vectors $\{(4,0), (3,1), (1,3), (0,4)\}$ as in Example 7.14.

7.7 Suppose that a principal ideal $\langle \mathbf{t}^{\mathbf{a}} \rangle$ has an associated prime P_F of codimension at least 2 in the affine semigroup ring $\Bbbk[Q]$. Must every other nonunit principal ideal $\langle \mathbf{t}^{\mathbf{b}} \rangle$ have an associated prime of codimension at least 2? What if $\mathbf{b} \notin F$?

7.8 Every affine semigroup ring $\Bbbk[Q]$ has a unique maximal $\mathbb{Z}^d$-graded ideal $\mathfrak{m}$, whether or not Q is pointed. Prove Nakayama's Lemma for $\Bbbk[Q]$: if M is a finitely generated $\mathbb{Z}^d$-graded module over $\Bbbk[Q]$ and $m_1, \ldots, m_r \in M$ are homogeneous elements whose images modulo $\mathfrak{m}$ generate $M/\mathfrak{m}M$, then $m_1, \ldots, m_r$ generate M. You may use a local or $\mathbb{Z}$-graded version of Nakayama's Lemma in your proof.

7.9 Prove that an affine semigroup Q is pointed if and only if the maximal $\mathbb{Z}^d$-graded ideal $\mathfrak{m}$ of $\Bbbk[Q]$ is maximal in the usual sense (so $\Bbbk[Q]/\mathfrak{m}$ is a field). In this case, show that a fixed set monomials in $\Bbbk[Q]$ generates $\Bbbk[Q]$ as a k-algebra if and only if it generates $\mathfrak{m}$ as an ideal. Now use Exercise 7.8 to prove Proposition 7.15.

7.10 Fix any ring $\Bbbk$ and any abelian group A. Prove that if $A' \subseteq A$ is any subgroup, then $\Bbbk[A]$ is free as a module over $\Bbbk[A']$, and any system $c_1, \ldots, c_\ell \in A$ of representatives for the cosets of A' yields a $\Bbbk[A']$-basis $\{\mathbf{t}^{c_1}, \ldots, \mathbf{t}^{c_\ell}\}$ for $\Bbbk[A]$.

7.11 Prove that a triangle in the plane $\mathbb{R}^2$ whose vertices are the only lattice points in it (so there are no other lattice points on the edges either) has area $1/2$. Using the tetrahedra with vertices $(0,0,0), (1,0,0), (0,1,0), (1,1,a)$ for $a > 0$, show that the analogous statement is false in $\mathbb{R}^n$ for $n \geq 3$.

7.12 (Generalize Theorem 7.21) For a lattice $L \subset \mathbb{Z}^n$, consider its *Lawrence ideal*

$$I_{\Lambda(L)} \;=\; \langle \mathbf{x}^{\mathbf{c}^+} \mathbf{y}^{\mathbf{c}^-} - \mathbf{x}^{\mathbf{c}^-} \mathbf{y}^{\mathbf{c}^+} \mid \mathbf{c} \in L \rangle \;\subset\; \Bbbk[x_1, \ldots, x_n, y_1, \ldots, y_n].$$

Prove that $I_\Lambda(L)$ has a unique set of minimal generators $\mathbf{x}^{\mathbf{c}^+} \mathbf{y}^{\mathbf{c}^-} - \mathbf{x}^{\mathbf{c}^-} \mathbf{y}^{\mathbf{c}^+}$. Give a combinatorial characterization of the vectors $\mathbf{c}$ occurring in these generators.

7.13 Let C be the cone in $\mathbb{R}^4$ generated by $(1,0,0,0), (0,1,0,0), (0,0,1,0)$, and $(1,2,3,5)$. Show that the Hilbert basis $\mathcal{H}_C$ contains the vector $(1,1,1,1)$.

7.14 Prove that the semigroup $C \cap A$ in Theorem 7.16 is saturated. Use the result to verify that if Q and Q' are saturated semigroups in $\mathbb{Z}^d$, then so is $Q \cap Q'$.

7.15 Let Q be an affine semigroup. The normalization $\Bbbk[Q_{\mathrm{sat}}]$ is a finitely generated module over $\Bbbk[Q]$ by [Eis95, Corollary 13.13]. Use this fact to prove that Q contains a translate of its saturation: $\mathbf{a} + Q_{\mathrm{sat}} \subseteq Q$ for some $\mathbf{a} \in Q$.

7.16 Given any positive weight vector w and any fixed ideal I of S, prove that there is a term order $<$ on S such that $\text{in}_<(\text{in}_w(I)) = \text{in}_<(I)$. In other words, some Gröbner basis for I with respect to w is also a Gröbner basis for I with respect to $<$. We say that $<$ **refines** w. Conclude using Exercise 2.5 that if w is generic for I, then I has a unique reduced Gröbner basis with respect to w.

7.17 Let L be the rank 2 sublattice of $\mathbb{Z}^6$ spanned by $(1,1,0,-1,-1,0)$ and $(0,1,1,0,-1,-1)$. Compute the ideal I_L and compute the polyhedron $\mathcal{P}_w$ for a generic choice of $w \in \mathbb{R}^6$. Show that $\Delta_w(I_L)$ is a triangulation of an octahedron.

7.18 Fix a $d \times n$ matrix $\mathbf{A}$ and a weight vector w. Form a $(d+1) \times (n+1)$ matrix $\hat{\mathbf{A}}_w$ by appending to $\mathbf{A}$ a top row w and then a left column $\mathbf{a}_0 = (1,0,\ldots,0)$. For the semigroup $\hat{Q}_w$ in $\mathbb{Z}^{d+1}$ generated by the columns of $\hat{\mathbf{A}}_w$ and the kernel L of $\mathbf{A}$, show that $\Bbbk[\hat{Q}_w]/\langle \mathbf{t}^{\mathbf{a}_0}\rangle \cong S/\text{in}_w(I_L)$. What is the relation with Theorem 7.33?

Notes

Lattice ideals are generalizations of the *toric ideals* discussed in [Stu96], which constitute the special class characterized in Theorem 7.4. The study of algebras presented by toric ideals has a long tradition in commutative algebra. We refer to Villarreal's book [Vil01] for additional information and references. The study of toric ideals is related to the theory of integer programming, as explained in [Stu96, Chapter 4]. The book by Schrijver [Sch86] is an excellent reference on integer programming and related topics relevant to this chapter, such as Smith normal form, linear programming duality, and the history of Gordan's Lemma.

Readers interested in software for computing with the objects we have discussed (Hilbert bases, lattice ideals, their Gröbner bases, and so on) are encouraged to try Hemmecke's program 4ti2, available at http://www.4ti2.de/. Various Gröbner basis packages also offer special features for working with lattice ideals.

The phenomena explored in Example 7.14, Exercise 7.6, and Exercise 7.7 concerning low-dimensional associated primes of principal ideals relate to the Cohen–Macaulay condition (and Serre's condition S_2). See Exercises 12.12 and 13.5 for further details from this point of view. Corollary 7.23 has been substantially refined by Thompson [Tho02, Theorem 1.2.7]. He shows that if Q is a pointed affine semigroup ("sharp finitely generated torsion-free commutative monoid"), then for any complete flag of faces in Q, there is an inclusion $Q \hookrightarrow \mathbb{N}^d$ taking the given complete flag to the standard complete flag in $\mathbb{N}^d$.

Theorem 7.27 first appeared in [BG99]. It is related to our discussion of Ehrhart polynomials in Chapter 12. For further reading on structural properties of semigroups and lattice polytopes, their triangulations, and connections to K-theory, we recommend the forthcoming book by Bruns and Gubeladze [BG05].

The material in Section 7.4 is taken from a paper by Sturmfels, Weismantel, and Ziegler [SWZ95, Section 7]. Theorem 7.33 generalizes the correspondence between regular triangulations and initial ideals of toric ideals [Stu96, Chapter 8].

Exercise 7.4 is based on [Ish87, Lemma 1.2]. Lawrence ideals (Exercise 7.12) are studied in [Stu96, Chapter 7]. Exercise 7.15 says that the region far interior to an unsaturated affine semigroup is devoid of "holes". This observation can be the starting point for induction arguments based on filtrations of $\mathbb{Z}^d$-graded modules in which successive quotients are Cohen–Macaulay.

Chapter 8

Multigraded polynomial rings

Having treated semigroup rings in the previous chapter, we now return to the polynomial ring $S = \Bbbk[x_1, \ldots, x_n]$, but this time with a grading by an arbitrary abelian group A. The notion of K-polynomial still works, although the notion of Hilbert series has to be revised for nonpositive gradings, where the graded pieces of S do not have finite dimension as vector spaces over $\Bbbk$. Multigradings give rise to a rich theory of *multidegrees*, directly generalizing the usual degree of a projective variety.

8.1 Multigradings

Definition 8.1 The polynomial ring S is called **multigraded by** A when it has been endowed with a **degree map** $\deg : \mathbb{Z}^n \to A$.

As in earlier chapters, monomials $\mathbf{x}^{\mathbf{u}} = x_1^{u_1} x_2^{u_2} \cdots x_n^{u_n}$ in S are identified with vectors $\mathbf{u} = (u_1, \ldots, u_n)$ in $\mathbb{N}^n$. Multigrading S by A is equivalent to fixing a semigroup homomorphism $\deg : \mathbb{N}^n \to A$ taking each monomial $\mathbf{x}^{\mathbf{u}}$ to its *degree* $\deg(\mathbf{u})$, which we also write as $\deg(\mathbf{x}^{\mathbf{u}})$. The distinguished system of n elements $\deg(x_1), \ldots, \deg(x_n)$ in A will be denoted by $\mathbf{a}_1, \ldots, \mathbf{a}_n$ throughout the chapter. In slightly different language, the grading on S is determined by an exact sequence

$$A \longleftarrow \mathbb{Z}^n \longleftarrow L \longleftarrow 0 \tag{8.1}$$

of abelian groups. The degree map $\mathbb{Z}^n \to A$ need not be surjective onto A, although often it will be. One such surjective case is when the grading of S is presented in the form of a basis for the sublattice L of $\mathbb{Z}^n$; the map $\deg$ is then taken to be projection onto $A = \mathbb{Z}^n / L$.

149

Example 8.2 Let $n = 3$ and let L be the rank 2 sublattice spanned by $(1,1,1)$ and $(1,3,5)$. Then $A = \mathbb{Z}^3/L$ is isomorphic to $\mathbb{Z} \oplus \mathbb{Z}/2\mathbb{Z}$, and the grading of $S = \Bbbk[x,y,z]$ defined by (8.1) can be expressed by regarding

$$\deg(x) = (1,1), \quad \deg(y) = (-2,1), \quad \text{and} \quad \deg(z) = (1,0)$$

as the three generators of $A = \mathbb{Z} \oplus \mathbb{Z}/2\mathbb{Z}$.   ◇

Throughout this chapter we use the symbol A to denote the group A together with the distinguished elements $\mathbf{a}_1, \ldots, \mathbf{a}_n$, and let $Q = \deg(\mathbb{N}^n)$ denote the subsemigroup of the group A generated by $\mathbf{a}_1, \ldots, \mathbf{a}_n$.

For $\mathbf{a} \in A$ let $S_{\mathbf{a}}$ denote the vector space (over the field $\Bbbk$) of homogeneous polynomials having degree $\mathbf{a}$ in the A-grading. A $\Bbbk$-linear basis of $S_{\mathbf{a}}$ is given by the (possibly infinite but also perhaps empty) set of all monomials $\mathbf{x}^{\mathbf{u}}$ satisfying $\deg(\mathbf{u}) = \mathbf{a}$, so S has the direct sum decomposition

$$S = \bigoplus_{\mathbf{a} \in A} S_{\mathbf{a}} \quad \text{satisfying} \quad S_{\mathbf{a}} \cdot S_{\mathbf{b}} \subseteq S_{\mathbf{a}+\mathbf{b}}.$$

Note that $S_{\mathbf{0}} = \Bbbk[L \cap \mathbb{N}^n]$ is a normal semigroup ring over $\Bbbk$, because $L \cap \mathbb{N}^n$ is a saturated semigroup (by Exercise 7.14 with $Q = L$ and $Q' = \mathbb{N}^n$).

Example 8.3 The Hilbert basis for $L \cap \mathbb{N}^3$ in Example 8.2 consists of the vectors $(4,2,0)$, $(1,1,1)$, and $(0,2,4)$. Hence the degree $\mathbf{0}$ component is

$$S_{(0,0)} \quad = \quad \Bbbk[x^4y^2, xyz, y^2z^4].$$

This normal affine semigroup ring is isomorphic to $\Bbbk[u,v,w]/\langle uw - v^4 \rangle$.   ◇

Proposition 8.4 *Each graded component $S_{\mathbf{a}}$ of a multigraded polynomial ring is a finitely generated module over the normal semigroup ring $S_{\mathbf{0}}$.*

Our proof will be algorithmic; see Exercise 8.1 for nonalgorithmic proof.

Proof. We may assume that $\mathbf{a} \in Q$ and that we are given one monomial $\mathbf{x}^{\mathbf{u}}$ lying in $S_{\mathbf{a}}$. Here is an algorithm that computes a finite generating set for $S_{\mathbf{a}}$ as a module over $S_{\mathbf{0}}$. Form the sublattice $L_{\mathbf{a}}$ of $\mathbb{Z}^{n+1} = \mathbb{Z}^n \oplus \mathbb{Z}$ generated by $(\mathbf{u},1)$ and all vectors $(\mathbf{b},0)$ for $\mathbf{b}$ running over a basis of L. Equivalently, $L_{\mathbf{a}}$ is the kernel of the abelian group homomorphism

$$\mathbb{Z}^{n+1} \to A \quad \text{sending} \quad (\mathbf{v},r) \mapsto \deg(\mathbf{v}) - r\mathbf{a}.$$

This shows that $L_{\mathbf{a}}$ is independent of the choice of $\mathbf{u}$.

Next compute the Hilbert basis $\mathcal{H}$ of the semigroup $L_{\mathbf{a}} \cap \mathbb{N}^{n+1}$. Let $\mathcal{H}_r$ denote the set of all vectors $\mathbf{v} \in \mathbb{N}^n$ that satisfy $(\mathbf{v},r) \in \mathcal{H}$. Then $\mathcal{H}_0$ is the Hilbert basis (considered above) of $L \cap \mathbb{N}^n$, and $\mathcal{H}_1$ is the desired finite generating set. Indeed, we can check directly that $\{\mathbf{x}^{\mathbf{v}} \mid (\mathbf{v},1) \in \mathcal{H}\}$ minimally generates $S_{\mathbf{a}}$ as a module over $S_{\mathbf{0}} = \Bbbk[\mathbf{x}^{\mathbf{v}} \mid (\mathbf{v},0) \in \mathcal{H}]$.   □

Example 8.5 Let us illustrate for Example 8.2 the procedure in the proof of Proposition 8.4 in degree $\mathbf{a} = (4,1)$. A typical monomial of that degree is $\mathbf{x}^{\mathbf{u}}$ with $\mathbf{u} = (5,2,3)$, so $L_{\mathbf{a}}$ is the rank 3 sublattice of $\mathbb{Z}^4$ spanned by $(1,1,1,0)$, $(1,3,5,0)$, and $(5,2,3,1)$. The Hilbert basis of $L_{\mathbf{a}} \cap \mathbb{N}^{3+1}$ consists of the nine vectors

$$(0,2,4,0), (4,2,0,0), (1,1,1,0), (0,1,6,1), (1,0,3,1),$$
$$(3,0,1,1), (6,1,0,1), (0,0,8,2), (8,0,0,2).$$

The first three of these vectors correspond to the monomial generators of $S_{(0,0)}$ as in Example 8.3. The next four vectors furnish the generators $\{yz^6, xz^3, x^3z, x^6y\}$ of $S_{(4,1)}$ as a module over the algebra $S_{(0,0)}$. ◇

In nice cases, the semigroup ring $S_{\mathbf{0}}$ is just $\Bbbk$ itself. There are many ways to characterize such good fortune.

Theorem 8.6 *The following conditions are equivalent for a polynomial ring S multigraded by A, with image semigroup $Q = \deg(\mathbb{N}^n)$.*

1. *There exists $\mathbf{a} \in Q$ such that the vector space $S_{\mathbf{a}}$ is finite-dimensional.*
2. *The only polynomials of degree $\mathbf{0}$ are the constants; i.e., $S_{\mathbf{0}} = \Bbbk$.*
3. *For all $\mathbf{a} \in A$, the $\Bbbk$-vector space $S_{\mathbf{a}}$ is finite-dimensional.*
4. *For all finitely generated graded modules M (see Definition 8.12) and degrees $\mathbf{a} \in A$, the $\Bbbk$-vector space $M_{\mathbf{a}}$ is finite-dimensional.*
5. *The only nonnegative vector in the lattice L is $\mathbf{0}$; i.e., $L \cap \mathbb{N}^n = \{\mathbf{0}\}$.*
6. *The semigroup Q has no units, and no variable x_i has degree zero.*

Proof. (1) ⇒ (2): For any $\mathbf{a} \in Q$ there exists some nonconstant monomial $\mathbf{x}^{\mathbf{u}}$ of degree $\mathbf{a}$. Multiplication by $\mathbf{x}^{\mathbf{u}}$ defines a monomorphism of vector spaces over $\Bbbk$ from $S_{\mathbf{0}}$ into $S_{\mathbf{a}}$, so $S_{\mathbf{0}}$ is a vector space of finite dimension. Hence $S_{\mathbf{0}}$ is the semigroup ring for an affine semigroup ring with finitely many elements. The only such affine semigroup is the zero semigroup.

(2) ⇒ (3): Proposition 8.4.

(3) ⇒ (4): The finiteness condition is clearly stable under taking finite direct sums of graded modules and quotients by graded submodules. Now take a graded free presentation of M—that is, a surjection from a finitely generated graded free module onto M—by choosing a graded generating set.

(4) ⇒ (1): Obvious.

(2) ⇔ (5): The degree zero component $S_{\mathbf{0}}$ is spanned as a $\Bbbk$-vector space by all monomials $\mathbf{x}^{\mathbf{u}}$, where $\mathbf{u}$ ranges over $L \cap \mathbb{N}^n$.

(2) ⇔ (6): $\deg(\mathbf{x}^{\mathbf{u}}) = \mathbf{0}$ if and only if one of the following occurs: (i) $\mathbf{u} = \mathbf{0}$; (ii) $\mathbf{x}^{\mathbf{u}} = x_i$ is one of the variables, in which case $\deg(x_i) = \mathbf{0}$; or (iii) $\mathbf{x}^{\mathbf{u}} = x_i \mathbf{x}^{\mathbf{v}}$ factors nontrivially to give an equation $\deg(x_i) = -\deg(\mathbf{x}^{\mathbf{v}})$ in Q, in which case $\deg(x_i)$ is a unit. □

Definition 8.7 If the equivalent conditions of Theorem 8.6 hold for a *torsion-free* abelian group A, then we call the grading by A **positive**, and we say that S is a **positively (multi)graded polynomial ring**.

Corollary 8.8 *If S is positively graded by A, then the semigroup $Q = \deg(\mathbb{N}^n)$ is pointed, and hence can be embedded in $\mathbb{N}^d$ for $d = \operatorname{rank}(A)$.*

Proof. This statement follows from Corollary 7.23 and Theorem 8.6.6.   □

It would have been nice to conclude immediately from the equivalent conditions of Theorem 8.6 that Q can be embedded in $\mathbb{N}^d$ for some d—in other words, that Q is affine. Unfortunately, this is false.

Example 8.9 Consider the subsemigroup Q generated by the two elements $\mathbf{a} = (1,1)$ and $\mathbf{b} = (1,0)$ in the group $A = \mathbb{Z} \oplus \mathbb{Z}/2\mathbb{Z}$, so that $\mathbf{a} - \mathbf{b}$ has order 2 in A. The degree map $\mathbb{N}^2 \to Q$ induces a surjection $\mathbb{Z}^2 \to A$ sending the basis vectors to $\mathbf{a}$ and $\mathbf{b}$. The kernel L of this surjection $\mathbb{Z}^2 \to A$ is generated by $(2, -2)$ and hence does not intersect the positive quadrant $\mathbb{N}^2$. Thus the conditions of Theorem 8.6 are satisfied. But there can be no embedding of Q into $\mathbb{N}^d$ for any d, because although no element of Q itself has finite order, the group A contains the torsion element $\mathbf{a} - \mathbf{b}$.   ◇

There is a fundamental difference between cases where the image of $\mathbb{Z}^n$ inside the grading group A has torsion and when it is torsion-free: when the image has torsion, there are graded ideals whose associated primes are not graded (as we will see in Example 8.10), whereas such bad behavior does not occur in the torsion-free case (as we will see in Proposition 8.11).

Example 8.10 In the situation of Example 8.2, the polynomial $y^2 z^4 - 1$ is homogeneous for the given grading by $\mathbb{Z} \oplus \mathbb{Z}/2\mathbb{Z}$. However, the ideal it generates equals the intersection

$$\langle y^2 z^4 - 1 \rangle \;\; = \;\; \langle yz^2 - 1 \rangle \cap \langle yz^2 + 1 \rangle$$

of two principal prime ideals neither of whose generators are homogeneous. Indeed, $\deg(yz^2) = (0,1)$, whereas $\deg(1) = (0,0)$.   ◇

Proposition 8.11 *Let S be multigraded by a torsion-free abelian group A. All associated primes of multigraded S-modules are multigraded.*

Proof. This is [Eis95, Exercise 3.5]. The proof, based on that of the corresponding $\mathbb{Z}$-graded statement in [Eis95, Section 3.5], is essentially presented in the aforementioned exercise from [Eis95]. It works because torsion-free grading groups $A \cong \mathbb{Z}^d$ can be totally ordered, for instance lexicographically (so $\mathbf{a} < \mathbf{b}$ when the earliest nonzero coordinate of $\mathbf{a} - \mathbf{b}$ is negative).   □

The proof fails for torsion groups because they admit no total orderings compatible with the group operation. However, Proposition 8.11 continues to hold for infinitely generated modules, because every associated prime is associated to a finitely generated submodule by definition.

The rest of this chapter concerns modules over multigraded polynomial rings. For the record, let us make a precise definition.

Definition 8.12 Let S be a polynomial ring multigraded by A. An S-module M is **multigraded by** A (sometimes we just say **graded**) if it has been endowed with a decomposition $M = \bigoplus_{\mathbf{a} \in A} M_{\mathbf{a}}$ as a direct sum of graded components such that $S_{\mathbf{a}} M_{\mathbf{b}} \subseteq M_{\mathbf{a}+\mathbf{b}}$ for all $\mathbf{a}, \mathbf{b} \in A$. Write $M(\mathbf{a})$ for the A-graded **translate** of M that satisfies $M(\mathbf{a})_{\mathbf{b}} = M_{\mathbf{a}+\mathbf{b}}$ for all $\mathbf{a}, \mathbf{b} \in A$.

The convention for A-graded translates makes the rank 1 free module $S(-\mathbf{a})$ into a copy of S generated in degree $\mathbf{a}$. The tensor product $M \otimes_S N$ of two multigraded modules is still multigraded, its degree $\mathbf{a}$ component being spanned by all elements $m_{\mathbf{b}} \otimes n_{\mathbf{a}-\mathbf{b}}$ such that $m_{\mathbf{b}} \in M_{\mathbf{b}}$ and $n_{\mathbf{a}-\mathbf{b}} \in N_{\mathbf{a}-\mathbf{b}}$. Consequently, $M(\mathbf{a}) = M \otimes_S S(\mathbf{a})$ is another way to express an A-graded translate of M. The notion of graded homomorphism makes sense as well: a map $\phi : M \to N$ is *graded (of degree $\mathbf{0}$)* if $\phi(M_{\mathbf{a}}) \subseteq N_{\mathbf{a}}$ for all $\mathbf{a} \in A$. Graded maps of graded modules have graded kernels, images, and cokernels.

Definition 8.12 does not assume that M is finitely generated, and indeed, we will see a variety of infinitely generated examples in Chapter 11. In addition, a graded module M might be nonzero in degrees from A that lie outside of the subgroup generated by $Q = \deg(\mathbb{N}^n)$.

Example 8.13 Let $S = \Bbbk[a, b, c, d]$ be multigraded by $\mathbb{Z}^2$, with $\deg(a) = (4, 0)$, $\deg(b) = (3, 1)$, $\deg(c) = (1, 3)$, and $\deg(d) = (0, 4)$. These vectors generate the semigroup $Q' \subset \mathbb{Z}^2$ in Fig. 7.2. Express the semigroup ring $\Bbbk[Q']$ as a quotient of S, as in Example 7.14. Although the semigroup Q' does not generate $\mathbb{Z}^2$ as a group, the $\mathbb{Z}^2$-graded translate $M = \Bbbk[Q'](-(1, 1))$ of the semigroup ring $\Bbbk[Q']$ is still a valid $\mathbb{Z}^2$-graded S-module, and its graded components $M_{\mathbf{a}}$ for $\mathbf{a} \in Q'$ are all zero. $\diamond$

8.2 Hilbert series and K-polynomials

Given a finitely generated graded module M over a positively multigraded polynomial ring, the dimensions $\dim_{\Bbbk}(M_{\mathbf{a}})$ are all finite, by Theorem 8.6. In the case where $M = S$ is the polynomial ring itself, the dimension of $S_{\mathbf{a}}$ is the cardinality of the fiber $(\mathbf{u} + L) \cap \mathbb{N}^n$ for any vector $\mathbf{u} \in \mathbb{Z}^n$ mapping to $\mathbf{a}$ under the degree map $\mathbb{Z}^n \to A$, where again, L is the kernel of the degree map as in (8.1). Geometrically, this cardinality is the number of lattice points in the polytope $(\mathbf{u} + \mathbb{R}L) \cap \mathbb{R}^n_{\geq 0}$. Just as in the "coarsely graded" case (where $A = \mathbb{Z}$; see Chapter 2 and Section 12.1, for example) and the "finely graded" case (where $A = \mathbb{Z}^n$; see Part I), the generating functions for the dimensions of the multigraded pieces of graded modules play a central role.

Definition 8.14 The **Hilbert function** of a finitely generated module M over a positively graded polynomial ring is the set map $A \to \mathbb{N}$ whose value at each group element $\mathbf{a} \in A$ is the vector space dimension $\dim_{\Bbbk}(M_{\mathbf{a}})$. The multigraded **Hilbert series** of M is the Laurent series

$$H(M; \mathbf{t}) = \sum_{\mathbf{a} \in A} \dim_{\Bbbk}(M_{\mathbf{a}}) \mathbf{t}^{\mathbf{a}} \quad \text{in the additive group} \quad \mathbb{Z}[[A]] = \prod_{\mathbf{a} \in A} \mathbb{Z} \cdot \mathbf{t}^{\mathbf{a}}.$$

Elements in the abelian group $\mathbb{Z}[[A]]$ are not really Laurent series, but just formal elements in the product, and the letter $\mathbf{t}$ here is a dummy variable. However, when we have an explicitly given inclusion $A \subseteq \mathbb{Z}^d$, so that $\mathbf{a}_1, \ldots, \mathbf{a}_n$ are vectors of length d, the symbol $\mathbf{t}$ can also stand for the list $t_1, \ldots, t_d$ of variables, so that $\mathbf{t^a} = t_1^{a_1} \cdots t_d^{a_d}$ as usual. This common special case lends the suggestive name to the elements of $\mathbb{Z}[[A]]$.

We would like to write Hilbert series as rational functions, just as we did in Corollary 4.20 for monomial ideals in the finely graded polynomial ring, where the degree map has image $Q = \mathbb{N}^n$. In order to accomplish this task in the current more general setting, we need an ambient algebraic structure in which to equate Laurent series with rational functions. To start, consider the semigroup ring $\mathbb{Z}[Q] = \bigoplus_{\mathbf{a} \in Q} \mathbb{Z} \cdot \mathbf{t^a}$ over the integers $\mathbb{Z}$. When Q is pointed (Definition 7.8), the ideal of $\mathbb{Z}[Q]$ generated by all monomials $\mathbf{t^a} \neq 1$ is a proper ideal. The completion of $\mathbb{Z}[Q]$ at this ideal [Eis95, Chapter 7] is the ring $\mathbb{Z}[[Q]]$ of *power series supported on* Q. Let us justify the name.

Lemma 8.15 *Elements in the completion $\mathbb{Z}[[Q]]$ for a pointed semigroup Q can be expressed uniquely as formal series $\sum_{\mathbf{a} \in Q} c_{\mathbf{a}} \mathbf{t^a}$ with $c_{\mathbf{a}} \in \mathbb{Z}$.*

Proof. The lemma is standard when $Q = \mathbb{N}^n$, as $\mathbb{Z}[[\mathbb{N}^n]] = \mathbb{Z}[[x_1, \ldots, x_n]]$ is an honest power series ring. For a general pointed semigroup Q, write $\mathbb{Z}[Q]$ as a module over $\mathbb{Z}[\mathbb{N}^n] = \mathbb{Z}[x_1, \ldots, x_n]$. Then $\mathbb{Z}[[Q]]$ is the completion of $\mathbb{Z}[Q]$ at the ideal $\mathfrak{m} = \langle x_1, \ldots, x_n \rangle \subset \mathbb{Z}[\mathbb{N}^n]$. It follows from condition 3 in Theorem 8.6 that every element in $\mathbb{Z}[[Q]]$ has *some* expression as a power series $p(\mathbf{t}) = \sum_{\mathbf{a} \in Q} c_{\mathbf{a}} \mathbf{t^a}$. To see that this expression is unique, we need only show that $p(\mathbf{t})$ is nonzero in $\mathbb{Z}[[Q]]$ whenever $c_{\mathbf{a}} \neq 0$ for some $\mathbf{a} \in Q$. We will use that the natural map $\mathbb{Z}[Q]/\mathfrak{m}^r\mathbb{Z}[Q] \to \mathbb{Z}[[Q]]/\mathfrak{m}^r\mathbb{Z}[[Q]]$ is an isomorphism for all $r \in \mathbb{N}$, which follows by definition of completion.

Choose a vector $\mathbf{u} \in \mathbb{N}^n$ mapping to $\mathbf{a}$. There is a positive integer r such that $c_{\mathbf{a}}\mathbf{x^u}$ lies outside of $\mathfrak{m}^r$, and for this choice of r, the image of the series $p(\mathbf{t})$ in $\mathbb{Z}[[Q]]/\mathfrak{m}^r\mathbb{Z}[[Q]] \cong \mathbb{Z}[Q]/\mathfrak{m}^r\mathbb{Z}[Q]$ is a nonzero *polynomial*. $\square$

Any element $p(\mathbf{t}) \in \mathbb{Z}[[Q]]$ with constant term ± 1 is a unit in $\mathbb{Z}[[Q]]$. Indeed, if $p(\mathbf{t}) = 1 - q(\mathbf{t})$ and q has constant term 0, then the inverse $p^{-1} = 1 + q + q^2 + \cdots$ is well-defined because $\mathbb{Z}[[Q]]$ is complete with respect to an ideal containing q. In particular, $1 - \mathbf{t^a}$ is invertible for all $\mathbf{a} \in Q$.

Lemma 8.16 *The Hilbert series for the multigraded polynomial ring S is*

$$H(S; \mathbf{t}) = \frac{1}{(1 - \mathbf{t^{a_1}}) \cdots (1 - \mathbf{t^{a_n}})}$$

as an element in $\mathbb{Z}[[Q]]$, if the multigrading on S is positive. $\square$

Proof. Let $H(S; \mathbf{x})$ be the Hilbert series of S in the fine multigrading by $\mathbb{N}^n$. Viewed as a power series, the image of $H(S; \mathbf{x})$ under the surjection $\mathbb{Z}[[\mathbb{N}^n]] \to \mathbb{Z}[[Q]]$ equals $H(S; \mathbf{t})$ by definition. On the other hand, $H(S; \mathbf{x}) = 1/(1 - x_1) \cdots (1 - x_n)$ by the first equation in Example 1.11. Now take the image of this equation in $\mathbb{Z}[[Q]]$. $\square$

As we saw in Example 8.13, the Hilbert series of a finitely generated module need not be supported on Q, so the completion $\mathbb{Z}[[Q]]$ is not big enough to hold all of the Hilbert series of finitely generated modules. Let us say that a Laurent series $p(\mathbf{t}) = \sum_{\mathbf{a} \in A} c_{\mathbf{a}} \mathbf{t}^{\mathbf{a}}$ is *supported on finitely many translates of Q* if there exist $\mathbf{b}_1, \ldots, \mathbf{b}_k \in A$ (depending on p) such that

$$c_{\mathbf{a}} \neq 0 \quad \Rightarrow \quad \mathbf{a} \in (\mathbf{b}_1 + Q) \cup \cdots \cup (\mathbf{b}_k + Q).$$

The set $\mathbb{Z}[[Q]][A]$ of such series is a ring; in fact, it equals the $\mathbb{Z}[[Q]]$-algebra

$$\mathbb{Z}[[Q]][A] \;\;=\;\; \mathbb{Z}[[Q]] \otimes_{\mathbb{Z}[Q]} \mathbb{Z}[A],$$

as outlined in Exercise 8.2. That exercise explains the role of the seemingly cumbersome tensor product, given that it amounts to a notation for inverting the monomials $t^{\mathbf{a}_1}, \ldots, t^{\mathbf{a}_n} \in \mathbb{Z}[[Q]]$ in the case that Q generates A.

Example 8.17 Let $A = \mathbb{Z}^2$, and consider the subsemigroup $Q \subset A$ generated by the four vectors $(4,0)$, $(3,1)$, $(1,3)$, and $(0,4)$. [This is the semigroup from Examples 7.14 and 8.13, where it was called Q'.] In this case,

$$\mathbb{Z}[[Q]] \;\cong\; \mathbb{Z}[[a,b,c,d]]/\langle bc - ad, c^3 - bd^2, ac^2 - b^2 d, b^3 - a^2 c \rangle$$

is the ring of power series supported on Q. The subgroup $A' \subset A$ generated by Q is a sublattice of index 4. The quotient group $\mathbb{Z}^2/A'$ is cyclic of order 4 and generated by the image of $(1,0)$. Letting s be an independent variable, we therefore find that $\mathbb{Z}[A] \cong \mathbb{Z}[A'][s]/\langle s^4 - a \rangle$, so

$$\mathbb{Z}[[Q]][A] \;\cong\; \mathbb{Z}[[Q]][a^{-1}, b^{-1}, c^{-1}, d^{-1}][s]/\langle s^4 - a \rangle.$$

This ring is free as a module over $\mathbb{Z}[[Q]][A']$, one reason being that the equation $s^4 - a$ is monic in s [Eis95, Proposition 4.1]. We can recover the "s, t" notation for $\mathbb{Z}[Q]$ as in Example 7.14 by setting $t = sa^{-1}b$, so that $\mathbb{Z}[Q]$ is the subring of $\mathbb{Z}[[Q]][A]$ generated by the monomials s^4, $s^3 t$, st^3, and t^4. $\diamond$

The main result of this section, Theorem 8.20, implies that the Hilbert series of any finitely generated module is supported on finitely many translates of Q and hence lies in $\mathbb{Z}[[Q]][A]$. Noting that $(1 - \mathbf{t}^{\mathbf{a}})$ is still invertible in this ring for $\mathbf{a} \in Q$, we find (for example) that the Hilbert series of the graded translate $S(-\mathbf{a})$ is $\mathbf{t}^{\mathbf{a}}/\prod_{j=1}^{n}(1 - \mathbf{t}^{\mathbf{a}_j})$. This tiny observation is the crux of Theorem 8.20, but we need free resolutions to see why.

Given our positive multigrading by A, we can choose a *coarsening* to a positive $\mathbb{Z}$-grading, under which S is $\mathbb{N}$-graded and the degrees of the variables are all strictly positive. Indeed, by Corollary 7.23 there is a linear map $A \to \mathbb{Z}$ that induces a morphism $Q \to \mathbb{N}$ of semigroups taking every nonzero element of Q to a strictly positive number. Using the fact that Nakayama's Lemma [Eis95, Corollary 4.8] holds in the positively $\mathbb{Z}$-graded case [Eis95, Exercise 4.6], we conclude that Nakayama's Lemma holds for any positively multigraded ring S. This observation has a number of consequences, the first being that every finitely generated multigraded module M has a well-defined number of minimal generators in each degree $\mathbf{b} \in A$.

Just as in [Eis95, Exercise 4.11a] for positive $\mathbb{Z}$-gradings, every positively multigraded free (or projective) S-module $\mathcal{F}$ of rank r has the form $\mathcal{F} \cong \bigoplus_{j=1}^{r} S(-\mathbf{b}_j)$ for some list $\mathbf{b}_1, \ldots, \mathbf{b}_r$ of degrees in A. Moreover, given a fixed degree $\mathbf{b} \in A$, the number of indices j satisfying $\mathbf{b}_j = \mathbf{b}$ is independent of the direct sum decomposition. As M is graded, there is a graded surjection $M \twoheadleftarrow \mathcal{F}_0$ onto M from a multigraded free module $\mathcal{F}_0$, and the kernel is a graded submodule of $\mathcal{F}_0$. Hence we can iterate this procedure to construct a free resolution $\mathcal{F}_\bullet$ of M. Let us record a more precise result, which we derive also for the useful case of nonpositive multigradings.

In the proof of the next proposition, we assume the Hilbert syzygy theorem for free resolutions of ungraded modules, which has an elementary proof using Gröbner bases (Schreyer's algorithm [Eis95, Corollary 15.11]).

Proposition 8.18 *Every finitely generated multigraded module M over S has a finite multigraded resolution by multigraded free modules of the form $S(-\mathbf{b}_1) \oplus \cdots \oplus S(-\mathbf{b}_r)$, even if the grading is not positive.*

Proof. Calculate a finite free resolution of M using Gröbner bases. The reduced Gröbner basis for a graded submodule of a multigraded free module is homogeneous for the given multigrading. Therefore each free module in the resolution has the desired form automatically. □

Questions about nonpositive multigradings will occupy our attention in Section 8.4. For now, let us focus again on the positively graded case.

Lemma 8.19 *If $0 \leftarrow M \leftarrow N_0 \leftarrow N_1 \leftarrow \cdots \leftarrow N_r \leftarrow 0$ is an exact sequence of finitely generated positively multigraded modules, then the Hilbert series of M equals the alternating sum of those for $N_0, \ldots, N_r$:*

$$H(M; \mathbf{t}) \;=\; \sum_{j=0}^{r} (-1)^j H(N_j; \mathbf{t}).$$

Proof. For each $\mathbf{a} \in A$, the degree $\mathbf{a}$ piece of the given exact sequence of modules is an exact sequence of finite-dimensional vector spaces over $\Bbbk$. The rank-nullity theorem from linear algebra says that the alternating sum of the dimensions of these vector spaces equals zero. □

Theorem 8.20 *The Hilbert series of a finitely generated graded module M over a polynomial ring positively multigraded by A is a Laurent series supported on finitely many translates of $Q = \deg(\mathbb{N}^n)$. More precisely, there is a unique Laurent polynomial $\mathcal{K}(M; \mathbf{t}) \in \mathbb{Z}[A]$ such that in $\mathbb{Z}[[Q]][A]$,*

$$H(M; \mathbf{t}) \;=\; \frac{\mathcal{K}(M; \mathbf{t})}{\prod_{i=1}^{n}(1 - \mathbf{t}^{\mathbf{a}_i})}.$$

Proof. By the obvious equality $H(M(-\mathbf{a}); \mathbf{t}) = \mathbf{t}^{\mathbf{a}} H(M; \mathbf{t})$ for graded translates of arbitrary finitely generated modules M and the fact that Hilbert series are additive on direct sums, we deduce from Lemma 8.16 that

$$H(S(-\mathbf{b}_1) \oplus \cdots \oplus S(-\mathbf{b}_r); \mathbf{t}) = \frac{\mathbf{t}^{\mathbf{b}_1} + \cdots + \mathbf{t}^{\mathbf{b}_r}}{\prod_{i=1}^{n}(1 - \mathbf{t}^{\mathbf{a}_i})}. \tag{8.2}$$

The existence of a Laurent polynomial $\mathcal{K}(M; \mathbf{t})$ satisfying the required conditions now follows by applying Lemma 8.19 to a free resolution, as in Proposition 8.18. The uniqueness of $\mathcal{K}(M; \mathbf{t})$ results from the fact that $\mathcal{K}(M; \mathbf{t}) = H(M; \mathbf{t}) \prod_{i=1}^{n}(1 - \mathbf{t}^{\mathbf{a}_i})$ in $\mathbb{Z}[[Q]][A]$. $\qquad\square$

Definition 8.21 The Laurent polynomial $\mathcal{K}(M; \mathbf{t})$ that is the numerator in Theorem 8.20 is called the K-**polynomial** of M.

The "K" here stands for "K-theory", but it also seems to fit in nicely with the terms "f-vector" and "h-polynomial", which are related notions.

8.3 Multigraded Betti numbers

As we have seen in the previous section, Nakayama's Lemma applies when the multigrading is positive. Consequently, as in Chapter 1, the notion of minimal generator extends to a notion of *minimal* graded free resolution, which is defined by the property that all of the differentials become zero when tensored with $\Bbbk = S/\mathfrak{m}$, where $\mathfrak{m} = \langle x_1, \ldots, x_m \rangle$. Equivalently, minimality means that the differentials can be represented by matrices whose nonzero entries are homogeneous of nonzero degree. Using the symmetry of Tor as in Section 1.5, we can count the number $\beta_{i,\mathbf{a}}(M)$ of summands $S(-\mathbf{a})$ in homological degree i of any minimal free resolution of M.

Definition 8.22 Let M be a graded module over a positively multigraded polynomial ring. The ith multigraded **Betti number** of M in degree $\mathbf{a}$ is the vector space dimension $\beta_{i,\mathbf{a}}(M) = \dim_{\Bbbk} \operatorname{Tor}_i^S(\Bbbk, M)_{\mathbf{a}}$.

Betti numbers fail to be well-defined when the grading is not positive, because the notion of minimality for free resolutions breaks. Viewed another way, although Definition 8.22 makes sense for any graded module M because $\operatorname{Tor}_i^S(\Bbbk, M)$ is graded, we can only expect $\beta_{i,\mathbf{a}}(M)$ to be finite when M is finitely generated and the multigrading is positive, via Theorem 8.6.

Analyzing the proof of Theorem 8.20 a little more closely, we can see how K-polynomials are built out of Betti numbers.

Proposition 8.23 *If M is a finitely generated positively multigraded module, then the K-polynomial records the alternating sum of its Betti numbers:*

$$\mathcal{K}(M; \mathbf{t}) = \sum_{\substack{\mathbf{a} \in A \\ i \geq 0}} (-1)^i \beta_{i,\mathbf{a}}(M) \, \mathbf{t}^{\mathbf{a}}.$$

Proof. Use the proof of Theorem 8.20 on a minimal free resolution of M: if the ith homological degree of this resolution is $S(-\mathbf{b}_1) \oplus \cdots \oplus S(-\mathbf{b}_r)$, then (8.2) contributes $(-1)^i \sum_{\mathbf{a} \in A} \beta_{i,\mathbf{a}}(M) \, \mathbf{t}^{\mathbf{a}} / \prod_{i=1}^{n}(1 - \mathbf{t}^{\mathbf{a}_i})$ to the Hilbert series of M.   □

Example 8.24 For any multigrading of S by A, the Hilbert series of the residue field $\Bbbk = S/\langle x_1, \ldots, x_n \rangle$ is just $1 \in \mathbb{Z}[[Q]][A]$. This agrees with the calculation of its K-polynomial from the Koszul complex, which yields

$$\mathcal{K}(S/\langle x_1, \ldots, x_n \rangle; \mathbf{t}) \;=\; \sum_{\Lambda \subseteq \{1,\ldots,n\}} (-1)^{|\Lambda|} \mathbf{t}^{\Lambda} \;=\; \prod_{i=1}^{n}(1 - \mathbf{t}^{\mathbf{a}_i}),$$

where we write $\mathbf{t}^{\Lambda} = \prod_{i \in \Lambda} \mathbf{t}^{\mathbf{a}_i}$ for any subset $\Lambda \subseteq \{1, \ldots, n\}$.   ◇

Some of the terms (or parts of terms) in the sum from Proposition 8.23 may cancel, even though the Betti numbers come from a minimal resolution. This phenomenon did not occur in Example 8.24, but we have already seen it in Examples 1.14 and 1.25.

Proposition 8.23 is often more useful as a means to bound Betti numbers than to compute K-polynomials, as the latter task is usually easier. Indeed, the calculation of K-polynomials can be reduced using Gröbner bases to the $\mathbb{N}^n$-graded case from Part I, where we have more combinatorial techniques at our disposal, because Hilbert series (and hence K-polynomials) are unchanged by taking initial submodules of graded free modules. [Proof: The *standard monomials* (those outside the initial submodule) are a vector space basis modulo both the original module and its initial submodule.]

Our purpose in the rest of this section is to show that Betti numbers can only increase under taking initial ideals. To prove this important statement in Theorem 8.29, we use a "lifting" construction, which turns the passage to an initial ideal or submodule into a continuous operation. Briefly, some power of a homogenizing parameter y is attached to each trailing term, and letting y approach zero yields the initial submodule, while setting $y = 1$ recovers the original module. Geometrically, this procedure yields a family of modules over the affine line, called a *Gröbner degeneration*, whose *special fiber* is the initial submodule.

To give the details, suppose that $\mathcal{F}$ is a free module with basis $\mathbf{e}_1, \ldots, \mathbf{e}_r$. A *weight order* on $\mathcal{F}$ is determined by a positive weight vector $w \in \mathbb{R}^n_{\geq 0}$ and integer weights $\varepsilon_1, \ldots, \varepsilon_r$ on the basis vectors of $\mathcal{F}$. Thus a weight order on $\mathcal{F}$ is a weight order on S together with *basis weights* ε. The terminology in Section 7.4 extends easily. For example, the *weight* of a term $c_{\mathbf{u},j} \mathbf{x}^{\mathbf{u}} \mathbf{e}_j$ is $w \cdot \mathbf{u} + \varepsilon_j$, and if K is a submodule of $\mathcal{F}$, then $\text{in}_{w,\varepsilon}(K)$ is its initial submodule with respect to w, ε.

Definition 8.25 (Homogenization) Fix a free module $\mathcal{F} = \bigoplus_{j=1}^{r} S \cdot \mathbf{e}_j$ over a polynomial ring S multigraded by A, with each basis vector $\mathbf{e}_j$ in degree $\mathbf{b}_j \in A$, and fix a weight order (w, ε) on $\mathcal{F}$. Introduce a new variable y, and let $S[y]$ be multigraded by $A \times \mathbb{Z}$, with $\deg(x_i) = (\mathbf{a}_i, w_i)$ for $i = 1, \ldots, n$, and $\deg(y) = (\mathbf{0}, 1)$. The **homogenization** of $\mathcal{F}$ is

$$\mathcal{F}[y] = \bigoplus_{j=1}^{r} S[y] \cdot \tilde{\mathbf{e}}_j \quad \text{with} \quad \deg(\tilde{\mathbf{e}}_j) = (\mathbf{b}_j, \varepsilon_j),$$

so $\mathcal{F}[y]$ is a graded free $S[y]$-module. Given $f = \sum c_{\mathbf{u},j} \mathbf{x}^{\mathbf{u}} \mathbf{e}_j \in \mathcal{F}$, let $\ell(f) = \max\{w \cdot \mathbf{u} + \varepsilon_j \mid c_{\mathbf{u},j} \neq 0\}$ be the weight of $\mathrm{in}_{w,\varepsilon}(f)$. The **homogenization**

$$\tilde{f} = y^{\ell(f)} \sum c_{\mathbf{u},j} (y^{-w \cdot \mathbf{u}} \mathbf{x}^{\mathbf{u}})(y^{-\varepsilon_j} \mathbf{e}_j)$$

is then a homogeneous element of $\mathcal{F}[y]$. Finally, if K is a submodule of $\mathcal{F}$, then its **homogenization** $\tilde{K} \subseteq \mathcal{F}[y]$ is generated by $\{\tilde{f} \mid f \in K\}$.

The multigrading in Definition 8.25 need not be positive, but if it is, then so is the multigrading on $S[y]$, by Theorem 8.6.6. For a concrete example of homogenization, including what is to come, see Exercise 8.3.

Suppose now that we are given a term order on $\mathcal{F}$ (Section 2.2; or see [Eis95, Section 15.2], where this is called a "monomial order") and a submodule $K \subseteq \mathcal{F}$. In what follows, we shall use without further comment that there is a weight order on $\mathcal{F}$ under which the initial submodule $\mathrm{in}_{w,\varepsilon}(K)$ equals the initial submodule $\mathrm{in}(K)$ for the given term order (see [Eis95, Exercises 15.11–15.13] for methods and references). In particular, using weight orders as in the forthcoming statement suffices to treat all term orders.

Proposition 8.26 *If $\tilde{K}$ is the homogenization of a submodule K of a free module $\mathcal{F}$ with respect to a weight order (w, ε), then $\tilde{K}/(y-1)\tilde{K} \cong K$ and $\tilde{K}/y\tilde{K} \cong \mathrm{in}_{w,\varepsilon}(K)$. Moreover, $\mathcal{F}[y]/\tilde{K}$ is free as a module over $\mathbb{k}[y]$.*

Proof. The isomorphisms are immediate from Definition 8.25. For freeness, the proof of [Eis95, Theorem 15.17] works here mutatis mutandis. $\square$

Proposition 8.26 says how to interpolate between a module and its initial module. The next two results develop a method to interpolate between free resolutions of these modules.

Lemma 8.27 *Let R be a ring and $\tilde{\mathcal{F}}_\bullet$ a free resolution of an $R[y]$-module M over $R[y]$. If y is a nonzerodivisor on M, then $\tilde{\mathcal{F}}_\bullet/y\tilde{\mathcal{F}}_\bullet$ is a free resolution of M/yM over R. The previous sentence also holds with $y-1$ in place of y.*

Proof. The homology of $\tilde{\mathcal{F}}_\bullet/y\tilde{\mathcal{F}}_\bullet$ is $\mathrm{Tor}_\bullet^{R[y]}(R, M)$, which can also be calculated by tensoring the $R[y]$-free resolution

$$0 \leftarrow R[y] \xleftarrow{\cdot y} R[y] \leftarrow 0$$

of R with M and taking the homology. The complex resulting after tensoring over $R[y]$ with M is $0 \leftarrow M \xleftarrow{\cdot y} M \leftarrow 0$, which has only the homology M/yM in homological degree zero, because y is a nonzerodivisor on M. The proof remains unchanged when y is replaced by $y - 1$ throughout. $\square$

Proposition 8.28 *Adopt all of the notation from Definition 8.25, and fix an $S[y]$-free resolution $\tilde{\mathcal{F}}_{\bullet}$ of $\mathcal{F}[y]/\tilde{K}$ that is multigraded by $A \times \mathbb{Z}$. Then*

1. *$\tilde{\mathcal{F}}_{\bullet}/(y-1)\tilde{\mathcal{F}}_{\bullet}$ is an S-free resolution of $\mathcal{F}/K$ that is multigraded by A;*
2. *$\tilde{\mathcal{F}}_{\bullet}/y\tilde{\mathcal{F}}_{\bullet}$ is an S-free resolution of $\mathcal{F}/\mathrm{in}_{w,\varepsilon}(K)$ multigraded by A; and*
3. *the numbers of degree $\mathbf{a}$ generators in each homological degree coincide.*

Proof. Both y and $y - 1$ are nonzerodivisors on $\mathcal{F}[y]/\tilde{K}$ by the freeness in Proposition 8.26. Therefore, ignoring the grading for the moment, $\tilde{\mathcal{F}}_{\bullet}/y\tilde{\mathcal{F}}_{\bullet}$ and $\tilde{\mathcal{F}}_{\bullet}/(y-1)\tilde{\mathcal{F}}_{\bullet}$ are S-free resolutions of the desired modules by Lemma 8.27 with $R = S$. For statements 1 and 2, it remains only to check that the differentials of $\tilde{\mathcal{F}}_{\bullet}/y\tilde{\mathcal{F}}_{\bullet}$ and $\tilde{\mathcal{F}}_{\bullet}/(y-1)\tilde{\mathcal{F}}_{\bullet}$ are multigraded by A.

The polynomial ring $S[y]$ carries another multigrading, namely a multigrading by A, by forgetting the $\mathbb{Z}$-coordinate of $A \times \mathbb{Z}$. This multigrading of $S[y]$ by A is never positive, but every $S[y]$-module graded by $A \times \mathbb{Z}$ is nevertheless naturally graded by A as well. Now simply note that y and $y - 1$ are homogeneous for this multigrading by A on $S[y]$.

For statement 3, consider for each homological degree i and degree $\mathbf{a} \in A$ the direct sum $\tilde{\mathcal{F}}_{i,\mathbf{a}}$ of all summands of $\tilde{\mathcal{F}}_i$ that are generated in degrees of the form $(\mathbf{a}, k) \in A \times \mathbb{Z}$. For both $\lambda = 0$ and $\lambda = 1$, the specialization $\tilde{\mathcal{F}}_{i,\mathbf{a}}/(y - \lambda)\tilde{\mathcal{F}}_{i,\mathbf{a}}$ at $y = \lambda$ is a graded free S-module generated in degree $\mathbf{a} \in A$, and in both cases its rank equals that of $\tilde{\mathcal{F}}_{i,\mathbf{a}}$. $\square$

Theorem 8.29 (Upper-semicontinuity) *Fix a graded submodule K of a graded free module $\mathcal{F}$ over a positively multigraded polynomial ring. If $\mathrm{in}(K)$ is the initial submodule of K for some term order or weight order, then*

$$\beta_{i,\mathbf{a}}(\mathcal{F}/K) \ \leq \ \beta_{i,\mathbf{a}}(\mathcal{F}/\mathrm{in}(K)) \quad \text{for all } i \in \mathbb{N} \text{ and } \mathbf{a} \in A.$$

Proof. Assume that the free resolution $\tilde{\mathcal{F}}_{\bullet}$ in Proposition 8.28 is minimal, which we can do because the multigrading on $S[y]$ is positive. Since y lies in the graded maximal ideal of $S[y]$, the free resolution $\tilde{\mathcal{F}}_{\bullet}/y\tilde{\mathcal{F}}_{\bullet}$ of $\mathcal{F}/\mathrm{in}(K)$ in Proposition 8.28 is minimal. Hence the (equal) numbers in Proposition 8.28.3, which can only exceed $\beta_{i,\mathbf{a}}(\mathcal{F}/K)$, are equal to $\beta_{i,\mathbf{a}}(\mathcal{F}/\mathrm{in}(K))$. $\square$

Remark 8.30 Since the Betti numbers of $\mathcal{F}/\mathrm{in}(K)$ can only jump up from those of $\mathcal{F}/K$, the excess must cancel (one by one, say) in pairs from different homological degrees. However, a stronger statement is true as a consequence of the proof and the $\mathbb{Z}$-graded analogue of Exercise 1.11: the jumping Betti numbers must cancel in *consecutive* pairs because they correspond to nonminimal summands in an honest free resolution of $\mathcal{F}/K$.

The projective dimension of a module M in Definition 5.52 is characterized by the vanishing of its Betti numbers past that homological degree, and M is Cohen–Macaulay if that homological degree is the codimension of M. Note that the (co)dimension of a positively multigraded module can be recovered from its Hilbert series, which is unchanged under taking initial submodules. Therefore Theorem 8.29 has the following consequence.

Corollary 8.31 *If $\mathcal{F}/K$ is a quotient of a positively multigraded free module and $\mathrm{in}(K)$ is the initial submodule of K for some term order on $\mathcal{F}$, then the projective dimension of $\mathcal{F}/K$ satisfies $\mathrm{pd}(\mathcal{F}/K) \leq \mathrm{pd}(\mathcal{F}/\mathrm{in}(K))$. In particular, $\mathcal{F}/K$ is Cohen–Macaulay if $\mathcal{F}/\mathrm{in}(K)$ is Cohen–Macaulay.* □

8.4 *K*-polynomials in nonpositive gradings

Theorem 8.20 immediately fails for nonpositive multigradings: the dimensions $\dim_{\Bbbk}(M_{\mathbf{a}})$ can be infinite by Theorem 8.6—even for finitely generated modules M. Nonetheless, there is a notion of K-polynomial that extends the positively multigraded concept of "Hilbert series numerator" to the nonpositive case. The idea is to take our cue from the proof of Theorem 8.20 and convert Proposition 8.23 into a definition.

Definition 8.32 If $\mathcal{F}$ is a multigraded free module that is isomorphic to $S(-\mathbf{b}_1) \oplus \cdots \oplus S(-\mathbf{b}_r)$, define the Laurent polynomial $[\mathcal{F}]_{\mathbf{t}} = \mathbf{t}^{\mathbf{b}_1} + \cdots + \mathbf{t}^{\mathbf{b}_r}$ in the group algebra $\mathbb{Z}[A]$. If M is a finitely generated multigraded module with a finite resolution $\mathcal{F}_{\bullet} \to M$ by free modules of that form, then define

$$\mathcal{K}(M; \mathbf{t}) \;=\; \sum_j (-1)^j [\mathcal{F}_j]_{\mathbf{t}}$$

to be the *K*-**polynomial** of M (relative to $\mathcal{F}_{\bullet}$).

If $\mathcal{F}$ is a free module, then the K-polynomial of $\mathcal{F}$ relative to its tautological (that is, length zero) free resolution is just $\mathcal{K}(\mathcal{F}; \mathbf{t}) = [\mathcal{F}]_{\mathbf{t}}$. For a less trivial example, the displayed equation in Example 8.24 calculates the K-polynomial of the residue field $\Bbbk$ relative to the Koszul complex, independently of whether or not the multigrading is positive.

Example 8.33 Everything still works when A has (or is) torsion, say when $A = \mathbb{Z}/3\mathbb{Z}$ and $S = \Bbbk[x]$, with $\deg(x) = 1 \pmod 3$. A quotient such as $\Bbbk[x]/\langle x^4 + 17x \rangle$ has K-polynomial $1 - t \in \mathbb{Z}[t]/\langle t^3 - 1 \rangle = \mathbb{Z}[\mathbb{Z}/3\mathbb{Z}]$, relative to the free resolution $0 \leftarrow S \xleftarrow{x^4+17x} S(-1) \leftarrow 0$, or your favorite other choice of free resolution. ◇

Proposition 8.18 guarantees that any given module possesses a multigraded free resolution $\mathcal{F}_{\bullet}$ from which to calculate a K-polynomial. When the multigrading is positive, we are justified in leaving $\mathcal{F}_{\bullet}$ out of the notation $\mathcal{K}(M; \mathbf{t})$ by Theorem 8.20 and Lemma 8.19. Although we are still justified in leaving out $\mathcal{F}_{\bullet}$ when the multigrading is nonpositive, meaning that K-polynomials do not depend on the resolutions used to calculate them, it is more difficult to prove without relying on Hilbert series.

Theorem 8.34 *If S is an arbitrary (perhaps not positively) multigraded polynomial ring and M is a finitely generated multigraded module, then the K-polynomials of M relative to any two finite free resolutions are equal.*

Although we will use this theorem implicitly in what follows, we only sketch its proof, which consists more or less of routine homological algebra.

Sketch of proof. Define a multigraded free resolution $\mathcal{F}_{\boldsymbol{\cdot}}$ to *dominate* another such resolution $\mathcal{F}'_{\boldsymbol{\cdot}}$ if there is a surjection $\mathcal{F}_{\boldsymbol{\cdot}} \twoheadrightarrow \mathcal{F}'_{\boldsymbol{\cdot}}$ of resolutions, in the sense that $\mathcal{F}_i$ maps surjectively to $\mathcal{F}'_i$ for all i, inducing an isomorphism on M. The kernel of such a morphism must necessarily be a free resolution of the zero-module. Since a resolution of zero is split exact, its K-polynomial relative to any finite free resolution equals 0. Therefore the K-polynomials of $\mathcal{F}_{\boldsymbol{\cdot}}$ and $\mathcal{F}'_{\boldsymbol{\cdot}}$ are equal if $\mathcal{F}_{\boldsymbol{\cdot}}$ dominates $\mathcal{F}'_{\boldsymbol{\cdot}}$.

For the remainder of the proof, one shows that given two finite free multigraded resolutions $\mathcal{F}'_{\boldsymbol{\cdot}}$ and $\mathcal{F}''_{\boldsymbol{\cdot}}$ of M, there is a third resolution $\mathcal{F}_{\boldsymbol{\cdot}}$ dominating both. One way to construct an $\mathcal{F}_{\boldsymbol{\cdot}}$ is to pick a chain map $\alpha : \mathcal{F}'_{\boldsymbol{\cdot}} \to \mathcal{F}''_{\boldsymbol{\cdot}}$ lifting the isomorphism $M \overset{\approx}{\to} M$ on cohomology, and then let $\mathcal{F}_{\boldsymbol{\cdot}}$ be the *mapping cocylinder* of α. [Mapping cocylinder is the dual notion to mapping cylinder; in our case, $\mathcal{F}_{\boldsymbol{\cdot}}$ can be alternately described as $\mathrm{Hom}(\mathcal{F}^{\bullet}, S)$, where $\mathcal{F}^{\bullet}$ is the usual mapping cylinder of $\alpha^* : \mathrm{Hom}(\mathcal{F}''_{\boldsymbol{\cdot}}, S) \to \mathrm{Hom}(\mathcal{F}'_{\boldsymbol{\cdot}}, S)$.] $\square$

The nontrivial issue in Theorem 8.34 is related to the fact that the failure of Nakayama's Lemma for general multigradings causes the notion of minimal free resolution to break.

Example 8.35 Take $S = \Bbbk[x, y]$ and set $A = \mathbb{Z}$, with $\deg(x) = 1$ and $\deg(y) = -1$. If $I = \langle xy - 1 \rangle$, which is homogeneous for this multigrading, then I can also be represented as the ideal $\langle x^2y^2 - xy, x^2y^2 - 1 \rangle$. Neither generator can be omitted, even though we know that I is a principal ideal. Therefore it just makes no sense to say that the first multigraded Betti number of S/I is 1 or 2. Nevertheless, the two free resolutions

$$0 \leftarrow S \xleftarrow{\;\left[\, xy-1 \,\right]\;} S \leftarrow 0 \quad \text{and} \quad 0 \leftarrow S \xleftarrow{\;\left[\, x^2y^2-xy \quad x^2y^2-1 \,\right]\;} S^2 \xleftarrow{\;\begin{bmatrix} xy+1 \\ xy \end{bmatrix}\;} S \leftarrow 0$$

yield the same K-polynomial $\mathcal{K}(S/I; \mathbf{t}) = 0$. Note that multiplication by the degree zero element $xy - 1$ induces a degree $\mathbf{0}$ multigraded isomorphism $S \to \langle xy - 1 \rangle$, so $\mathcal{K}(\langle xy - 1 \rangle; t) = \mathcal{K}(S; t) = 1$ should be expected. $\diamond$

One of our reasons for developing the homogenization techniques in Section 8.3 is that they work even for nonpositive multigradings. In particular, we get that K-polynomials are still invariant under Gröbner degeneration.

Theorem 8.36 *Fix a term order on a multigraded free module $\mathcal{F}$. If K is a graded submodule of $\mathcal{F}$ then $\mathcal{K}(\mathcal{F}/K; \mathbf{t}) = \mathcal{K}(\mathcal{F}/\mathrm{in}(K); \mathbf{t})$.*

Proof. Immediate from Proposition 8.28 and Theorem 8.34. $\square$

K-polynomials in nonpositive gradings still retain some of the enumerative data that they express via Hilbert series in Theorem 8.20 for positive gradings. The way out of the obvious problem that arbitrary modules—even finitely generated ones—need not have Hilbert series is to define ourselves out, by restricting our attention to modules that do have Hilbert series. Such modules arise in a number of applications, notably to homogeneous coordinate rings of toric varieties.

Definition 8.37 Let $S = \Bbbk[\mathbf{x}]$ be a polynomial ring multigraded by A. A multigraded module M is **modest** if $\dim_{\Bbbk}(M_{\mathbf{a}})$ is finite for all $\mathbf{a} \in A$.

Although the grading group A can have torsion, the notation $\mathbb{Z}[[A]]$ still makes sense for the additive group of functions $A \to \mathbb{Z}$, which we still call *Laurent series*. Modest modules are precisely those graded modules M that have well-defined Hilbert series, which we again denote by $H(M; \mathbf{t}) \in \mathbb{Z}[[A]]$.

Example 8.38 Let $I = \langle xy, yz \rangle = \langle y \rangle \cap \langle x, z \rangle$, and consider the module $M = \Bbbk[x, y, z]/I$, with the multigrading as in Example 8.2. This module is modest, and the Hilbert function $\mathbb{Z} \oplus \mathbb{Z}/2\mathbb{Z} \to \mathbb{N}$ of M can be represented as a $2 \times \infty$ array of integers. In degrees ranging from $(-4, 0)$ in the lower left to $(4, 1)$ in the upper right, the Hilbert function is

0	0	1	0	0	1	1	2	2
1	0	0	0	1	1	2	2	3

$\leftrightarrow$

			y			x	xz	xz^2	xz^3
								x^3	x^3z
y^2					1	z	x^2	x^2z	$\cdots$
							z^2	z^3	

.

The Hilbert series of M lies in the group $\mathbb{Z}[[\mathbb{Z} \oplus \mathbb{Z}/2\mathbb{Z}]] = \mathbb{Z}[[s, s^{-1}, t]]/\langle t^2 - 1 \rangle$ of Laurent series supported on A. The subgroups $\mathbb{Z}[[s^{-1}]][t]/\langle t^2 - 1 \rangle$ and $\mathbb{Z}[[s]][t]/\langle t^2 - 1 \rangle$ are rings in which all elements with constant coefficients ± 1 are units. The Hilbert series of M can therefore be expressed as the sum

$$H(M; s, t) = \frac{s^{-2}t}{1 - s^{-2}t} + \frac{1}{(1 - st)(1 - s)},$$

where the two ratios—which are viewed as lying in the two subgroups above—sum the positive powers of y and the monomials in $\Bbbk[x, z]$. $\diamond$

The Laurent series $\mathbb{Z}[[A]]$ naturally constitute a module over the group algebra $\mathbb{Z}[A]$, whose elements we still call *Laurent polynomials*. To see the module structure, observe that for a Laurent series $p(\mathbf{t}) \in \mathbb{Z}[[A]]$ and a Laurent polynomial $f(\mathbf{t}) \in \mathbb{Z}[A]$, the coefficient of $\mathbf{t}^{\mathbf{a}}$ in $p(\mathbf{t})f(\mathbf{t})$ depends on only finitely many coefficients of p and f. The Laurent series module $\mathbb{Z}[[A]]$ contains the group algebra $\mathbb{Z}[A]$ as a submodule. This observation allows us to reexamine the concept of "rational function".

Definition 8.39 A Laurent series $p(\mathbf{t}) \in \mathbb{Z}[[A]]$ is **summable** if there is a Laurent polynomial $f(\mathbf{t}) \in \mathbb{Z}[A]$ whose product with p lies in $\mathbb{Z}[A]$. In this case, $f(\mathbf{t})p(\mathbf{t}) = K(\mathbf{t})$ is called the **sum** of p with respect to f, and we write

$$p(\mathbf{t}) \equiv \frac{K(\mathbf{t})}{f(\mathbf{t})}.$$

The ratio $K(\mathbf{t})/f(\mathbf{t})$ is formal and should not be viewed as lying in an ambient structure; indeed, no single ring could contain all such ratios, as $f(\mathbf{t})$ might be a zerodivisor in $\mathbb{Z}[A]$. The congruence symbol "$\equiv$" rather than simple equality "$=$" is meant to indicate that distinct summable Laurent series can sum to the same Laurent polynomial. Thus p is congruent to its sum modulo the series that sum to zero with respect to f. Such zero-sums can even occur when A has no torsion; for example, the Laurent series $\sum_{\mathbf{a}\in\mathbb{Z}^d}\mathbf{t}^\mathbf{a}$ sums to zero with respect to $(1-t_1)\cdots(1-t_d)$. In what follows, we will be particularly interested in the case where $f(\mathbf{t}) = \prod_{i=1}^{n}(1-\mathbf{t}^{\mathbf{a}_i})$.

Lemma 8.40 *Let S/I be a modest monomial quotient. Then*

$$H(S/I; \mathbf{t}) \ \equiv \ \frac{\mathcal{K}(S/I; \mathbf{t})}{\prod_{i=1}^{n}(1 - \mathbf{t}^{\mathbf{a}_i})}.$$

Proof. Let us use variables $\mathbf{s} = s_1, \ldots, s_n$ to write monomials in the additive group $\mathbb{Z}[[\mathbb{Z}^n]]$, to contrast with the formal variable $\mathbf{t}$ for $\mathbb{Z}[[A]]$. Call a series $p(\mathbf{s})$ in $\mathbb{Z}[[\mathbb{Z}^n]]$ *modest* if for each $\mathbf{a} \in A$, only finitely many monomials $\mathbf{s}^\mathbf{u}$ satisfying $\deg(\mathbf{u}) = \mathbf{a}$ have nonzero coefficient in p. Denote the set of modest Laurent series in $\mathbb{Z}[[\mathbb{Z}^n]]$ by $\mathbb{Z}[[\mathbb{Z}^n]]_A$. Modest monomial quotients S/I are precisely those having modest $\mathbb{Z}^n$-graded Hilbert series $H(S/I; \mathbf{s})$.

The modest formal series $\mathbb{Z}[[\mathbb{Z}^n]]_A \subset \mathbb{Z}[[\mathbb{Z}^n]]$ clearly form a group under addition and are closed under multiplication by arbitrary Laurent monomials $\mathbf{s}^\mathbf{u} \in \mathbb{Z}[\mathbb{Z}^n]$, which act as shift operators. Therefore, the modest series $\mathbb{Z}[[\mathbb{Z}^n]]_A \subset \mathbb{Z}[[\mathbb{Z}^n]]$ form a $\mathbb{Z}[\mathbb{Z}^n]$-submodule. On the other hand, the formal series $\mathbb{Z}[[A]]$ also constitute a module over $\mathbb{Z}[\mathbb{Z}^n]$, with the monomial $\mathbf{s}^\mathbf{u} \in \mathbb{Z}[\mathbb{Z}^n]$ acting as multiplication by $\mathbf{t}^{\deg(\mathbf{u})}$. The specialization map

$$\mathbb{Z}[[\mathbb{Z}^n]]_A \ \to \ \mathbb{Z}[[A]] \tag{8.3}$$

sending $\mathbf{s}^\mathbf{u}$ to $\mathbf{t}^{\deg(\mathbf{u})}$ defines a homomorphism of $\mathbb{Z}[\mathbb{Z}^n]$-modules.

Suppose that $p(\mathbf{s}) \in \mathbb{Z}[[\mathbb{Z}^n]]_A$ is summable with respect to $f(\mathbf{s})$. Then its image $\bar{p}(\mathbf{t})$ in $\mathbb{Z}[[A]]$ is summable with respect to the image $\bar{f}(\mathbf{t})$ of f in $\mathbb{Z}[A]$, and its sum $f(\mathbf{s})p(\mathbf{s}) \in \mathbb{Z}[\mathbb{Z}^n]$ maps to the sum $\bar{f}(\mathbf{t})\bar{p}(\mathbf{t}) \in \mathbb{Z}[A]$ of the series $\bar{p}(\mathbf{t})$, because (8.3) is a homomorphism of $\mathbb{Z}[\mathbb{Z}^n]$-modules. On the other hand, the $\mathbb{Z}^n$-graded K-polynomial of S/I maps by definition to the A-graded K-polynomial of S/I under the specialization $\mathbb{Z}[\mathbb{Z}^n] \to \mathbb{Z}[A]$ induced by (8.3). Now apply Theorem 8.20 to the grading by $\mathbb{Z}^n$, which is positive, and use (8.3) to specialize the result to the A-grading. $\square$

Theorem 8.41 *If M is finitely generated and modest, then*

$$H(M; \mathbf{t}) \ \equiv \ \frac{\mathcal{K}(M; \mathbf{t})}{\prod_{i=1}^{n}(1 - \mathbf{t}^{\mathbf{a}_i})}.$$

Proof. Express M as a quotient $\mathcal{F}/N$ of a multigraded free module. The Hilbert series of the modest quotient $\mathcal{F}/N$ equals that of $\mathcal{F}/\mathrm{in}(N)$ under any term order, because the standard monomials for the initial submodule in(N) form a vector space basis for both $\mathcal{F}/N$ and $\mathcal{F}/\text{in}(N)$. Since $\mathcal{F}/\text{in}(N)$ is a direct sum of multigraded translates of monomial quotients of S, and K-polynomials as well as Hilbert series are additive on direct sums, it suffices by Lemma 8.40 to know that K-polynomials are preserved under passing to quotients by initial submodules. This last bit is Theorem 8.36. $\square$

Example 8.42 Putting the Hilbert series in Example 8.38 over a common denominator yields the $\mathbb{Z} \oplus \mathbb{Z}/2\mathbb{Z}$-graded K-polynomial of M, namely
$$\mathcal{K}(M; s, t) = s^{-2}t(1 - st)(1 - s) + 1 - s^{-2}t = 1 - s^{-1}t - s^{-1}t^2 + t^2. \qquad \diamond$$

Two open problems concerning Hilbert series and K-polynomials are

(i) how to represent the kernel of the homomorphism of abelian groups {Hilbert series of modest modules} $\to$ {K-polynomials}; and

(ii) how to write down Hilbert series for immodest modules.

The map {Hilbert series} $\to$ {K-polynomials} in (i) is never injective when the grading is nonpositive: there can be many modest modules, with very different-looking Hilbert series, that nonetheless have equal K-polynomials. Such ambiguity does not occur in the positively graded case by Theorem 8.20, because the rational functions that represent positively graded Hilbert series lie in an ambient power series ring that is an integral domain.

8.5 Multidegrees

We saw in Part I that free resolutions in the finest possible multigrading are essentially combinatorial in nature. For coarser gradings, in contrast, combinatorial data can usually be extracted only after a certain amount of condensation. Although K-polynomials can sometimes suffice for this purpose, even they might end up carrying an overload of information. In such cases we prefer a multigraded generalization of the degree of a $\mathbb{Z}$-graded ideal. The characterizing properties of these *multidegrees* give them enormous potential to encapsulate finely textured combinatorics, as we will see in the cases of Schubert and quiver polynomials in Chapters 15–17.

The ordinary $\mathbb{Z}$-graded degree of a module is usually defined via the leading coefficient of its Hilbert polynomial (see Exercise 8.14). However, as Hilbert polynomials do not directly extend to multigraded situations, we must instead rely on a different characterization. In the next definition, the symbol $\text{mult}_\mathfrak{p}(M)$ denotes the *multiplicity* of a module M at the prime $\mathfrak{p}$, which by definition equals the length of the largest finite-length submodule in the localization of M at $\mathfrak{p}$ [Eis95, Section 3.6].

Definition 8.43 Let S be a polynomial ring multigraded by a subgroup $A \subseteq \mathbb{Z}^d$ (so in particular, A is torsion-free). Let $\mathcal{C} : M \mapsto \mathcal{C}(M; \mathbf{t})$ be a function from finitely generated graded S-modules to $\mathbb{Z}[t_1, \ldots, t_d]$.

1. The function C is **additive** if for all modules M,

$$C(M; \mathbf{t}) \;=\; \sum_{k=1}^{r} \mathrm{mult}_{\mathfrak{p}_k}(M) \cdot C(S/\mathfrak{p}_k; \mathbf{t}),$$

where $\mathfrak{p}_1, \ldots, \mathfrak{p}_r$ are the maximal dimensional associated primes of M.

2. The function C is **degenerative** if whenever $M = \mathcal{F}/K$ is a graded free presentation and $\mathrm{in}(M) := \mathcal{F}/\mathrm{in}(K)$ for some term or weight order,

$$C(M; \mathbf{t}) \;=\; C(\mathrm{in}(M); \mathbf{t}).$$

The definition of additivity implicitly uses the fact (Proposition 8.11) that $\mathfrak{p}_1, \ldots, \mathfrak{p}_r$ are themselves multigraded, given that M is. This subtlety is why we restrict to torsion-free gradings when dealing with multidegrees.

Readers who have previously seen the notion of $\mathbb{Z}$-graded degree will recognize it as being both additive and degenerative. The main goal of this section is to produce a multigraded analogue of degree that also satisfies these properties. The very fact that $\mathbb{Z}$-graded degrees already satisfy them means that we will need extra information to distinguish multidegrees from the usual $\mathbb{Z}$-graded degree. To that end, we have the following uniqueness statement, our main result of the section.

Theorem 8.44 *Exactly one additive degenerative function C satisfies*

$$C(S/\langle x_{i_1}, \ldots, x_{i_r}\rangle; \mathbf{t}) \;=\; \langle \mathbf{a}_{i_1}, \mathbf{t}\rangle \cdots \langle \mathbf{a}_{i_r}, \mathbf{t}\rangle$$

for all prime monomial ideals $\langle x_{i_1}, \ldots, x_{i_r}\rangle$, *where* $\langle \mathbf{a}, \mathbf{t}\rangle = a_1 t_1 + \cdots + a_d t_d$.

Proof. If M is a finitely generated graded module, then let $M \cong \mathcal{F}/K$ be a graded presentation, and pick a term order on $\mathcal{F}$. Set $M' = \mathcal{F}/\mathrm{in}(K)$, so $C(M; \mathbf{t}) = C(M'; \mathbf{t})$ because C is degenerative. Note that M' is a direct sum of A-graded translates of monomial quotients of S. Since all of the associated primes of M' are therefore monomial primes, the values of C on the prime monomial quotients of S determine $C(M'; \mathbf{t})$ by additivity. This proves uniqueness. Existence will follow from Corollary 8.47, Proposition 8.49, and Theorem 8.53. ☐

From the perspective of uniqueness, we could as well have put *any* function of $\mathbf{t}$ on the right-hand side of Theorem 8.44. However, it takes very special such right-hand sides to guarantee that a function C actually exists. Indeed, C is severely overdetermined, because most submodules $K \subseteq \mathcal{F}$ have many distinct initial submodules $\mathrm{in}(K)$, as the term order varies. This brings us to our main definition: to prove the existence of a function C satisfying Theorem 8.44, we shall explicitly construct one from K-polynomials.

We work exclusively with torsion-free gradings, and therefore we assume $A \subseteq \mathbb{Z}^d$. The symbol $\mathbf{t^a}$ thus stands for an honest Laurent monomial

$t_1^{a_1} \cdots t_d^{a_d}$, allowing us to substitute $1 - t_j$ for every occurrence of t_j. Doing so yields a rational function $(1 - t_1)^{a_1} \cdots (1 - t_d)^{a_d}$, which can be expanded (even if some of the a_j are less than zero) as a well-defined power series

$$\prod_{j=1}^{d}(1 - t_j)^{a_j} \;=\; \prod_{j=1}^{d}\left(1 - a_j t_j + \frac{a_j(a_j - 1)}{2}t_j^2 - \cdots\right)$$

in $\mathbb{Z}[[t_1, \ldots, t_d]]$. Doing the same for each monomial in an arbitrary Laurent polynomial $K(\mathbf{t})$ results in a power series $K[[\mathbf{1} - \mathbf{t}]]$. If $K(\mathbf{t})$ is a polynomial, so $\mathbf{a} \in \mathbb{N}^d$ whenever $\mathbf{t}^{\mathbf{a}}$ appears in $K(\mathbf{t})$, then $K[[\mathbf{1} - \mathbf{t}]]$ is a polynomial.

Definition 8.45 The **multidegree** of a $\mathbb{Z}^d$-graded S-module M is the sum $\mathcal{C}(M; \mathbf{t}) \in \mathbb{Z}[t_1, \ldots, t_d]$ of all terms in $\mathcal{K}[[M; \mathbf{1} - \mathbf{t}]]$ having total degree $\mathrm{codim}(M) = n - \dim(M)$. When $M = S/I$ is the coordinate ring of a subvariety $X \subseteq \mathbb{k}^n$, we may also write $[X]_A$ or $\mathcal{C}(X; \mathbf{t})$ to mean $\mathcal{C}(M; \mathbf{t})$.

Example 8.46 Let $S = \mathbb{k}[a, b, c, d]$ be multigraded by $\mathbb{Z}^2$, with

$$\deg(a) = (2, -1), \quad \deg(b) = (1, 0), \quad \deg(c) = (0, 1), \quad \text{and} \quad \deg(d) = (-1, 2).$$

If M is the module $S/\langle b^2, bc, c^2 \rangle$, then

$$\begin{aligned}
\mathcal{K}(M; \mathbf{t}) &= 1 - \mathbf{t}^{\deg(b^2)} - \mathbf{t}^{\deg(bc)} - \mathbf{t}^{\deg(c^2)} + \mathbf{t}^{\deg(b^2 c)} + \mathbf{t}^{\deg(bc^2)} \\
&= 1 - t_1^2 - t_1 t_2 - t_2^2 + t_1^2 t_2 + t_1 t_2^2
\end{aligned}$$

because of the Scarf resolution. Gathering $-t_1^2 + t_1^2 t_2 = -t_1^2(1 - t_2)$, we get

$$\begin{aligned}
\mathcal{K}[[M; \mathbf{1} - \mathbf{t}]] &= 1 - (1 - t_1)^2 t_2 - t_1(1 - t_2)^2 - (1 - t_1)(1 - t_2) \\
&= 3 t_1 t_2 - t_1^2 t_2 - t_1 t_2^2,
\end{aligned}$$

so

$$\mathcal{C}(M; \mathbf{t}) = 3 t_1 t_2$$

is the sum of degree $2 = \mathrm{codim}(M)$ terms in $\mathcal{K}[[M; \mathbf{1} - \mathbf{t}]]$.   ◇

Immediately from Definition 8.45 and the invariance of K-polynomials under Gröbner degeneration in Theorem 8.36, we get the analogous invariance of multidegrees under Gröbner degeneration.

Corollary 8.47 *The multidegree function $M \mapsto \mathcal{C}(M; \mathbf{t})$ is degenerative.*

Next let us verify that the multidegrees of quotients by monomial primes satisfy the formula in Theorem 8.44. For this we need a lemma.

Lemma 8.48 *Let $\mathbf{b} \in \mathbb{Z}^d$. If $K(\mathbf{t}) = 1 - \mathbf{t}^{\mathbf{b}} = 1 - t_1^{b_1} \cdots t_d^{b_d}$, then substituting $1 - t_j$ for each occurrence of t_j yields $K[[\mathbf{1} - \mathbf{t}]] = b_1 t_1 + \cdots + b_d t_d + O(\mathbf{t}^2)$, where $O(\mathbf{t}^2)$ denotes a sum of terms each of which has total degree at least 2.*

Proof. $K[[1 - t]] = 1 - \prod_{j=1}^{d}(1 - t_j)^{b_j} = 1 - \prod_{j=1}^{d}\left(1 - b_j t_j + O(t_j^2)\right)$,
and this equals $1 - \left(1 - \sum_{j=1}^{d}(b_j t_j) + O(t^2)\right) = \left(\sum_{j=1}^{d} b_j t_j\right) + O(t^2)$. $\square$

The linear form $b_1 t_1 + \cdots + b_d t_d$ in Lemma 8.48 can also be expressed as the inner product $\langle b, t \rangle$ of the vector b with the vector $t = (t_1, \ldots, t_d)$. It can be useful to think of this as the logarithm of the Laurent monomial t^b.

Proposition 8.49 *K-polynomials of prime monomial quotients satisfy*

$$\mathcal{K}[[S/\langle x_{i_1}, \ldots, x_{i_r}\rangle; 1 - t]] = \left(\prod_{\ell=1}^{r}\langle a_{i_\ell}, t\rangle\right) + O(t^{r+1}),$$

where $O(t^{r+1})$ is a sum of forms each of which has total degree at least $r + 1$. In particular, the multidegree of a prime monomial quotient of S is

$$\mathcal{C}(S/\langle x_{i_1}, \ldots, x_{i_r}\rangle; t) = \langle a_{i_1}, t\rangle \cdots \langle a_{i_r}, t\rangle.$$

Proof. Using the Koszul complex, the K-polynomial of the quotient module $M = S/\langle x_{i_1}, \ldots, x_{i_r}\rangle$ is computed to be $\mathcal{K}(M; t) = (1 - t^{a_{i_1}}) \cdots (1 - t^{a_{i_r}})$. Now apply Lemma 8.48 to each of the r factors in this product. $\square$

Example 8.50 Consider the ring $S = \Bbbk[a, b, c, d]$ multigraded by $\mathbb{Z}^2$ with

$$\deg(a) = (3, 0), \; \deg(b) = (2, 1), \; \deg(c) = (1, 2), \; \text{and } \deg(d) = (0, 3).$$

Then, using variables $s = s_1, s_2$, we have multidegrees

$$\begin{array}{rcccl}
\mathcal{C}(S/\langle a, b\rangle, s) & = & (3s_1)(2s_1 + s_2) & = & 6s_1^2 + 3s_1 s_2, \\
\mathcal{C}(S/\langle a, d\rangle, s) & = & (3s_1)(3s_2) & = & 9s_1 s_2, \\
\mathcal{C}(S/\langle c, d\rangle, s) & = & (s_1 + 2s_2)(3s_2) & = & 3s_1 s_2 + 6s_2^2.
\end{array}$$

Note that these multidegrees all lie inside the ring

$$\mathbb{Z}[3s_1, s_1 + 2s_2, 2s_1 + s_2, 3s_2] = \mathbb{Z}[s_1 + 2s_2, 2s_1 + s_2]$$

and not just $\mathbb{Z}[s_1, s_2]$, since the group A is the proper subgroup of $\mathbb{Z}s_1 \oplus \mathbb{Z}s_2$ generated by $s_1 + 2s_2$ and $2s_1 + s_2$. Nonetheless, the definition of multidegree via $\mathcal{K}[[M; 1 - s]]$ still works verbatim. $\Diamond$

Remark 8.51 Warning: Proposition 8.49 does not state that the product $\langle a_{i_1}, t\rangle \cdots \langle a_{i_r}, t\rangle$ is nonzero. Indeed, it can very easily be zero, if $a_{i_\ell} = 0$ for some ℓ. However, this is the only way to get zero, as the product of linear forms takes place in a polynomial ring over $\mathbb{Z}$, which has no zerodivisors.

Proposition 8.49 implies that multidegrees of quotients of S by monomial prime ideals are insensitive to multigraded translation.

Corollary 8.52 *If $M = S/\langle x_{i_1}, \ldots, x_{i_r} \rangle$ then $\mathcal{C}(M(-\mathbf{b}); \mathbf{t}) = \mathcal{C}(M; \mathbf{t})$.*

Proof. Shifting by $\mathbf{b}$ multiplies the K-polynomial by $\mathbf{t}^{\mathbf{b}}$, so $\mathcal{K}(M(-\mathbf{b}); \mathbf{t}) = \mathbf{t}^{\mathbf{b}} \mathcal{K}(M; \mathbf{t})$. The degree r form in $\mathcal{K}[[M(-\mathbf{b}); \mathbf{1} - \mathbf{t}]]$ is the product of the lowest degree forms in $\mathcal{K}[[M; \mathbf{1} - \mathbf{t}]]$ and $(\mathbf{1} - \mathbf{t})^{\mathbf{b}}$, the latter of which is 1. $\square$

In view of Corollary 8.47 and Proposition 8.49, our final result of the chapter completes the characterization of multidegrees in Theorem 8.44.

Theorem 8.53 *The multidegree function $\mathcal{C} : M \mapsto \mathcal{C}(M; \mathbf{t})$ is additive.*

Proof. Let $M = M_\ell \supset M_{\ell-1} \supset \cdots \supset M_1 \supset M_0 = 0$ be a filtration of M in which $M_j/M_{j-1} \cong (S/\mathfrak{p}_j)(-\mathbf{b}_j)$ for multigraded primes $\mathfrak{p}_j$ and vectors $\mathbf{b}_j \in A$. Such a filtration exists because we can choose a homogeneous associated prime $\mathfrak{p}_1 = \mathrm{ann}(m_1)$ by Proposition 8.11, set $M_1 = \langle m_1 \rangle \cong (S/\mathfrak{p}_1)(-\deg(m_1))$, and then continue by Noetherian induction on M/M_1.

The quotients $S/\mathfrak{p}_j$ all have dimension at most $\dim(M)$, and if $S/\mathfrak{p}$ has dimension exactly $\dim(M)$, then $\mathfrak{p} = \mathfrak{p}_j$ for exactly $\mathrm{mult}_\mathfrak{p}(M)$ values of j (localize the filtration at $\mathfrak{p}$ to see this). Also, additivity of K-polynomials on short exact sequences implies that $\mathcal{K}(M; \mathbf{t}) = \sum_{j=1}^\ell \mathcal{K}(M_j/M_{j-1}; \mathbf{t})$.

Assume for the moment that M is a direct sum of multigraded translates of quotients of S by monomial ideals. Then all the primes $\mathfrak{p}_j$ are monomial primes. Therefore the only power series $\mathcal{K}[[M_j/M_{j-1}; \mathbf{1} - \mathbf{t}]]$ contributing terms of degree $\mathrm{codim}(M)$ to $\mathcal{K}[[M; \mathbf{1} - \mathbf{t}]]$ are those for which $M_j/M_{j-1} \cong (S/\mathfrak{p}_j)(-\mathbf{b}_j)$ has maximal dimension, by Proposition 8.49 and Corollary 8.52. Thus the theorem holds for such M.

Before continuing with the case of general modules M, let us generalize Proposition 8.49, and hence Corollary 8.52, to arbitrary modules.

Claim 8.54 *If M has codimension r, then $\mathcal{K}[[M; \mathbf{1} - \mathbf{t}]] = \mathcal{C}(M; \mathbf{t}) + O(\mathbf{t}^{r+1})$. In particular, $\mathcal{C}(M(-\mathbf{b}); \mathbf{t}) = \mathcal{C}(M; \mathbf{t})$ for arbitrary modules M.*

Proof. We have just finished showing that the first statement holds for direct sums of multigraded shifts of monomial quotients of S. By Corollary 8.47, every module $M \cong \mathcal{F}/K$ of codimension r has the same multidegree as such a direct sum, namely $\mathcal{F}/\mathrm{in}(K)$, whose codimension is also r. The second statement follows as in the proof of Corollary 8.52. $\square$

Now the argument before Claim 8.54 works for arbitrary modules M and primes $\mathfrak{p}_j$, using the Claim in place of Proposition 8.49 and Corollary 8.52. $\square$

Example 8.55 Let $S = \Bbbk[a, b, c, d]$ and let $I = \langle b^2 - ac, bc - ad, c^2 - bd \rangle$ be the *twisted cubic* ideal. Then I has initial ideal $\mathrm{in}(I) = \langle b^2, bc, c^2 \rangle$ under the reverse lexicographic term order with $a > b > c > d$. Since $\mathrm{in}(I)$ is supported on $\langle b, c \rangle$ with multiplicity 3, the multidegree of S/I under the $\mathbb{Z}^2$-grading from Example 8.46 is $\mathcal{C}(S/I; t_1, t_2) = 3\langle \deg(b), \mathbf{t} \rangle \langle \deg(c), \mathbf{t} \rangle = 3t_1 t_2$. This agrees with the multidegree in Example 8.46, as it should by

Theorem 8.53. It also equals the multidegree $\mathcal{C}(S/I; \mathbf{t})$ of the twisted cubic, by Corollary 8.47.

On the other hand, the twisted cubic ideal I has initial ideal $\operatorname{in}(I) = \langle ac, ad, bd \rangle = \langle a, b \rangle \cap \langle a, d \rangle \cap \langle c, d \rangle$ under the lexicographic term order with $a > b > c > d$. Using the multigrading and notation from Example 8.50, additivity in Theorem 8.53 implies that

$$
\begin{aligned}
\mathcal{C}(S/\langle ac, ad, bd \rangle; \mathbf{s}) &= \mathcal{C}(S/\langle a, b \rangle; \mathbf{s}) + \mathcal{C}(S/\langle a, d \rangle; \mathbf{s}) + \mathcal{C}(S/\langle c, d \rangle; \mathbf{s}) \\
&= 6s_1^2 + 15s_1 s_2 + 6s_2^2.
\end{aligned}
$$

We conclude by Corollary 8.47 that $\mathcal{C}(S/I; \mathbf{s}) = 6s_1^2 + 15s_1 s_2 + 6s_2^2$. This multidegree also equals $3\langle \deg(b), \mathbf{s} \rangle \langle \deg(c), \mathbf{s} \rangle = 3(s_1 + 2s_2)(2s_1 + s_2)$. $\diamond$

Exercises

8.1 Prove that $S_{\mathbf{a}}$ is generated as a module over S_0 by any set of monomials that generates the ideal $\langle S_{\mathbf{a}} \rangle$ inside of S.

8.2 Let Q be a pointed affine semigroup in $A \cong \mathbb{Z}^d$, and let $A' \subseteq A$ be the subgroup generated by Q. Write $\mathbb{Z}[[Q]][A] = \mathbb{Z}[[Q]] \otimes_{\mathbb{Z}[Q]} \mathbb{Z}[A]$. Note that when $A' = A$, the ring $\mathbb{Z}[[Q]][A]$ equals the localization $\mathbb{Z}[[Q]][\mathbf{t}^{-\mathbf{a}_1}, \ldots, \mathbf{t}^{-\mathbf{a}_n}]$.

(a) Show that if $A' = A$, every element in $\mathbb{Z}[[Q]][A]$ can be represented uniquely by a series $\sum_{\mathbf{a} \in A} c_{\mathbf{a}} \mathbf{t}^{\mathbf{a}}$ supported on a union of finitely many translates of Q.
(b) In the situation of part (a), prove that every series supported on a union of finitely many translates of Q lies in $\mathbb{Z}[[Q]][A]$.
(c) Use Exercise 7.10 to verify parts (a) and (b) when A' does not equal A.

8.3 Consider the twisted cubic ideal I in Example 8.55, and let $w = (0, 1, 3, 2)$.

(a) Prove that the homogenization of I with respect to the weight vector w is the ideal $\tilde{I} = \langle ac - b^2 y, bc - ady^2, c^2 - bdy^3, b^3 - a^2 dy \rangle$ in $S[y]$.
(b) Compute a minimal free resolution of $\tilde{I}$ graded by $\mathbb{Z}^2 \times \mathbb{Z}$, where the multi-grading of $\mathbb{k}[a, b, c, d]$ by $\mathbb{Z}^2$ is as in either Example 8.46 or Example 8.50.
(c) Verify Proposition 8.28, Theorem 8.29, and Corollary 8.31 in this case by plugging $y = 0$ and $y = 1$ into matrices for the maps in the resolution from (b) and exhibiting the consecutive pairs as described in Remark 8.30.

8.4 Express Exercise 8.3(a) as an instance of Exercise 7.18. More generally, express Exercise 7.18 as an instance of Proposition 8.26.

8.5 Let $S = \mathbb{k}[\mathbf{x}]$ for $\mathbf{x} = \{x_{ij} \mid i, j = 1, \ldots, 4\}$. With $|\cdot| = \det(\cdot)$, set

$$
I = \left\langle \begin{vmatrix} x_{11} & x_{12} \\ x_{21} & x_{22} \end{vmatrix}, \begin{vmatrix} x_{11} & x_{12} \\ x_{31} & x_{32} \end{vmatrix}, \begin{vmatrix} x_{21} & x_{22} \\ x_{31} & x_{32} \end{vmatrix}, \begin{vmatrix} x_{11} & x_{13} & x_{14} \\ x_{21} & x_{23} & x_{24} \\ x_{31} & x_{33} & x_{34} \end{vmatrix}, \begin{vmatrix} x_{12} & x_{13} & x_{14} \\ x_{22} & x_{23} & x_{24} \\ x_{32} & x_{33} & x_{34} \end{vmatrix} \right\rangle.
$$

Compute the K-polynomials and multidegrees of the quotient S/I in the multi-gradings by $\mathbb{Z}^4$ in which (i) $\deg(x_{ij}) = t_i$ and (ii) $\deg(x_{ij}) = s_j$.

8.6 Make arbitrarily long lists of polynomials generating the ideal $\langle xy - 1 \rangle \subset \mathbb{k}[x, y]$, none of which can be left off. (Example 8.35 has lists of length 1 and 2.) Confirm that the corresponding free resolutions all give the same K-polynomial.

8.7 Write down formulas for the K-polynomial and multidegree of the quotient of S by an irreducible monomial ideal (i.e. generated by powers of variables).

8.8 Show that arbitrary multidegrees are nonnegative, in the following sense: the multidegree of any module of dimension $n - r$ over S is a nonnegative sum of "squarefree" homogeneous forms $\langle \mathbf{a}_{i_1}, \mathbf{t} \rangle \cdots \langle \mathbf{a}_{i_r}, \mathbf{t} \rangle$ of degree r with $i_1 < \cdots < i_r$.

8.9 If the linear forms $\langle \mathbf{a}_1, \mathbf{t} \rangle, \ldots, \langle \mathbf{a}_n, \mathbf{t} \rangle$ are nonzero and generate a pointed affine semigroup in $\mathbb{Z}^d$, deduce that no product $\langle \mathbf{a}_{i_1}, \mathbf{t} \rangle \cdots \langle \mathbf{a}_{i_r}, \mathbf{t} \rangle$ for $i_1 < \cdots < i_r$ is zero and that all of these forms together generate a pointed semigroup in $\mathbb{Z}^R$, where $R = \binom{n+r-1}{n}$ is the number of monomials of degree r in n variables.

8.10 Prove that the positivity in Exercise 8.8 holds for positive multigradings in the stronger sense that any nonempty nonnegative sum of forms $\langle \mathbf{a}_{i_1}, \mathbf{t} \rangle \cdots \langle \mathbf{a}_{i_r}, \mathbf{t} \rangle$ is nonzero. Conclude that $\mathcal{C}(M; \mathbf{t}) \neq 0$ if $M \neq 0$ is positively graded.

8.11 Let M and M' be two $\mathbb{Z}^n$-graded S-modules. Give examples demonstrating that the product of the multidegrees of M and M' need not be expressible as the multidegree of a $\mathbb{Z}^n$-graded module. Can you find sufficient conditions on M and M' to guarantee that $\mathcal{C}(M; \mathbf{t})\mathcal{C}(M'; \mathbf{t}) = \mathcal{C}(M \otimes_S M'; \mathbf{t})$?

8.12 Let M be a multigraded module and $z \in S$ a homogeneous nonzerodivisor on M of degree $\mathbf{b}$. Prove that

(a) $\mathcal{K}(M/zM; \mathbf{t}) = (1 - \mathbf{t}^{\mathbf{b}})\mathcal{K}(M; \mathbf{t})$, and

(b) $\mathcal{C}(M/zM; \mathbf{t}) = \langle \mathbf{b}, \mathbf{t} \rangle \mathcal{C}(M; \mathbf{t})$.

8.13 Let $I \subseteq S$ be multidgraded for a positive $\mathbb{Z}^d$-grading. Suppose that J is a $\mathbb{Z}^d$-graded radical ideal contained inside I and that J is equidimensional (also known as *pure*: all of its associated primes have the same dimension). If S/I and S/J have equal multidegrees, deduce that $I = J$. Hint: Use Exercise 8.10.

8.14 Let S be $\mathbb{Z}$-graded in the standard way, with $\deg(x_i) = 1$ for $i = 1, \ldots, n$. The usual $\mathbb{Z}$-graded degree $e(M)$ of a graded module M is usually defined as $(r - 1)!$ times the leading coefficient of the Hilbert polynomial of M, where $r = \dim(M)$. Prove that the multidegree of M is $e(M)t^{n-r}$.

8.15 Suppose S is multigraded by A, with $\deg(x_i) = \mathbf{a}_i \in A$, and suppose $A \to A'$ is a homomorphism of abelian groups sending $\mathbf{a}_i$ to $\mathbf{a}_i' \in A'$. Prove the following:

(a) The homomorphism $A \to A'$ induces a new multigrading $\deg'$ on S, in which $\deg'(x_i) = \mathbf{a}_i'$.

(b) If a module M is multigraded by A, then M is also multigraded by A'.

(c) If $\mathcal{K}(M; \mathbf{t})$ and $\mathcal{K}(M; \mathbf{s})$ are the K-polynomials of M under the multigradings by A and A', then $\mathcal{K}(M; \mathbf{t})$ maps to $\mathcal{K}(M; \mathbf{s})$ under the homomorphism $\mathbb{Z}[A] \to \mathbb{Z}[A']$ of group algebras. In particular, this sends $\mathbf{t}^{\mathbf{a}_i}$ to $\mathbf{s}^{\mathbf{a}_i'}$.

(d) If $\mathcal{C}(M; \mathbf{t})$ and $\mathcal{C}(M; \mathbf{s})$ are the multidegrees of M under the multigradings by A and A', then $\mathcal{C}(M; \mathbf{t})$ maps to $\mathcal{C}(M; \mathbf{s})$ under the homomorphism $\mathbb{Z}[\mathbf{t}] \to \mathbb{Z}[\mathbf{s}]$ of polynomial rings, which sends $\langle \mathbf{a}_i, \mathbf{t} \rangle$ to $\langle \mathbf{a}_i', \mathbf{s} \rangle$.

8.16 Verify functoriality of K-polynomials and multidegrees for the twisted cubic $\Bbbk[a, b, c, d]/\langle b^2 - ac, ad - bc, c^2 - bd \rangle$ under the two multigradings in Examples 8.46, 8.50, and 8.55. The morphism of gradings sends $t_1 \mapsto 2s_1 - s_2$ and $t_2 \mapsto 2s_2 - s_1$.

8.17 Prove that an ideal I inside an a priori ungraded polynomial ring $\Bbbk[\mathbf{x}]$ is homogeneous for a weight vector $w \in \mathbb{Z}^n$ if and only if some (and hence every) reduced Gröbner basis for I is homogeneous for w. Conclude that there is a unique finest I-**universal grading** on $\Bbbk[\mathbf{x}]$ in which the ideal I is homogeneous.

8.18 For any polynomial $g \in \Bbbk[\mathbf{x}]$, let $\log(g)$ be the set of exponent vectors on monomials having nonzero coefficient in g. Suppose that $\mathcal{G}$ is the reduced Gröbner basis of I for some term order. If L is the sublattice of $\mathbb{Z}^n$ generated by the sets $\log(g) - \log(\mathrm{in}(g))$ for $g \in \mathcal{G}$, show that the $\mathbb{Z}^n/L$-grading on $\Bbbk[\mathbf{x}]$ is universal for I.

8.19 Prove that if $L \subseteq \mathbb{Z}^n$ is a sublattice, then the universal grading for the lattice ideal I_L is the multigrading by $\mathbb{Z}^n/L$.

8.20 Define the **universal K-polynomial** and **universal multidegree** of the quotient $\Bbbk[\mathbf{x}]/I$ to be its K-polynomial and multidegree in the I-universal grading. Compute the universal K-polynomial and multidegree of S/I from Exercise 8.5.

Notes

Geometrically, a multigrading on a polynomial ring comes from the action of an algebraic torus times a finite abelian group. The importance of this point of view has surged in recent years due to its connections with toric varieties (see Chapter 10). Multigraded $\Bbbk[x_1, \ldots, x_n]$-modules correspond to *torus-equivariant sheaves* on the vector space $\Bbbk^n$. The K-polynomial of a module is precisely the class represented by the corresponding sheaf in the *equivariant K-theory* of $\Bbbk^n$; this is the content of Theorem 8.34. The degenerative property of K-polynomials in Theorem 8.36 is an instance of the constancy of K-theory classes in flat families. See [BG05] for more on K-theory in the toric context.

The increase of Betti numbers in Theorem 8.29 can be interpreted in terms of associated graded modules for filtrations [Vas98, Section B.2], or as an instance of a more general upper-semicontinuity for flat families [Har77, Theorem III.12.8].

The notion of multidegree, essentially in the form of Definition 8.45, seems to be due to Borho and Brylinski [BB82, BB85] as well as to Joseph [Jos84]. The *equivariant multiplicities* used by Rossmann [Ros89] in complex-analytic contexts are equivalent. Multidegrees are called *T-equivariant Hilbert polynomials* in [CG97, Section 6.6], where they are proved to be additive as well as homogeneous of degree equal to the codimension (the name is confusing when compared with usual Hilbert polynomials). Elementary proofs of these facts appear also in [BB82]. Multidegrees are algebraic reformulations of the geometric *torus-equivariant Chow classes* (or *equivariant cohomology classes* when $\Bbbk = \mathbb{C}$) of varieties in $\Bbbk^n$ [Tot99, EG98]; this is proved in [KMS04, Proposition 1.19]. The transition from K-polynomials to multidegrees is a manifestation of the Grothendieck–Riemann–Roch Theorem.

Exercises 8.8–8.10 come from [KnM04b, Section 1.7], where the positively multigraded case of the characterization in Theorem 8.44 (that is, including the degenerative property) was noted. Exercise 8.13 appears in [Mar03, Section 12] and [KnM04b, Lemma 1.7.5]. The t-multidegree in Exercise 8.5 is a Schur function, since I is a Grassmannian Schubert determinantal ideal (Exercises 15.2 and 16.9).

Chapter 9

Syzygies of lattice ideals

The Hilbert series of a pointed affine semigroup $Q = \mathbb{N}\{\mathbf{a}_1, \ldots, \mathbf{a}_n\}$ inside the group $A = \mathbb{Z}^d$ is defined as the formal sum of all monomials $\mathbf{t}^{\mathbf{b}} = t_1^{b_1} \cdots t_d^{b_d}$, where $\mathbf{b}$ runs over the vectors in Q. The Hilbert series of Q is a rational generating function of the form

$$\sum_{\mathbf{b} \in Q} \mathbf{t}^{\mathbf{b}} = \frac{\mathcal{K}_Q(t_1, \ldots, t_d)}{(1 - \mathbf{t}^{\mathbf{a}_1})(1 - \mathbf{t}^{\mathbf{a}_2}) \cdots (1 - \mathbf{t}^{\mathbf{a}_n})}, \tag{9.1}$$

where $\mathcal{K}_Q(t_1, \ldots, t_d)$ is a polynomial with integer coefficients. In fact, writing $\Bbbk[Q]$ as the quotient S/I_L of a polynomial ring $S = \Bbbk[x_1, \ldots, x_n]$ multigraded by A, the polynomial $\mathcal{K}_Q$ is the K-polynomial of S/I_L.

We have seen in Proposition 8.23 that $\mathcal{K}_Q$ records the alternating sum of the multigraded Betti numbers of S/I_L, occurring in a free resolution over S. In this chapter we study the minimal free resolution of the lattice ideal I_L and resulting formulas for the polynomial $\mathcal{K}_Q$. Our main goal is to give a geometric construction for the generators and all higher syzygies of I_L. Generalizing our results for monomial ideals in Part I of this book, we introduce the Scarf complex and the hull complex of a lattice ideal, and we interpret $\mathcal{K}_Q$ as the graded Euler characteristic of the hull complex.

9.1 Betti numbers

Throughout this chapter we fix a pointed affine semigroup Q generating the group $A = \mathbb{Z}^d$, and we assume that $\mathbf{a}_1, \ldots, \mathbf{a}_n$ are the unique minimal generators of Q. The polynomial ring $S = \Bbbk[x_1, \ldots, x_n]$ is multigraded by A via $\deg(x_i) = \mathbf{a}_i$, and this grading of S is positive by Theorem 8.6.6. Hence every finitely generated graded S-module has a minimal generating set and a minimal free resolution. In addition, its Hilbert function is faithfully represented by either its Hilbert series or its K-polynomial (Theorem 8.20).

Recall that the A-graded translate $S(-\mathbf{b})$ is the free S-module with one generator in degree $\mathbf{b} \in Q$. Equivalently, $S(-\mathbf{b})$ is isomorphic to the principal ideal $\langle \mathbf{x}^{\mathbf{u}} \rangle$, where $\mathbf{x}^{\mathbf{u}}$ is any monomial of degree $\mathbf{b}$. With this notation, the minimal free resolution of I_L as an S-module looks like

$$0 \leftarrow I_L \leftarrow \bigoplus_{\mathbf{b} \in Q} S(-\mathbf{b})^{\beta_{0,\mathbf{b}}} \leftarrow \bigoplus_{\mathbf{b} \in Q} S(-\mathbf{b})^{\beta_{1,\mathbf{b}}} \leftarrow \cdots \leftarrow \bigoplus_{\mathbf{b} \in Q} S(-\mathbf{b})^{\beta_{r,\mathbf{b}}} \leftarrow 0.$$

Here, $r \leq n - 1$ because S/I_L has projective dimension at most n and all the Betti numbers $\beta_{i,\mathbf{b}}$ are simultaneously minimized. The latter condition is equivalent to requiring that no nonzero scalars appear in the matrices representing the differentials. (Monomial matrix notation does not lend enough advantage for gradings as coarse as the A-grading on S to warrant its use in this context.) The Betti number $\beta_{i,\mathbf{b}} = \beta_{i,\mathbf{b}}(I_L)$ is the number of minimal i^{th} syzygies of the lattice ideal I_L in degree $\mathbf{b}$. In particular, the number of minimal generators of I_L in degree $\mathbf{b}$ equals $\beta_{0,\mathbf{b}}$.

Example 9.1 (Syzygies of the twisted cubic curve) If Q is the subsemigroup of $\mathbb{Z}^2$ generated by $\{(1,0),(1,1),(1,2),(1,3)\}$, then

$$\Bbbk[Q] \;=\; \Bbbk[s, st, st^2, st^3] \;=\; \Bbbk[a,b,c,d]/I_L$$

is the coordinate ring of the twisted cubic curve in projective 3-space. The ideal I_L has three minimal generators and two minimal first syzygies, making five nonzero Betti numbers: $\beta_{0,(2,2)} = \beta_{0,(2,3)} = \beta_{0,(2,4)} = \beta_{1,(3,4)} = \beta_{1,(3,5)} = 1$. In multigraded notation, the minimal free resolution is

$$
\begin{array}{ccccccc}
& & S(-(2,2)) & & & & \\
& & \oplus & & S(-(3,4)) & & \\
0 \longleftarrow I_L \longleftarrow & & S(-(2,3)) & \longleftarrow & \oplus & \longleftarrow & 0. \\
& & \oplus & & S(-(3,5)) & & \\
& & S(-(2,4)) & & & &
\end{array}
$$

The differential $S^3 \leftarrow S^2$ is given by the matrix $\begin{bmatrix} a & b \\ b & c \\ c & d \end{bmatrix}$, whose 2×2 minors minimally generate I_L, as in Example 8.55. Students of commutative algebra will recognize this as an instance of the *Hilbert–Burch Theorem* for Cohen–Macaulay rings of codimension 2. From the resolution (and using different notation for monomials in $\mathbb{Z}[[Q]]$ than in Example 8.55), we get

$$\mathcal{K}_Q \;=\; 1 - s^2 t^2 - s^2 t^3 - s^2 t^4 + s^3 t^4 + s^3 t^5,$$

so

$$\frac{\mathcal{K}_Q(s,t)}{(1-s)(1-st)(1-st^2)(1-st^3)} \;=\; \frac{1 + st + st^2}{(1-s)(1-st^3)}$$

is the Hilbert series of the semigroup Q. $\diamond$

We will express the Betti numbers $\beta_{i,\mathbf{b}}$ in terms of a certain simplicial complex on the vertex set $\{1, \ldots, n\}$. For any degree $\mathbf{b} \in Q$ define

$$\Delta_{\mathbf{b}} = \Big\{ I \subseteq \{1, \ldots, n\} \,\Big|\, \mathbf{b} - \sum_{i \in I} \mathbf{a}_i \text{ lies in } Q \Big\}.$$

A subset I of $\{1, \ldots, n\}$ lies in $\Delta_{\mathbf{b}}$ if and only if $I \subseteq \mathrm{supp}(\mathbf{x}^{\mathbf{u}})$ for some monomial $\mathbf{x}^{\mathbf{u}}$ of degree $u_1 \mathbf{a}_1 + u_2 \mathbf{a}_2 + \cdots + u_n \mathbf{a}_n = \mathbf{b}$. In other words, $\Delta_{\mathbf{b}}$ is the simplicial complex generated by the collection $\{\mathrm{supp}(\mathbf{x}^{\mathbf{u}}) \mid \mathbf{x}^{\mathbf{u}} \in S_{\mathbf{b}}\}$ of subsets of $\{1, \ldots, n\}$. Note that the vector space $S_{\mathbf{b}}$ spanned by all monomials $\mathbf{x}^{\mathbf{u}}$ of degree $\mathbf{b}$ has finite dimension by Theorem 8.6.

Theorem 9.2 *The Betti number $\beta_{j,\mathbf{b}}$ of I_L equals the dimension over $\Bbbk$ of the jth reduced homology group $\widetilde{H}_j(\Delta_{\mathbf{b}}; \Bbbk)$.*

Proof. The proof has the same structure as many of the proofs in Part I: find a multigraded complex of free S-modules with the appropriate homology, and identify the graded pieces of this complex as the desired reduced chain complexes. As in Lemma 1.32, the desired Betti number can be expressed as $\beta_{j,\mathbf{b}} = \dim_{\Bbbk} \mathrm{Tor}_S^{j+1}(\Bbbk, \Bbbk[Q])_{\mathbf{b}}$, where the field $\Bbbk$ is given the structure of an S-module via $\Bbbk \cong S/\mathfrak{m}$ for the maximal ideal $\mathfrak{m} = \langle x_1, \ldots, x_n \rangle$. This allows us to compute $\beta_{j,\mathbf{b}}$ by tensoring the Koszul complex $\mathbb{K}_{\boldsymbol{\cdot}}$ with $\Bbbk[Q]$ and then taking the jth homology module of the resulting complex. Note that $\mathbb{K}_{\boldsymbol{\cdot}}$ is graded by Q, with the summand $S(-\sigma)$ for the face σ generated in degree $\sum_{i \in \sigma} \mathbf{a}_i$.

The tensor product $\Bbbk[Q] \otimes_S \mathbb{K}_{\boldsymbol{\cdot}}$ is obtained from $\mathbb{K}_{\boldsymbol{\cdot}}$ simply by replacing $S(-\sigma)$ with $\Bbbk[Q](-\sum_{i \in \sigma} \mathbf{a}_i)$. The summands of this complex contributing to its component in degree $\mathbf{b}$ are precisely those summands $\Bbbk[Q](-\sum_{i \in \sigma} \mathbf{a}_i)$ such that $\mathbf{b} - \sum_{i \in \sigma} \mathbf{a}_i$ lies in Q. Moreover, each such summand contributes the 1-dimensional vector space $\Bbbk\{\mathbf{t}^{\mathbf{b} - \sum_{i \in \sigma} \mathbf{a}_i}\}$. Hence, just as in the proof of Proposition 1.28, we conclude that $(\Bbbk[Q] \otimes_S \mathbb{K}_{\boldsymbol{\cdot}})_{\mathbf{b}}$ is the reduced chain complex of $\Delta_{\mathbf{b}}$, but with the empty face $\varnothing$ in homological degree 0 instead of -1. It follows that its $(j+1)^{\mathrm{st}}$ homology is $\widetilde{H}_j(\Delta_{\mathbf{b}}; \Bbbk)$, as desired. $\square$

Corollary 9.3 *The lattice ideal I_L has a minimal generator in degree $\mathbf{b}$ if and only if the simplicial complex $\Delta_{\mathbf{b}}$ is disconnected.*

As an application of Theorem 9.2, we give a bound on the projective dimension of I_L—that is, on the length of its minimal resolution. Recall that this is the largest integer r satisfying $\beta_{r,\mathbf{b}} \neq 0$ for some $\mathbf{b} \in Q$. When $n \gg d$, the following is better than the Hilbert Syzygy Theorem bound $r \leq n - 1$.

Corollary 9.4 *The projective dimension of I_L is at most $2^{n-d} - 2$.*

Proof. Let $F_1, F_2, \ldots, F_s$ denote the distinct facets (maximal faces) of $\Delta_{\mathbf{b}}$. There exist monomials $\mathbf{x}^{\mathbf{u}_1}, \mathbf{x}^{\mathbf{u}_2}, \ldots, \mathbf{x}^{\mathbf{u}_s}$ of degree $\mathbf{b}$ such that $\mathrm{supp}(\mathbf{u}_i) =$

F_i for $i = 1, \ldots, s$. We claim that $s \leq 2^{n-d}$. Otherwise, there exist two vectors $\mathbf{u}_i$ and $\mathbf{u}_j$ such that $\mathrm{mod}_2(\mathbf{u}_i) = \mathrm{mod}_2(\mathbf{u}_j)$ in $(\mathbb{Z}/2\mathbb{Z})^n$, where $\mathrm{mod}_2(\_)$ denotes the operation of taking all coordinates modulo 2. The midpoint $\frac{1}{2}(\mathbf{u}_i + \mathbf{u}_j)$ is a nonnegative vector in $\mathbb{N}^n$ of degree $\mathbf{b}$. Its support $\mathrm{supp}(\frac{1}{2}(\mathbf{u}_i + \mathbf{u}_j)) = F_i \cup F_j$ is a face of $\Delta_\mathbf{b}$, and it properly contains both F_i and F_j. This is a contradiction to our choice that F_i and F_j are facets. It has been shown that $\Delta_\mathbf{b}$ has at most 2^{n-d} facets. Computing the homology of $\Delta_\mathbf{b}$ by the nerve of the cover by its facets (as we did in Theorem 5.37), we see that the homology of $\Delta_\mathbf{b}$ vanishes in dimension $2^{n-d} - 1$ and higher. $\square$

It can be shown that the upper bound in Corollary 9.4 is tight: for every $m \geq 1$ there exists a lattice L of rank m in $\mathbb{Z}^{2^m}$ such that the projective dimension of I_L equals $2^m - 2$. We demonstrate the construction for $m = 3$.

Example 9.5 Choose the lattice L in $\mathbb{Z}^8$ with basis given by the rows of

$$
\mathbf{L} = \begin{bmatrix} 1 & 1 & 1 & 2 & -2 & -1 & -1 & -1 \\ 1 & 2 & -2 & -1 & 1 & 1 & -1 & -1 \\ 1 & -1 & 1 & -1 & 1 & -1 & 2 & -2 \end{bmatrix}.
$$

This matrix has the properties that all eight sign patterns appear among its columns and its maximal minors are relatively prime. The latter condition ensures that $Q = \mathbb{N}^8 / L$ is an affine semigroup. The ideal I_L has 13 minimal generators, and its minimal free resolution looks like

$$
0 \leftarrow I_L \leftarrow S^{13} \leftarrow S^{44} \leftarrow S^{67} \leftarrow S^{56} \leftarrow S^{28} \leftarrow S^8 \leftarrow S^1 \leftarrow 0.
$$

Hence I_L has projective dimension 6, the maximal number allowed by Corollary 9.4. The unique minimal sixth syzygy occurs in the degree $\mathbf{b} = (3, 3, 2, 2, 2, 1, 2, 0) \pmod{L}$. This vector is the column sum of all positive entries in the matrix $\mathbf{L}$. There are precisely eight monomials in degree $\mathbf{b}$:

$$x_1^3 x_2^3 x_3^2 x_4^2 x_5^2 x_6 x_7^2, \; x_1^2 x_2^4 x_3 x_4^3 x_5 x_6^2 x_8^2, \; x_1^2 x_2 x_3^4 x_4^3 x_5 x_7^3 x_8, \; x_1 x_2^2 x_3^3 x_4^4 x_6 x_7 x_8^3,$$

$$x_1^2 x_2^2 x_3 x_5^3 x_6^2 x_7^3 x_8, \; x_1 x_2^3 x_4 x_5^2 x_6^3 x_7 x_8^4, \; x_1 x_3^3 x_4 x_5^2 x_6 x_7^4 x_8^3, \; x_2 x_3^2 x_4^2 x_5 x_6^2 x_7^2 x_8^5.$$

Each monomial misses a different variable. This means that $\Delta_\mathbf{b}$ is the boundary of the 7-dimensional simplex, so $\tilde{H}_6(\Delta_\mathbf{b}; \Bbbk) = \Bbbk^1$. $\diamond$

9.2   Laurent monomial modules

The formula for Betti numbers in the previous section suggests that resolutions of lattice ideals are similar to resolutions of monomial ideals. In the remainder of this chapter we will make this similarity precise by showing that lattice ideals can be regarded as "infinite periodic monomial ideals".

Definition 9.6 Let $T = S[x_1^{-1}, \ldots, x_n^{-1}]$ be the Laurent polynomial ring. An S-submodule M of T generated by Laurent monomials $\mathbf{x}^\mathbf{u}$ with $\mathbf{u} \in \mathbb{Z}^n$ is called a **Laurent monomial module**.

In general, a Laurent monomial module M need not be generated by its subset of minimal monomials (with respect to divisibility). For instance, let $n = 2$ and choose M to be the Laurent monomial module over $\Bbbk[x, y]$ spanned by $x^u y^v$ for (u, v) satisfying

$$u \geq 2 \quad \text{or} \quad (u \geq 1 \text{ and } v \geq 1).$$

Then M has only one minimal monomial, namely xy, but this element does not generate M: the monomial $x^2 y^{-17}$ lies in M but is not divisible by xy.

In what follows we only consider Laurent monomial modules M that are generated by their minimal monomials. If the set of minimal monomial generators of M is finite, then M is a $\mathbb{Z}^n$-graded translate of a monomial ideal of S. Hence we will be mainly interested in Laurent monomial modules whose generating sets are infinite. We can still draw pictures, but the usual staircase diagrams for monomial ideals become infinite staircases for Laurent monomial modules.

Example 9.7 Consider the Laurent monomial module in $\Bbbk[x, y][x^{-1}, y^{-1}]$ generated by the Laurent monomials $\left(\frac{x}{y}\right)^i$ for $i \in \mathbb{Z}$. The staircase diagram really is a staircase, but an infinite one:

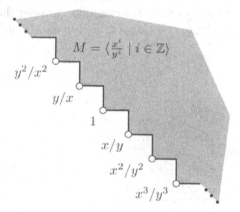

This Laurent monomial module is a model for all powers of the maximal ideal in $\Bbbk[x, y]$: intersecting it with any shift of $S = \Bbbk[x, y]$ produces the ideal $\langle x, y \rangle^r$ for some r. In fact, any Laurent monomial module can be thought of as the limit of the monomial ideals obtained by intersecting it with shifted positive orthants. $\diamond$

The construction of the hull complex in Chapter 4.4 works mutatis mutandis for Laurent monomial modules, using infinitely generated free modules. Given a Laurent monomial module M, we fix a real number $\lambda \gg 0$ and form the unbounded n-dimensional convex polyhedron

$$
\begin{aligned}
\mathcal{P}_\lambda &= \text{conv}\{\lambda^{\mathbf{u}} \mid \mathbf{x}^{\mathbf{u}} \in M\} \\
&= \text{conv}\{(\lambda^{u_1}, \lambda^{u_2}, \ldots, \lambda^{u_n}) \mid x_1^{u_1} x_2^{u_2} \cdots x_n^{u_n} \in M\}.
\end{aligned}
$$

The face poset of $\mathcal{P}_\lambda$ is independent of the large real number λ, and its vertices are precisely the minimal generators of M (this is why we are assuming M has minimal generators). The *hull complex* hull(M) is the polyhedral cell complex consisting of the bounded faces of $\mathcal{P}_\lambda$. The vertices of hull(M) are labeled by monomials. As in Chapter 4, a complex of free modules $\mathcal{F}_{\text{hull}(M)}$ is defined, and the result of Theorem 4.17 still holds.

Theorem 9.8 *The complex $\mathcal{F}_{\text{hull}(M)}$ is a $\mathbb{Z}^n$-graded free resolution of the Laurent monomial module M. This **hull resolution** has length at most n.*

The length bound says that hull resolutions respect the syzygy theorem.

Example 9.9 The hull complex hull(M) for the Laurent monomial module in Example 9.7 is the real line with a vertex at each integer point.   ◇

The results on Betti numbers of monomial ideals proved in the first part of this book also remain valid for Laurent monomial modules, except that now we may have minimal syzygies in infinitely many degrees. Here is another kind of infinite behavior we have to watch out for.

Example 9.10 A polyhedral cell complex is **locally finite** if every face meets finitely many others. In general, hull complexes of Laurent monomial modules need not be locally finite. For example, consider the Laurent monomial module M over $\Bbbk[x, y, z]$ generated by y/x and $(z/y)^i$ for all $i \in \mathbb{Z}$:

$$M = \left\langle \frac{y}{x} \right\rangle + \left\langle \left(\frac{y}{z} \right)^i \,\middle|\, i \in \mathbb{Z} \right\rangle.$$

The vertex y/x lies on infinitely many edges of hull(M). Only one of these edges is needed in the minimal free resolution of M over $\Bbbk[x, y, z]$, though. ◇

The connection with lattice ideals and semigroup rings arises from Laurent monomial modules whose generating Laurent monomials form a group under multiplication. Let $L \subset \mathbb{Z}^n$ be a sublattice whose intersection with $\mathbb{N}^n$ is $\{0\}$. This condition ensures the existence of a linear functional with strictly positive coordinates that vanishes on L, a hypothesis satisfied when L is the lattice associated with a pointed affine semigroup Q.

Definition 9.11 Given a lattice L whose intersection with $\mathbb{N}^n$ is $\{\mathbf{0}\}$, the **lattice module** M_L is the S-submodule of the Laurent polynomial ring $T = S[x_1^{-1}, \ldots, x_n^{-1}]$ generated by $\{\mathbf{x}^\mathbf{u} \mid \mathbf{u} \in L\}$.

The hypothesis on L guarantees that the elements of L form a *minimal* generating set for M_L.

Example 9.12 The Laurent monomial module in Example 9.7 is the lattice module M_L for the lattice $L = \ker(1,1) = \{(u, -u) \in \mathbb{Z}^2 \mid u \in \mathbb{Z}\}$. More generally, consider the lattice $L = \ker(1, 1, \ldots, 1)$, which consists of all vectors in $\mathbb{Z}^n$ with zero coordinate sum. The corresponding lattice module M_L is generated by all Laurent monomials of total degree 0, and it is the limit of powers of the maximal ideal $\langle x_1, \ldots, x_n \rangle$. Indeed, any intersection of M_L with a $\mathbb{Z}^n$-translate of $S = \mathbb{k}[x_1, \ldots, x_n]$ produces the ideal $\langle x_1, \ldots, x_n \rangle^r$ for some r. A picture of a finite part of this staircase for $n = 3$ looks as follows:

The white dots in this picture are the integer vectors in the lattice L. ◇

Let us write a lattice module M_L in terms of generators and relations. There is one generator $\mathbf{e}_\mathbf{u}$ for each element $\mathbf{u}$ in the lattice L, and M_L is the free S-module on the generators $\{\mathbf{e}_\mathbf{u} \mid \mathbf{u} \in L\}$ modulo the relations

$$\mathbf{x}^\mathbf{w} \cdot \mathbf{e}_\mathbf{u} - \mathbf{x}^{\mathbf{w}'} \cdot \mathbf{e}_\mathbf{v} = 0$$

for all $\mathbf{u}, \mathbf{v} \in L$ and $\mathbf{w}, \mathbf{w}' \in \mathbb{N}^n$ satisfying $\mathbf{w} + \mathbf{u} = \mathbf{w}' + \mathbf{v}$. This set of relations is far from minimal; an improvement is to consider only those relations where $\mathbf{x}^\mathbf{w}$ and $\mathbf{x}^{\mathbf{w}'}$ are relatively prime. Write $\mathbf{u} = \mathbf{v} + \mathbf{w}$ and decompose $\mathbf{w}$ into positive and negative parts, so $\mathbf{w} = \mathbf{w}_+ - \mathbf{w}_-$. Then we can express M_L as the free S-module on $\{\mathbf{e}_\mathbf{u} \mid \mathbf{u} \in L\}$ modulo the relations

$$\mathbf{x}^{\mathbf{w}_-} \cdot \mathbf{e}_{\mathbf{v}+\mathbf{w}} - \mathbf{x}^{\mathbf{w}_+} \cdot \mathbf{e}_\mathbf{v} = 0 \quad \text{for all} \quad \mathbf{v}, \mathbf{w} \in L. \tag{9.2}$$

The abelian group L acts freely on the generators of the lattice module M_L. The presentation of M_L by the syzygies (9.2) is nonminimal but invariant under the action of L. It would be nice to identify a *finite* set of first syzygies

$$\mathbf{x}^{\mathbf{w}_-} \cdot \mathbf{e}_\mathbf{w} - \mathbf{x}^{\mathbf{w}_+} \cdot \mathbf{e}_\mathbf{0} = 0 \quad \text{with} \quad \mathbf{w} \in L \tag{9.3}$$

such that M_L is presented by their translates (9.2) as $\mathbf{v}$ ranges over L. For instance, the lattice module for $L = \ker(1, \ldots, 1)$ in Example 9.12 is minimally presented by the lattice translates of the relations (9.3) for $\mathbf{w}$ in the set $\{\mathbf{e}_1 - \mathbf{e}_2, \mathbf{e}_2 - \mathbf{e}_3, \ldots, \mathbf{e}_{n-1} - \mathbf{e}_n\}$ of $n - 1$ differences of unit vectors.

It would be really nice to find a whole free resolution of M_L that is acted on by L. Such an *equivariant free resolution* is provided by the hull resolution. The point is that the lattice L permutes the faces of hull(M_L).

Example 9.13 The hull complex of the lattice $L = \ker(1, \ldots, 1)$ is an infinite periodic subdivision of an $(n-1)$-dimensional Euclidean space. It is isomorphic to the face poset of the infinite hyperplane arrangement consisting of all points in $L \otimes \mathbb{R}$ possessing two coordinates whose difference is an integer. The complex hull(M_L) has $n-1$ maximal faces modulo the lattice action; they are called *hypersimplices*. In three variables, the hull complex is the tessellation of the plane $\mathbb{R} \otimes L$ by two classes of triangles: "up" triangles and "down" triangles. Part of this tessellation is depicted in Example 4.22. This hull complex has three edges modulo the action of $L = \ker(1, 1, 1)$. They correspond to **c** being one of the three vectors $\mathbf{e}_1 - \mathbf{e}_2$, $\mathbf{e}_1 - \mathbf{e}_3$, and $\mathbf{e}_2 - \mathbf{e}_3$. Of the resulting three first syzygies (9.3) it suffices to take only two for a minimal presentation of M_L.   ◇

Calculating the hull complex hull(M_L) is a finite algorithmic problem, even though it has infinitely many cells. This is because of a minor miracle, to the effect that the phenomenon of Example 9.10 will not happen for a Laurent monomial module M_L arising from a lattice L.

Theorem 9.14 *The hull complex of a lattice module is locally finite.*

Proof. We claim that the vertex $\mathbf{0} \in L$ is incident to only finitely many edges of hull(M_L). This claim implies the theorem because (i) the lattice L acts transitively on the vertices of hull(M_L), so it suffices to consider the vertex $\mathbf{0}$, and (ii) every face of hull(M_L) containing $\mathbf{0}$ is uniquely determined by the edges containing $\mathbf{0}$, so $\mathbf{0} \in L$ lies in only finitely many faces.

To prove the claim we introduce the following definition. A nonzero vector $\mathbf{u} = \mathbf{u}_+ - \mathbf{u}_-$ in our lattice L is called *primitive* if there is no other vector $\mathbf{v} \in L \setminus \{\mathbf{u}, \mathbf{0}\}$ such that $\mathbf{v}_+ \leq \mathbf{u}_+$ and $\mathbf{v}_- \leq \mathbf{u}_-$. The primitive vectors in L can be computed as follows. Fix any sign pattern in $\{-1, +1\}^n$ and consider the pointed affine semigroup consisting of all vectors in L whose nonzero entries are consistent with the chosen sign pattern. A vector in L is primitive if and only if it lies in the Hilbert basis of the semigroup associated to its sign pattern. Each of these Hilbert bases is finite by Theorem 7.16, and by taking the union over all sign patterns, we conclude that the set of primitive vectors in L is finite.

We will now prove that for any edge $\{\mathbf{0}, \mathbf{u}\}$ of the hull complex hull(M_L), the vector $\mathbf{u}$ is primitive. As the set of primitive vectors is finite, this proves the claim and hence the theorem. Suppose that $\mathbf{u} \in L \setminus \{\mathbf{0}\}$ is not primitive, and choose $\mathbf{v} \in L \setminus \{\mathbf{u}, \mathbf{0}\}$ such that $\mathbf{v}_+ \leq \mathbf{u}_+$ and $\mathbf{v}_- \leq \mathbf{u}_-$. This implies $\lambda^{v_i} + \lambda^{u_i - v_i} \leq 1 + \lambda^{u_i}$ for all $i \in \{1, \ldots, n\}$ and $\lambda \gg 0$. In other words, for $\lambda \gg 0$, the vector $\lambda^{\mathbf{v}} + \lambda^{\mathbf{u}-\mathbf{v}}$ is componentwise smaller than or equal to the vector $\lambda^{\mathbf{0}} + \lambda^{\mathbf{u}}$. We conclude that the midpoint of the segment conv$\{\lambda^0, \lambda^u\}$ lies in conv$\{\lambda^v, \lambda^{u-v}\} + \mathbb{R}^n_{\geq 0}$, and hence conv$\{\lambda^0, \lambda^u\}$ is not an edge of the polyhedron $\mathcal{P}_\lambda = \text{conv}\{\lambda^{\mathbf{w}} \mid \mathbf{w} \in L\} + \mathbb{R}^n_{\geq 0}$. □

The lattice L acts on the set of faces of the hull complex. Two faces are considered *equivalent modulo* L if they lie in the same orbit. By definition, any two vertices of hull(M_L) are equivalent modulo L.

Corollary 9.15 *There are only finitely many equivalence classes modulo* L *of faces in the hull complex* hull(M_L).

From the proof of Theorem 9.14 we derive the following general algorithm for computing the hull complex. The first step is to find all primitive vectors in L. A convenient way to do this is described in Exercise 7.12. Next, compute the link of $\mathbf{0}$ in hull(M_L) by computing the faces of the polyhedral cone spanned by the vectors

$$\lambda^{\mathbf{u}} - \lambda^{\mathbf{0}} = (\lambda^{u_1} - 1, \ldots, \lambda^{u_n} - 1),$$

where $\mathbf{u}$ runs over all primitive vectors. Typically, many of the vectors $\lambda^{\mathbf{u}} - \lambda^{\mathbf{0}}$ here are not extreme rays of the cone. Those primitive vectors $\mathbf{u}$ are discarded, as they do not correspond to edges of hull(M_L). Finally, identify faces of the link of $\mathbf{0}$ that correspond to the same face of hull(M_L). This is done by translating the link of $\mathbf{0}$ to the various neighbors $\mathbf{u}$.

Example 9.16 An interesting lattice module, to be discussed in greater detail in Example 9.26, is the one given by the rank 3 sublattice $L = \ker([20\ 24\ 25\ 31])$ in $\mathbb{Z}^4$. This lattice has 75 primitive vectors, but only 7 of them are edges of hull(M_L). Thus the module M_L is minimally presented by 7 classes of first syzygies as in (9.3). It has 12 second syzygies and 6 third syzygies, modulo the action of L. ◇

9.3 Free resolutions of lattice ideals

Fix a lattice $L \subset \mathbb{Z}^n$ satisfying $L \cap \mathbb{N}^n = \{\mathbf{0}\}$. We wish to determine the following fundamental objects concerning the lattice ideal I_L and the semigroup ring S/I_L:

1. generators for I_L;
2. the $\mathbb{Z}^n/L$-graded Hilbert series of S/I_L, as a rational function; and
3. a (minimal) free resolution of S/I_L over S.

Of course, $3 \Rightarrow 2 \Rightarrow 1$, so we will aim for free resolutions.

The essential idea is to express the semigroup ring S/I_L as a quotient of the lattice module M_L by the action of L. In order to do that, let us formalize the action by introducing the *group algebra* $S[L]$ of the abelian group L over the polynomial ring S. Explicitly, this is the subalgebra

$$S[L] = \Bbbk[\mathbf{x}^{\mathbf{u}}\mathbf{z}^{\mathbf{v}} \mid \mathbf{u} \in \mathbb{N}^n \text{ and } \mathbf{v} \in L]$$

of the Laurent polynomial algebra $S[z_1^{\pm 1}, \ldots, z_n^{\pm 1}]$. The group ring $S[L]$ carries a $\mathbb{Z}^n$-grading via $\deg(\mathbf{x}^{\mathbf{u}} \mathbf{z}^{\mathbf{v}}) = \mathbf{u} + \mathbf{v}$. In the previous section we considered the following transitive action of L on the monomials in M_L:

$$\mathbf{v} \cdot \mathbf{x}^{\mathbf{u}} = \mathbf{x}^{\mathbf{u}+\mathbf{v}} \quad \text{for} \quad \mathbf{v} \in L \text{ and } \mathbf{x}^{\mathbf{u}} \in M_L.$$

This action is reformulated using $S[L]$ by stipulating that $\mathbf{x}^{\mathbf{u}} \mathbf{z}^{\mathbf{v}} = \mathbf{x}^{\mathbf{u}+\mathbf{v}}$. Thus the lattice module M_L becomes a $\mathbb{Z}^n$-graded cyclic $S[L]$-module:

$$M_L \;\cong\; S[L]/\langle \mathbf{x}^{\mathbf{u}} - \mathbf{x}^{\mathbf{v}} \mathbf{z}^{\mathbf{u}-\mathbf{v}} \mid \mathbf{u}, \mathbf{v} \in \mathbb{N}^n \text{ and } \mathbf{u} - \mathbf{v} \in L \rangle. \qquad (9.4)$$

In fact, any $\mathbb{Z}^n$-graded S-module with an equivariant action of L such that $\mathbf{v} \in L$ acts as a homomorphism of degree $\mathbf{v}$ is naturally an $S[L]$-module. Here, "equivariant" means that the homomorphisms $\mathbf{v}$ commute with the action of S. Another way of making the same statement is this: The category of L-equivariant $\mathbb{Z}^n$-graded S-modules is isomorphic to the category

$$\mathcal{A} \;=\; \{\mathbb{Z}^n\text{-graded } S[L]\text{-modules}\}.$$

Consider any object M in $\mathcal{A}$. How do we define the quotient of the L-equivariant module M by the action of L? We wish to identify $m \in M$ with $\mathbf{z}^{\mathbf{v}} \cdot m$ whenever $\mathbf{v} \in L$, so that the quotient is an S-module whose elements are orbits of the action of L on M_L. When $M = S[L]$ itself, this quotient is

$$\begin{aligned}
S[L]/L &= S[L]/\langle \mathbf{x}^{\mathbf{u}} \mathbf{z}^{\mathbf{v}} - \mathbf{x}^{\mathbf{u}} \mid \mathbf{u} \in \mathbb{N}^n \text{ and } \mathbf{v} \in L \rangle \\
&= S[L]/\langle \mathbf{z}^{\mathbf{v}} - 1 \mid \mathbf{v} \in L \rangle \\
&\cong S.
\end{aligned}$$

However, this copy of S is no longer $\mathbb{Z}^n$-graded, because $\mathbf{x}^{\mathbf{u}}$ and $\mathbf{x}^{\mathbf{u}} \mathbf{z}^{\mathbf{v}}$, which have different $\mathbb{Z}^n$-graded degrees $\mathbf{u}$ and $\mathbf{u} + \mathbf{v}$, map to the same element $\mathbf{x}^{\mathbf{u}}$. On the other hand, all of the preimages in $S[L]$ of $\mathbf{x}^{\mathbf{u}} \in S$ have $\mathbb{Z}^n$-graded degrees that are congruent modulo L. We conclude that the above copy of the polynomial ring S is $\mathbb{Z}^n/L$-graded, with $\mathbf{x}^{\mathbf{u}}$ having degree $\mathbf{u} \pmod{L}$.

For an arbitrary $\mathbb{Z}^n$-graded $S[L]$-module M, our quotient M/L will similarly be obtained by "setting $\mathbf{z}^{\mathbf{v}} = 1$ for all $\mathbf{v} \in L$". Algebraically, this is just tensoring M over $S[L]$ with $S = S[L]/\langle \mathbf{z}^{\mathbf{v}} - 1 \mid \mathbf{v} \in L \rangle$, yielding

$$M/L \;=\; M \otimes_{S[L]} S[L]/L \;=\; M \otimes_{S[L]} S.$$

As with $S[L]/L$, the quotient M/L is no longer $\mathbb{Z}^n$-graded, but only $\mathbb{Z}^n/L$-graded. This tensor product therefore defines a functor of categories

$$\pi : \mathcal{A} \to \mathcal{B} \;=\; \{\mathbb{Z}^n/L\text{-graded } S\text{-modules}\}.$$

The great thing about the functor π is that it forgets nothing significant. In particular, it is *exact*: it maps exact sequences to exact sequences.

Theorem 9.17 *The functor $\pi : \mathcal{A} \to \mathcal{B}$ sending M to M/L is an equivalence of categories.*

Proof. By condition (iii) of [MacL98, Theorem IV.4.1], we must show that

- π is *fully faithful*, meaning that π induces a natural identification $\mathrm{Hom}_{\mathcal{A}}(M, M') = \mathrm{Hom}_{\mathcal{B}}(\pi(M), \pi(M'))$; and
- every object $N \in \mathcal{B}$ is isomorphic to $\pi(M)$ for some object $M \in \mathcal{A}$.

Each module $M \in \mathcal{A}$ is $\mathbb{Z}^n$-graded, so the lattice $L \subset S[L]$ acts on M as a group of S-equivariant automorphisms. For each $\mathbf{a} \in \mathbb{Z}^n/L$, the functor π identifies the spaces $M_{\mathbf{u}}$ for $\mathbf{u}$ mapping to $\mathbf{a} \pmod{L}$ as the single space $\pi(M)_{\mathbf{a}}$. A morphism $f : M \to M'$ in $\mathcal{A}$ is a collection of $\Bbbk$-linear maps $f_{\mathbf{u}} : M_{\mathbf{u}} \to M'_{\mathbf{u}}$ compatible with the action by L and with multiplication by each variable x_i. A morphism $g : \pi(M) \to \pi(M')$ in $\mathcal{B}$ is a collection of $\Bbbk$-linear maps $g_{\mathbf{a}} : \pi(M)_{\mathbf{a}} \to \pi(M')_{\mathbf{a}}$ compatible with multiplication by each variable x_i. Given $\mathbf{a} \in \mathbb{Z}^n/L$, the functor π identifies the maps $f_{\mathbf{u}}$ for $\mathbf{u}$ mapping to $\mathbf{a} \pmod{L}$ as the single map $\pi(f)_{\mathbf{a}}$.

The above discussion implies that π takes distinct morphisms to distinct morphisms (so π is *faithful*); now we must show that there are no remaining morphisms between $\pi(M)$ and $\pi(M')$ in $\mathcal{B}$ (so π is *full*). Given a morphism $g \in \mathrm{Hom}_{\mathcal{B}}(\pi(M), \pi(M'))$, define a morphism $f \in \mathrm{Hom}_{\mathcal{A}}(M, M')$ by the rule $f_{\mathbf{u}} = g_{\mathbf{a}}$ whenever $\mathbf{u}$ maps to $\mathbf{a} \pmod{L}$. Then $\pi(f) = g$, establishing the desired identification of Hom groups.

Finally, we define an inverse to π by constructing the "universal cover" of any given object $N = \bigoplus_{\mathbf{a} \in \mathbb{Z}^n/L} N_{\mathbf{a}}$ in $\mathcal{B}$. Define the $\Bbbk$-vector space $M = \bigoplus_{\mathbf{u} \in \mathbb{Z}^n} M_{\mathbf{u}}$ by setting $M_{\mathbf{u}} = N_{\mathbf{a}}$ whenever $\mathbf{u}$ maps to $\mathbf{a} \pmod{L}$. For every vector $\mathbf{u} \in \mathbb{Z}^n$ mapping to $\mathbf{a} \pmod{L}$, lift each multiplication map $N_{\mathbf{a}} \xrightarrow{\cdot x_i} N_{\mathbf{a}+\mathbf{a}_i}$ to a map $M_{\mathbf{u}} \xrightarrow{\cdot x_i} M_{\mathbf{u}+\mathbf{e}_i}$, and let $\mathbf{z}^{\mathbf{v}}$ for $\mathbf{v} \in L$ act on M as the identity map from $M_{\mathbf{u}}$ to $M_{\mathbf{u}+\mathbf{v}}$. These multiplication maps make the vector space M into a module over $S[L]$ satisfying $\pi(M) = N$. $\square$

We now apply this functor π to the lattice module M_L. By definition,

$$\pi(M_L) = M_L \otimes_{S[L]} S.$$

The tensor product means that in the presentation (9.4) of M_L, we replace $S[L]$ by S and set all occurrences of any $\mathbf{z}$-monomial $\mathbf{z}^{\mathbf{v}}$ to 1. Thus

$$\begin{aligned}
\pi(M_L) &= S/\langle \mathbf{x}^{\mathbf{u}} - \mathbf{x}^{\mathbf{v}} \mid \mathbf{u}, \mathbf{v} \in \mathbb{N}^n \text{ and } \mathbf{u} - \mathbf{v} \in L \rangle \\
&= S/I_L.
\end{aligned}$$

We now have achieved our goal of writing S/I_L as the quotient of M_L by the action of L. Next, we can use the functoriality of π and Theorem 9.17 to translate free resolutions of M_L in $\mathcal{A}$ to free resolutions of S/I_L in $\mathcal{B}$.

Corollary 9.18 *If $\mathcal{F}.$ is any $\mathbb{Z}^n$-graded free resolution of M_L over $S[L]$, then $\pi(\mathcal{F}.)$ is a $\mathbb{Z}^n/L$-graded free resolution of S/I_L over S. Moreover, $\mathcal{F}.$ is a minimal resolution if and only if $\pi(\mathcal{F}.)$ is a minimal resolution.*

What is a resolution of M_L over $S[L]$? It is just a resolution of M_L as an S-module along with an action of L that is *free*, meaning that no element of L has a fixed point. These exist because π is an equivalence. We have constructed an explicit such resolution of M_L with a free L-action in the previous section. This was the hull resolution $\mathcal{F}_{\bullet} = \mathcal{F}_{\mathrm{hull}(M_L)}$. The point is that we can now write hull(M_L) as an exact sequence

$$\mathrm{hull}(M_L): \quad 0 \leftarrow S[L] \leftarrow S[L]^{\beta_1} \leftarrow S[L]^{\beta_2} \leftarrow S[L]^{\beta_3} \leftarrow \cdots \qquad (9.5)$$

in which β_i is the number of L-equivalence classes of i-dimensional faces of hull(M_L), and the differentials involve monomials in both $\mathbf{x}$ and $\mathbf{z}$. The $\mathbf{z}$-monomials take care of any ambiguity in choosing representatives for the faces of the hull complex. We shall see an explicit example shortly.

Definition 9.19 The **hull resolution** of the semigroup ring S/I_L equals $\pi(\mathcal{F}_{\mathrm{hull}(M_L)})$. It is gotten from (9.5) by replacing $S[L]$ with S and $\mathbf{z}$ with 1.

Theorem 9.20 *The hull resolution of the semigroup ring S/I_L is a finite $\mathbb{Z}^n/L$-graded free resolution of length $\leq n$.*

Proof. The lattice L acts freely on hull(M_L), which implies that $\mathcal{F}_{\mathrm{hull}(M_L)}$ is a free $S[L]$-module. Since π(free $S[L]$-module) is a free S-module, the hull resolution of S/I_L is a resolution by free S-modules. The finiteness holds because of Corollary 9.15. The length of $\mathcal{F}_{\mathrm{hull}(M_L)}$ is at most n because hull(M_L) is the set of bounded faces of a polyhedron inside $\mathbb{R}^n$.   □

Example 9.21 Consider the monomial curve $t \mapsto (t^4, t^3, t^5)$ in affine 3-space. Its defining prime ideal in $S = \Bbbk[x_1, x_2, x_3]$ is the lattice ideal

$$I_L = \langle x_1 x_2^2 - x_3^2, x_1 x_3 - x_2^3, x_2 x_3 - x_1^2 \rangle,$$

for the kernel L of the matrix $[4\ 3\ 5]$. The corresponding lattice module

$$M_L = \langle x_1^u x_2^v x_3^w \mid 4u + 3v + 5w = 0 \rangle$$

in $\Bbbk[x_1^{\pm 1}, x_2^{\pm 1}, x_3^{\pm 1}]$ is pictured at the top of Fig. 9.1. The hull complex below it triangulates $\mathbb{R} \otimes L$ using L for vertices. The labeling on every pair of up and down triangles is obtained from the representative labeling

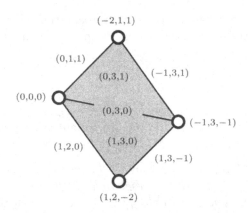

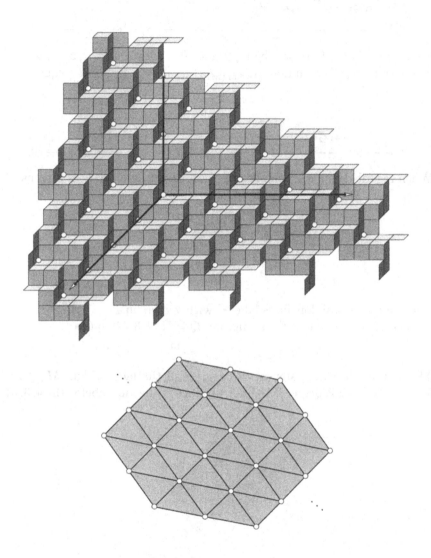

Figure 9.1: A lattice module and its hull complex

by adding some vector in L to all of the labels. As in (9.5) we write the hull resolution $\mathcal{F}_{\text{hull}(M_L)}$ as an exact sequence over the group algebra $S[L]$:

$$0 \leftarrow S[L] \xleftarrow{\left[x_1 x_2^2 - \frac{x_3^2 z_1 z_2^2}{z_3^2} \quad x_1 x_3 - \frac{x_2^3 z_1 z_3}{z_2^3} \quad x_2 x_3 - \frac{x_1^2 z_2 z_3}{z_1^2}\right]} S[L]^3 \xleftarrow{\begin{bmatrix} x_2 & \frac{x_1 z_2 z_3}{z_1^2} \\ x_1 & x_3 \\ x_3 & \frac{x_2^2 z_3}{z_1 z_2^2} \end{bmatrix}} S[L]^2 \leftarrow 0.$$

We now apply the functor π by replacing $S[L]$ with S and z_1, z_2, z_3 with 1. The resulting hull resolution $\pi(\mathcal{F}_{\text{hull}(M_L)})$ of $S/I_L = \Bbbk[t^4, t^3, t^5]$ equals

$$0 \longleftarrow S \xleftarrow{\left[x_1 x_2^2 - x_3^2 \quad x_1 x_3 - x_2^3 \quad x_2 x_3 - x_1^2\right]} S^3 \xleftarrow{\begin{bmatrix} x_2 & x_1 \\ x_1 & x_3 \\ x_3 & x_2^2 \end{bmatrix}} S^2 \longleftarrow 0.$$

When regarded as a cell complex, the hull resolution of S/I_L is a torus

whose fundamental domain is labeled with vectors in L.

The K-polynomial of the semigroup $Q = \mathbb{N}\{4, 3, 5\}$ equals

$$\mathcal{K}_Q(t) \;=\; 1 - t^8 - t^9 - t^{10} + t^{13} + t^{14}.$$

This is the alternating sum of the degrees of the faces of $\text{hull}(M_L)/L$ in $\mathbb{Z}^3/L \cong \mathbb{Z}$. Each $\mathbb{Z}$-degree is the dot product of the face label with $(4, 3, 5)$:

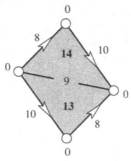

The Hilbert series of S/I_L is obtained from the K-polynomial by dividing by the appropriate denominator:

$$\frac{1 - t^8 - t^9 - t^{10} + t^{13} + t^{14}}{(1 - t^4)(1 - t^3)(1 - t^5)} \;=\; \frac{1}{1 - t} - t - t^2 \;=\; \sum\{t^a \mid a \in \mathbb{N}\{4, 3, 5\}\}.$$

The denominator comes from the Hilbert series $\frac{1}{(1-t^4)(1-t^3)(1-t^5)}$ of S. $\diamond$

Example 9.22 Suppose that L is a *unimodular lattice*. This means that for all subsets $\sigma \subseteq \{1, \ldots, n\}$, the group $\mathbb{Z}^n / (L + \sum_{i \in \sigma} \mathbb{Z}e_i)$ is torsion-free. This property holds for an affine semigroup $Q = \mathbb{N}\{a_1, \ldots, a_n\}$ in $\mathbb{Z}^d$ if every linearly independent d-element subset of $\{a_1, \ldots, a_n\}$ is a basis of $\mathbb{Z}^d$.

Consider the *Lawrence lifting* $\Lambda(L) = \{(\mathbf{u}, -\mathbf{u}) \in \mathbb{Z}^{2n} \mid \mathbf{u} \in L\}$, which is also a unimodular lattice, but now in $\mathbb{Z}^{2n}$. Its corresponding lattice ideal is

$$I_{\Lambda(L)} = \langle \mathbf{x}^{\mathbf{u}}\mathbf{y}^{\mathbf{v}} - \mathbf{x}^{\mathbf{u}}\mathbf{y}^{\mathbf{v}} \mid \mathbf{u} - \mathbf{v} \in L \rangle \subset \mathbb{k}[x_1, \ldots, x_n, y_1, \ldots, y_n].$$

These *unimodular Lawrence ideals* have the characteristic property that all of their initial monomial ideals are squarefree [Stu96, Remark 8.10].

The hull resolution of $I_{\Lambda(L)}$ is not necessarily minimal, even if L is unimodular. However, the minimal resolution does come from a cellular resolution of $M_{\Lambda(L)}$ and is described by a combinatorial construction:

Step 1. Take the infinite hyperplane arrangement

$$\{x_i = j \mid i = 1, \ldots, n \text{ and } j \in \mathbb{Z}\}.$$

Step 2. Let $\mathcal{H}_L$ be its intersection with $L \otimes \mathbb{R}$.

Step 3. Form the quotient $\mathcal{H}_L/L$.

The lattice L acts on the cells of the arrangement $\mathcal{H}_L$ with finitely many orbits. The vertices of $\mathcal{H}_L$ are labeled by the elements of $\Lambda(L)$. The corresponding algebraic complex $\mathcal{F}_{\mathcal{H}_L}$ is an L-equivariant minimal free resolution of the lattice module $M_{\Lambda(L)}$. The quotient complex $\mathcal{H}_L/L$ is a finite cell complex. By Corollary 9.18, the minimal i^{th} syzygies of $I_{\Lambda(L)}$ are in bijection with the i-dimensional faces of $\mathcal{H}_L/L$.

A particular example of the minimal resolution described here is the Eagon–Northcott complex for the 2×2 minors of a generic $2 \times n$ matrix. Another example is featured in Exercise 9.9. $\diamond$

9.4 Genericity and the Scarf complex

Definition 9.23 A Laurent monomial module M in $T = \mathbb{k}[x_1^{\pm 1}, \ldots, x_n^{\pm 1}]$ is called **generic** if all its minimal first syzygies $\mathbf{x}^{\mathbf{u}}e_i - \mathbf{x}^{\mathbf{v}}e_j$ have full support.

This condition means that every variable x_ℓ appears either in $\mathbf{x}^{\mathbf{u}}$ or in $\mathbf{x}^{\mathbf{v}}$. This definition is the essence behind genericity for monomial ideals, although for ideals there are "boundary effects" coming from the fact that $\mathbb{N}^n$ is a special subset of $\mathbb{Z}^n$. To be precise, the genericity condition on the minimal first syzygies $\mathbf{x}^{\mathbf{u}}e_i - \mathbf{x}^{\mathbf{v}}e_j$ of an ideal requires only that $\mathrm{supp}(\mathbf{x}^{\mathbf{u}+\mathbf{v}}) = \mathrm{supp}(\mathrm{lcm}(m_i, m_j))$, as opposed to $\mathrm{supp}(\mathbf{x}^{\mathbf{u}+\mathbf{v}}) = \{1, \ldots, n\}$ for Laurent monomial modules. This definition allows us to treat the boundary exponent 0 differently than the strictly positive exponents coming from the interior of $\mathbb{N}^n$. Just like the hull complex, the Scarf complex defined earlier for monomial ideals makes sense for Laurent monomial modules, too, as does the theorem on free resolutions of generic objects.

Theorem 9.24 *For generic Laurent monomial modules M, the following coincide:*

1. *The Scarf complex of M*
2. *The hull resolution of M*
3. *The minimal free resolution of M*

Proof. The proof of Theorem 6.13 carries over from monomial ideals to Laurent monomial modules.   □

A lattice L in $\mathbb{Z}^n$ is called *generic* if its associated lattice module M_L is generic. Equivalently, the lattice L is generic if the lattice ideal I_L is generated by binomials $\mathbf{x}^u - \mathbf{x}^v$ with full support, so every variable x_ℓ appears in every minimal generator of I_L. Applying Corollary 9.18 to the minimal free resolution in Theorem 9.24, we get the following result.

Corollary 9.25 *The minimal free resolution of a generic lattice ideal I_L is its **Scarf complex**, which is the image under π of the Scarf complex of M_L.*

The lattice L in Example 9.21 is generic because all three generators of

$$I_L \;=\; \langle x_1 x_2^2 - x_3^2, x_1 x_3 - x_2^3, x_2 x_3 - x_1^2 \rangle$$

have full support. The Scarf complex of M_L coincides with the hull complex depicted in Fig. 9.1. The Scarf complex of I_L is a minimal free resolution. Geometrically, it is a subdivision of the torus with two triangles.

Example 9.26 Things become much more complicated in four dimensions. The smallest codimension 1 generic lattice module in four variables is determined by the lattice $L = \ker([20\ 24\ 25\ 31]) \subset \mathbb{Z}^4$. The lattice ideal I_L is the ideal of the monomial curve $t \mapsto (t^{20}, t^{24}, t^{25}, t^{31})$ in affine 4-space. The group algebra is $S[L] = \Bbbk[a, b, c, d][\mathbf{z}^v \mid \mathbf{v} \in L]$, and

$$\begin{aligned}
M_L \;=\; S[L]/\langle & a^4 - bcd\,\mathbf{z}^*, a^3 c^2 - b^2 d^2\,\mathbf{z}^*, a^2 b^3 - c^2 d^2\,\mathbf{z}^*, ab^2 c - d^3\,\mathbf{z}^*, \\
& b^4 - a^2 cd\,\mathbf{z}^*, b^3 c^2 - a^3 d^2\,\mathbf{z}^*, c^3 - abd\,\mathbf{z}^* \rangle,
\end{aligned}$$

where, for instance, the $*$ in $a^4 - bcd\,\mathbf{z}^*$ is the vector in L that is 4 times the first generator minus 1 times each of the second, third, and fourth generators. The hull = Scarf = minimal resolution of S/I_L has the form

$$0 \longleftarrow S \longleftarrow S^7 \longleftarrow S^{12} \longleftarrow S^6 \longleftarrow 0.$$

Up to the action of L, there are 6 tetrahedra corresponding to the second syzygies and 12 triangles corresponding to the first syzygies.   ◇

In Theorem 6.26 we described what it means for a monomial ideal to be generic. Similar equivalences hold for monomial modules M. In particular, M is generic if and only if its Scarf complex is unchanged by arbitrary deformations. It would nice to make a similar statement also for deformations in the subclass of lattice modules. Here, the situation is more complicated, but it is the case that generic lattices deserve to called "generic" among all lattices: they are "abundant" in a sense that we are about to make precise.

Consider the set $\mathcal{S}_{d,n}$ of all rational $d \times n$ matrices $\mathbf{L}$ such that the row span of $\mathbf{L}$ meets $\mathbb{N}^n$ only in the origin. Each such matrix $\mathbf{L}$ defines a rank d sublattice $L = \text{rowspan}_{\mathbb{Q}}(\mathbf{L}) \cap \mathbb{Z}^n$. Let $\mathcal{T}_{d,n}$ be the subset of all matrices $\mathbf{L}$ in $\mathcal{S}_{d,n}$ such that the corresponding lattice L is not generic.

Theorem 9.27 (Barany and Scarf) *The closure of $\mathcal{T}_{d,n}$ has measure zero in the closure of $\mathcal{S}_{d,n}$ in $\mathbb{R}^{d \times n}$.*

Proof. Condition (A3) in the article [BaS96] by Barany and Scarf describes an open set of matrices $\mathbf{L}$ that represent generic lattices. Theorem 1 in [BaS96] shows that the set of all generic lattices with a fixed Scarf complex is an open polyhedral cone. The union of these cones is a dense subset in the closure of $\mathcal{S}_{d,n}$. □

Theorem 9.27 means in practice that if the rational matrix $\mathbf{L}$ is chosen at random, with respect to any reasonable distribution on rational matrices, then the corresponding lattice ideal will be generic. What is puzzling is that virtually all lattice ideals one encounters in commutative algebra seem to be nongeneric; i.e., they lie in the measure zero subset $\mathcal{T}_{d,n}$. The deterministic construction of generic lattice ideals with prescribed properties (such as Betti numbers) is an open problem that appears to be difficult. It is also not known how to "deform" a lattice ideal to a "nearby" generic lattice ideal.

Exercises

9.1 Let Q be the affine semigroup in $\mathbb{Z}^d$ spanned by the vectors $\mathbf{e}_i + \mathbf{e}_j$, where $1 \leq i < j \leq d$. In other words, Q is spanned by all zero-one vectors with precisely two ones. Determine the K-polynomial $\mathcal{K}_Q(t_1, \ldots, t_d)$ of the semigroup Q.

9.2 Let M be the Laurent monomial module generated by $\{x^u y^v z^w \mid u+v+w = 0$ and not all three coordinates of (u, v, w) are even$\}$. Draw a picture of M. Find a cellular minimal free resolution of M over $\Bbbk[x, y, z]$.

9.3 Let L be the kernel of the matrix $\begin{bmatrix} 3 & 2 & 1 & 0 \\ 0 & 1 & 2 & 3 \end{bmatrix}$. Show that the hull resolution of the Laurent monomial module M_L is minimal. What happens modulo the action by the lattice L? Answer: Depicted after the last exercise in this chapter.

9.4 What projective dimensions are possible for ideals I_L of pointed affine semigroups spanned by six vectors in $\mathbb{Z}^3$? Give an explicit example for each value.

9.5 Compute the hull resolution for the ideal of 2×2 minors in a 2×4 matrix.

9.6 Consider the lattice ideal generated by all the 2×2 minors of a generic 4×4 matrix, and compute its minimal free resolution. Classify all syzygies up to symmetry, and determine the corresponding simplicial complexes $\Delta_\mathbf{b}$.

9.7 Compute the hull resolution of the ideal of 2×2 minors of a generic 3×3 matrix, and compare it with the minimal free resolution of that same ideal.

9.8 Compute the hull complex hull(M_L) of the sublattice of $\mathbb{Z}^5$ spanned by three vectors $(1, -2, 1, 0, 0)$, $(0, 1, -2, 1, 0)$, and $(0, 0, 1, -2, 1)$.

9.9 Let Q be the subsemigroup of $\mathbb{Z}^3$ generated by the six vectors $(1, 0, 0)$, $(0, 1, 0)$, $(0, 0, 1)$, $(-1, 1, 0)$, $(-1, 0, 1)$, and $(0, -1, 1)$. Determine the corresponding lattice L and show that it is unimodular. Then compute (a) generators for the Lawrence ideal $I_{\Lambda(L)}$, (b) the three-dimensional cell complex $\mathcal{H}_L/L$ as in Example 9.22, and (c) the minimal free resolution of $I_{\Lambda(L)}$.

9.10 Determine the K-polynomial $\mathcal{K}_Q(t)$ of the semigroup $Q = \mathbb{N}\{20, 24, 25, 31\}$ in Example 9.16.

9.11 Find an explicit generic lattice L of codimension 1 in $\mathbb{Z}^5$. List the faces of the Scarf complex of your lattice and describe the minimal free resolution of I_L.

Answer to Exercise 9.3 Translates of the left picture by L constitute hull(M_L):

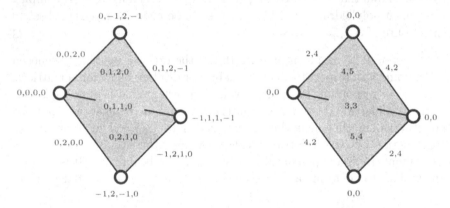

Opposite edge labels become equal if the matrix defining L is applied, as at right.

Notes

The bulk of the presentation in this chapter is based on [BS98]. In particular, the Laurent monomial module point of view originated there, as did Theorems 9.8, 9.14, 9.17, 9.20, and 9.24, as well as Corollaries 9.15 and 9.18. Corollary 9.25 had previously appeared in [PS98a, Theorem 4.2].

A more general version of Theorem 9.2 appeared in [Sta96, Theorem I.7.9], attributed to Stanley, Hochster, "and perhaps others". The consequence in Corollary 9.4 was derived in [PS98b, Theorem 2.3]. The unimodular Lawrence ideals in Example 9.22 were expounded upon greatly throughout [BPS01].

The Scarf complex of a lattice was introduced by the mathematical economist Herbert Scarf [Sca86]. This article, which also explains the connection to integer programming, was the original inspiration for the work of Bayer, Peeva, and Sturmfels [BPS98]. Theorem 9.27 is due to Barany and Scarf [BaS96]. A translation into the language of commutative algebra was given in [PS98a, Section 4]. The generic lattice ker([20 24 25 31]) in Example 9.26 is the "smallest" generic lattice, and it was found by exhaustive search in Maple for [PS98a, Example 4.5].

It is an open problem to characterize the Betti numbers of generic lattice ideals. Partial progress in this direction has been made by Björner [Bj00], but we expect further restrictions along the lines of Section 6.4.

Chapter 10

Toric varieties

Just as standard $\mathbb{N}$-graded polynomial rings give rise to projective geometry, multigraded polynomial rings give rise to toric geometry. The purpose of this chapter is to make sense of this statement. We begin by explaining how the geometry and representation theory of abelian group actions on vector spaces gives rise to multigradings on polynomial rings and how the affine quotients by such actions are reflected algebraically. Then we treat the projective case, which considers an additional grading by $\mathbb{Z}$. The main point comes next: a toric variety is characterized by the data of a multigraded polynomial ring and a squarefree monomial ideal that is in a precise sense compatible with the multigrading. Through the geometry of invariant theory, we relate this *homogeneous coordinate ring* perspective to the more classical constructions of toric varieties from fans and polytopes.

For simplicity, we work here over the field $\Bbbk = \mathbb{C}$ of complex numbers.

10.1 Abelian group actions

Toric varieties are quotients X of certain open subsets of the n-dimensional vector space $\mathbb{C}^n$ by actions of subgroups G of the standard n-torus $(\mathbb{C}^*)^n$. By virtue of the inclusion $(\mathbb{C}^*)^n \subset \mathbb{C}^n$, the quotient group $T = (\mathbb{C}^*)^n/G$ is always a subvariety of X, and the action of T on itself extends to an action on all of X. In this section and the next we present algebraic constructions of those quotients X regarded as the nicest, namely affine and projective toric varieties; we postpone the general definition and construction of toric varieties as quotients via homogeneous coordinate rings until Section 10.3.

Suppose we are given an exact sequence of (additive) abelian groups

$$0 \longleftarrow A \longleftarrow \mathbb{Z}^n \longleftarrow L \longleftarrow 0 \tag{10.1}$$

defining a multigrading on $S = \mathbb{C}[\mathbf{x}]$ as in Chapter 8 (note that $A \longleftarrow \mathbb{Z}^n$ is surjective here). Considering the nonzero complex numbers $\mathbb{C}^*$ as a multiplicative abelian group, we get a corresponding map from $(\mathbb{C}^*)^n = \mathrm{Hom}(\mathbb{Z}^n, \mathbb{C}^*)$ to $\mathrm{Hom}(L, \mathbb{C}^*)$, induced by taking homomorphisms into $\mathbb{C}^*$. Since the group $\mathbb{C}^*$ is *divisible*, meaning that every element has an mth root for every $m \in \mathbb{N}$, the homomorphism $(\mathbb{C}^*)^n \to \mathrm{Hom}(L, \mathbb{C}^*)$ is surjective. In other words, the exact sequence (10.1) dualizes to an exact sequence

$$1 \longrightarrow G \longrightarrow (\mathbb{C}^*)^n \longrightarrow \mathrm{Hom}(L, \mathbb{C}^*) \longrightarrow 1 \qquad (10.2)$$

of multiplicative abelian groups, where $G = \mathrm{Hom}(A, \mathbb{C}^*)$ is the *character group* of A. Thus (10.2) defines an embedding of G into the group $(\mathbb{C}^*)^n$ of diagonal invertible $n \times n$ matrices. The sequence (10.1) gives a *presentation* of the (additive) group A, whereas the sequence (10.2) gives a *representation* of the (multiplicative) group G.

Example 10.1 Recall Example 8.3, where $n = 3$ and L is the lattice spanned by $(1, 1, 1)$ and $(1, 3, 5)$ inside $\mathbb{Z}^3$, so that $A = \mathbb{Z}^3/L \cong \mathbb{Z} \oplus \mathbb{Z}/2\mathbb{Z}$. The group G is the kernel of the multiplicative group homomorphism

$$(\mathbb{C}^*)^3 \to \mathrm{Hom}(L, \mathbb{C}^*) \quad \text{sending} \quad (z_1, z_2, z_3) \mapsto (z_1 z_2 z_3, z_1 z_2^3 z_3^5).$$

Hence G equals the subvariety of $(\mathbb{C}^*)^3$ cut out by the lattice ideal

$$
\begin{aligned}
I_L &= \langle z_1 z_2 z_3 - 1, \, z_1 z_2^3 z_3^5 - 1 \rangle \\
&= \langle z_1 - z_3, \, z_2 z_3^2 - 1 \rangle \cap \langle z_1 + z_3, \, z_2 z_3^2 + 1 \rangle.
\end{aligned}
$$

The two components of I_L correspond to the torsion $\mathbb{Z}_2 = \{\pm 1\}$ of $G = \mathrm{Hom}(\mathbb{Z} \oplus \mathbb{Z}/2\mathbb{Z}, \mathbb{C}^*) \cong \mathbb{C}^* \times \mathbb{Z}_2$. This group G acts on the vector space $\mathbb{C}^3$ by sending $(\alpha, \beta) \in G$ to the diagonal matrix with entries $(\alpha\beta, \alpha^{-2}\beta, \alpha)$. $\diamond$

In general, let $z_1, \ldots, z_n$ denote the coordinates on the torus $(\mathbb{C}^*)^n$, so as to distinguish them from the coordinates $x_1, \ldots, x_n$ on the affine space $\mathbb{C}^n$. The subgroup G of $(\mathbb{C}^*)^n$ is the common zero set of the lattice ideal I_L, which is regarded here as an ideal in the Laurent polynomial ring $\mathbb{C}[z_1^{\pm 1}, \ldots, z_n^{\pm 1}]$. The torus $(\mathbb{C}^*)^n$ acts on the polynomial ring $S = \mathbb{C}[x_1, \ldots, x_n]$ by scaling variables: $(\zeta_1, \ldots, \zeta_n) \in (\mathbb{C}^*)^n$ sends the variable x_i to $\zeta_i x_i$. This action of $(\mathbb{C}^*)^n$ restricts to an action of $G = \mathcal{V}(I_L)$ on S.

Lemma 10.2 *A polynomial $f \in S$ is a common eigenvector for G if and only if it is homogeneous under the multigrading by A. In particular, $f \in S$ is fixed by G if and only if it is homogeneous of degree $\mathbf{0}$, so $\deg(f) \in L \cap \mathbb{N}^n$.*

Proof. If $\zeta = (\zeta_1, \ldots, \zeta_n)$ represents an element in G, then the image of a polynomial $f(x_1, \ldots, x_n) = \sum c_{\mathbf{u}} \mathbf{x}^{\mathbf{u}}$ under ζ can be computed as follows:

$$f(\zeta_1 x_1, \ldots, \zeta_n x_n) = \sum c_{\mathbf{u}} \cdot \zeta^{\mathbf{u}} \cdot \mathbf{x}^{\mathbf{u}}.$$

By definition of G, we get $\zeta^{\mathbf{u}} = \zeta^{\mathbf{v}}$ for all $\zeta \in G$ if and only if $\mathbf{u} \equiv \mathbf{v} \pmod{L}$. Hence f is an eigenvector if and only if all vectors $\mathbf{u}$ with $c_{\mathbf{u}} \neq 0$ have the same image $\mathbf{a}$ in $\mathbb{Z}^n/L = A$, or equivalently, if f is homogeneous of degree $\mathbf{a}$. The second statement concerns the special case $\mathbf{a} = \mathbf{0}$. $\square$

Lemma 10.3 *An ideal I inside S is stable under the action of G (that is, $G \cdot I = I$) if and only if I is homogeneous for the multigrading by A.*

Proof. Every homogeneous ideal is generated by homogeneous polynomials, which are simultaneous eigenvectors for all of G by Lemma 10.2. Therefore such ideals are stable under the action of G. For the converse, suppose that I is a G-stable ideal and $f \in I$. It suffices to prove that every homogeneous component of f lies in I. Write $f = \sum_{\mathbf{a} \in A'} c_{\mathbf{a}} \cdot f_{(\mathbf{a})}$, where A' is a finite subset of A and $f_{(\mathbf{a})}$ is homogeneous of degree $\mathbf{a}$. A basic result in representation theory states that the characters of the finitely generated abelian group A are $\mathbb{C}$-linearly independent. We can therefore find a subset G' of G such that the complex matrix $[\sigma(\mathbf{a})]_{\sigma \in G', \mathbf{a} \in A'}$ is square and invertible. This implies that the graded components $f_{(\mathbf{a})}$ are $\mathbb{C}$-linear combinations of the images of f under the group elements σ that lie in G'. Each of these images lies in I, and therefore each $f_{(\mathbf{a})}$ lies in I. $\square$

Note that Lemma 10.3 generalizes Proposition 2.1.

The most basic construction of a quotient in algebraic geometry is via the ring of invariant polynomial functions. In our abelian setting, the ring S^G of invariants equals the normal semigroup ring $S_0 = \Bbbk[L \cap \mathbb{N}^n]$. This is the second statement in Lemma 10.2. The elements of $S^G = S_0$ are precisely those polynomials that are constant along all orbits of G on $\mathbb{C}^n$.

Example 10.4 In Example 10.1, the invariant ring for the action of $G \cong \mathbb{C}^* \times \mathbb{Z}_2$ equals

$$\mathbb{C}[x_1, x_2, x_3]^G = \mathbb{C}[x_1^4 x_2^2, x_1 x_2 x_3, x_2^2 x_3^4] \cong \mathbb{C}[u, v, w]/\langle uw - v^4 \rangle.$$

The inclusion of this ring into $\mathbb{C}[x_1, x_2, x_3]$ defines a morphism of affine varieties from $\mathbb{C}^3$ onto the surface $uw = v^4$ inside $\mathbb{C}^3$. Each G-orbit in $\mathbb{C}^3$ is mapped to a unique point under this morphism. Moreover, distinct G-orbits are mapped to distinct points on the surface, provided the orbits are sufficiently general. The surface $uw = v^4$ is the quotient, denoted by $\mathbb{C}^3 /\!/ G$. $\diamond$

Definition 10.5 The **affine GIT quotient** of $\mathbb{C}^n$ modulo G is the affine toric variety $\operatorname{Spec}(S^G)$ whose coordinate ring is the invariant ring S^G:

$$\mathbb{C}^n /\!/ G := \operatorname{Spec}(S^G) = \operatorname{Spec}(S_0) = \operatorname{Spec}(\mathbb{C}[Q]).$$

where $Q = \mathbb{N}^n \cap L$ is the saturated pointed semigroup in degree $\mathbf{0}$. The acronym GIT stands for *Geometric Invariant Theory*.

Officially, the *spectrum* $\operatorname{Spec}(S^G)$ of the ring S^G is the set of all prime ideals in S^G together with the Zariski topology on this set. However, since S^G is an integral domain that is generated as a $\mathbb{C}$-algebra by a finite set of monomials, namely those corresponding to the Hilbert basis $\mathcal{H}_Q$ of Q, we can identify $\operatorname{Spec}(S^G)$ with the closure of the variety parametrized by those monomials. In particular, $\operatorname{Spec}(S^G)$ is an irreducible affine subvariety of a complex vector space whose basis is in bijection with the Hilbert basis $\mathcal{H}_Q$. Observe that by Proposition 7.20, every saturated affine semigroup Q can be expressed as $Q = L \cap \mathbb{N}^n$, so the spectrum of every normal affine semigroup ring $\mathbb{C}[Q]$ is an affine toric variety.

This construction of the quotient $\mathbb{C}^n /\!\!/ G$ is fully satisfactory when G is a finite group. Note that in this case, the two groups $(G, *)$ and $(A, +)$ are actually isomorphic. Indeed, every cyclic group is isomorphic to its character group, and this property is preserved under taking direct sums.

Let us work out an important family of examples of cyclic group actions.

Example 10.6 (Veronese rings) Fix a positive integer p and let L denote the sublattice of $\mathbb{Z}^n$ consisting of all vectors whose coordinate sum is divisible by p. Then $A = \mathbb{Z}^n/L$ is isomorphic to the cyclic group $\mathbb{Z}/p\mathbb{Z}$, and the grading of $S = \mathbb{C}[x_1, \dots, x_n]$ is given by total degree modulo p. The multiplicative group $G \cong \mathbb{Z}/p\mathbb{Z}$ acts on $\mathbb{C}^n$ via $(x_1, \dots, x_n) \mapsto (\zeta x_1, \dots, \zeta x_n)$, where ζ is a primitive p^{th} root of unity. The invariant ring $S^G = S_0$ is the $\Bbbk$-linear span of all monomials $x_1^{i_1} x_2^{i_2} \cdots x_n^{i_n}$ with the property that p divides $i_1 + i_2 + \cdots + i_n$. It is minimally generated as a $\Bbbk$-algebra by those monomials with $i_1 + \cdots + i_n = p$. Equivalently, the Hilbert basis of $Q = L \cap \mathbb{N}^n$ is

$$\mathcal{H}_Q = \{(i_1, i_2, \dots, i_n) \in \mathbb{N}^n \mid i_1 + i_2 + \cdots + i_n = p\}.$$

The ring S^G is the pth *Veronese subring* of the polynomial ring S. $\Diamond$

10.2   Projective quotients

A major drawback of the affine GIT quotient is that $\mathbb{C}^n /\!\!/ G$ is often only a point. Indeed, the spectrum of S^G is a point if and only if S^G consists just of the ground field $\mathbb{C}$, or equivalently, when the only polynomials constant along all G-orbits are the constant polynomials. In view of our characterization of positive gradings in Theorem 8.6, we reach the following conclusion.

Corollary 10.7 *The A-grading is positive if and only if $\mathbb{C}^n /\!\!/ G$ is a point.*

To fix this problem, we now introduce *projective GIT quotients*. These quotients are toric varieties that are not affine, so their description is a bit more tricky. In particular, more data are needed than simply the action of G on $\mathbb{C}^n$: we must fix an element $\mathbf{a}$ in the grading group A.

Consider the graded components $S_{r\mathbf{a}}$ where r runs over all nonnegative integers, and take their (generally infinite) direct sum

$$S_{(\mathbf{a})} = S_0 \oplus S_{\mathbf{a}} \oplus S_{2\mathbf{a}} \oplus S_{3\mathbf{a}} \oplus \cdots . \tag{10.3}$$

This graded S_0-module, each of whose graded pieces $S_{r\mathbf{a}}$ is a finitely generated over S_0 by Proposition 8.4, is actually an S_0-subalgebra of S. Indeed, the product of an element in $S_{r\mathbf{a}}$ and an element in $S_{r'\mathbf{a}}$ lies in $S_{(r+r')\mathbf{a}}$

by definition. Of course, every S_0-algebra is automatically a $\mathbb{C}$-algebra as well. In what follows it will be crucial to distinguish the S_0-algebra structure on $S_{(\mathbf{a})}$ from its $\mathbb{C}$-algebra structure. The S_0-algebra structure carries a natural $\mathbb{N}$-grading, which we emphasize by introducing an auxilliary grading variable γ that allows us to write

$$S_{(\mathbf{a})} \;=\; \bigoplus_{r=0}^{\infty} \gamma^r S_{r\mathbf{a}} \;=\; S_0 \oplus \gamma S_{\mathbf{a}} \oplus \gamma^2 S_{2\mathbf{a}} \oplus \gamma^3 S_{3\mathbf{a}} \oplus \cdots. \quad (10.4)$$

Definition 10.8 The **projective GIT quotient** of $\mathbb{C}^n$ modulo G at $\mathbf{a}$ is the projective spectrum $\mathbb{C}^n /\!\!/_{\mathbf{a}} G$ of the $\mathbb{N}$-graded S_0-algebra $S_{(\mathbf{a})}$:

$$\mathbb{C}^n /\!\!/_{\mathbf{a}} G \;=\; \mathrm{Proj}(S_{(\mathbf{a})}) \;=\; \mathrm{Proj}\Big(\bigoplus_{r=0}^{\infty} \gamma^r S_{r\mathbf{a}}\Big).$$

Officially, the toric variety $\mathrm{Proj}(S_{(\mathbf{a})})$ consists of all prime ideals in $S_{(\mathbf{a})}$ homogeneous with respect to γ and not containing the *irrelevant ideal*

$$S_{(\mathbf{a})}^{+} \;=\; \bigoplus_{r=1}^{\infty} \gamma^r S_{r\mathbf{a}} \;=\; \gamma S_{\mathbf{a}} \oplus \gamma^2 S_{2\mathbf{a}} \oplus \gamma^3 S_{3\mathbf{a}} \oplus \cdots.$$

If P is such a homogeneous prime ideal in $S_{(\mathbf{a})}$, then $P \cap S_0$ is a prime ideal in S_0. This statement is more commonly phrased in geometric language.

Proposition 10.9 *The map $P \mapsto P \cap S_0$ defines a projective morphism from the projective GIT quotient $\mathbb{C}^n /\!\!/_{\mathbf{a}} G$ to the affine GIT quotient $\mathbb{C}^n /\!\!/ G$. $\mathbb{C}^n /\!\!/_{\mathbf{a}} G$ is a projective toric variety if and only if S is positively graded by A.*

Proof. The canonical map from the projective spectrum of an $\mathbb{N}$-graded ring to the spectrum of its $\mathbb{N}$-graded degree zero part is a projective morphism by definition, proving the first statement. For the second, a complex variety is projective over $\mathbb{C}$ if and only if it admits a projective morphism to the point $\mathrm{Spec}(\mathbb{C})$. Thus the "if" direction is a consequence of Theorem 8.6 and Corollary 10.7. For the "only if" direction, note that $\mathbb{C}^n /\!\!/_{\mathbf{a}} G \to \mathbb{C}^n /\!\!/ G$ is a surjective morphism to $\mathrm{Spec}(S_0)$. Since projective varieties admit only constant maps to affine varieties, the affine variety $\mathrm{Spec}(S_0)$ must be a point. $\square$

The ring $S_{(\mathbf{a})}$ and the quotient $\mathbb{C}^n /\!\!/_{\mathbf{a}} G$ can be computed using the algorithm in the proof of Proposition 8.4: compute the Hilbert basis $\mathcal{H}$ for the saturated semigroup $L_{\mathbf{a}} \cap \mathbb{N}^{n+1}$, where $L_{\mathbf{a}}$ is the kernel of $\mathbb{Z}^{n+1} \to A$ under the morphism sending $(\mathbf{v}, r)$ to $(\mathbf{v} \pmod{L}) - r \cdot \mathbf{a}$. Let $\mathcal{H}_0$ be the set of elements in $\mathcal{H}$ having last coordinate zero, and set $\mathcal{H}_+ = \mathcal{H} \smallsetminus \mathcal{H}_0$.

Proposition 10.10 *The S_0-algebra $S_{(\mathbf{a})}$ is minimally generated over S_0 by the monomials $\mathbf{x}^{\mathbf{u}} \gamma^r$, where $(\mathbf{u}, r)$ runs over all vectors in $\mathcal{H}_+$.*

Proof. In the proof of Proposition 8.4, we saw that S_0 is minimally generated as a $\mathbb{C}$-algebra by the monomials $\mathbf{x}^{\mathbf{u}}$ for $\mathbf{u}$ in $\mathcal{H}_0$. Likewise, the ring $S_{(\mathbf{a})}$ is minimally generated as a $\mathbb{C}$-algebra by the monomials $\mathbf{x}^{\mathbf{u}}\gamma^r$ for $(\mathbf{u}, r)$ in $\mathcal{H}$. It follows that the monomials $\mathbf{x}^{\mathbf{u}}\gamma^r$ with $(\mathbf{u}, r) \in \mathcal{H}_+$ generate $S_{(\mathbf{a})}$ as an S_0-algebra. None of these monomials can be omitted. $\square$

The toric variety $\mathbb{C}^n /\!\!/_{\mathbf{a}} G$ is covered by affine open subsets $\mathcal{U}(\mathbf{x}^{\mathbf{u}}\gamma^r)$, one for each generator $\mathbf{x}^{\mathbf{u}}\gamma^r$ of $S_{(\mathbf{a})}$ over S_0. This affine open subset consists of all points in $\mathbb{C}^n /\!\!/_{\mathbf{a}} G$ for which the coordinate $\mathbf{x}^{\mathbf{u}}\gamma^r$ is nonzero. More precisely, $\mathcal{U}(\mathbf{x}^{\mathbf{u}}\gamma^r)$ is by definition the spectrum of the $\mathbb{C}$-algebra consisting of elements of γ-degree 0 in the localization of $S_{(\mathbf{a})}$ inverting $\mathbf{x}^{\mathbf{u}}\gamma^r$.

Proposition 10.11 *The affine toric variety* $\mathcal{U}(\mathbf{x}^{\mathbf{u}}\gamma^r)$ *is the spectrum of the semigroup ring over $\mathbb{C}$ for the semigroup* $\{\mathbf{w} \in L \mid (\mathbf{w} + \mathbb{N}\mathbf{u}) \cap \mathbb{N}^n \neq \varnothing\}$ *of vectors $\mathbf{w}$ in L that can be made positive by adding high multiples of $\mathbf{u}$.*

Proof. The γ-degree 0 part of the localization $S_{(\mathbf{a})}[\mathbf{x}^{-\mathbf{u}}\gamma^{-r}]$ is spanned by all monomials $\mathbf{x}^{\mathbf{v}-s\mathbf{u}}$ for nonnegative integers s and monomials $\mathbf{x}^{\mathbf{v}}$ of degree $rs \cdot \mathbf{a}$. The monomial $\mathbf{x}^{\mathbf{w}}$ for $\mathbf{w} = \mathbf{v} - s\mathbf{u}$ satisfies $\mathbf{v} \in (\mathbf{w} + \mathbb{N}\mathbf{u}) \cap \mathbb{N}^n$. $\square$

Example 10.12 (The two resolutions of the cone over the quadric)
Consider the action of $G = \mathbb{C}^*$ on affine 4-space given by $(x_1, x_2, x_3, x_4) \mapsto (zx_1, zx_2, z^{-1}x_3, z^{-1}x_4)$. This notation should be thought of as indicating the map $\mathbb{C}[x_1, x_2, x_3, x_4] \to \mathbb{C}[z, z^{-1}] \otimes_{\mathbb{C}} \mathbb{C}[x_1, x_2, x_3, x_4]$ on coordinate rings reflecting the morphism $\mathbb{C}^* \times \mathbb{C}^4 \to \mathbb{C}^4$; the variables $\{x_1, x_2, x_3, x_4\}$ go to the tensor products $\{z \otimes x_1, z \otimes x_2, z^{-1} \otimes x_3, z^{-1} \otimes x_4\}$.

Here $A = \mathbb{Z}$, the variables x_1 and x_2 have degree 1, and the variables x_3 and x_4 have degree -1. The affine 3-fold

$$\mathbb{C}^4 /\!\!/ G \;=\; \mathrm{Spec}(S_0) \;=\; \mathrm{Spec}(\mathbb{C}[x_1x_3, x_1x_4, x_2x_3, x_2x_4])$$

is the cone over the quadric. It has an isolated singularity at the origin. There are two natural ways to resolve the singularity of $\mathbb{C}^4 /\!\!/ G$. They are given by the map in Proposition 10.9 for $\mathbf{a} = -1$ and $\mathbf{a} = 1$, respectively:

$$\mathbb{C}^4 /\!\!/_{-1} G \qquad\qquad \mathbb{C}^4 /\!\!/_1 G$$
$$\searrow \qquad\qquad \swarrow$$
$$\mathbb{C}^4 /\!\!/ G$$

Let us compute the map for $\mathbf{a} = 1$ in more detail. The ring $S_{(\mathbf{a})}$ is

$$S_{(1)} \;=\; S_0[\gamma x_1, \gamma x_2] \;=\; \mathbb{C}[x_1x_3, x_1x_4, x_2x_3, x_2x_4, \gamma x_1, \gamma x_2].$$

The projective spectrum of this ring with respect to the γ-grading is the projective GIT quotient $\mathbb{C}^4 /\!\!/_1 G$. It has a cover consisting of two affine spaces:

$$\mathcal{U}(\gamma x_1) \;=\; \mathrm{Spec}(\mathbb{C}[x_2/x_1, x_1x_3, x_1x_4]) \;\cong\; \mathbb{C}^3,$$
$$\mathcal{U}(\gamma x_2) \;=\; \mathrm{Spec}(\mathbb{C}[x_1/x_2, x_2x_3, x_2x_4]) \;\cong\; \mathbb{C}^3.$$

We conclude that $\mathbb{C}^4 /\!\!/_1 G$ and (by symmetry) $\mathbb{C}^4 /\!\!/_{-1} G$ are smooth. $\diamond$

Example 10.13 (Toric quiver varieties) Fix a finite directed graph on the vertex set $V = \{1, \ldots, d\}$. The edge set E is a subset of $V \times V$. Loops and multiple edges are allowed. The torus $(\mathbb{C}^*)^V$ with coordinates z_i for $i \in V$ acts on the vector space $\mathbb{C}^E$ with coordinates x_{ij} for $(i, j) \in E$ via

$$x_{ij} \mapsto z_i z_j^{-1} \cdot x_{ij}.$$

The grading group A is the codimension 1 sublattice of $\mathbb{Z}^V$ consisting of vectors with zero coordinate sum. We are interested in the affine quotient $\mathbb{C}^E /\!\!/ (\mathbb{C}^*)^V$ and the projective quotients $\mathbb{C}^E /\!\!/_\mathbf{a} (\mathbb{C}^*)^V$ for $\mathbf{a} \in A$.

Every directed cycle $i_1, i_2, \ldots, i_r, i_1$ gives a monomial of degree $\mathbf{0}$,

$$x_{i_1 i_2} x_{i_2 i_3} x_{i_3 i_4} \cdots x_{i_r i_1},$$

and these monomials minimally generate the semigroup ring $S_\mathbf{0} = K[x_{ij}]_\mathbf{0}$. Thus $\mathbb{C}^E /\!\!/ (\mathbb{C}^*)^V$ is the variety parametrized by these cycle monomials. This affine toric variety is generally singular. The algebra $S_{(\mathbf{a})}$ is generated over $S_\mathbf{0}$ by its monomials of degree $\mathbf{a}$, and the minimal generators are those monomials whose support is a forest. If $\mathbf{a}$ is sufficiently generic, then these forests are spanning trees and $\mathbb{C}^E /\!\!/_\mathbf{a} (\mathbb{C}^*)^V$ is smooth. Example 10.12 is the case where $V = \{1, 2\}$ with edges $(1, 2)$, $(1, 2)$, $(2, 1)$, and $(2, 1)$.

If the given graph is acyclic then $\mathbb{C}^E /\!\!/_\mathbf{a} (\mathbb{C}^*)^V$ is the projective variety parametrized by all monomials $\gamma \cdot \mathbf{x}^\mathbf{u}$, where $\mathbf{u} \in \mathbb{N}^E$ is a flow on a tree having A-degree $\mathbf{a} \in \mathbb{Z}^V$. For instance, let $d = 5$, take $E = \{1, 2\} \times \{3, 4, 5\}$ to be the (acyclically directed) complete bipartite graph $K_{2,3}$, and let $\mathbf{a} = (-3, -3, 2, 2, 2)$. The $S_\mathbf{0}$-algebra $S_{(\mathbf{a})}$ is generated by the seven monomials in $S_\mathbf{a}$. They correspond to the vertices of a regular hexagon plus one interior point. The projective variety $\mathbb{C}^E /\!\!/_\mathbf{a} (\mathbb{C}^*)^V$ is the projective plane blown up at three points. $\diamond$

Every lattice polytope $\mathcal{P}$ of dimension $n - d$ gives rise to a projective toric variety $X_\mathcal{P}$. In Example 10.13 we encountered $X_\mathcal{P}$ for $\mathcal{P}$ a regular hexagon. The general construction proceeds via the map ν from Proposition 7.20, which adapts just as well for polytopes as it does for cones. To be precise, suppose the polytope $\mathcal{P}$ has n facets with primitive integer inner normal vectors $\nu_1, \ldots, \nu_n$. Then $\mathcal{P}$ is defined by inequalities $\nu_i \cdot \mathcal{P} \geq -w_i$ for some vector $\mathbf{w} \in \mathbb{Z}^n$. The map $\nu : \mathbb{R}^{n-d} \to \mathbb{R}^n$ sending $\mathbf{u} \in \mathbb{R}^{n-d}$ to $(\nu_1 \cdot \mathbf{u}, \ldots, \nu_n \cdot \mathbf{u})$ takes $\mathbb{R}^{n-d}$ to a subspace $V \subseteq \mathbb{R}^n$. The map ν is injective because $\mathcal{P}$ has a vertex, and its restriction to $\mathcal{P}$ is an isomorphism

$$\mathcal{P} \;\cong\; \nu(\mathcal{P}) \;=\; \mathbb{R}^n_{\geq -\mathbf{w}} \cap V, \tag{10.5}$$

where $\mathbb{R}^n_{\geq -\mathbf{w}} = \{\mathbf{v} \in \mathbb{R}^n \mid v_i \geq -w_i \text{ for all } i\}$. Set $L = V \cap \mathbb{Z}^n$, and denote by $\mathbf{a}$ the coset in $\mathbb{Z}^n / L$ containing $\mathbf{w}$. With $G = \mathrm{Hom}(\mathbb{Z}^n / L, \mathbb{C}^*)$ as before,

$$X_\mathcal{P} \;:=\; \mathbb{C}^n /\!\!/_\mathbf{a} G \tag{10.6}$$

is the projective toric variety associated with the lattice polytope $\mathcal{P}$. It is reasonable that $X_{\mathcal{P}}$ depends on $\mathbf{a}$ rather than $\mathbf{w}$, since the polytope $\mathbb{R}^n_{\geq -\mathbf{w}} \cap V$ is a lattice translate of $\mathbb{R}^n_{\geq -\mathbf{v}} \cap V$ whenever $\mathbf{v} \equiv \mathbf{w} \pmod{L}$.

Example 10.14 (The 3-dimensional cube) We construct the toric variety $X_{\mathcal{P}}$ associated with the standard 3-dimensional cube $\mathcal{P} = \operatorname{conv}\{0,1\}^3$. For the representation (10.5) with $n = 6$, we take $\mathbf{w} = (1,1,1,0,0,0)$ and

$$L \;=\; \mathbb{Z} \cdot \{(1,0,0,-1,0,0),\ (0,1,0,0,-1,0),\ (0,0,1,0,0,-1)\}.$$

This lattice induces the action of $G = (\mathbb{C}^*)^3$ on $\mathbb{C}^6$ via

$$(x_1, x_2, x_3, x_4, x_5, x_6) \;\longmapsto\; (z_1 x_1, z_2 x_2, z_3 x_3, z_1 x_4, z_2 x_5, z_3 x_6). \qquad (10.7)$$

Hence $X_{\mathcal{P}}$ equals $\mathbb{P}^1 \times \mathbb{P}^1 \times \mathbb{P}^1$, the product of three projective lines. Points on $X_{\mathcal{P}}$ are represented by vectors in $\mathbb{C}^6$ modulo scaling (10.7). However, some vectors in $\mathbb{C}^6$ are not allowed. They are the zeros of the irrelevant ideal

$$\langle S_{\mathbf{a}} \rangle \;=\; \langle x_1, x_4 \rangle \cap \langle x_2, x_5 \rangle \cap \langle x_3, x_6 \rangle.$$

Irrelevant ideals of general toric varieties will appear in the next section. $\diamond$

10.3   Constructing toric varieties

A general toric variety is constructed from a fan in a lattice. To be consistent with earlier notation, we take this lattice to be $L^\vee = \operatorname{Hom}(L, \mathbb{Z})$, the lattice dual to L. Its relation to the grading group A comes from applying the contravariant functor $\operatorname{Hom}(\_, \mathbb{Z})$ to the sequence (10.1):

$$0 \longrightarrow \operatorname{Hom}(A, \mathbb{Z}) \longrightarrow \mathbb{Z}^n \longrightarrow L^\vee \longrightarrow \operatorname{Ext}^1(A, \mathbb{Z}) \longrightarrow 0. \qquad (10.8)$$

Let $\nu_1, \ldots, \nu_n$ denote the images in $L^\vee$ of the unit vectors in $\mathbb{Z}^n$ under the middle morphism of (10.8), so the embedding $L \hookrightarrow \mathbb{Z}^n$ is given by $\mathbf{u} \mapsto (\nu_1 \cdot \mathbf{u}, \ldots, \nu_n \cdot \mathbf{u})$. We write $C = \mathbb{R}_{\geq 0}\{\nu_1, \ldots, \nu_n\}$ for the cone generated by these n vectors in the real vector space $L^\vee \otimes \mathbb{R}$. This cone C may be pointed, but frequently (when we have better luck) it is not.

Lemma 10.15 *The A-grading is positive if and only if C equals $L^\vee \otimes \mathbb{R}$.*

Proof. The A-grading is not positive if and only if there exists a nonzero vector $\mathbf{u}$ in $L \cap \mathbb{N}^n$, by Theorem 8.6. On the other hand, the cone C fails to equal $L^\vee \otimes \mathbb{R}$ if and only if all functionals ν_i lie on one side of a hyperplane in $L^\vee$. This hyperplane is orthogonal to some nonzero vector $\mathbf{u}$ in L, which we may choose to satisfy $\nu_i \cdot \mathbf{u} \geq 0$ for $i = 1, \ldots, n$. The image of $\mathbf{u}$ in $\mathbb{Z}^n$ under the inclusion $L \hookrightarrow \mathbb{Z}^n$ is $(\nu_1 \cdot \mathbf{u}, \ldots, \nu_n \cdot \mathbf{u}) \in L \cap \mathbb{N}^n$ by definition. $\square$

Definition 10.16 Fix a cone C inside $L^\vee \otimes \mathbb{R}$. A **fan** in $L^\vee$ is a collection Σ of subcones $\sigma \subseteq C$ satisfying the following properties:

- Every cone $\sigma \in \Sigma$ is pointed.
- Every face of a cone in Σ is also in Σ.
- The intersection of two cones in Σ is a common face of each cone.

The fan is **compatible** with the multigrading by A if $C = \mathbb{R}_{\geq 0}\{\nu_1, \ldots, \nu_n\}$ is the cone defined after (10.8), and also:

- The cones in Σ are generated by images of unit vectors under $\mathbb{Z}^n \to L^\vee$:

$$\sigma \;=\; \mathbb{R}_{\geq 0}\{\nu_{i_1}, \ldots, \nu_{i_s}\} \text{ for } \sigma \in \Sigma.$$

The fan Σ is **complete** if every point of $L^\vee$ lies in some cone of Σ (so $C = L^\vee \otimes \mathbb{R}$ as in Lemma 10.15 if Σ is compatible). If every cone in Σ is generated by part of a $\mathbb{Z}$-basis for $L^\vee$, then Σ is called **smooth**; if the generators are merely linearly independent, then Σ is called **simplicial**.

Example 10.17 (The normal fan of a lattice polytope) Consider a lattice polytope $\mathcal{P}$ as in (10.5). The multigrading by A is positive because $\mathcal{P}$ is bounded. The vectors in $C = L^\vee \otimes \mathbb{R}$ are linear functionals on the polytope $\mathcal{P}$. Stipulating that two such functionals ν and ν' are equivalent if they are minimized on the same face of $\mathcal{P}$, the set of closures of the equivalence classes is a fan $\Sigma(\mathcal{P})$, compatible in the sense of Definition 10.16 (see Theorem 10.30). The fan $\Sigma(\mathcal{P})$ is called the **(inner) normal fan** of $\mathcal{P}$. (See [Zie95, Example 7.3] for a nice picture of an outer normal fan.) $\diamond$

All of the fans we encounter will be compatible, and these can be encoded by squarefree monomial ideals, given the homomorphism $\mathbb{Z}^n \to L^\vee$.

Definition 10.18 The **irrelevant ideal** of a compatible fan Σ is the squarefree monomial ideal

$$B_\Sigma \;=\; \langle x_{j_1} \cdots x_{j_s} \mid \{\nu_1, \ldots, \nu_n\} \smallsetminus \{\nu_{j_1}, \ldots, \nu_{j_s}\} \text{ spans a cone of } \Sigma \rangle$$

in $S = \mathbb{C}[\mathbf{x}]$. Equivalently, the Alexander dual of the irrelevant ideal is

$$I_\Sigma \;=\; \langle x_{i_1} \cdots x_{i_r} \mid \nu_{i_1}, \ldots, \nu_{i_r} \text{ do not lie in a common cone of } \Sigma \rangle.$$

The Stanley–Reisner simplicial complex of the ideal I_Σ can be identified with the variety $\mathcal{V}(I_\Sigma)$. Its facets are those subsets of $\{1, \ldots, n\}$ that index the maximal faces of Σ. If Σ is a simplicial fan, then this simplicial complex is precisely Σ itself. The facets of the simplicial complex associated with B_Σ are those subsets of $\{1, \ldots, n\}$ complementary to minimal nonfaces of Σ. Consequently, the variety $\mathcal{V}(B_\Sigma)$ is usually harder to visualize in terms of the fan Σ; but see Theorem 10.30 for the projective GIT case, where $\mathcal{V}(B_\Sigma)$ has a simple geometric description.

Example 10.19 (The 3-cube revisited) For the normal fan Σ of the 3-cube in Example 10.14, the simplicial complex of I_Σ is the boundary of the octahedron. The simplicial complex of B_Σ consists of three tetrahedra. $\diamond$

The variety $\mathcal{V}(B_\Sigma)$ of the irrelevant ideal consists of coordinate subspaces in the vector space $\mathbb{C}^n$. We will be interested in the G-orbits on the complement $\mathcal{U}_\Sigma = \mathbb{C}^n \smallsetminus \mathcal{V}(B_\Sigma)$ of this subspace arrangement. To begin, $\mathcal{U}_\Sigma$ is the union over all cones $\sigma \in \Sigma$ of open subsets, each of which is defined by the nonvanishing of a monomial $\mathbf{x}^{\overline{\sigma}} = x_{j_1} \cdots x_{j_s}$ in B_Σ:

$$\mathcal{U}_\Sigma = \bigcup_{\sigma \in \Sigma} \mathcal{U}_\sigma, \quad \text{where} \quad \mathcal{U}_\sigma = \mathbb{C}^n \smallsetminus \mathcal{V}(\mathbf{x}^{\overline{\sigma}}) = \mathrm{Spec}(S[\mathbf{x}^{-\overline{\sigma}}]). \quad (10.9)$$

We have seen in Definition 10.5 that taking degree $\mathbf{0}$ pieces can be interpreted geometrically as taking the quotient by the action of the torus G. Therefore, we have a collection of affine GIT quotients

$$X_\sigma \;=\; \mathcal{U}_\sigma /\!/ G \;=\; \mathrm{Spec}(S[\mathbf{x}^{-\overline{\sigma}}]_{\mathbf{0}}); \quad (10.10)$$

these affine toric varieties arise from a multigraded generalization of the procedure in Proposition 10.11. For notation, let $\mathbf{e}_{\overline{\sigma}} = \mathbf{e}_{j_1} + \cdots + \mathbf{e}_{j_s}$ be the sum of all the unit vectors in $\mathbb{Z}^n$ corresponding to rays $\nu_{j_1}, \ldots, \nu_{j_s}$ outside σ.

Lemma 10.20 *The affine toric variety X_σ equals the spectrum of the semigroup ring over $\mathbb{C}$ for the semigroup $\{\mathbf{w} \in L \mid (\mathbf{w} + \mathbb{N}\mathbf{e}_{\overline{\sigma}}) \cap \mathbb{N}^n \neq \varnothing\}$ of vectors $\mathbf{w}$ in L that can be made positive by adding high multiples of $\mathbf{e}_{\overline{\sigma}}$.*

Proof. A Laurent monomial $\mathbf{x}^{\mathbf{w}}$ lies in the localization $S[\mathbf{x}^{-\overline{\sigma}}]$ if and only if $\mathbf{w} = \mathbf{v} - r\mathbf{e}_{\overline{\sigma}}$ for some $\mathbf{v} \in \mathbb{N}^n$ and $r \in \mathbb{N}$; in other words, $\mathbf{w} + r\mathbf{e}_{\overline{\sigma}} = \mathbf{v} \in \mathbb{N}^n$. On the other hand, $\mathbf{x}^{\mathbf{w}}$ has degree $\mathbf{0}$ if and only if $\mathbf{w} \in L$. $\square$

Proposition 10.21 *The semigroup $\{\mathbf{w} \in L \mid (\mathbf{w} + \mathbb{N}\mathbf{e}_{\overline{\sigma}}) \cap \mathbb{N}^n \neq \varnothing\}$ from Lemma 10.20 equals the semigroup $\sigma^\vee \cap L$, where $\sigma^\vee$ is the cone in $L \otimes \mathbb{R}$ dual to σ, consisting of linear functionals taking nonnegative values on σ.*

Proof. Suppose σ is generated as a real cone by $\nu_{i_1}, \ldots, \nu_{i_r}$. Let $\mathbf{e}_i^*$ be the basis vector of $\mathbb{Z}^n$ mapping to ν_i. The subset of L on which the linear functionals $\mathbf{e}_{i_1}^*, \ldots, \mathbf{e}_{i_r}^*$ take nonnegative values is by definition the set of lattice points in the real cone $\sigma^\vee \subseteq L \otimes \mathbb{R}$ dual to σ. On the other hand, the subset of $\mathbb{Z}^n$ on which $\mathbf{e}_{i_1}^*, \ldots, \mathbf{e}_{i_r}^*$ take nonnegative values is precisely the semigroup $\mathbb{N}^n - \mathbb{N}\mathbf{e}_{\overline{\sigma}}$ where coordinates not corresponding to generators of σ are allowed to be negative. Intersecting this semigroup with L again yields the part of L where the functionals $\mathbf{e}_{i_1}^*, \ldots, \mathbf{e}_{i_r}^*$ take nonnegative values. $\square$

Example 10.22 For the toric variety $\mathbb{P}^2$, the lattice $L \subset \mathbb{Z}^3$ is the kernel of $[1, 1, 1]$. The three semigroups $\sigma^\vee \cap L$ result from the intersection of L with $\mathbb{N}^3 - \mathbb{N}\mathbf{e}_{\overline{\sigma}}$, where $\overline{\sigma}$ is the singleton $\{i\}$ for $i = 1, 2$, or 3. The cones $\sigma^\vee$ in $\mathbb{R}L$ are the "shadows" of $\mathbb{R}^3_{\geq 0}$ obtained by projecting it along the coordinate directions to $\mathbb{R}L$; see the illustration in Fig. 10.1. $\diamond$

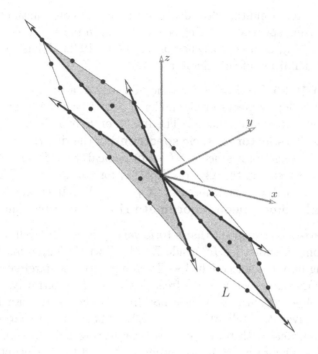

Figure 10.1: The three semigroups $\sigma^\vee \cap L$ for $\mathbb{P}^2$ as shadows of $\mathbb{N}^3$ in L

Geometrically then, the spectrum of $\mathbb{C}[\sigma^\vee \cap L]$ is the affine GIT quotient X_σ resulting from the action of G on the affine variety $\mathcal{U}_\sigma$. Since we are interested in quotienting all of $\mathcal{U}_\Sigma = \mathbb{C}^n \smallsetminus \mathcal{V}(B_\Sigma)$ and not just an open affine subvariety, we need to know how to glue the affine GIT quotients from different cones in Σ.

Corollary 10.23 *If τ is a face of a cone $\sigma \in \Sigma$, then X_τ is an open affine toric subvariety of X_σ. More precisely, $S[\mathbf{x}^{-\overline{\tau}}]_0$ is a localization of $S[\mathbf{x}^{-\overline{\sigma}}]_0$.*

Proof. The ring $S[\mathbf{x}^{-\overline{\tau}}]_0$ is obtained from $S[\mathbf{x}^{-\overline{\sigma}}]_0$ by inverting all monomials $\mathbf{x}^{\mathbf{w}}$ for which the linear functional $\mathbf{w}$ vanishes on τ. $\square$

Intersecting two open subsets $\mathcal{U}_{\sigma_1}$ and $\mathcal{U}_{\sigma_2}$ yields the open subset $\mathcal{U}_\tau$ for the cone $\tau = \sigma_1 \cap \sigma_2$, which lies in the fan Σ by definition of fan. More importantly, Corollary 10.23 says that this remains true if we take the quotient by G, thereby replacing $\mathcal{U}$ by X: the affine variety X_τ is naturally an open affine subvariety of both X_{σ_1} and X_{σ_2}. Hence we can glue them along X_τ. Doing this for all cones in Σ yields a variety X_Σ.

Lemma 10.24 *The open subvariety $\mathcal{U}_\Sigma = \mathbb{C}^n \smallsetminus \mathcal{V}(B_\Sigma)$ of $\mathbb{C}^n$ comes endowed with a morphism $\mathcal{U}_\Sigma \to X_\Sigma$ of varieties.*

Proof. The gluing used to define $\mathcal{U}_\Sigma$ and X_Σ from their open affines $\mathcal{U}_\sigma$ and X_σ commutes with the projections $\mathcal{U}_\sigma \to X_\sigma$ by Corollary 10.23. $\square$

Had we fixed a multigraded ideal I inside S, we could have carried out the gluing using spectra Y_σ of rings $(S[\mathbf{x}^{-\overline{\sigma}}]/I)_0$ in place of the spectra X_σ of rings $S[\mathbf{x}^{-\overline{\sigma}}]_0$, as a consequence of Corollary 10.23. Thus we arrive at the central definition of this chapter.

Definition 10.25 Let $R = S/I$ be the quotient of a multigraded polynomial ring by a homogeneous ideal, and let $B = \langle \mathbf{x}^{\overline{\sigma}} \mid \sigma \in \Sigma \rangle$ be an irrelevant ideal for some compatible fan Σ. The image of B in R is the **irrelevant ideal** of R. The **spector** (or **toric spectrum**) of the ring R with irrelevant ideal B is the variety (or scheme if I is not radical) $\mathrm{SpecTor}(R, B)$ covered by the affine spectra of the algebras $R[\mathbf{x}^{-\overline{\sigma}}]_0$ for cones $\sigma \in \Sigma$. The spector of $R = S$, where $I = 0$, is denoted by X_B or by X_Σ. It is called the **toric variety** with **homogeneous coordinate ring** S and irrelevant ideal B.

Let us stress at this point that a toric variety is equally well determined by giving only a sublattice $L^\vee$ inside $\mathbb{Z}^n$ along with a fan Σ inside $L \otimes \mathbb{R}$, or by giving only the surjection $A \twoheadleftarrow \mathbb{Z}^n$ along with a squarefree monomial ideal B. Of course, we are not free to choose B arbitrarily, given the surjection $A \twoheadleftarrow \mathbb{Z}^n$, just as we are not free to choose the fan in $L \otimes \mathbb{R}$ arbitrarily, given the sublattice $L^\vee$ in $\mathbb{Z}^n$. The point is that we could, if we desired, deal with toric varieties by referring only to combinatorial commutative algebra of the multigrading by A and the irrelevant ideal B.

Example 10.26 Here is a concrete example demonstrating how the spector of a multigraded ring can depend on the choice of irrelevant ideal. For positive integers r and s, consider the polytope $\mathcal{P}_{r,s}$ beneath the planes $z = y$ and $z = x$, above the xy-plane, and satisfying $x \leq r$ and $y \leq s$. The polytope $\mathcal{P}_{r,s}$ is defined by inequalities $\nu_i \cdot \mathbf{u} \geq \mathbf{w}$ for $\mathbf{w} = (0, 0, 0, -r, -s)$, where the linear functionals $\nu_1, \ldots, \nu_5$ are the rows of the matrix $\mathbf{L}$ below:

$$\mathbf{A} = \begin{bmatrix} 0 & 1 & 1 & 1 & 0 \\ 1 & 0 & 1 & 0 & 1 \end{bmatrix} \quad \text{and} \quad \mathbf{L} = \begin{bmatrix} 0 & 1 & -1 \\ 1 & 0 & -1 \\ 0 & 0 & 1 \\ -1 & 0 & 0 \\ 0 & -1 & 0 \end{bmatrix}.$$

The weights of the variables in the multigrading by $A = \mathbb{N}^2$ are the columns of $\mathbf{A}$, and $\mathbf{a} = (r, s)$ is the image of $-\mathbf{w} = (0, 0, 0, r, s)$ in A, so our notation agrees with (10.5). The distinction we will make is between the two cases $r > s$ and $r < s$, yielding $\mathcal{P}_{r,s}$ in the left and right pictures, respectively:

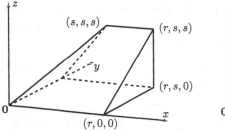
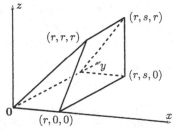

Given the multigrading, the only extra information we need to define a toric variety is an irrelevant ideal in $S = \Bbbk[x_1, x_2, x_3, x_4, x_5]$. When $r > s$, the corresponding polytope $\mathcal{P}_{r>s}$ has toric variety

$$X_{\mathcal{P}_{r>s}} = \operatorname{SpecTor}(S, \langle x_2, x_4 \rangle \cap \langle x_1, x_3, x_5 \rangle),$$

whereas for $r < s$, the corresponding polytope $\mathcal{P}_{r<s}$ has toric variety

$$X_{\mathcal{P}_{r<s}} = \operatorname{SpecTor}(S, \langle x_1, x_5 \rangle \cap \langle x_2, x_3, x_4 \rangle).$$

The reader can verify these claims directly or apply Theorem 10.30. When $r > s$, a monomial lies in the irrelevant ideal if and only if the degree of some monomial with the same support lies interior to the "chamber" $a_1 \geq a_2$ in $A = \mathbb{N}^2$. The analogous statement holds for $a_1 \leq a_2$ when $r < s$. ◇

10.4 Toric varieties as quotients

Now that we have seen how the spectra of the affine semigroup rings $\mathbb{C}[\sigma^\vee \cap L] = S[\mathbf{x}^{-\overline{\sigma}}]_0$ cover the toric variety X_Σ and how this information is recorded globally via the homogeneous coordinate ring, we would like to ascertain what kind of "quotientlike" properties are enjoyed by $\operatorname{SpecTor}(R, B_\Sigma)$, at least when $R = S$. To this end, a variety X is called the *categorical quotient* of a variety $\mathcal{U}$ modulo the action by an algebraic group G if there is a G-equivariant morphism $\mathcal{U} \to X$, in which X carries the trivial G-action, with the property that any G-equivariant morphism from $\mathcal{U}$ to a variety Y with trivial G-action factors uniquely as $\mathcal{U} \to X \to Y$.

Theorem 10.27 *The toric spectrum* $X_\Sigma = \operatorname{SpecTor}(S, B_\Sigma)$ *is the categorical quotient of* $\mathcal{U}_\Sigma = \mathbb{C}^n \setminus \mathcal{V}(B_\Sigma)$ *by* G.

Proof. Suppose $\mathcal{U}_\Sigma \to Y$ is a G-equivariant morphism. Then any local function on Y induces a *G-invariant* function on an open subset U of the variety $\mathcal{U}_\Sigma$. Any G-invariant function on U is locally given by elements in a localization of $S[\mathbf{x}^{-\overline{\sigma}}]_0 = \mathbb{C}[\sigma^\vee \cap L]$. This describes the local maps of structure sheaves $\mathcal{O}_Y \to \mathcal{O}_{X_\Sigma}$, giving the desired morphism $X_\Sigma \to Y$. □

Note that when the fan Σ has only one maximal cone, I_Σ is the zero ideal and B_Σ is the unit ideal. In this case, $\mathcal{V}(B_\Sigma)$ is empty, so $\mathcal{U}_\Sigma = \mathbb{C}^n$ and X_Σ is simply the affine GIT quotient $\mathbb{C}^n /\!/ G$.

The disadvantage of categorical quotients is that sometimes many orbits get lumped together, so the geometric fibers of the morphism $\mathcal{U}_\Sigma \to X_\Sigma$ need not be single orbits.

Example 10.28 Consider the situation of Example 10.12, where Σ is the 3-dimensional fan consisting of a quadrangular cone and its faces. The fiber of the morphism $\mathbb{C}^4 \to X_\Sigma = \mathbb{C}^4 /\!/ G$ over the origin consists of the 2-planes $x_1 = x_2 = 0$ and $x_3 = x_4 = 0$. Hence there are infinitely many G-orbits mapping to the origin under the quotient morphism (G has dimension 1). ◇

In general terms, the categorical quotient in the above example fails to be a so-called *geometric quotient*, where by definition the fibers of the quotient are exactly the orbits. The reason why is that the fan is not simplicial.

Theorem 10.29 *If the fan Σ is simplicial, then the toric spectrum $X_\Sigma = \mathrm{SpecTor}(S, B_\Sigma)$ is a geometric quotient of $\mathcal{U}_\Sigma = \mathbb{C}^n \smallsetminus \mathcal{V}(B_\Sigma)$ by G; that is, the fibers of the morphism $\mathcal{U}_\Sigma \to X_\Sigma$ are precisely the G-orbits on $\mathcal{U}_\Sigma$.*

Proof. We must prove that if $\sigma \in \Sigma$ is a simplicial cone, then the morphism on spectra induced by inclusion of $\mathbb{C}[\sigma^\vee \cap L]$ as the degree $\mathbf{0}$ piece of $S[\mathbf{x}^{-\overline{\sigma}}]$ has fibers that are orbits of G. For ease of notation, set $\check{\sigma} = \sigma^\vee \cap L$, write $\mathbb{Z}^\lambda$ for $\lambda \subseteq \{1, \ldots, n\}$ to mean the subgroup of $\mathbb{Z}^n$ generated by those basis vectors $\mathbf{e}_\ell$ for $\ell \in \lambda$, and recall the setup from (10.9) and (10.10).

Orthogonal projection of $\mathbb{R}^n$ with kernel $\mathbb{R}^{\overline{\sigma}} = \mathbb{Z}^{\overline{\sigma}} \otimes \mathbb{R}$ onto the subspace $\mathbb{R}^\sigma$ induces a *surjection* from the real cone $\sigma^\vee$ to the orthant $\mathbb{R}^\sigma_{\geq 0}$; indeed, this is equivalent to σ being a simplicial cone. This projection maps $L \otimes \mathbb{R}$ surjectively to $\mathbb{R}^\sigma$. Consequently, letting $d = \dim(G)$ as usual, some choice of d basis vectors $\mathbf{e}_{\ell_1}, \ldots, \mathbf{e}_{\ell_d}$ in $\mathbb{R}^n$ whose corresponding rays $\nu_{\ell_1}, \ldots, \nu_{\ell_d}$ lie outside σ are independent modulo $L \otimes \mathbb{R}$. It follows that $L + \mathbb{Z}^\lambda$ has finite index in $\mathbb{Z}^n$, where $\lambda = \{\ell_1, \ldots, \ell_d\}$.

Set $G' = \mathrm{Hom}(\mathbb{Z}^n/(L + \mathbb{Z}^\lambda), \mathbb{C}^*)$, and consider the semigroup $\check{\sigma} + \mathbb{Z}^\lambda$ inside $\mathbb{Z}^n$ generated by $\check{\sigma}$ and the basis vectors $\pm\mathbf{e}_{\ell_1}, \ldots, \pm\mathbf{e}_{\ell_d}$. Then G' is finite, and the inclusion $\mathbb{C}[\check{\sigma} + \mathbb{Z}^\lambda] \hookrightarrow S[\mathbf{x}^{-\overline{\sigma}}]$ induces the quotient morphism $\mathcal{U}_\sigma \to \mathcal{U}_\sigma /\!\!/ G'$. On the other hand, the inclusion $\mathbb{C}[\check{\sigma}] \hookrightarrow \mathbb{C}[\check{\sigma} + \mathbb{Z}^{\overline{\sigma}}]$ induces the projection from $\mathcal{U}_\sigma = (\mathbb{C}^*)^\lambda \times X_\sigma$ onto X_σ.

In summary, the morphism $\mathcal{U}_\sigma \to X_\sigma$ factors as a composite

$$\mathcal{U}_\sigma \longrightarrow \mathcal{U}_\sigma /\!\!/ G' = (\mathbb{C}^*)^\lambda \times X_\sigma \longrightarrow X_\sigma$$

of two quotients. Both of these quotients are geometric, since quotients by finite group actions are always geometric, as are projections of products. We conclude that $\mathcal{U}_\sigma \to X_\sigma$ is a geometric quotient, since the composition of two geometric quotients is again a geometric quotient. $\square$

We will next relate the projective GIT quotients to our definition of toric variety. Let $\mathbf{a} \in A = \mathbb{Z}^n/L$. The set of points in $\mathbb{R}^n_{\geq 0}$ mapping to $\mathbf{a}$ under the projection $\mathbb{R}^n \to A \otimes \mathbb{R}$ is a polyhedron $\mathcal{P}_\mathbf{a}$. Equivalently, picking a representative $\mathbf{w} \in \mathbb{Z}^n$ for the class $\mathbf{a}$ recovers the polyhedron

$$\mathcal{P}_\mathbf{a} \;=\; \mathbb{R}^n_{\geq 0} \cap (\mathbf{w} + L \otimes \mathbb{R}) \tag{10.11}$$

as an intersection of the orthant $\mathbb{R}^n_{\geq 0}$ with the affine translate of the subspace $L \otimes \mathbb{R}$ by $\mathbf{w}$. The geometry of $\mathcal{P}_\mathbf{a}$ in (10.11) is precisely the same as that of $\mathcal{P}_\mathbf{w}$ after (7.9), except that for $\mathcal{P}_\mathbf{a}$, the roles of the various lattices and vector spaces have changed: picking an element $\mathbf{u} \in L$ specifies a lift of each vector $\nu_i \in L^\vee$ to height $w_i + u_i$ inside of $L^\vee_\mathbb{R} \times \mathbb{R}$. We will assume that $\mathcal{P}_\mathbf{a}$ has full dimension $n - d = \mathrm{rank}(L)$. The polyhedron $\mathcal{P}_\mathbf{a}$ only intersects some of the faces of $\mathbb{R}^n_{\geq 0}$; we say that $\mathcal{P}_\mathbf{a}$ *misses* the other faces of $\mathbb{R}^n_{\geq 0}$.

Theorem 10.30 *The normal fan $\Sigma = \Sigma(\mathcal{P}_\mathbf{a})$ is compatible, and its irrelevant variety $\mathcal{V}(B_\Sigma)$ corresponds to the simplicial complex of faces of $\mathbb{R}^n_{\geq 0}$ missed by $\mathcal{P}_\mathbf{a}$. The projective GIT quotient $\mathbb{C}^n /\!\!/_\mathbf{a} G$ is the toric variety X_Σ.*

Proof. As functions of a positive integer r, both the normal fan $\Sigma(\mathcal{P}_{r\mathbf{a}})$ and the projective GIT quotient $\mathbb{C}^n /\!\!/_{r\mathbf{a}} G$ are constant. Indeed, the normal fan does not change under scaling the polyhedron by positive real multiples, and the projective spectrum of an $\mathbb{N}$-graded ring does not change under taking Veronese subrings. Since $\mathcal{P}_\mathbf{a}$ has rational vertices, we therefore assume—after replacing $\mathcal{P}_\mathbf{a}$ by $\mathcal{P}_{r\mathbf{a}}$ for some large positive integer r, perhaps—that every face of $\mathcal{P}_\mathbf{a}$ has an integer point in its relative interior.

The fan Σ is compatible because, by Farkas' Lemma [Zie95, Proposition 1.9], every functional maximized along a face F of $\mathcal{P}_\mathbf{a}$ is a nonnegative combination of the outer normals to the facets containing F.

Next we identify the irrelevant ideal of Σ. A subset ν_σ of the *primitive* (meaning shortest) integer vectors $\nu_1, \ldots, \nu_n$ along the rays in Σ equals the subset lying in a single cone $\sigma \in \Sigma$ if and only if ν_σ is precisely the subset minimized along a face F of $\mathcal{P}_\mathbf{a}$. This occurs if and only if there is a lattice point $\mathbf{u} \in F$ such that $\nu_i \cdot \mathbf{u}$ is nonzero precisely for ν_i in the complement $\nu_{\bar{\sigma}}$ of ν_σ, or equivalently, there is a monomial $\mathbf{x}^\mathbf{u}$ of degree $\mathbf{a}$ with support $\{i \mid \nu_i \in \nu_{\bar{\sigma}}\}$. The lattice point $\mathbf{u}$ is a witness for the fact that $\mathcal{P}_\mathbf{a}$ intersects the face F of $\mathbb{R}^n_{\geq 0}$ with support $\{i \mid \nu_i \in \nu_{\bar{\sigma}}\}$. Since the monomials with support $\{i \mid \nu_i \in \nu_{\bar{\sigma}}\}$ for cones $\sigma \in \Sigma$ are exactly those in the irrelevant ideal B_Σ, a monomial $\mathbf{x}^\mathbf{v}$ lies in B_Σ if and only if its support $\mathrm{supp}(\mathbf{x}^\mathbf{v})$ corresponds to a face of $\mathbb{R}^n_{\geq 0}$ intersecting $\mathcal{P}_\mathbf{a}$. Hence $\mathbf{x}^\mathbf{v}$ lies outside of B_Σ precisely when $\mathrm{supp}(\mathbf{x}^\mathbf{v})$ corresponds to a face of $\mathbb{R}^n_{\geq 0}$ missed by $\mathcal{P}_\mathbf{a}$. The conclusion about $\mathcal{V}(B_\Sigma)$ follows as a consequence.

The reason why X_Σ coincides with $\mathbb{C}^n /\!\!/_\mathbf{a} G$ is that when $\deg(\mathbf{x}^\mathbf{u}) = \mathbf{a}$ and $\mathrm{supp}(\mathbf{x}^\mathbf{u}) = \{i \mid \nu_i \in \nu_{\bar{\sigma}}\}$, the $\mathbb{Z}$-graded degree piece of the localization $S_{(\mathbf{a})}[\mathbf{x}^{-\mathbf{u}}]$ is isomorphic to $\mathbb{C}[\sigma^\vee \cap L]$. Thus X_Σ and $\mathbb{C}^n /\!\!/_\mathbf{a} G$ have covers by isomorphic open affines that agree on the overlaps. To see the isomorphism

$$S_{(\mathbf{a})}[\mathbf{x}^{-\mathbf{u}}] \cong \mathbb{C}[\sigma^\vee \cap L] \quad \text{when} \quad \mathrm{supp}(\mathbf{x}^\mathbf{u}) = \{i \mid \nu_i \in \nu_{\bar{\sigma}}\},$$

observe that $S_{(\mathbf{a})}[\mathbf{x}^{-\mathbf{u}}]$ is spanned as a vector space over $\mathbb{C}$ by the (Laurent) monomials expressible as $\mathbf{x}^\mathbf{v}/\mathbf{x}^{m\mathbf{u}} = \mathbf{x}^{\mathbf{v}-m\mathbf{u}}$ for $\mathbf{v} \in \mathcal{P}_{m\mathbf{a}}$ and $m \in \mathbb{N}$. Writing $\mathbf{v} = \mathbf{v}_1 + \cdots + \mathbf{v}_m$ as a sum of m lattice points in $\mathcal{P}_\mathbf{a}$, we find that $\mathbf{x}^{\mathbf{v}-m\mathbf{u}} = \mathbf{x}^{\mathbf{v}_1-\mathbf{u}} \cdots \mathbf{x}^{\mathbf{v}_m-\mathbf{u}}$, so $S_{(\mathbf{a})}[\mathbf{x}^{-\mathbf{u}}]$ is the semigroup ring for the semigroup generated by the lattice points in the translate of $\mathcal{P}_\mathbf{a}$ by $-\mathbf{u}$ (so $\mathbf{u}$ is moved to the origin). This semigroup consists of the lattice points in L on which the vectors in σ are nonnegative, which is $\sigma^\vee \cap L$ by definition. $\square$

Remark 10.31 Our identification of the irrelevant ideal B_Σ in the proof of Theorem 10.30 actually showed that B_Σ equals the radical of the ideal $\langle S_{r\mathbf{a}} \rangle$ generated by all monomials of degree $r\mathbf{a}$, for any sufficiently large $r \in \mathbb{N}$. Alternatively, if choosing a large integer r seems unnatural, we could think of $S_{(a)}$ as a subring of S, so its irrelevant ideal $S^+_{(a)}$ is a subset of S. Then B_Σ is the radical of the ideal $\langle S^+_{(a)} \rangle$ generated by the irrelevant ideal of $S_{(a)}$.

Example 10.32 (Cubes yet again) The three tetrahedra at the end of Example 10.19 arise by embedding the cube into a simplex of dimension 5: each tetrahedron is the intersection of two codimension 1 simplices corresponding to opposite faces of the cube. These nonintersecting pairs of faces correspond to pairs of coordinate hyperplanes in $\mathbb{R}^6_{\geq 0}$ intersecting in a face of $\mathbb{R}^6_{\geq 0}$ that misses the cone over the cube.

The picture is somewhat simpler one dimension lower down, where the square is expressed as the intersection of a tetrahedron with a 2-plane E:

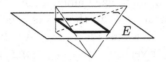

When two facets of the tetrahedron intersect E in opposite edges of the square, they intersect at an edge of the tetrahedron missed by the square. ◇

Example 10.33 (The five varieties of 2×2 minors of a 2×3 matrix) Let S be the polynomial ring generated by the entries of a 2×3 matrix $X = (x_{ij})$ of variables and consider the ideal of 2×2 minors

$$ I \;=\; \langle x_{11}x_{22} - x_{12}x_{21},\; x_{11}x_{32} - x_{12}x_{31}, x_{21}x_{32} - x_{22}x_{31} \rangle. $$

There are five different ways, all very natural, of associating to the prime ideal I a subvariety Y of a toric variety X. In each case, the inclusion of $Y = \mathrm{SpecTor}(S/I, B)$ in $X = \mathrm{SpecTor}(S, B)$ is specified by the irrelevant ideal B and a sublattice L of the lattice $\mathbb{Z}^{2 \times 3}$ of integer 2×3 matrices.

1. If $L = \mathbb{Z}^{2 \times 3}$ consists of all integer 2×3 matrices and $B = \langle 1 \rangle$, then $Y \subset X = \mathbb{C}^6$ is the cone over the Segre variety $\mathbb{P}^1 \times \mathbb{P}^2$. See case 2.

2. If L consists of all matrices whose entries sum to zero and $B = \langle x_{11}, x_{12}, x_{13}, x_{21}, x_{22}, x_{23} \rangle$, then $Y = \mathbb{P}^1 \times \mathbb{P}^2$ is the variety in $X = \mathbb{P}^5$.

3. If L consists of all matrices with zero row sums and $B = \langle x_{11}, x_{12}, x_{13} \rangle$ $\cap \langle x_{21}, x_{22}, x_{23} \rangle$, then $Y = \mathbb{P}^2$ is the diagonal in $X = \mathbb{P}^2 \times \mathbb{P}^2$.

4. If L consists of all matrices with zero column sums and $B = \langle x_{11}, x_{21} \rangle$ $\cap \langle x_{12}, x_{22} \rangle \cap \langle x_{13}, x_{23} \rangle$, then $Y = \mathbb{P}^1$ is the small diagonal in $X = \mathbb{P}^1 \times \mathbb{P}^1 \times \mathbb{P}^1$.

5. Let L be all matrices with zero row and column sums and B the ideal of monomials whose support involves both rows and all three columns. Then Y is the distinguished point (the identity element of the dense torus) of a smooth toric surface X, namely the blowup of $\mathbb{P}^2$ at three points. We encountered this surface in Example 10.13.   ◇

Exercises

10.1 Any sublattice $L \subseteq \mathbb{Z}^n$ determines an affine toric variety $\operatorname{Spec}(\mathbb{C}[L \cap \mathbb{Z}^n])$ with a homogeneous coordinate ring graded by $\mathbb{Z}^n/L$. What is its irrelevant ideal?

10.2 Decompose the polynomial $f = (x_1 + x_2 + x_3)^4$ into homogeneous components, $f = \sum_{\mathbf{a} \in A'} c_{\mathbf{a}} \cdot f_{(\mathbf{a})}$, with respect to the A-grading in Example 10.1. Write each $f_{(\mathbf{a})}$ as a $\mathbb{C}$-linear combination of the images of f under various $\zeta \in G$.

10.3 Pick an algebraic geometry textbook and review the definitions of Spec and Proj. Draw a picture of the real points of $\operatorname{Proj}(\mathbb{C}[x, y, z]/\langle x^3 - y^2 z \rangle)$.

10.4 Consider the special case of Example 10.13 in which the graph is the complete bipartite graph $K_{r,s}$, directed from one partite set to the other. Show that A is the subgroup of $\mathbb{Z}^r \times \mathbb{Z}^s$ consisting of all pairs $(\mathbf{a}, \mathbf{b}) = (a_1, \ldots, a_r, b_1, \ldots, b_s)$ satisfying $a_1 + \cdots + a_r = b_1 + \cdots + b_s$, and that the polytope $\mathcal{P}_{(\mathbf{a}, \mathbf{b})}$ is the *transportation polytope* consisting of all nonnegative real $r \times s$ matrices with row sums $\mathbf{a}$ and column sums $\mathbf{b}$. Prove that the projective toric variety corresponding to $\mathcal{P}_{(\mathbf{a}, \mathbf{b})}$ is smooth when $\mathbf{a}$ and $\mathbf{b}$ lie outside of finitely many hyperplanes.

10.5 Let L be the column span of the 3×6 matrix $\mathbf{L}$ in (7.1) and pick a nonzero pair $(\mathbf{u}, r)$ as in Proposition 10.11. Compute the affine toric variety $\mathcal{U}(\mathbf{x}^{\mathbf{u}} \gamma^r)$.

10.6 Consider the complete graph K_5 on five nodes. Characterize the vectors $\mathbf{a}$ such that the toric quiver variety $\mathbb{C}^E /\!\!/_{\mathbf{a}} (\mathbb{C}^*)^V$ is smooth. What is its dimension?

10.7 Consider the action of $G = \mathbb{C}^*$ on $\mathbb{C}^4$ in Example 10.12. Set $B = \langle x_1, x_2 \rangle$ and $X_B = \operatorname{SpecTor}(\mathbb{C}[x_1, x_2, x_3, x_4], B)$. Is X_B a geometric quotient of $\mathbb{C}^4$?

10.8 Let I be a homogeneous ideal of S and B an irrelevant ideal.

(a) Explain why $\operatorname{SpecTor}(S/I, B)$ is naturally a subvariety (or subscheme, if I is not a radical ideal) of $\operatorname{SpecTor}(S, B)$.
(b) Prove that two ideals I and I' define the same subvariety (and even the same subscheme) if their saturations with respect to B are equal.
(c) Prove the converse of (b). Hint: See [FM05].

10.9 What changes (if any) must be made to Theorem 10.30 when $\mathcal{P}_{\mathbf{a}}$ does not have full dimension $n - d$, so that the cones in $\Sigma(\mathcal{P}_{\mathbf{a}})$ are no longer pointed?

10.10 Prove the converse to Theorem 10.29 when the quotient is affine: If the cone $L \cap \mathbb{N}^n$ generates a group of rank $n - d$ but has more than $n - d$ facets, then some fiber of quotient morphism $\mathbb{C}^n \twoheadrightarrow \mathbb{C}^n /\!\!/ G$ contains infinitely many orbits of G. Hint: Think of the fiber over the origin as the zero set of an irrelevant ideal, and check that one of its components must have dimension at least $d+1 = \dim(G)+1$.

10.11 Prove the converse to Theorem 10.29 in general: If Σ is not simplicial, then some fiber of the quotient morphism $\mathcal{U}_\Sigma \to X_\Sigma$ contains infinitely many G-orbits.

10.12 In Example 10.26, what is the irrelevant ideal for the toric variety when $r = s$? How does Example 10.26 relate to Example 10.12?

10.13 (The diagonal embedding of a toric variety) Consider any toric variety $X_\Sigma = \operatorname{SpecTor}(S, B)$, where S is graded by the abelian group A. Show that $X_\Sigma \times X_\Sigma$ equals the variety $\operatorname{SpecTor}(S', B')$ where $S' = S \otimes_{\mathbb{C}} S$ for a suitable ideal B' and grading on S'. Determine the homogeneous prime ideal in S' whose variety is the diagonal embedding $X_\Sigma \subset X_\Sigma \times X_\Sigma$. Hint: See Exercise 7.12.

Notes

Books can be written—and have been written [Oda88, Ful93, Ewa96, BP02], and still are being written [BG05, FM05]—about toric varieties. Our main goal has been to give an idea of the extent to which one can understand various parts of the subject almost entirely from the perspective of multigraded commutative algebra. In fact, it is possible to go quite a bit further. For example, one can give an elementary definition of *sheaf* and *equivariant sheaf* for toric varieties using only multigraded algebra [Cox95, Mus02], without going into the technicalities of sheaf theory. A number of cohomology rings associated to a smooth projective toric variety, including the ordinary and torus-equivariant cohomology and K-rings, can also be treated in the context of combinatorial commutative algebra.

The homogeneous coordinate ring of a toric variety (Definition 10.25) was discovered by Audin [Aud91], Cox [Cox95], and Musson [Mus94]. Theorem 10.27 and Theorem 10.29 both appear in Cox's article.

It is probably possible to define the spector of any pair (R, B) in which R is a commutative algebra (over an algebraically closed field $\Bbbk$) graded by a finitely generated abelian group A and $B \subseteq R$ is a graded ideal. Any such construction would essentially output the GIT quotient of $\mathrm{Spec}(R) \smallsetminus \mathcal{V}(B)$ by $G = \mathrm{Hom}(A, \Bbbk^*)$. From this perspective, the upshot of Section 10.3, and Definition 10.25 in particular, is that when B is the irrelevant ideal for a compatible fan, we can get an explicit combinatorial handle on the quotient, $\mathrm{SpecTor}(R, B)$, including an open affine cover. Readers interested in learning more about the generalities of quotients by algebraic group actions should start with the fundamental reference [MFK94].

The toric quiver varieties in Example 10.13 were introduced by Hille [Hil98] and discussed further by Altmann and Hille [AH99]. Such varieties are special cases of *quiver varieties*, where one associates a linear map to each directed edge, but the vector spaces at the vertices need not have dimension 1. The group acting on the space of such *quiver representations* is a product of general linear groups (one for each vertex), and the quiver variety is obtained as the quotient by this action. Although this setup is the same one underlying Chapter 17, there are only finitely many orbits there, and it is these orbits in Chapter 17 that interests us, rather than the moduli space of orbits as in Example 10.13.

Fig. 10.1 was inspired by the cover art of [Hof79] (we should have named the cones G, E, and B).

Chapter 11

Irreducible and injective resolutions

Let $Q \subseteq \mathbb{Z}^d$ be an affine semigroup (throughout this chapter, we do not require Q to be pointed or to generate $\mathbb{Z}^d$). Every monomial ideal $I \subseteq \Bbbk[Q]$ has a resolution by Q-graded free modules—that is, $\mathbb{Z}^d$-graded free $\Bbbk[Q]$-modules with summands generated in degrees $\mathbf{a} \in Q$. Each summand $\Bbbk[Q](-\mathbf{a})$ in such a free module can be thought of alternatively as the principal ideal $\langle \mathbf{t^a} \rangle \subseteq \Bbbk[Q]$, so *a Q-graded free resolution of I is a resolution of I by principal monomial ideals.* When $Q \ncong \mathbb{N}^d$, the ring $\Bbbk[Q]$ is not regular, so a classical theorem of Serre implies that there are ideals of $\Bbbk[Q]$ whose free resolutions over $\Bbbk[Q]$ cannot be made finite.

This infiniteness has a number of disadvantages. For starters, we have little hope of actually writing down the whole free resolution. Even granted that we can somehow "know" the whole resolution, we have to be careful when using infinite resolutions to write the Hilbert series of $\Bbbk[Q]/I$ as an alternating sum. Furthermore, free resolutions best capture the kinds of algebraic data associated to I expressible in terms of generators and relations; geometric data such as associated primes call for a different construction.

In this chapter we show that every monomial ideal I in an affine semigroup ring $\Bbbk[Q]$ has a finite resolution in terms of irreducible monomial ideals. This construction is closely related to injective modules and injective resolutions. We characterize these injective objects combinatorially, and we demonstrate that they can be computed quite explicitly.

11.1 Irreducible resolutions

In this chapter we use the term "ideal" to mean a monomial ideal in the semigroup ring $\Bbbk[Q]$, unless otherwise stated. An ideal I is principal if and only if $I = I_1 + I_2$ implies $I \in \{I_1, I_2\}$. Thus principal ideals correspond

Figure 11.1: An irreducible ideal

to semigroup ideals of Q that are "primitive for unions", in the sense that they cannot be written nontrivially as unions of ideals. Dually:

Definition 11.1 An ideal $W \subseteq \Bbbk[Q]$ is **irreducible** if every expression $W = W_1 \cap W_2$ of W as an intersection of ideals implies that $W \in \{W_1, W_2\}$.

Thus irreducible ideals are "primitive for intersections". We use the symbol "W" for irreducible ideals because that is how they look (Fig. 11.1). This geometric picture will be made precise at the end of Section 11.2, on injective modules. Instead of resolving a given ideal I using principal ideals as before, we now resolve $M = \Bbbk[Q]/I$ using quotients by irreducible ideals.

Definition 11.2 The quotient $\overline{W} = \Bbbk[Q]/W$ of the semigroup ring $\Bbbk[Q]$ modulo an irreducible ideal W is called an **irreducible quotient**. Such a module is $\mathbb{Z}^d$-graded with its generator in degree $\mathbf{0}$. An **irreducible resolution** $\overline{W}^{\cdot}$ of a $\mathbb{Z}^d$-graded module M over $\Bbbk[Q]$ is a graded exact sequence

$$0 \to M \to \overline{W}^0 \to \overline{W}^1 \to \overline{W}^2 \to \cdots \qquad \text{with} \qquad \overline{W}^i = \bigoplus_{j=1}^{\mu^i} \overline{W}^{ij},$$

where each W^{ij} is an irreducible ideal of $\Bbbk[Q]$. The irreducible resolution is **minimal** if the numbers μ^i are all simultaneously minimized (among irreducible resolutions of M). We say that the irreducible resolution is **finite** if each μ^i is finite and $\overline{W}^i = 0$ for $i \gg 0$.

Example 11.3 Let $Q = \mathbb{N}^2$ and consider the ideal $I = \langle x^4; x^2 y^2, y^4 \rangle$ in $\Bbbk[Q] = \Bbbk[x, y]$. The following sequence is a minimal irreducible resolution:

$$0 \to \Bbbk[x, y]/I \to \Bbbk[x, y]/\langle x^4, y^2 \rangle \oplus \Bbbk[x, y]/\langle x^2, y^4 \rangle \to \Bbbk[x, y]/\langle x^2, y^2 \rangle \to 0.$$

It corresponds to the "exclusion–inclusion"

that expresses the set of monomials outside of I in terms of "boxes".   ◇

In order for a finitely generated $\mathbb{Z}^d$-graded $\Bbbk[Q]$-module M to have an irreducible resolution, a necessary condition is that $M_{\mathbf{a}} = 0$ for $\mathbf{a} \in \mathbb{Z}^d \smallsetminus Q$; that is, the module M has to be Q-graded. This condition is also sufficient.

Theorem 11.4 *Every finitely generated Q-graded module M has a finite minimal irreducible resolution, and it is unique up to isomorphism.*

This theorem applies in particular to ideals I and their quotients $\Bbbk[Q]/I$. An immediate consequence is the following combinatorial statement, which we have already seen in action in Example 7.14.

Corollary 11.5 *Every monomial ideal $I \subseteq \Bbbk[Q]$ has a unique irredundant expression $I = W_1 \cap \cdots \cap W_r$ as an intersection of irreducible ideals W_j.*

Proof. If $\overline{W}^{\bullet}$ is a minimal irreducible resolution of $\Bbbk[Q]/I$, then choose $r = \mu^0$ and $W_j = W^{0j}$. The kernel of the composite homomorphism $\Bbbk[Q] \to \Bbbk[Q]/I \to \bigoplus_{j=1}^{r} \overline{W}_j$ is the intersection of ideals $W_1 \cap \cdots \cap W_r$. Existence follows because $\Bbbk[Q]/I \to \overline{W}^0$ is an inclusion. Uniqueness follows from the uniqueness of minimal irreducible resolutions in Theorem 11.4. $\square$

It is worth pausing at this juncture to remark that, although the ideals W_1 and W_2 in Definition 11.2 are required to be monomial ideals by the conventions of this chapter, such monomial ideals are always irreducible in the ungraded sense of Remark 5.17 anyway. The proof of this statement requires some facts about $\mathbb{Z}^d$-graded irreducible ideals, so we postpone it until Proposition 11.41, at the end of the chapter.

We take Corollary 11.5 as the motivation for the rest of this chapter, whose eventual aim is to prove Theorem 11.4 (after Example 11.40). Along the way, we will see how *injective modules* and *injective resolutions* arise naturally, allowing their well-behaved homological behavior to rub off onto irreducible resolutions. Also, we will attempt to dispel the common belief that injective modules must necessarily be unwieldy behemoths, by describing them combinatorially in the context of affine semigroup rings.

Let us illustrate the difference between free resolutions and injective resolutions for the ideal $I = \langle x^4, x^2y^2, y^4 \rangle$ from Example 11.3. The free resolution of $\Bbbk[x,y]/I$ (i) covers the set of standard monomials modulo I with all of $\mathbb{N}^2$, (ii) uncovers the monomials in I using translated copies of the positive quadrant $\mathbb{N}^2$, and finally, (iii) excludes the monomials in I that were uncovered too many times:

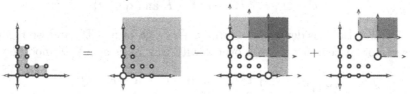

In contrast, an injective resolution of $\Bbbk[x,y]/I$ starts by covering the set of standard monomials using translated copies of the *negative* quadrant $-\mathbb{N}^2$.

It then subtracts off those monomials that were covered too many times—including those outside of the positive quadrant. Finally, the injective resolution adds back in those monomials subtracted off too many times:

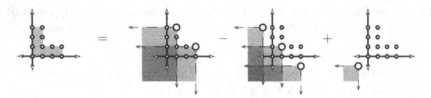

We recover the irreducible resolution in Example 11.3 from this injective resolution by ignoring $\mathbb{Z}^2$-graded degrees outside the semigroup $Q = \mathbb{N}^2$. That this works for any semigroup Q will be the content of Proposition 11.39.

Irreducible monomials ideals in a polynomial ring $\Bbbk[x_1, \ldots, x_d]$ are easy to recognize because they are generated by powers of the variables. However, when $Q \not\cong \mathbb{N}^d$, there seems to be no simple way of telling an irreducible ideals from its generators. Testing irreducibility, computing irreducible decompositions, and computing irreducible resolutions are challenging algorithmic problems. The computationally inclined reader may wish to think about Exercise 11.1 before moving on to the next section.

Example 11.6 Fix the semigroup $Q = \mathbb{N}\{(-1,1),(0,1),(1,1)\} \subset \mathbb{Z}^2$. A typical example of an irreducible ideal in $\Bbbk[Q] = \Bbbk[x,y,z]/\langle xz - y^2\rangle$ is

$$W \;=\; \langle x^3, x^2 y, yz^2, z^3 \rangle.$$

We can show that W is irreducible by noting that

$$\overline{W} \;=\; \Bbbk\{Q \cap ((0,4) - Q)\}.$$

Can you find an irreducible ideal in Q with more than four generators? ◇

11.2   Injective modules

Recall that F is a face of the semigroup $Q \subset \mathbb{Z}^d$ if $Q \smallsetminus F$ is a prime ideal.

Definition 11.7 The **injective hull** of the face F of Q is the subset

$$F - Q \;=\; \{\mathbf{f} - \mathbf{q} \mid \mathbf{f} \in F \text{ and } \mathbf{q} \in Q\}$$

of $\mathbb{Z}^d$. We also consider its translates $\mathbf{a} + F - Q$ for $\mathbf{a} \in \mathbb{Z}^d$, and we regard the vector space $\Bbbk\{\mathbf{a} + F - Q\}$ over $\Bbbk$ with that basis as a $\Bbbk[Q]$-module via

$$t^{\mathbf{q}} \cdot t^{\mathbf{u}} \;=\; \begin{cases} t^{\mathbf{q}+\mathbf{u}} & \text{if } \mathbf{q} + \mathbf{u} \in \mathbf{a} + F - Q \\ 0 & \text{if otherwise.} \end{cases}$$

The module $\Bbbk\{\mathbf{a} + F - Q\}$ is called an **indecomposable injective** of Q.

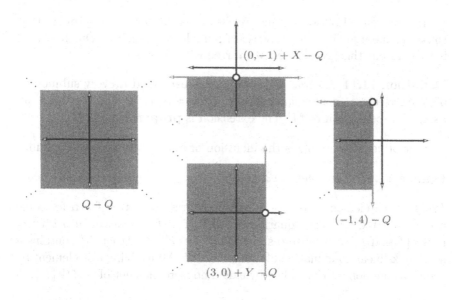

Figure 11.2: $\mathbb{Z}^2$-graded translates of injective hulls for $Q = \mathbb{N}^2$

Example 11.8 When $Q = \mathbb{N}^2$, so $\Bbbk[Q] = \Bbbk[x, y]$ is the polynomial ring in two variables, there are four faces: the trivial face $\mathcal{O} = \{\mathbf{0}\}$, the x-axis X, the y-axis Y, and the whole semigroup $\mathbb{N}^2$. Examples of subsets $\mathbf{a} + F - Q$ appear in Fig. 11.2, each dot lying at the appropriate $\mathbf{a} \in \mathbb{Z}^2$. $\diamond$

Although arbitrary $\mathbb{Z}^d$-graded translates of indecomposable injectives are allowed, sometimes there are homogeneous isomorphisms *of degree* $\mathbf{0}$ between two different translates. It is instructive to check the following proposition for the faces F in Fig. 11.2.

Proposition 11.9 *An indecomposable injective* $\Bbbk\{\mathbf{a}+F-Q\}$ *is isomorphic to* $\Bbbk\{\mathbf{b} + F - Q\}$ *as a* $\Bbbk[Q]$*-module if and only if* $\mathbf{a} + \mathbb{Z}F = \mathbf{b} + \mathbb{Z}F$.

Proof. The two modules are isomorphic if and only if $\mathbf{a} \in \mathbf{b} + F - Q$ and $\mathbf{b} \in \mathbf{a} + F - Q$. This condition is equivalent to $\mathbf{a} - \mathbf{b} \in F - Q$ and $\mathbf{b} - \mathbf{a} \in F - Q$, which is the same as $\mathbf{a} - \mathbf{b} \in (F - Q) \cap (Q - F) = \mathbb{Z}F$. $\square$

Definition 11.10 An **injective module** over $\Bbbk[Q]$ is any direct sum of indecomposable injectives:

$$ E \;=\; \bigoplus_{k \in K} \Bbbk\{\mathbf{a}_k + F^k - Q\}. $$

Here, K is an index set that can be infinite, the vectors $\mathbf{a}_k$ lie in $\mathbb{Z}^d$, and the faces F^k of Q can be repeated.

In Theorem 11.30 we will justify the name "injective" by proving that these are the injectives in the sense of homological algebra. Our first goal is to work out their connection with irreducible ideals.

Definition 11.11 A submodule $M \subseteq N$ is **essential** if every submodule of N intersects M nontrivially: $0 \neq N' \subseteq N \Rightarrow N' \cap M \neq 0$. We call N an **essential extension** of M. The extension is **proper** if $N \neq M$.

Our principal example is the inclusion of a face into its injective hull.

Lemma 11.12 *The inclusion $\Bbbk\{F\} \subset \Bbbk\{F - Q\}$ is an essential extension.*

Proof. Each element $\mathbf{u} \in F - Q$ can be expressed as $\mathbf{u} = \mathbf{f} - \mathbf{a}$ for some $\mathbf{a} \in Q$ and $\mathbf{f} \in F$. The equation $\mathbf{a} + \mathbf{u} = \mathbf{f} \in F$ translates into $\mathbf{t^a t^u} = \mathbf{t^f} \in \Bbbk\{F\}$. If N' is a nonzero submodule of $\Bbbk\{F - Q\}$, then N' contains a nonzero $\Bbbk$-linear combination of monomials $\mathbf{t^u}$. Multiplying this element by a suitable monomial $\mathbf{t^a}$ as above yields a nonzero element of $N' \cap \Bbbk\{F\}$. $\square$

The most common argument using an essential extension $M \subseteq N$ says: If a homomorphism $N \to N'$ induces an inclusion $M \to N'$, then $N \to N'$ is also an inclusion. The proof of the "if" part of our next result uses this argument. For notation, the Q-*graded part* of a module M is the submodule $M_Q = \bigoplus_{\mathbf{a} \in Q} M_{\mathbf{a}}$ obtained by ignoring all $\mathbb{Z}^d$-graded degrees outside of Q.

Theorem 11.13 *A monomial ideal W is irreducible if and only if the Q-graded part of some indecomposable injective module E satisfies $E_Q = \overline{W}$.*

Proof. First we prove the "if" direction. The multiplication rule in Definition 11.7 implies that $\Bbbk\{\mathbf{a} + Q - F\}_Q$ is isomorphic to $\overline{W}$ for *some* ideal W. Supposing that $W \neq \Bbbk[Q]$, we may as well assume $\mathbf{a} \in Q$ by Proposition 11.9 (add an element way inside F), so that $\mathbf{t^a} \in \overline{W}$ generates an essential submodule $\Bbbk\{\mathbf{a} + F\}$. Suppose $W = I_1 \cap I_2$. The copy of $\Bbbk\{\mathbf{a} + F\}$ inside $\overline{W}$ must include into $\Bbbk[Q]/I_j$ for $j = 1$ or 2; indeed, if both induced maps $\Bbbk\{\mathbf{a} + F\} \to \Bbbk[Q]/I_j$ have nonzero kernels, then these kernels intersect in a nonzero submodule of $\Bbbk[\mathbf{a} + F]$ because $\Bbbk[F]$ is a domain. Essentiality of $\Bbbk\{\mathbf{a} + F\} \subseteq \overline{W}$ forces $\overline{W} \to \Bbbk[Q]/I_j$ to be an inclusion for some j, so W contains—and hence equals—this ideal I_j. Thus W is irreducible.

Now we prove the "only if" direction. Since W is irreducible, its radical is the unique prime ideal $P_F = \Bbbk\{Q \smallsetminus F\}$ associated to $\overline{W}$. Let N be the span $\Bbbk\{\mathbf{t^u} \in \Bbbk[Q] \mid (W : \mathbf{t^u}) = P_F\}$ of all monomials in $\overline{W}$ with annihilator equal to P_F, which is a $\Bbbk[Q]$-submodule of $\overline{W}$. Define U to be the exponent vectors on a finite set of monomials generating N. Given $\mathbf{u} \in U$, we have $\mathbf{t^{u+f}} \notin W$ for $\mathbf{f} \in F$. Consequently, all monomials with exponents in $Q \cap (\mathbf{u} + F - Q)$ lie outside W, because W is an ideal. Thus the ideal $W^{\mathbf{u}}$ defined by $\overline{W}^{\mathbf{u}} = \Bbbk\{\mathbf{u} + F - Q\}_Q$ contains W. But every monomial in $\Bbbk[Q] \smallsetminus W$ has a monomial multiple whose annihilator equals P_F, whence $W = \bigcap_{\mathbf{u} \in U} W^{\mathbf{u}}$. Irreducibility of W implies that $W = W^{\mathbf{u}}$ for some $\mathbf{u}$. $\square$

Theorem 11.13 says approximately that the standard monomials for an irreducible monomial ideal lie in the intersection of a cone and a translate of its negative, justifying the heuristic illustration in Fig. 11.1.

11.3 Monomial matrices revisited

Earlier in this book, we used monomial matrices as a convenient notational device to write down complexes of free modules over $\mathbb{Z}^n$-graded polynomial rings. Now we extend this construction to injective $\Bbbk[Q]$-modules.

When we defined monomial matrices in Section 1.4, we tacitly assumed a full understanding of the $\mathbb{N}^n$-graded homomorphisms $S(-\mathbf{b}) \to S(-\mathbf{c})$ between a pair of copies of $S = \Bbbk[x_1, \ldots, x_n]$. Of course, such a homomorphism is completely determined by the image of the generator $1_\mathbf{b}$ of $S(-\mathbf{b})$: either the map is zero or it takes $1_\mathbf{b}$ to a nonzero scalar multiple of the monomial $\mathbf{x}^{\mathbf{b}-\mathbf{c}} \cdot 1_\mathbf{c}$, which sits in degree $\mathbf{b}$ of $S(-\mathbf{c})$.

To justify using monomial matrices here, we need to get a handle on homomorphisms between (indecomposable) injectives. For this purpose, let us review the notion of homogeneous homomorphism in more detail. In what follows, $\mathbb{Z}^d$-graded $\Bbbk$-algebras R always have $\Bbbk$ contained in the degree zero piece $R_\mathbf{0}$. The principal examples to think of are $R = \Bbbk[Q]$ and $R = \Bbbk$.

Definition 11.14 Let R be a $\mathbb{Z}^d$-graded $\Bbbk$-algebra. A map $\phi : M \to N$ of graded R-modules is **homogeneous of degree** $\mathbf{b} \in \mathbb{Z}^d$ (or just **homogeneous** when $\mathbf{b} = \mathbf{0}$) if $\phi(M_\mathbf{a}) \subseteq N_{\mathbf{a}+\mathbf{b}}$. For fixed $\mathbf{b} \in \mathbb{Z}^d$, the set of such maps is a $\Bbbk$-vector space denoted by

$$
\begin{aligned}
\underline{\mathrm{Hom}}_R(M, N)_\mathbf{b} &= \text{degree } \mathbf{b} \text{ homogeneous maps } M \to N \\
&= \text{homogeneous maps } M \to N(\mathbf{b}) \\
&= \text{homogeneous maps } M(-\mathbf{b}) \to N.
\end{aligned}
$$

As the notation suggests, if R is either $\Bbbk$ or $\Bbbk[Q]$, and M is a $\Bbbk[Q]$-module,

$$
\underline{\mathrm{Hom}}_R(M, N) = \bigoplus_{\mathbf{b} \in \mathbb{Z}^d} \underline{\mathrm{Hom}}_R(M, N)_\mathbf{b}
$$

is a $\mathbb{Z}^d$-graded $\Bbbk[Q]$-module, with $\mathbf{x}^\mathbf{a}\phi$ defined by $(\mathbf{x}^\mathbf{a}\phi)(m) = \phi(\mathbf{x}^\mathbf{a} m)$.

When $R = \Bbbk[Q]$, we write $\underline{\mathrm{Hom}}(M, N) = \underline{\mathrm{Hom}}_{\Bbbk[Q]}(M, N)$ if no confusion can result. The graded module $\underline{\mathrm{Hom}}(M, N)$ is isomorphic to the $\mathbb{Z}$-graded and ungraded versions whenever M is finitely generated (all versions can be calculated using the same graded free presentation of M).

The obvious combinatorial relation between the localization $\Bbbk[Q - F]$ and the injective hull $\Bbbk\{F - Q\}$ underlies a deeper algebraic duality. To pinpoint it, we "turn modules upside down" algebraically.

Definition 11.15 The **Matlis dual** of a graded $\Bbbk[Q]$-module M is the $\Bbbk[Q]$-module $M^\vee = \underline{\operatorname{Hom}}_\Bbbk(M, \Bbbk)$. In other words, $M^\vee$ is defined by

$$(M^\vee)_{-\mathbf{u}} \;=\; \operatorname{Hom}_\Bbbk(M_\mathbf{u}, \Bbbk),$$

the multiplication $(M^\vee)_{-\mathbf{u}} \xrightarrow{t^\mathbf{a}} (M^\vee)_{\mathbf{a}-\mathbf{u}}$ being transpose to $M_{\mathbf{u}-\mathbf{a}} \xrightarrow{t^\mathbf{a}} M_\mathbf{u}$. Observe that $(M^\vee)^\vee = M$, as long as $\dim_\Bbbk(M_\mathbf{b})$ is finite for all $\mathbf{b} \in \mathbb{Z}^d$.

Note that the Matlis dual of the localization $\Bbbk[Q-F]$ of $\Bbbk[Q]$ along F is the injective hull $\Bbbk\{F-Q\}$ of $\Bbbk[F]$. In symbols, $\Bbbk\{F-Q\} = \Bbbk[Q-F]^\vee$.

Matlis duality behaves well with respect to Hom and tensor product:

Lemma 11.16 $\underline{\operatorname{Hom}}(M, N^\vee) = (M \otimes N)^\vee$.

Proof. The result is a consequence of the adjointness between $\underline{\operatorname{Hom}}$ and $\otimes$ that holds for arbitrary $\mathbb{Z}^d$-graded $\Bbbk$-algebras R and R-modules M, N:

$$\underline{\operatorname{Hom}}_\Bbbk(M \otimes_R N, \Bbbk) \;=\; \underline{\operatorname{Hom}}_R(M, \underline{\operatorname{Hom}}_\Bbbk(N, \Bbbk)) \;=\; \underline{\operatorname{Hom}}_R(M, N^\vee).$$

Here, the base ring $\Bbbk$ does not even need to be a field. $\square$

A map between injective modules can be represented by a matrix each of whose entries is a degree $\mathbf{0}$ homomorphism $\Bbbk\{\mathbf{a}+F-Q\} \to \Bbbk\{\mathbf{b}+G-Q\}$ between indecomposable injectives. Just as with free modules over $\Bbbk[\mathbb{N}^n]$, it is therefore crucial to know (i) that the vector space of such maps is either $\Bbbk$ or zero, and (ii) the conditions on F, G, $\mathbf{a}$, and $\mathbf{b}$ that force zero.

Proposition 11.17 *The $\Bbbk$-vector space* $\underline{\operatorname{Hom}}\left(\Bbbk\{\mathbf{a}+F-Q\}, \Bbbk\{\mathbf{b}+G-Q\}\right)_\mathbf{0}$ *is either zero or 1-dimensional. The following conditions are equivalent.*

1. $\underline{\operatorname{Hom}}\left(\Bbbk\{\mathbf{a}+F-Q\}, \Bbbk\{\mathbf{b}+G-Q\}\right)_\mathbf{0} = \Bbbk$.
2. $\mathbf{a}+F-Q \supseteq \mathbf{b}+G-Q$.
3. $F \supseteq G$ *and* $\mathbf{b} \in \mathbf{a}+F-Q$.

Proof. Lemma 11.16 implies the first equality below. The second uses the same lemma with the roles of M and N switched:

$$\begin{aligned}
\underline{\operatorname{Hom}}\left(\Bbbk\{\mathbf{a}+F-Q\}, \Bbbk\{\mathbf{b}+G-Q\}\right) & \\
&= (\Bbbk\{\mathbf{a}+F-Q\} \otimes \Bbbk\{-\mathbf{b}+Q-G\})^\vee \\
&= \underline{\operatorname{Hom}}\left(\Bbbk\{-\mathbf{b}+Q-G\}, \Bbbk\{-\mathbf{a}+Q-F\}\right).
\end{aligned}$$

Any nonzero degree $\mathbf{0}$ homomorphism between such localizations must be an injection, induced by an inclusion $-\mathbf{b}+Q-G \subseteq -\mathbf{a}+Q-F$. The statement about 0 or $\Bbbk$ is immediate. Taking negatives yields the criterion of part 2. The equivalence of part 2 and part 3 can be checked directly. $\square$

Consider elements in direct sums of indecomposable injectives as *row vectors*, so a matrix acts on the *right* side of a vector and the arrows in cochain complexes of injectives go to the right.

Definition 11.18 An **(injective) monomial matrix** is a matrix of constants $\lambda_{qp} \in \Bbbk$ such that:

1. Each row is labeled by a vector $\mathbf{a}_q \in \mathbb{Z}^d$ and a face F^q of Q.
2. Each column is labeled by a vector $\mathbf{a}_p \in \mathbb{Z}^d$ and a face F^p of Q.
3. $\lambda_{qp} = 0$ unless $F^p \subseteq F^q$ and $\mathbf{a}_p \in \mathbf{a}_q + F^q - Q$.

Sometimes we use *monomial labels* $\mathbf{t}^{\mathbf{a}_q}$ and $\mathbf{t}^{\mathbf{a}_p}$ in place of the *vector labels* $\mathbf{a}_q$ and $\mathbf{a}_p$.

Theorem 11.19 *Monomial matrices represent maps of injective modules:*

$$
\begin{array}{c}
\cdots\ F^p\ \cdots \\
\cdots\ \mathbf{a}_p\ \cdots \\
F^q\ \mathbf{a}_q
\begin{bmatrix}
& & \\
& \lambda_{qp} & \\
& &
\end{bmatrix}
\end{array}
$$

$$
\bigoplus_q \Bbbk\{\mathbf{a}_q + F^q - Q\} \xrightarrow{\hspace{3cm}} \bigoplus_p \Bbbk\{\mathbf{a}_p + F^p - Q\}.
$$

Two monomial matrices represent the same map of injectives (with fixed direct sum decompositions) if and only if (i) their scalar entries are equal, (ii) the corresponding faces F^r are equal, where $r = p, q$, and (iii) the corresponding vectors $\mathbf{a}_r$ are congruent modulo $\mathbb{Z}F^r$.

Proof. Proposition 11.17 immediately implies the first sentence. The second sentence is the content of Proposition 11.9. $\square$

Definition 11.18 really does constitute an extension of the notion of monomial matrix from Section 1.4. All that we have done here is added *face labels* to the data of the row and column labels and changed the condition for λ_{qp} to be nonzero accordingly. The reader should check that when $Q = \mathbb{N}^n$ and $F^q = F^p = \{\mathbf{0}\}$ for all q and p, the only surviving condition on λ_{qp} is $\mathbf{a}_q \succeq \mathbf{a}_p$, and this is precisely the condition on $-\mathbf{a}_q$ and $-\mathbf{a}_p$ stipulated by Definition 1.23. (The negatives on $\mathbf{a}_q$ and $\mathbf{a}_p$ stem from Matlis duality.)

As with cellular monomial matrices for complexes of free modules, cellular injective monomial matrices can be specified simply by labeling the cell complex with the appropriate face and vector labels.

Example 11.20 Resume the notation from Example 11.8. The following sequence of maps is cellular, supported on a line segment. The vector labels are all zero. The vertices have face labels X and Y, the interior has face label $\mathbb{N}^2$, and the empty set has face label $\mathcal{O}$.

$$
0 \to \Bbbk\{\mathbb{Z}^2\} \xrightarrow{\ \mathbb{N}^2\ \mathbf{0}\begin{bmatrix} X & Y \\ 0 & 0 \\ 1 & 1 \end{bmatrix}\ } \Bbbk\{X - \mathbb{N}^2\} \oplus \Bbbk\{Y - \mathbb{N}^2\} \xrightarrow{\ \begin{matrix} & \mathcal{O} \\ & 0 \\ X\ 0 \\ Y\ 0 \end{matrix}\begin{bmatrix} -1 \\ 1 \end{bmatrix}\ } \Bbbk\{\mathcal{O} - \mathbb{N}^2\} \to 0
$$

This sequence of maps is actually a complex, and it would be exact except that the kernel of the first map $\Bbbk\{\mathbb{Z}^2\} \to \Bbbk\{X - \mathbb{N}^2\} \oplus \Bbbk\{Y - \mathbb{N}^2\}$ is isomorphic to $\Bbbk\{(1,1) + \mathbb{N}^2\}$.

The same cell complex also supports a completely different complex of injectives. Here, monomials $\mathbf{t^a}$ replace the vector labels $\mathbf{a}$:

$$
0 \to \Bbbk\{-\mathbb{N}^2\} \xrightarrow{\mathcal{O}\, 1 \begin{smallmatrix} \mathcal{O} & \mathcal{O} \\ y^{-1} & x^{-1} \\ [\ 1 & 1\ ] \end{smallmatrix}} \begin{matrix} \Bbbk\{(0,-1) - \mathbb{N}^2\} \\ \oplus \\ \Bbbk\{(-1,0) - \mathbb{N}^2\} \end{matrix} \xrightarrow{\begin{smallmatrix} & \mathcal{O} \\ & x^{-1}y^{-1} \\ \mathcal{O}\, y^{-1} \\ \mathcal{O}\, x^{-1} \end{smallmatrix} \begin{bmatrix} -1 \\ 1 \end{bmatrix}} \Bbbk\{(-1,-1) - \mathbb{N}^2\} \to 0
$$

This complex is also exact except at the left, where the kernel is just $\Bbbk$ in $\mathbb{Z}^2$-graded degree $\mathbf{0}$. In fact, this is just the Matlis dual of the Koszul complex in two variables (Definition 1.26). $\diamond$

11.4   Essential properties of injectives

In more general commutative algebraic settings, injectives are important because of their simple homological behavior, in analogy with free modules.

Definition 11.21 A graded $\Bbbk[Q]$-module J is called **homologically injective** if $M \mapsto \underline{\mathrm{Hom}}_{\Bbbk[Q]}(M, J)$ takes exact sequences to exact sequences.

In other words, if $0 \to M \to N \to P \to 0$ is exact, then so is

$$0 \leftarrow \underline{\mathrm{Hom}}(M, J) \leftarrow \underline{\mathrm{Hom}}(N, J) \leftarrow \underline{\mathrm{Hom}}(P, J) \leftarrow 0.$$

For (10.2) in Chapter 10 we exploited this valuable property in the context of (ungraded) $\mathbb{Z}$-modules, otherwise known as abelian groups: divisible groups, such as $\mathbb{C}^*$, are homologically injective. In general, only the surjectivity of $\underline{\mathrm{Hom}}(M, J) \leftarrow \underline{\mathrm{Hom}}(N, J)$ can fail, even for arbitrary J. The surjectivity for homologically injective J can be read equivalently as follows.

Lemma 11.22 *J is homologically injective if whenever $M \subseteq N$ and $\phi : M \to J$ are given, some map $\psi : N \to J$ extends ϕ; that is, $\psi|_M = \phi$.*

Judging from what we have already called the modules $\Bbbk\{F - Q\}$ and their direct sums in Definition 11.10, we had better reconcile our combinatorial definition of injective module with the usual homological one. The goal of this section is to accomplish just that, in Theorem 11.30.

Recall that a module N is *flat* if tensoring any exact sequence with N yields another exact sequence. The examples of flat modules to keep in mind are the localizations $\Bbbk[Q - F]$. In fact, localizations are pretty much the only examples that can come up in the context of graded modules over affine semigroup rings (cf. the next lemma and Theorem 11.30).

Lemma 11.23 *N is flat if and only if $N^\vee$ is homologically injective.*

Proof. $M \mapsto M \otimes N$ is exact if and only if $M \mapsto (M \otimes N)^\vee$ is. Now use the equality $(M \otimes N)^\vee = \underline{\mathrm{Hom}}\,(M, N^\vee)$ of Lemma 11.16. □

Thus "flat" and "injective" are Matlis dual conditions. Heuristically, a module $\Bbbk\{T\}$ is flat if T is an intersection of positive half-spaces for facets of Q, whereas $\Bbbk[T]$ is injective if T is an intersection of negative half-spaces.

Proposition 11.24 *Indecomposable injectives are homologically injective.*

Proof. Since $\Bbbk[Q - F]^\vee = \Bbbk\{F - Q\}$, this follows from Lemma 11.23. □

For any $\mathbb{Z}^d$-graded module M, the Matlis dual can be expressed as $M^\vee = \underline{\mathrm{Hom}}_{\Bbbk[Q]}(M, \Bbbk[Q]^\vee)$ by Lemma 11.16 with $N = \Bbbk[Q]$. Proposition 11.24 says in this case that Matlis duality is exact, which is obvious from the fact that $\Bbbk$ is a field, because taking vector space duals is exact. Taking $\underline{\mathrm{Hom}}$ into $\Bbbk[Q]^\vee$ (= the injective hull of $\Bbbk$) provides a better algebraic formulation of Matlis duality than Definition 11.15, by avoiding degree-by-degree vector space duals. It should convince you that dualization with respect to injective modules can have concrete combinatorial interpretations.

Homological injectivity behaves very well with respect to (categorical) direct products of modules. Unfortunately, the usual product of infinitely many $\mathbb{Z}^d$-graded modules $(M^p)_{p \in P}$ is not necessarily $\mathbb{Z}^d$-graded. Indeed, there may be sequences $(y_p)_{p \in P} \in \prod_{p \in P} M^p$ of homogeneous elements that have distinct degrees, in which case $\prod_{p \in P} M^p$ fails to be the direct sum of its graded components. Such poor behavior occurs even in the simplest of cases, in the presence of only one variable x (so $Q = \mathbb{N}$): the product $\prod_{i=0}^\infty \Bbbk[x]$ of infinitely many copies of $\Bbbk[x]$ has an element $(1, x, x^2, \ldots)$ that is not expressible as a finite sum of homogeneous elements. The remedy is to take the largest $\mathbb{Z}^d$-graded submodule of the usual product.

Definition 11.25 The $\mathbb{Z}^d$-**graded product** $^*\!\!\prod_{p \in P} M^p$ is the submodule of the usual product generated by arbitrary products of homogeneous elements of the same degree. Explicitly, this is the module that has

$$\left({}^*\!\!\prod_{p \in P} M^p \right)_{\mathbf{b}} = \prod_{p \in P} M_{\mathbf{b}}^p$$

as its component in $\mathbb{Z}^d$-graded degree $\mathbf{b}$.

Lemma 11.26 *Arbitrary $\mathbb{Z}^d$-graded products of homologically injective modules are homologically injective.*

Proof. The natural map $\underline{\mathrm{Hom}}\,(N, {}^*\!\!\prod_{p \in P} M^p) \to {}^*\!\!\prod_{p \in P} \underline{\mathrm{Hom}}\,(N, M^p)$ is an isomorphism (write out carefully what it means to be a homogeneous element of degree $\mathbf{a}$ on each side). Apply Definition 11.21 to the case where each M^p is homologically injective. □

It is very easy to produce (in an abstract sense) nonzero maps from arbitrary modules to homological injectives. The next result capitalizes on this ease: we can stick a module injectively into a product of indecomposable injectives by explicitly making sure that no element maps to zero.

Proposition 11.27 *Every module M is isomorphic to a submodule of a homologically injective module. If M is finitely generated, then M is isomorphic to a submodule of a finite direct sum of indecomposable injectives.*

Proof. Homogeneous elements $y \in M$ generate finitely generated submodules. Using Proposition 8.11 and Lemma 7.10, pick a face F such that P_F is associated to M, so $\langle \mathbf{t^a} y \rangle \cong \Bbbk\{\mathbf{u}_y + F^y\}$ for some $\mathbf{a} \in Q$ and some vector $\mathbf{u}_y \in \mathbb{Z}^d$. The corresponding inclusion $\langle \mathbf{t^a} y \rangle \hookrightarrow \Bbbk\{\mathbf{u}_y + F^y - Q\}$ extends to a map $\phi_y : M \to \Bbbk\{\mathbf{u}_y + F^y - Q\}$ by homological injectivity of the latter. The graded product of such maps over $y \in M$ is a homomorphism $(\phi_y)_{y \in M} : M \to {}^* \prod_y \Bbbk\{\mathbf{u}_y + F^y - Q\}$ to a homologically injective module (Lemma 11.26 and Proposition 11.24) that is an inclusion by construction.

When M is finitely generated, each of the finitely many submodules $(0 :_M P_F)$ annihilated by a monomial prime ideal is itself finitely generated. Using the above construction, it suffices to take the graded product over all y in a finite set containing generators for each of the modules $(0 :_M P_F)$. This finite product is a direct sum.   □

Lemma 11.28 *Let J be homologically injective and E any module.*

1. *If E is a direct summand of J, then E is homologically injective.*
2. *If $J \subseteq E$, then J is a direct summand of E.*

Proof. To prove the first part, let $J = J' \oplus J''$ and apply $\underline{\mathrm{Hom}}\,(\_, J) = \underline{\mathrm{Hom}}\,(\_, J') \oplus \underline{\mathrm{Hom}}\,(\_, J'')$ to any exact sequence. For the second part, the surjection $\underline{\mathrm{Hom}}\,(J, J) \twoheadleftarrow \underline{\mathrm{Hom}}\,(E, J)$ produces a homomorphism $E \to J$ mapping to id_J, which is by definition a splitting of the inclusion $J \hookrightarrow E$. □

Proposition 11.29 *A module J is homologically injective if and only if J has no proper essential extensions.*

Proof. First assume J is homologically injective. If $J \subseteq M$ is an essential extension, then writing $M = J \oplus N$ for some N by the second part of Lemma 11.28, it must be that $N = 0$, so $J = M$.

Now assume J has no proper essential extension. Use Proposition 11.27 to find an inclusion $J \hookrightarrow E$ into a homologically injective module E. The set of submodules of E trivially intersecting J has a maximal element M by Zorn's Lemma. The natural map $J \to E/M$ makes the quotient E/M into essential extension of J by construction, so $J \cong E/M$. Thus $E = J \oplus M$. Homological injectivity of J is the first part of Lemma 11.28.   □

Theorem 11.30 *A module is homologically injective if and only if it is injective in the combinatorial sense of Definition 11.10.*

Proof. Finite direct sums of indecomposable injectives are homologically injective by Proposition 11.24 and Lemma 11.26. Now let J be an arbitrary direct sum of indecomposable injectives, and suppose that $J \subseteq E$ is an essential extension. If $x \in E$, then $\langle x \rangle \cap J$ is isomorphic to (a $\mathbb{Z}^d$-graded translate of) an ideal of $\Bbbk[Q]$, so it is finitely generated because $\Bbbk[Q]$ is Noetherian. Since every generator involves only finitely many indecomposable summands of J, the submodule $\langle x \rangle \cap J$ lies in a direct sum $J' \subset J$ of finitely many summands of J. By construction, $J' + \langle x \rangle$ is an essential extension of J', so $x \in J'$ by Proposition 11.29 and the first sentence of this paragraph. Apply Proposition 11.29 again to conclude that J is homologically injective.

Now suppose that J is homologically injective, and let $\mathcal{E}$ be the set of indecomposable injective submodules of J. Among all subsets of $\mathcal{E}$, consider the subsets whose elements pairwise intersect in 0. These subsets form a poset $\mathcal{P}$ (under inclusion) that has a maximal element $\mathcal{E}' \in \mathcal{P}$ by Zorn's Lemma. The sum of the modules in $\mathcal{E}'$ is a homologically injective submodule $J' \subseteq J$ by the previous paragraph, and we can write $J = J' \oplus J''$ as a direct sum in which J'' is also homologically injective, by Lemma 11.28.

Suppose $J'' \neq 0$. Then it has an associated prime, which has the form P_F by Proposition 8.11 and Lemma 7.10, so some element $x \in J''$ generates a submodule isomorphic to $\Bbbk[F](-\mathbf{a})$ for some $\mathbf{a} \in \mathbb{Z}^d$. The inclusion $\langle x \rangle \subset J''$ can be extended to a map $\Bbbk\{\mathbf{a} + F - Q\} \to J''$ by Lemma 11.22, and this map is also an inclusion, because $\Bbbk[F] \subset \Bbbk\{F - Q\}$ is an essential extension. Denoting the image by $M \subseteq J''$, we find that $\mathcal{E}' \subsetneq \mathcal{E}' \cup \{M\} \in \mathcal{P}$ contradicts maximality of $\mathcal{E}'$, thereby proving $J = J'$. □

Every result in this chapter therefore holds for the injective modules in Definition 11.10, and we can forget the term "homologically injective".

11.5 Injective hulls and resolutions

Proposition 11.27 has about the same value as its dual statement for free modules: "Every module has a generating set." Well, of course it does. Much more useful is the analogue to "Every module has a *minimal* generating set."

Definition 11.31 An **injective hull** of a module M is an injective module $E(M)$ containing M as an essential submodule.

Note, for example, that the indecomposable injective $\Bbbk\{F - Q\}$ has been called the injective hull of $\Bbbk[F]$ ever since Definition 11.10.

Theorem 11.32 *Injective hulls exist and are unique up to isomorphism.*

Proof. Existence: Choose an injection $M \hookrightarrow J$ with J injective using Proposition 11.27, and let $E \subseteq J$ be maximal among essential extensions of M contained in J; these exist by Zorn's Lemma. Suppose $E \subseteq E'$ is an essential extension. Lemma 11.22 produces a homomorphism $E' \to J$ whose image contains E. Since the image cannot strictly contain E by maximality of E, and the kernel is zero by essentiality of $E \subseteq E'$, it must be that $E = E'$. Hence E is injective by Proposition 11.29.

Uniqueness: Let $M \subseteq E$ and $M \subseteq E'$ be injective hulls. Lemma 11.22 produces a map $E \to E'$ whose image contains M. The kernel of this map trivially intersects M and is hence zero because $M \subseteq E$ is an essential extension. This forces the image to be an injective module and therefore a summand of E'. Since $M \subseteq E'$ is an essential extension, the image is E'. $\square$

Do not read more into Theorem 11.32 than it states: injective hulls are not unique up to *canonical* isomorphism. In other words, there may be many isomorphisms between two injective hulls of M. Minimal generating sets have the same (manageable) problem, stemming from the fact that vector spaces do not always come with canonical bases.

An *irreducible hull* of M is an essential extension of M that is a direct sum of irreducible quotients. Theorem 11.32 immediately implies the corresponding result for irreducible hulls, using Theorem 11.13.

Corollary 11.33 *Irreducible hulls of Q-graded modules exist, and they are unique up to isomorphism. The irreducible hull of a Q-graded module is the Q-graded part of its injective hull.*

Another consequence of Theorem 11.32 is that every module has a special sort of resolution by injective modules.

Definition 11.34 An **injective resolution** of M is an exact sequence

$$J^\bullet : 0 \to M \to J^0 \xrightarrow{\lambda^0} J^1 \xrightarrow{\lambda^1} J^2 \xrightarrow{\lambda^2} \cdots$$

with all J^j injective. $J^\bullet$ is **minimal** if $J^0 = E(M)$ is the injective hull of M and $J^{j+1} = E(\lambda^j(J^j))$ is the injective hull of the image of λ^j for all $j \geq 0$.

Corollary 11.35 *Every module has an injective resolution. Minimal injective resolutions are unique up to isomorphism; in fact, if $J^\bullet$ and $E^\bullet$ are injective resolutions of M with $J^\bullet$ minimal, $E^\bullet$ contains $J^\bullet$ as a subcomplex.*

Proof. Use Theorem 11.32 and Lemma 11.28 to show by induction on cohomological degree j that $E^j \cong J^j \oplus \tilde{E}^j$ for an injective resolution $\tilde{E}^\bullet$ of 0. $\square$

Any module inherits numerical invariants from the generating degrees of the free modules in its minimal free resolution, namely the Betti numbers. Likewise, if injective modules possess numerical invariants, then they will be passed on to arbitrary modules as homological invariants by taking minimal injective resolutions. The question becomes: How unique is the decomposition of an injective module as a direct sum of indecomposables?

For arbitrary modules M, let $M[\mathbb{Z}F] = M \otimes_{\Bbbk[Q]} \Bbbk[Q - F]$ be the homogeneous localization of M along the face F.

Theorem 11.36 *If J is injective, then the localization $\underline{\mathrm{Hom}}\,(\Bbbk[F], J)[\mathbb{Z}F]$ is a free module over $\Bbbk[\mathbb{Z}F]$. Its $\mathbb{Z}^d$-graded piece in degree $\mathbf{a}$ satisfies*

$$\dim_{\Bbbk} \underline{\mathrm{Hom}}\,(\Bbbk[F], J)[\mathbb{Z}F]_{\mathbf{a}} \quad = \quad \#\text{summands isomorphic to } \Bbbk\{\mathbf{a} + F - Q\}$$

in any decomposition of J into a direct sum of indecomposable injectives.

Proof. The submodule $N = \underline{\mathrm{Hom}}\,(\Bbbk[F], \Bbbk\{\mathbf{a} + G - Q\})$ of elements inside $\Bbbk\{\mathbf{a}+G-Q\}$ annihilated by $\Bbbk[F]$ is zero unless $F \subseteq G$. Subsequently localizing N at F yields zero unless $F = G$, in which case $N[\mathbb{Z}F] = \Bbbk[\mathbb{Z}F](-\mathbf{a})$. It follows that if $J = \bigoplus_{k \in K} \Bbbk\{\mathbf{a}_k + F^k - Q\}$, then

$$\underline{\mathrm{Hom}}\,(\Bbbk[F], J)[\mathbb{Z}F] \quad = \quad \bigoplus_{F^k = F} \Bbbk[\mathbb{Z}F](-\mathbf{a}_k).$$

Using Proposition 11.9, which implies that $\mathbf{a}_k$ is only defined modulo $\mathbb{Z}F^k$, the result follows by taking degree $\mathbf{a}$ pieces. $\square$

Of course, the vector space dimensions and numbers need not be finite; the statement is then that they have the same cardinality. Have no fear, though: almost every injective module in this book has only finitely many summands. There are cases of combinatorial interest, however, where infinitely many summands do occur (see Example 13.17 in Chapter 13.2, for instance), although there are usually still finitely many that have been translated by any fixed $\mathbb{Z}^d$-graded degree.

To explain why $\mathbf{a} \in \mathbb{Z}^d/\mathbb{Z}F$ in what follows, recall Proposition 11.9.

Definition 11.37 *The jth **Bass number** of M along the face F in degree $\mathbf{a} \in \mathbb{Z}^d/\mathbb{Z}F$ is the number $\mu_F^{j,\mathbf{a}}(M)$ of summands isomorphic to $\Bbbk\{\mathbf{a}+F-Q\}$ appearing in J^j, for any minimal injective resolution $J^{\bullet}$ of M.*

The higher Bass numbers of M are no more abstract than the higher Betti numbers of M. Moreover, we will see after Proposition 11.39 that zeroth Bass numbers (which are finite for finitely generated modules by Proposition 11.27) measure characteristics of modules that are as tangible as minimal generators—namely irreducible components.

Our last main goal is to complete the proof of Theorem 11.4. Given any desired irreducible resolution, we begin by reconstructing an injective resolution whose Q-graded part is that irreducible resolution.

Lemma 11.38 *Any irreducible resolution $\overline{W}^{\bullet}$ of a Q-graded module M can be expressed as the Q-graded part $J_Q^{\bullet}$ of an injective resolution $J^{\bullet}$ of M.*

Proof. Since M is Q-graded, $M \hookrightarrow W^0 \hookrightarrow E(W^0)$ and $E(W^0)_Q = W^0$. Having chosen $\overline{W}^i \hookrightarrow J^i$ such that $\overline{W}^i = J_Q^i$, let $N = J^i/J^{i-1}$ and $J^{i+1} = E(N_Q) \oplus E(K)$. Choose a map $N \to E(N_Q)$ by applying Lemma 11.22 to the inclusion $N_Q \subseteq N$, and let $K \subseteq N$ be the kernel. Choosing a map $N \to E(K)$ extending $K \to E(K)$, we get a monomorphism $N \hookrightarrow E(N_Q) \oplus E(K)$. Since $K_Q = 0$, we have $E(K)_Q \cap K = 0$, so $E(K)_Q = 0$ by essentiality of $K \subseteq E(K)$. Then $N_Q = J_Q^{i+1} = \overline{W}^{i+1}$ by construction. $\square$

Now we extract minimal irreducible resolutions from injective resolutions.

Proposition 11.39 *Let M be a finitely generated Q-graded module. The Q-graded part of a minimal injective resolution of M is a finite minimal irreducible resolution of M.*

Proof. Let $J^{\bullet}$ be a minimal injective resolution of M. That $\overline{W}^{\bullet} = (J^{\bullet})_Q$ is an irreducible resolution follows from Theorem 11.13, so it remains to demonstrate minimality. For each j, the number of indecomposable summands in $\overline{W}^j$ equals the number of summands in J^j having a nonzero Q-graded part. This number is well-defined by Theorem 11.36, and is no larger than in any other injective resolution of M by Corollary 11.35. By Lemma 11.38, it is enough to show that $(J^j)_Q$ is finitely generated for each cohomological degree $j \geq 0$ (to get finiteness of the μ^i in Definition 11.2) and zero for all $j \gg 0$ (to get $\overline{W}^i = 0$ for $i \gg 0$ in Definition 11.2).

Corollary 11.33 implies that M has an irreducible resolution. By Proposition 11.27 and induction on cohomological degree, we may construct it so that every cohomological degree is finitely generated. Now construct an injective resolution $E^{\bullet}$ whose Q-graded part is this irreducible resolution using Lemma 11.38, and conclude from Corollary 11.35 that $(J^j)_Q \subseteq (E^j)_Q$ is finitely generated for each j.

Finally, for length-finiteness, consider for each Q-graded module N the set $V(N)$ of degrees $\mathbf{a} \in Q$ such that $N_{\mathbf{b}}$ vanishes for all $\mathbf{b} \in \mathbf{a} + Q$. The vector space $\Bbbk\{V(N)\}$ is naturally an ideal in $\Bbbk[Q]$. We leave it as an exercise for the reader to check that $V(M) \subsetneq V(\overline{W}/M)$ whenever $\overline{W}$ is the Q-graded part of an injective hull of M and $M \neq 0$ (that is, $V(M) \neq Q$). Noetherianity of $\Bbbk[Q]$ plus this strict containment force the sequence

$$\Bbbk\{V(M)\} \subseteq \Bbbk\{V(\overline{W}^0/M)\} \subseteq \Bbbk\{V(\overline{W}^1/\mathrm{image}(\overline{W}^0))\} \subseteq \cdots$$

of ideals to stabilize at the unit ideal of $\Bbbk[Q]$ after finitely many steps. $\square$

Proposition 11.39 for ideals says that zeroth Bass numbers precisely locate irreducible components.

Example 11.40 Look back at the illustration for $I = \langle x^4, x^2y^2, y^4 \rangle \subset \Bbbk[x, y]$ in Example 11.3. The injective hull of $\Bbbk[x, y]/I$ is the direct sum $\Bbbk[x, y]^{\vee}(-1, -3) \oplus \Bbbk[x, y]^{\vee}(-3, -1)$ appearing at the first stage of the injective resolution. The $\mathbb{N}^2$-graded part of $\Bbbk[x, y]^{\vee}(-1, -3)$ is $\Bbbk[x, y]/\langle x^2, y^4 \rangle$, so $\langle x^4, x^2y^2, y^4 \rangle = \langle x^2, y^4 \rangle \cap \langle x^4, y^2 \rangle$. $\diamond$

Proof of Theorem 11.4. Proposition 11.39 says that minimal irreducible resolutions exist as Q-graded parts of minimal injective resolutions. By Lemma 11.38 and Corollary 11.35, every minimal irreducible resolution can be expressed this way. $\square$

Finally, we prove that our irreducibility agrees with the usual notion.

Proposition 11.41 *If a monomial ideal W in $\Bbbk[Q]$ is irreducible in the sense of Definition 11.2, then W cannot be expressed as the intersection of two strictly larger ideals, even if nonmonomial ideals are allowed.*

Proof. In this proof, ideals are not assumed to be monomial ideals unless otherwise stated. Assume that the monomial ideal W is irreducible in the sense of Definition 11.2. It suffices to show that for any expression $W = W_1 \cap \cdots \cap W_m$ in which W_i is irreducible in the sense of Remark 5.17 for all i, we must have $W = W_i$ for some i. Indeed, we reduce to this situation by intersecting irreducible decompositions of any pair of ideals whose intersection equals W.

Write W as the Q-graded part of $\Bbbk\{\mathbf{a} + F - Q\}$ by Theorem 11.13. Thus $\overline{W} = \Bbbk[Q]/W$ has a unique associated prime P_F. Let R be the localization of $\Bbbk[Q]$ at P_F, in the category of (not necessarily graded) modules over $\Bbbk[Q]$, and let $\mathfrak{p}$ be the maximal ideal of R. Then $\overline{W}_\mathfrak{p} = (W_1)_\mathfrak{p} \cap \cdots \cap (W_m)_\mathfrak{p}$ is still an irreducible decomposition, although perhaps one that is more redundant than before localization. Assume that all intersectands W_i have the unique associated prime P_F, by omitting the rest if necessary.

Theorem 11.36 implies that the socle of $\overline{W}_\mathfrak{p}$, which is by definition the submodule $\mathrm{soc}(\overline{W}_\mathfrak{p})$ of elements in $\overline{W}_\mathfrak{p}$ annihilated by $\mathfrak{p}$, is a one-dimensional vector space over the residue field $R/\mathfrak{p}$. But $\mathrm{soc}(\overline{W}_\mathfrak{p})$ maps injectively to the socle $\bigoplus_{i=1}^m \mathrm{soc}(R/(W_i)_\mathfrak{p})$ of the module $\bigoplus_{i=1}^m R/(W_i)_\mathfrak{p}$. Consequently, the homomorphism $\mathrm{soc}(\overline{W}_\mathfrak{p}) \to \mathrm{soc}(R/(W_i)_\mathfrak{p})$ induced by the natural map $\overline{W}_\mathfrak{p} \to R/(Q_i)_\mathfrak{p}$ is injective for some i. Since $\mathrm{soc}(\overline{W}_\mathfrak{p})$ is an essential submodule of $\overline{W}_\mathfrak{p}$ by definition (every element of $\overline{W}_\mathfrak{p}$ has some nonzero R-multiple that is killed by $\mathfrak{p}$), it must be that the homomorphism $\overline{W}_\mathfrak{p} \to R/(W_i)_\mathfrak{p}$ is injective, so $W_\mathfrak{p} \supseteq (W_i)_\mathfrak{p}$. Since $\overline{W}$ and $\Bbbk[Q]/W_i$ both have unique associated prime P_F, we deduce that $W \supseteq W_i$ and therefore that $W = W_i$. □

In terms of gradings, Proposition 11.41 says that ideals irreducible *in the category of $\mathbb{Z}^d$-graded ideals* over $\Bbbk[Q]$ are irreducible in the category of all (not necessarily monomial) ideals in $\Bbbk[Q]$. This statement fails to hold when the $\mathbb{Z}^d$-grading is replaced by a grading with torsion. For example, consider the univariate polynomial ring $\Bbbk[x]$ graded by $\mathbb{Z}/2\mathbb{Z}$, with $\deg x \neq 0$. The ideal $\langle x^2 - 1 \rangle$ is irreducible (and in fact maximal) in the category of $\mathbb{Z}/2\mathbb{Z}$-graded ideals, but not in the category of all ideals. Indeed, $\langle x^2 - 1 \rangle = \langle x + 1 \rangle \cap \langle x - 1 \rangle$, but the intersectands are not $\mathbb{Z}/2\mathbb{Z}$-graded.

Exercises

11.1 Describe a combinatorial algorithm for testing whether an ideal in a subsemigroup Q of the two-dimensional lattice $\mathbb{Z}^2$ is irreducible.

11.2 For vectors $\mathbf{a}, \mathbf{b}, \mathbf{c} \in \mathbb{N}^n$ satisfying $\mathbf{c} \preceq \mathbf{b} \preceq \mathbf{a}$, define the **Alexander dual** of the homogeneous degree $\mathbf{0}$ injection $S(-\mathbf{b}) \to S(-\mathbf{c})$ with respect to $\mathbf{a}$ to be the surjection $S/\mathfrak{m}^{\mathbf{a} \smallsetminus \mathbf{c}} \to S/\mathfrak{m}^{\mathbf{a} \smallsetminus \mathbf{b}}$. Show that the Alexander dual with respect to $\mathbf{a}$ of a minimal free resolution of I is a minimal irreducible resolution of $S/I^{[\mathbf{a}]}$.

11.3 Using the setup from Example 7.14, compute a minimal irreducible resolution of the quotient $\mathbb{k}[Q']/\langle b \rangle$. Then calculate a minimal *injective* resolution of $\mathbb{k}[Q']/\langle b \rangle$ through cohomological degree 3. Do both of these tasks for $\mathbb{k}[Q']/\langle ab \rangle$.

11.4 Let Δ be a simplicial complex with n vertices, and set $\mathfrak{m}^{[2]} = \langle x_1^2, \dots, x_n^2 \rangle$. Prove the following relation between Matlis duality and Alexander duality:

$$\left((I_\Delta + \mathfrak{m}^{[2]})/\mathfrak{m}^{[2]} \right)^\vee = \left(\mathbb{k}[x_1, \dots, x_n]/(I_{\Delta^\star} + \mathfrak{m}^{[2]}) \right)(\mathbf{1}),$$

where the $\mathbb{Z}^n$-graded translate on the right-hand side is by $\mathbf{1} = (1, \dots, 1)$.

11.5 Generalize Exercise 11.4 to arbitrary monomial ideals in $S = \mathbb{k}[x_1, \dots, x_n]$:

$$\left((I + \mathfrak{m}^{\mathbf{a}+1})/\mathfrak{m}^{\mathbf{a}+1} \right)^\vee = \left(S/(I^{[\mathbf{a}]} + \mathfrak{m}^{\mathbf{a}+1}) \right)(\mathbf{a}).$$

11.6 Let $J^\bullet$ be a minimal injective resolution of a finitely generated $\mathbb{Z}^d$-graded $\mathbb{k}[Q]$-module M. Prove that every injective resolution of M is isomorphic to the direct sum of $J^\bullet$ with some number of cohomological shifts of trivial complexes having the form $0 \to \mathbb{k}\{\mathbf{a} + F - Q\} \cong \mathbb{k}\{\mathbf{a} + F - Q\} \to 0$.

11.7 Consider the labeled cell complex X at right below, where $\bar{1}$ is short for -1.

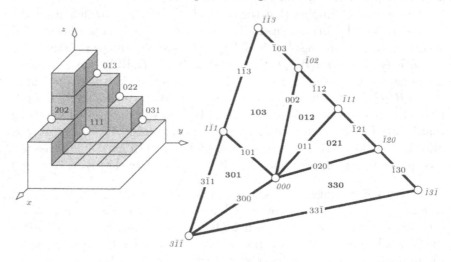

Endow each cell F with a face label $\sigma_F = \{i \mid (\mathbf{a}_F)_i = 3\}$ corresponding to the face of $\mathbb{N}^3$ generated by the basis vectors $\mathbf{e}_i$ such that $\mathbf{a}_F$ has i^{th} coordinate 3.

(a) Visually verify the bijection between the indecomposable injectives corresponding to the vector and face labels on the facets of X and the irreducible components of the ideal I whose staircase diagram is above. The rest of this exercise shows that X supports a minimal injective resolution of $\mathbb{k}[x, y, z]/I$.

(b) Pick orientations for the cells in X, and write down the transposes of the three (scalar) matrices of differentials for the boundary complex of the underlying complex $\underline{X}$. List these matrices from left to right, with the transpose of the map from 2-cells to 1-cells on the left.

(c) Label the rows and columns of the scalar matrices in (b) with the corresponding vector and face labels $\mathbf{a}_F, \sigma_F$ from X.

(d) Explain why parts (b) and (c) result in injective monomial matrices for a complex $J^{\bullet}$ of injectives, where again, we consider elements in direct sums of indecomposable injectives as *row* vectors (this is why we took transposes).

(e) By ascertaining which indecomposable summands contribute to each graded degree, check that the complex $J^{\bullet}$ in (d) is exact in some representative $\mathbb{Z}^3$-graded degrees. Be sure to choose some $\mathbb{Z}^3$-graded degrees that have one or more negative coordinates. (See Exercise 11.11 for more generality.)

(f) Conclude from (e) that $J^{\bullet}$ is a minimal injective resolution of $\mathbb{k}[x, y, z]/I$, where $I = \langle x^2 z^2, yz^3, y^2 z^2, y^3 z, xyz \rangle$.

11.8 How would you change the labeling on X in the picture from Exercise 11.7 to support a minimal free resolution of $\mathbb{k}[x, y, z]/(I + \langle x^4, y^4, z^4 \rangle)$? If you wanted a minimal free resolution of $\mathbb{k}[x, y, z]/I$, which faces of X would you ignore?

11.9 How would you use the labeled cell complex X from Exercise 11.7 to represent a minimal irreducible resolution of $\mathbb{k}[x, y, z]/I$? Which faces of X would you need to ignore?

11.10 Let Λ be a monomial matrix as in Definition 1.23, and fix $\mathbf{a} \in \mathbb{N}^n$. Denote by $\sigma \subseteq \{1, \ldots, n\}$ the face of $\mathbb{N}^n$ generated by $\{\mathbf{e}_i \mid i \in \sigma\}$. Given $\mathbf{b} \in \mathbb{Z}^n$, let $\sigma(\mathbf{b}) = \{i \mid b_i > a_i\}$. Show that appending the face labels $\sigma(\mathbf{a}_q)$ and $\sigma(\mathbf{a}_p)$ to the row labels $\mathbf{a}_q$ and column labels $\mathbf{a}_p$ of Λ yields an injective monomial matrix.

11.11 Fix a monomial ideal I generated in degrees $\preceq \mathbf{a}$ and a labeled cell complex X (with labels $\mathbf{a}_F$) supporting a minimal free resolution of $S/(I + \mathfrak{m}^{\mathbf{a}+\mathbf{1}})$. With notation as in Exercise 11.10, let $\tilde{X}$ be the *injectively labeled cell complex* with vector labels $\mathbf{a}_F - \mathbf{1}$ and face labels $\sigma(\mathbf{a}_F)$ on each cell $F \in \underline{X}$.

(a) Show that $\tilde{X}$ determines a complex of injectives whose injective monomial matrices have scalar entries that constitute the transposes of the boundary maps of X, listed from left to right so that the facets of X are on the left.

(b) Prove that the complex in part (a) is a minimal injective resolution of S/I. (One possibility is to adapt the proof of Theorem 5.37. Another is to reduce to the *statement* of Theorem 5.37 by applying Matlis duality, $\mathbb{Z}^n$-translating by $\mathbf{a}$, and taking $\mathbb{N}^n$-graded parts; note the exactness of these operations.)

(c) In what sense does this minimal injective resolution not depend on $\mathbf{a}$?

11.12 Use Exercise 11.11 to define the *Scarf triangulation* for a generic ideal I from the minimal injective resolution of $\mathbb{k}[\mathbf{x}]/I$, without referring to $I + \mathfrak{m}^{\mathbf{a}+\mathbf{1}}$.

11.13 Let $V(N)$ for a Q-graded module N consist of the degrees $\mathbf{a} \in Q$ such that $N_\mathbf{b}$ vanishes for all $\mathbf{b} \in \mathbf{a} + Q$. Prove the fact used in the proof of Proposition 11.39: For a nonzero finitely generated Q-graded module M, the set $V(M)$ is strictly contained inside $V(\overline{W}/M)$ if $\overline{W}$ is the Q-graded part of an injective hull of M.

Notes

It was not our goal in this chapter to give the most general account of graded injective modules over graded Noetherian rings; the interested reader should start with [BH98, Sections 3.1 and 3.2], on which some of the exposition in Sections 11.4–11.5 is based. The main point is that minimal injective resolutions exist in the category of graded modules over any arbitrarily-graded Noetherian ring. Essential extensions, indecomposable injectives, injective hulls, and irreducible decompositions generalize without alteration, although injective modules are themselves not so explicit, even in contexts as nice as positively multigraded polynomial rings.

Although injective resolutions carry over to more general graded Noetherian rings, irreducible resolutions are special to the finely graded case, where the vector space dimensions of the graded pieces of the ring have dimension 1. The reason is that the fine grading forces submodules of indecomposable injectives to be uniquely determined by the set of graded degrees in which they are nonzero. Finely graded injective modules were introduced by Goto and Watanabe [GW78] for the study of semigroup rings. Irreducible resolutions were introduced in [Mil02c] for the purpose of generalizing the Eagon–Reiner Theorem (Theorem 5.56) from the polynomial ring to affine semigroup rings.

There exist algorithms to compute irreducible decompositions and resolutions over normal semigroup rings [HM04], although it is still open to do so for unsaturated semigroups. The use of Bass numbers to compute irreducible components—even algorithmically—works in more general settings [Vas98, pp. 66–68].

Exercise 11.2 can be used to generalize Alexander duality to arbitrary $\mathbb{N}^n$-graded modules (see also Exercise 13.10): Taking the Alexander dual of a free resolution of any $\mathbb{N}^n$-graded module M whose generators and relations lie in degrees $\preceq \mathbf{a}$ yields an irreducible resolution for the *Alexander dual* of M with respect to $\mathbf{a}$. This and many of the exercises in this chapter are, to varying degrees, based on the content of [Mil00a]. The exercises in question include Exercises 11.7–11.10, which are loosely based on [Mil00a], and Exercises 11.2, 11.4, 11.5, and 11.11, whose solutions can be found more directly in specific results. In particular, Exercise 11.11 is one of the cellular cases of the general duality for resolutions; see [Mil00a, Corollary 4.9], for example. Exercise 11.13 is equivalent to [Mil02c, Lemma 2.3].

From the point of view in Exercise 11.12, part (d) of Theorem 6.26 is a rephrasing of the statement that the zeroth Bass numbers of S/I (Definition 11.37) are determined by the Scarf complex Δ_{I^*}, and part (b) of Theorem 6.26 says that the entire $\mathbb{Z}^n$-graded injective resolution of S/I is determined by Δ_{I^*}.

Future uses of irreducible resolutions could include applications to finely graded Hilbert series of monomial ideals in semigroup rings. For this, one would need to get a handle on the Hilbert series of irreducible quotients, which turns out to be a subtle lattice-points-in-polyhedra geometry problem. Describing explicit geometric or combinatorial irreducible resolutions (in the sense of Part I and Chapter 9) of monomial ideals in arbitrary semigroup rings remains a tantalizing open problem, even in the saturated case (Definition 7.24). A combinatorial or geometric solution to the dual problem of how to find generators for the intersection of two principal ideals in a semigroup ring would be a good start.

Chapter 12

Ehrhart polynomials

This chapter is concerned with counting the lattice points in a convex polytope $\mathcal{P}$. If the vertices of the polytope are lattice points themselves, then the number of lattice points in integer multiples $m\mathcal{P}$ of the given polytope is a polynomial function $E_{\mathcal{P}}(m)$ whose degree is the dimension of $\mathcal{P}$. The polynomial $E_{\mathcal{P}}$ was studied by Eugène Ehrhart in the 1960s and is called the *Ehrhart polynomial* of the polytope $\mathcal{P}$. We present a proof of Ehrhart's Theorem and also of Brion's Formula, which expresses the set of lattice points in $\mathcal{P}$ (rather than the number of them) as a rational function in several variables. The presentation highlights the interaction between the arithmetic aspects of polyhedra and multigraded commutative algebra. We conclude with a discussion of Barvinok's polynomial-time algorithm for computing Ehrhart polynomials of polytopes in fixed dimension. The algorithm is based on encoding lattice points in polytopes and certain multivariate Hilbert series in terms of short rational generating functions.

12.1 Ehrhart from Hilbert

Let $\mathcal{P}$ be a d-dimensional lattice polytope—that is, a full-dimensional convex polytope in $\mathbb{R}^d$ all of whose vertices lie in $\mathbb{Z}^d$. For any integer $m \geq 0$, the multiple $m \cdot \mathcal{P}$ is also a lattice polytope, and we can count its lattice points.

Definition 12.1 The function taking each integer $m \in \mathbb{N}$ to the number

$$E_{\mathcal{P}}(m) \;=\; \#\big((m \cdot \mathcal{P}) \cap \mathbb{Z}^d\big)$$

of lattice points in the polytope $m \cdot \mathcal{P}$ is the **Ehrhart polynomial** of $\mathcal{P}$.

The aim of this section is to prove the following theorem due to Ehrhart, which justifies the terminology in Definition 12.1.

Theorem 12.2 *The function $E_{\mathcal{P}} : \mathbb{N} \to \mathbb{N}$ is a polynomial of degree d.*

The leading term of the Ehrhart polynomial equals m^d times the volume of the polytope $\mathcal{P}$. Similarly, the second coefficient (after the leading term) equals $1/2$ times the sum of volumes of each facet, each normalized with respect to the sublattice in the hyperplane spanned by the facet. One might guess that we understand all of the coefficients similarly, but this is not the case: the constant coefficient equals 1, but the intervening coefficients are less well understood.

Example 12.3 The Ehrhart polynomial of the unit 3-cube $\text{conv}(\{0,1\}^3)$ is the cube of the Ehrhart polynomial of the unit segment:

$$E_{\text{cube}}(m) = m^3 + 3m^2 + 3m + 1 = (m+1)^3.$$

To get an octahedron with Ehrhart polynomial

$$E_{\text{octahedron}}(m) = \frac{2}{3}m^3 + 2m^2 + \frac{7}{3}m + 1$$

remove two antipodal vertices of the cube and take the convex hull of the remaining six vertices.   ◇

We will present a proof of Theorem 12.2 that exhibits an $\mathbb{N}$-graded polynomial ring (with all variables of degree 1) and a suitable module over it whose Hilbert polynomial equals $E_{\mathcal{P}}$. As in [Eis95, Theorem 1.11], the *Hilbert polynomial* of a module M is the polynomial whose values at large integers m equals the coefficient $\dim_{\Bbbk}(M_m)$ on t^m in the Hilbert series of M.

Let C be the cone in $\mathbb{R} \times \mathbb{R}^d$ generated by the points $(1, \mathbf{a})$ for lattice points $\mathbf{a}$ in the polytope $\mathcal{P}$. Although the cone C equals the convex hull of the semigroup Q generated by the lattice points $\{(1, \mathbf{a}) \mid \mathbf{a} \in \mathcal{P} \cap \mathbb{Z}^d\}$ in the copy of $\mathcal{P}$ "at height 1", the semigroup Q need not be saturated. Nonetheless, the semigroup ring $\Bbbk[Q_{\text{sat}}]$ for the saturation $Q_{\text{sat}} = C \cap \mathbb{Z}^{1+d}$ is a finitely generated module over the semigroup ring $\Bbbk[Q]$, by Proposition 7.25 and the finiteness of normalization [Eis95, Corollary 13.13].

The semigroup ring $\Bbbk[Q]$ is $\mathbb{Z} \times \mathbb{Z}^d$-graded, but for the moment, we will consider its Hilbert series in the coarser $\mathbb{Z}$-grading given by t_0-degree. The finer grading will arise in Section 12.3. For $m \in \mathbb{N}$, we write $\Bbbk[Q_{\text{sat}}]_m$ for the $\mathbb{Z}$-graded piece of the $\Bbbk[Q]$-module $\Bbbk[Q_{\text{sat}}]$ in degree m.

Lemma 12.4 $E_{\mathcal{P}}$ *is the $\mathbb{N}$-graded Hilbert function of* $\Bbbk[Q_{\text{sat}}]$:

$$E_{\mathcal{P}}(m) = \dim_{\Bbbk}(\Bbbk[Q_{\text{sat}}]_m).$$

Proof. The intersection of the cone C with the hyperplane at height m is a copy of $m \cdot \mathcal{P}$ by construction. The lattice points in this copy of $m \cdot \mathcal{P}$ correspond to the monomials of degree m in $\Bbbk[Q_{\text{sat}}]$ by Definition 7.24. $\square$

It is irrelevant for the statement of the previous lemma whether $\Bbbk[Q_{\text{sat}}]$ is considered as a module over itself, or over $\Bbbk[Q]$, or over some other $\mathbb{N}$-graded $\Bbbk$-algebra. The same comment applies to the next lemma, although its proof exploits a carefully chosen module structure.

Lemma 12.5 *If the polytope $\mathcal{P}$ is a lattice simplex, then the Hilbert function of the $\mathbb{N}$-graded module $\Bbbk[Q_{\mathrm{sat}}]$ equals its Hilbert polynomial; that is, $E_{\mathcal{P}}(m)$ is a polynomial for all nonnegative integers $m \in \mathbb{N}$, even small ones.*

Proof. Let $\mathcal{P}$ be the simplex with vertices $\mathbf{a}_1, \ldots, \mathbf{a}_{d+1}$ in $\mathbb{Z}^d$, and define L as the sublattice of $\mathbb{Z}^{d+1}$ spanned by $(1, \mathbf{a}_1), \ldots, (1, \mathbf{a}_{d+1})$. This lattice L has finite index inside $\mathbb{Z}^{d+1}$; in fact, its index $s = [\mathbb{Z}^{d+1} : L]$ is the volume of the half-open parallelepiped

$$B = \left\{ \sum_{i=1}^{d+1} \lambda_i \cdot (1, \mathbf{a}_i) \mid 0 \leq \lambda_i < 1 \right\}.$$

Every vector in $\mathbb{Z}^{d+1}$ lies inside precisely one translate of B by a lattice vector from L. Hence the set

$$B \cap \mathbb{Z}^{d+1} = \{\mathbf{b}_1, \mathbf{b}_2, \ldots, \mathbf{b}_s\}$$

of lattice points in B is a complete set of representatives for the cosets of $\mathbb{Z}^{d+1}$ modulo L. Moreover, Q_{sat} is the disjoint union

$$Q_{\mathrm{sat}} = \bigcup_{j=1}^{s} \{\mathbf{b}_j + \nu_1(1, \mathbf{a}_1) + \cdots + \nu_{d+1}(1, \mathbf{a}_{d+1}) \mid \nu_1, \ldots, \nu_{d+1} \in \mathbb{N}\}.$$

Setting $x_i = t_0 \mathbf{t}^{\mathbf{a}_i}$ inside the Laurent polynomial ring $\mathbb{Z}[\mathbf{t}^{\pm 1}][t_0]$, we conclude that $\Bbbk[Q_{\mathrm{sat}}]$ is the free $\Bbbk[x_1, \ldots, x_{d+1}]$-module of rank s with basis $\{\mathbf{t}^{\mathbf{b}_1}, \ldots, \mathbf{t}^{\mathbf{b}_s}\}$, where $\mathbf{t}^{\mathbf{b}}$ here means $t_0^{b_0} t_1^{b_1} \cdots t_d^{b_d}$. The $\mathbb{N}$-graded degree of the monomial $\mathbf{t}^{\mathbf{b}_j}$ is the first coordinate b_{j0}. The definition of B shows that

$$b_{j0} \leq d \quad \text{for} \quad j = 1, 2, \ldots, s,$$

so the Hilbert function of the free module with basis $\{\mathbf{t}^{\mathbf{b}_1}, \mathbf{t}^{\mathbf{b}_2}, \ldots, \mathbf{t}^{\mathbf{b}_s}\}$ is

$$E_{\mathcal{P}}(m) = \sum_{j=1}^{s} \binom{(d - b_{j0}) + m}{d}. \tag{12.1}$$

This expression is a polynomial in m, completing the proof. $\square$

Proof of Theorem 12.2. The normalization $\Bbbk[Q_{\mathrm{sat}}]$ is finitely generated as a module over the semigroup ring $\Bbbk[Q]$, which is itself generated in $\mathbb{N}$-graded degree 1. Lemma 12.4, along with a standard result [Eis95, Theorem 1.11] on Hilbert functions, shows that $E_{\mathcal{P}}(m)$ is a polynomial function in m for $m \gg 0$. The degree of the Hilbert polynomial of $\Bbbk[Q_{\mathrm{sat}}]$ equals its Krull dimension minus 1. Since $\Bbbk[Q_{\mathrm{sat}}]$ has the same dimension as $\Bbbk[Q]$, this Krull dimension is $d + 1$, so the Hilbert polynomial has degree d.

It remains to show that the Hilbert function of $\Bbbk[Q_{\mathrm{sat}}]$ equals its Hilbert polynomial. This nontrivial fact is precisely Lemma 12.5 when $\mathcal{P}$ is a simplex. A general lattice polytope $\mathcal{P}$ can be triangulated into lattice simplices (for example, by using a regular subdivision defined by any generic lifting [DRS04]), and we get the Ehrhart polynomial of P by taking an integer sum of the Ehrhart polynomials of all the simplices of various dimensions in the triangulation of P. (The integer coefficients are determined by the Möbius function of the poset of faces in the triangulation [Sta97, Section 4.6].) □

Many lattice polytopes arising in combinatorial problems enjoy the property that the semigroup Q is already saturated. Equivalently,

$$(m \cdot P) \cap \mathbb{Z}^d \;=\; m \cdot (P \cap \mathbb{Z}^d) \quad \text{for all } m \in \mathbb{N}.$$

If this holds, then we say that the polytope P is *normal*. If P is normal, then $E_P(m)$ equals the number of elements in

$$m \cdot (P \cap \mathbb{Z}^d) \;:=\; (P \cap \mathbb{Z}^d) + \cdots + (P \cap \mathbb{Z}^d).$$

The reader should take care in making the distinction between $m \cdot (P \cap \mathbb{Z}^d)$ and $(m \cdot P) \cap \mathbb{Z}^d$. All lattice polygons ($d = 2$) are normal. However, there exist nonnormal polytopes in dimensions $d \geq 3$. The following example also illustrates formula (12.1).

Example 12.6 The lattice tetrahedron P with vertex set

$$A \;=\; \{(0,0,0),\, (1,0,0),\, (0,1,0),\, (1,1,2)\}$$

is not normal, since $(1,1,1)$ lies in $2P$ but not in $2(P \cap \mathbb{Z}^d)$. The semigroup ring $\Bbbk[Q_{\mathrm{sat}}]$ is minimally generated by the five monomials t_0, $t_0 t_1$, $t_0 t_2$, $t_0 t_1 t_2 t_3^2$, and $t_0^2 t_1 t_2 t_3$. Over the polynomial ring $\Bbbk[t_0 \mathbf{t}^{\mathbf{a}} \mid \mathbf{a} \in A]$, the module $\Bbbk[Q_{\mathrm{sat}}]$ is free of rank 2. It has one generator in degree 0, namely the unit monomial 1, and one generator in degree 2, namely $t_0^2 t_1 t_2 t_3$. The Hilbert series of this graded module is

$$E_P(m) \;=\; \binom{m+3}{3} + \binom{m+1}{3} \;=\; \frac{1}{3} m^3 + m^2 + \frac{5}{3} m + 1,$$

and this is the Ehrhart polynomial of the tetrahedron P.   ◇

12.2 Dualizing complexes

In this section we provide the key ingredient for our proof of an elegant formula for the sum of the Laurent monomials corresponding to the lattice points in a polytope (Theorem 12.13). The ingredient is a canonical cellular injective resolution over a normal semigroup ring. Although we will in fact construct the appropriate cellular complex of injectives more generally for an arbitrary affine semigroup $Q \subseteq \mathbb{Z}^d$, its exactness requires certain hypotheses (that always hold when Q is saturated); see Section 13.4.

The cone $C = \mathbb{R}_{\geq 0}Q$ over a given affine semigroup $Q \subseteq \mathbb{Z}^d$ can be expressed as the Cartesian product of its lineality space $C \cap (-C)$ and the cone over a polytope $\mathcal{P}$. When Q is pointed, one way to construct a suitable polytope $\mathcal{P}$ is to take a transverse hyperplane section of C. In general, when C has lineality of dimension ℓ, a transverse affine-linear section of codimension $\ell + 1$ can be used. In any case, the faces of Q correspond bijectively to those of $\mathcal{P}$ by Lemma 7.12. For instance, the minimal face of Q corresponds to the empty face of $\mathcal{P}$. This allows us to define a cellular injective monomial matrix supported on $\mathcal{P}$, with scalar entries forming its reduced chain complex.

Definition 12.7 Label each face of the polytope $\mathcal{P}$ by the vector $\mathbf{0} \in \mathbb{Z}^d$ along with the corresponding face F of Q. The resulting cellular complex $\Omega_Q^{\bullet}$ of injective $\Bbbk[Q]$-modules is called the **dualizing complex**.

When Q is pointed, the dualizing complex of $\Bbbk[Q]$ therefore looks like

$$0 \to \Bbbk\{\mathbb{Z}^d\} \to \underbrace{\bigoplus \Bbbk\{T_F\}}_{\substack{\text{facets } F \\ \text{of } Q}} \to \underbrace{\bigoplus \Bbbk\{T_R\}}_{\substack{\text{ridges } R \\ \text{of } Q}} \to \cdots \to \underbrace{\bigoplus \Bbbk\{T_L\}}_{\substack{\text{rays } L \\ \text{of } Q}} \to \Bbbk\{-Q\} \to 0,$$

where $T_G = G - Q$ is the injective hull of the face G of Q. (A *ridge* is a face of codimension 2.) Although scalar matrices for the differential in the dualizing complex come from the reduced chain complex of $\mathcal{P}$ (as opposed to the reduced *co*chain complex), the differential of $\Omega_Q^{\bullet}$ is a coboundary map, which raises indices. Unfortunately, there is no single best cohomological shift for the dualizing complex. The two standard choices are as follows:

(i) Put $\Bbbk[\mathbb{Z}^d]$ in cohomological degree $-1 - \dim(\mathcal{P})$ and $\Bbbk\{-Q\}$ in cohomological degree zero.

(ii) Put $\Bbbk[\mathbb{Z}^d]$ in cohomological degree 0 and $\Bbbk\{-Q\}$ in cohomological degree $1 + \dim(\mathcal{P})$.

The first choice is sometimes called the *normalized* dualizing complex; it is more often used in the context of local duality [Har66b]. The second choice is more natural from the point of view of injective resolutions, as Theorem 12.11 will attest.

Example 12.8 Let Q be the integer points in the cone over a square $\mathcal{P}$ as in Example 7.13. Monomial matrices for the dualizing complex of the associated semigroup ring are depicted in Fig. 12.1, along with the vector- and face-labeled square $\mathcal{P}$ that supports it. The row and column labels composed of letters (such as ab, bc, cd, da) are simply the names of the faces. Every occurrence of $\mathbf{0} \in \mathbb{Z}^3$ is a vector label. To simplify notation, we use $T_F = F - Q$ for the injective hull of the face F of Q, which coincides with the integer points in the outer tangent cone at any point interior to F. $\diamond$

Definition 12.9 Let Q be a saturated affine semigroup. The **canonical module** of $\Bbbk[Q]$, denoted ω_Q, is the ideal spanned by all monomials $\mathbf{t}^{\mathbf{a}}$ such that $\mathbf{a}$ is interior to C, or equivalently, $\mathbf{a}$ does not lie on a proper face of Q.

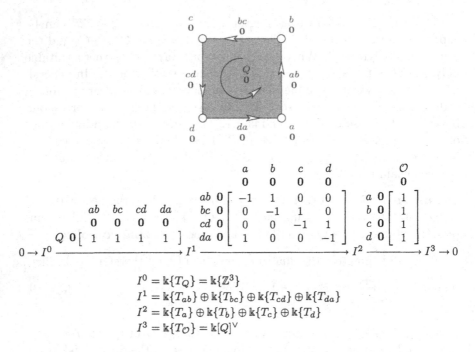

$$I^0 = \Bbbk\{T_Q\} = \Bbbk\{\mathbb{Z}^3\}$$
$$I^1 = \Bbbk\{T_{ab}\} \oplus \Bbbk\{T_{bc}\} \oplus \Bbbk\{T_{cd}\} \oplus \Bbbk\{T_{da}\}$$
$$I^2 = \Bbbk\{T_a\} \oplus \Bbbk\{T_b\} \oplus \Bbbk\{T_c\} \oplus \Bbbk\{T_d\}$$
$$I^3 = \Bbbk\{T_{\mathcal{O}}\} = \Bbbk[Q]^\vee$$

Figure 12.1: Dualizing complex for the cone over a square

We will need a geometric lemma to prove that the dualizing complex resolves the canonical module when Q is saturated. For each vector $\mathbf{a} \in \mathbb{R}^d$, define the subcomplex $\mathcal{P}_\mathbf{a}$ of the polytope $\mathcal{P}$ by

$$\mathcal{P}_\mathbf{a} = \{\text{faces of } \mathcal{P} \text{ corresponding to faces } F \text{ of } Q \text{ with } \mathbf{a} \notin F - C\}. \quad (12.2)$$

When $\mathbf{a}$ lies in the affine span of $\mathcal{P}$, the complex $\mathcal{P}_\mathbf{a}$ consists of the faces F of $\mathcal{P}$ such that $\mathbf{a} \notin F - T_F$, where the *inner tangent cone* T_F is generated by $\mathcal{P} - F$. Informally, $\mathcal{P}_\mathbf{a}$ is the closure of the set of faces of $\mathcal{P}$ whose interiors cannot be seen from $\mathbf{a}$. An example of this crucial case is illustrated below, where the subcomplex $\mathcal{P}_\mathbf{a}$ is the thickened union of two line segments.

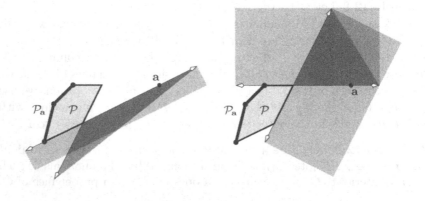

The left-hand figure shows that **a** lies in the (translated) outer tangent cones for the most southern edge and its northern vertex in $\mathcal{P}$, while the right-hand figure shows that the same point **a** lies in the outer tangent cones for the northeast vertex of $\mathcal{P}$ and the two edges of $\mathcal{P}$ containing it. Notice that **a** has to look through the interior relint($\mathcal{P}$) to see the relative interiors of the segments in $\mathcal{P}_\mathbf{a}$.

Lemma 12.10 $\mathcal{P}_\mathbf{a}$ *is a contractible union of facets whenever* $\mathbf{a} \notin \text{relint}(C)$.

Proof. Since $\mathcal{P}_\mathbf{a}$ is the void complex (no faces at all, not even the empty face $\varnothing$) when $\mathbf{a} \in -C$, we assume $\mathbf{a} \notin -C$. Note that $\mathcal{P}_\mathbf{a} = \mathcal{P}_{\mathbf{a}+\mathbf{b}}$ whenever $\mathbf{b}$ is in the lineality space $C \cap (-C)$, because C is invariant under translation by elements in $C \cap (-C)$. Therefore, decomposing C as its lineality space times a pointed subcone $\bar{C}$, we can assume that **a** lies in the linear span L of $\bar{C}$. Next, having assumed $\mathbf{a} \notin -C$, we can by Farkas' Lemma [Zie95, Proposition 1.8] find a hyperplane H in L transverse to $\bar{C}$ and containing **a**, so that $H \cap \bar{C} = \mathcal{P}$ is a valid choice for $\mathcal{P}$. If **a** lies relative interior to some face G of $\mathcal{P}$, then $\mathcal{P}_\mathbf{a} = \mathcal{P}_{\mathbf{a}-\mathbf{c}}$ for any small **c** in the inner tangent cone T_G, so we may as well assume that $\mathbf{a} \notin \mathcal{P}$.

Consider the polytope $\mathcal{P}'_\mathbf{a}$ that is the convex hull of $\mathcal{P}$ and **a**. There are two kinds of facets of $\mathcal{P}'_\mathbf{a}$: those containing **a** and those that are also facets of $\mathcal{P}$. The latter are exactly the facets G of $\mathcal{P}$ such that the linear functional ν_G minimized along G satisfies $\nu_G(\mathbf{a}) > \nu_G(G)$. On the other hand, the condition $\mathbf{a} \notin F - T_F$ for a face F of $\mathcal{P}$ to lie in $\mathcal{P}_\mathbf{a}$ is equivalent to $\nu_G(\mathbf{a}) > \nu_G(F) = \nu_G(G)$ for some facet G of $\mathcal{P}$ containing F. Therefore $\mathcal{P}_\mathbf{a}$ equals the union of the facets of $\mathcal{P}'_\mathbf{a}$ not meeting **a**. It follows from Exercise 4.8 that $\mathcal{P}_\mathbf{a}$ equals the entire subcomplex of $\mathcal{P}'_\mathbf{a}$ consisting of the faces not meeting **a**, whence $\mathcal{P}_\mathbf{a}$ is contractible by Lemma 4.18. $\square$

The contractibility implies the following important property of normal semigroup rings, which will resurface in different language in Section 13.5.

Theorem 12.11 *If* Q *is a saturated affine semigroup, then the dualizing complex* $\Omega_Q^\bullet$ *is a minimal injective resolution of the canonical module* ω_Q.

Proof. Write $\Omega^\bullet = \Omega_Q^\bullet$ for the dualizing complex of $\Bbbk[Q]$. In any $\mathbb{Z}^d$-graded degree $\mathbf{a} \in \text{relint}(C)$, where $C = \mathbb{R}_{\geq 0} Q$, the degree **a** piece of $\Omega^\bullet$ is

$$\Omega_\mathbf{a}^\bullet : \quad 0 \longrightarrow \Bbbk \longrightarrow 0 \longrightarrow 0 \longrightarrow \cdots$$

with the copy of $\Bbbk$ in cohomological degree 0. It remains only to show that $\Omega_\mathbf{a}^\bullet$ has no cohomology when $\mathbf{a} \in \mathbb{Z}^d \smallsetminus \text{relint}(C)$. We will in fact show that $\Omega_\mathbf{a}^\bullet$ agrees with (a homological shift of) the relative chain complex $\widetilde{C}.(\mathcal{P}, \mathcal{P}_\mathbf{a})$ with coefficients in $\Bbbk$, for the polyhedral subcomplex $\mathcal{P}_\mathbf{a}$ in Lemma 12.10. This suffices because of the long exact homology sequence

$$\cdots \longrightarrow \widetilde{H}_i(\mathcal{P}) \longrightarrow \widetilde{H}_i(\mathcal{P}, \mathcal{P}_\mathbf{a}) \longrightarrow \widetilde{H}_{i-1}(\mathcal{P}_\mathbf{a}) \longrightarrow \cdots$$

in which all terms $\tilde{H}.(\mathcal{P})$ and $\tilde{H}.(\mathcal{P}_{\mathbf{a}})$ are zero by contractibility.

Writing $\bar{F}$ for the face of $\mathcal{P}$ corresponding to $F \subseteq Q$, we have

$$
\begin{aligned}
\Omega_{\mathbf{a}}^{\cdot} \;&=\; \Bbbk \cdot \{\text{faces } \bar{F} \text{ of } \mathcal{P} \text{ satisfying } \Bbbk\{F-Q\}_{\mathbf{a}} \neq 0\} \\
&=\; \Bbbk \cdot \{\text{faces } \bar{F} \text{ of } \mathcal{P} \text{ satisfying } \mathbf{a} \in F - Q\} \\
&=\; \tilde{\mathcal{C}}.(\mathcal{P})/\tilde{\mathcal{C}}.(\text{faces } \bar{F} \text{ of } \mathcal{P} \text{ satisfying } \mathbf{a} \notin F - Q), \quad (12.3)
\end{aligned}
$$

so the proof is complete by the next lemma. $\qquad\qquad\qquad\qquad\square$

Lemma 12.12 *If $Q \subseteq \mathbb{Z}^d$ is a saturated semigroup and $\mathbf{a} \in \mathbb{Z}^d$, then $\bar{F} \in \mathcal{P}_{\mathbf{a}}$ if and only if $\mathbf{a} \notin F - Q$.*

Proof. Note the contrast with (12.2), which has the condition $\mathbf{a} \notin F - C$ rather than $\mathbf{a} \notin F - Q$. Replacing $\mathbb{Z}^d$ by the subgroup that Q generates, we may assume that Q generates $\mathbb{Z}^d$. For facets G of Q, we then have $G - Q = (G - C) \cap \mathbb{Z}^d$ because Q is saturated. From this we can conclude that $\mathbf{a} \in F - C$ if and only if $\mathbf{a} \in F - Q$ by expressing $F - C$ and $F - Q$ as intersections over the set of facets G containing F. $\qquad\square$

Theorem 12.11 fails for dualizing complexes associated to general affine semigroup rings, but the saturation hypothesis is much stronger than necessary, given the appropriate generalization of canonical module to the unsaturated case. These issues are central to Section 13.4.

12.3   Brion's Formula

Instead of merely counting the lattice points in a lattice polytope $\mathcal{P}$, as we did in Section 12.1, we can list them all, by writing a Laurent polynomial that records each lattice point separately:

$$
\Phi_{\mathcal{P}}(\mathbf{t}) \;:=\; \sum_{\mathbf{a} \in \mathcal{P} \cap \mathbb{Z}^d} \mathbf{t}^{\mathbf{a}}. \qquad\qquad (12.4)
$$

In addition to the *lattice point enumerator* in (12.4), we might also be interested in those lattice points in $\mathcal{P}$ not lying on the boundary of $\mathcal{P}$. Such lattice points contribute the terms to the *interior lattice point enumerator*

$$
\Phi_{-\mathcal{P}}(\mathbf{t}) \;:=\; (-1)^{\dim \mathcal{P}} \sum_{\mathbf{a} \in \mathrm{relint}(-1 \cdot \mathcal{P}) \cap \mathbb{Z}^d} \mathbf{t}^{\mathbf{a}}.
$$

In the above notation, we make a distinction between the formal symbol $-\mathcal{P}$ appearing in the subscript of Φ and the negated polytope

$$
-1 \cdot \mathcal{P} \;=\; \{\mathbf{a} \in \mathbb{R}^d \mid -\mathbf{a} \in \mathcal{P}\}.
$$

In particular, the notation $\Phi_{-m\mathcal{P}}(\mathbf{t})$ for $m \in \mathbb{N}$ is meant to be parsed as the interior enumerator $\Phi_{-(m \cdot \mathcal{P})}(\mathbf{t})$, which sums (up to a global sign) the lattice points interior to the polytope $-m \cdot \mathcal{P}$ obtained by negating $m \cdot \mathcal{P}$. For integers $m < 0$ we use the convention that $\Phi_{m\mathcal{P}}(\mathbf{t}) = \Phi_{-(-m \cdot \mathcal{P})}$ is an interior enumerator (note that $-m \cdot \mathcal{P}$ is a positive integer scaling of $\mathcal{P}$ here). Our goal in this section is to find a concise formula for the lattice point enumerators $\Phi_{m\mathcal{P}}$, simultaneously for all $m \in \mathbb{Z}$.

Associated to each vertex $\mathbf{v}$ of $\mathcal{P}$ is the *inner tangent cone* to $\mathcal{P}$ at $\mathbf{v}$:

$$C_{\mathbf{v}} \quad := \quad \text{the real cone generated by } \mathcal{P} - \mathbf{v}.$$

We already saw this and the outer tangent cone $-C_{\mathbf{v}}$ in the previous section. The *vertex semigroup* $Q_{\mathbf{v}} = C_{\mathbf{v}} \cap \mathbb{Z}^d$ is pointed because $\mathbf{v}$ is a vertex of $\mathcal{P}$. Let $\mathcal{H}_{\mathbf{v}} := \mathcal{H}_{C_{\mathbf{v}}}$ be the Hilbert basis of this semigroup. The following figure

illustrates some of these definitions; the tangent cone $C_{\mathbf{v}}$ consists of the real points in the shaded region, while $Q_{\mathbf{v}}$ consists of the lattice points there, and the Hilbert basis $\mathcal{H}_{\mathbf{v}}$ consists of the white dots. In general, $\mathcal{H}_{\mathbf{v}}$ determines a presentation of the *vertex semigroup ring* $\Bbbk[Q_{\mathbf{v}}]$ as a quotient of a polynomial ring:

$$S_{\mathbf{v}} \quad := \quad \Bbbk[x_{\mathbf{a}} \mid \mathbf{a} \in \mathcal{H}_{\mathbf{v}}] \quad \twoheadrightarrow \quad \Bbbk[Q_{\mathbf{v}}].$$

Given this presentation, the $S_{\mathbf{v}}$-module $\Bbbk[Q_{\mathbf{v}}]$ has a *vertex K-polynomial*

$$K_{\mathbf{v}}(\mathbf{t}) \quad := \quad \mathcal{K}(\Bbbk[Q_{\mathbf{v}}]; \mathbf{t})$$

and a *vertex denominator*

$$D_{\mathbf{v}}(\mathbf{t}) \quad := \quad \prod_{\mathbf{a} \in \mathcal{H}_{\mathbf{v}}} (1 - \mathbf{t}^{\mathbf{a}}).$$

The $\mathbb{Z}^d$-graded Hilbert series of $\Bbbk[Q_{\mathbf{v}}]$ is the rational generating function

$$\frac{K_{\mathbf{v}}(\mathbf{t})}{D_{\mathbf{v}}(\mathbf{t})} \quad = \quad \sum_{\mathbf{a} \in Q_{\mathbf{v}}} \mathbf{t}^{\mathbf{a}}.$$

Note that the Hilbert series of $\Bbbk\{-Q_{\mathbf{v}}\}$ is $K_{\mathbf{v}}(\mathbf{t}^{-1})/D_{\mathbf{v}}(\mathbf{t}^{-1})$. This fact will arise in the proof of the following theorem, which is the goal of this section.

Theorem 12.13 (Brion's Formula) *For all lattice polytopes $\mathcal{P}$ and $m \in \mathbb{Z}$,*

$$\Phi_{m\mathcal{P}}(\mathbf{t}) \quad = \quad \sum_{\substack{\text{vertices} \\ \mathbf{v} \in \mathcal{P}}} \left(\mathbf{t}^{m\mathbf{v}} \cdot \frac{K_{\mathbf{v}}(\mathbf{t})}{D_{\mathbf{v}}(\mathbf{t})} \right)$$

as rational functions of $\mathbf{t}$ with coefficients in $\mathbb{Q}$. In particular, the right-hand side sums to a Laurent polynomial whose nonzero coefficients are all 1.

The reader seeing this result for the first time should be shocked. It says that if you add together the lattice points in the inner tangent cones at all vertices of $\mathcal{P}$, you get precisely the sum of all the lattice points inside $\mathcal{P}$! The case $m < 0$ is even weirder: if you add together the lattice points in the outer tangent cones at all vertices of $\mathcal{P}$, you get (up to a sign) the sum of all lattice points interior to $\mathcal{P}$. None of these Laurent monomials appeared in the original sum! The case of one-dimensional polytopes is instructive.

Example 12.14 Let $d = 1$ and let $\mathcal{P} = [2, 3] \subset \mathbb{R}$ be the unit line segment from 2 to 3. The two vertices of this 1-dimensional polytope give

$$K_2(t) = 1 \quad \text{and} \quad D_2(t) = 1 - t,$$
$$K_3(t) = 1 \quad \text{and} \quad D_3(t) = 1 - 1/t.$$

The right-hand side of Brion's Formula equals

$$t^{2m} \cdot \frac{1}{1-t} + t^{3m} \cdot \frac{1}{1 - 1/t}. \tag{12.6}$$

For integers m of small absolute value, this rational function is

m	$\Phi_{m\mathcal{P}}(t)$
-4	$-t^{-9} - t^{-10} - t^{-11}$
-3	$-t^{-7} - t^{-8}$
-2	$-t^{-5}$
-1	0
0	t^0
1	$t^3 + t^2$
2	$t^6 + t^5 + t^4$
3	$t^9 + t^8 + t^7 + t^6$

which equals the desired sum of Laurent monomials. $\diamond$

For the proof of Brion's Formula, let C be the cone in $\mathbb{R} \times \mathbb{R}^d$ generated by the points $(1, \mathbf{a})$ for $\mathbf{a} \in \mathcal{P}$. Consider the saturated semigroup $Q = C \cap \mathbb{Z}^{1+d}$, which has $\mathcal{P}$ as a transverse hyperplane slice. [This semigroup Q coincides with the semigroup called Q_{sat} in Section 12.1.] Thus the faces of Q correspond bijectively to those of $\mathcal{P}$. In particular, the vertices $\mathbf{v} \in \mathcal{P}$ correspond to the extreme rays of Q, and the empty face of $\mathcal{P}$ corresponds to the face $\{\mathbf{0}\}$ of Q.

As in Section 12.1, the semigroup ring $\Bbbk[Q]$ is $\mathbb{Z} \times \mathbb{Z}^d$-graded. The Hilbert series of arbitrary modest $\mathbb{Z} \times \mathbb{Z}^d$-graded $\Bbbk[Q]$-modules can be expressed as formal doubly infinite series in t_0 with coefficients that are themselves formal series $p(t_1, \ldots, t_d) \in \mathbb{Z}[[\mathbb{Z}^d]]$. Call such a series $\sum_{m \in \mathbb{Z}} t_0^m p_m(\mathbf{t})$ *summable* if there is a *single* Laurent polynomial $f(\mathbf{t}) \in \mathbb{Z}[\mathbb{Z}^d]$ such that p_m is summable with respect to f for all $m \in \mathbb{Z}$ (Definition 8.39); in other words, $p_m(\mathbf{t})f(\mathbf{t})$ is a Laurent polynomial in $t_1, \ldots, t_d$ for every m.

Lemma 12.15 *Let F be a nonempty face of $\mathcal{P}$. If T_F denotes the injective hull of the face of Q corresponding to F, then the Hilbert series of $\Bbbk\{T_F\}$ is summable. More precisely, the Hilbert series $H(\Bbbk\{T_F\}; t_0, \mathbf{t})$ satisfies*

$$D_\mathbf{v}(\mathbf{t}^{-1}) \cdot H(\Bbbk\{T_\mathbf{v}\}; t_0, \mathbf{t}) \;\; = \;\; \sum_{m \in \mathbb{Z}} t_0^m \mathbf{t}^{m\mathbf{v}} \cdot K_\mathbf{v}(\mathbf{t}^{-1}) \qquad (12.7)$$

if $F = \mathbf{v}$ is a vertex of $\mathcal{P}$ and

$$D_\mathbf{v}(\mathbf{t}^{-1}) \cdot H(\Bbbk\{T_F\}; t_0, \mathbf{t}) \;\; = \;\; 0 \qquad (12.8)$$

for any vertex $\mathbf{v} \in F$ if $\dim F \geq 1$.

Proof. Translation by $m \cdot (1, \mathbf{w})$ for any vector $\mathbf{w} \in F$ gives a bijection $T_F \cap (0 \times \mathbb{Z}^d) \to T_F \cap (m \times \mathbb{Z}^d)$ between the parts of T_F at levels 0 and m. Thus $H(\Bbbk\{T_F\}; t_0, \mathbf{t}) = \sum_{m \in \mathbb{Z}} t_0^m \mathbf{t}^{m\mathbf{w}} H_{F,0}$ for $\mathbf{w} \in F$, where $H_{F,0} \in \mathbb{Z}[[\mathbb{Z}^d]]$ is the coefficient on $1 = t_0^0$.

If $F = \mathbf{v}$ is a vertex, then $T_\mathbf{v} \cap (1 \times \mathbb{Z}^d)$ is by definition the translate $(1, \mathbf{v}) - Q_\mathbf{v}$ of the "outer" vertex semigroup of $\mathcal{P}$ at $\mathbf{v}$. Thus $H_{\mathbf{v},0}$ is the Hilbert series of $\Bbbk\{-Q_\mathbf{v}\}$, proving (12.7) by Theorem 8.20. If $\dim F \geq 1$, we can choose a vertex $\mathbf{v}$ of F, along with another vector $\mathbf{w} \in F$ such that $\mathbf{w} - \mathbf{v}$ is a Hilbert basis vector of $C_\mathbf{v}$. Since $(1, \mathbf{w}) + T_F = (1, \mathbf{v}) + T_F$, it follows that $\mathbf{t}^{\mathbf{v} - \mathbf{w}} H_{F,0} = H_{F,0}$. Thus $1 - \mathbf{t}^{\mathbf{v} - \mathbf{w}}$ annihilates $H_{F,0}$. The final equation (12.8) follows from the fact that $1 - \mathbf{t}^{\mathbf{v} - \mathbf{w}}$ is a factor of $D_\mathbf{v}(\mathbf{t}^{-1})$. $\square$

The previous lemma hinged on the fact that the Hilbert series of an affine semigroup with nontrivial units sums to zero because the series is equal to its translates along its directions of lineality. The essence of the forthcoming proof will be that most of the indecomposable summands in the dualizing complex are cones with nonzero lineality. The exactness of the dualizing complex therefore results in an expression of the rational function 0 as an alternating sum of formal series that almost all sum to zero. The terms surviving with nonzero sums are those contributing to Brion's Formula.

Proof of Theorem 12.13. As in Lemma 12.15, let $Q \subset \mathbb{Z} \times \mathbb{Z}^d$ be generated by the vectors $\{(1, \mathbf{a}) \mid \mathbf{a} \in \mathcal{P} \cap \mathbb{Z}^d\}$. By Theorem 12.11, the Hilbert series of the canonical module ω_Q equals the alternating sum of the Hilbert series of the injective modules in the dualizing complex:

$$H(\omega_Q; t_0, \mathbf{t}) \;\; = \;\; \sum_{\text{faces } F \in \mathcal{P}} (-1)^{d - \dim F} H(\Bbbk\{T_F\}; t_0, \mathbf{t}). \qquad (12.9)$$

Again, we have used T_F to denote the injective hull of the face of Q corresponding to F. The sign $(-1)^{d - \dim F}$ occurs because $\Bbbk\{\mathbb{Z}^d\}$ lies in cohomological degree 0.

The left-hand side equals $\sum_{m < 0} t_0^{-m} \cdot (-1)^d \cdot \Phi_{m\mathcal{P}}(\mathbf{t}^{-1})$ because we are enumerating interior lattice points in $m \cdot \mathcal{P}$ instead of $-m \cdot \mathcal{P}$. Similarly, the term for the empty face $F = \varnothing$ on the right hand side of (12.9) equals

$$(-1)^{d+1} \cdot H(\Bbbk\{T_\varnothing\}; t_0, \mathbf{t}) \;=\; \sum_{m \in \mathbb{N}} t_0^{-m} \cdot (-1)^{d+1} \cdot \Phi_{m\mathcal{P}}(\mathbf{t}^{-1}) \quad (12.10)$$

because we are enumerating all the lattice points in $-m \cdot \mathcal{P}$ instead of $m \cdot \mathcal{P}$.

Let $\mathbf{D} = \prod_{\mathbf{v}} D_{\mathbf{v}}(\mathbf{t}^{-1})$ be the product of all the vertex denominators for the semigroups $-Q_{\mathbf{v}}$, where $\mathbf{v}$ ranges over all vertices of $\mathcal{P}$. Move the empty-face summand in (12.9) to the left-hand side, and multiply the resulting equation by $(-1)^d \cdot \mathbf{D}$. By Lemma 12.15, this multiplication kills every term in which F is a positive-dimensional face of $\mathcal{P}$, and it yields by (12.10) the identity

$$\mathbf{D} \cdot \sum_{m \in \mathbb{Z}} t_0^{-m} \cdot \Phi_{m\mathcal{P}}(\mathbf{t}^{-1}) \;=\; \sum_{\mathbf{v} \in \mathcal{P}} \sum_{m \in \mathbb{Z}} t_0^m \cdot \mathbf{t}^{m\mathbf{v}} \cdot \frac{\mathbf{D}}{D_{\mathbf{v}}(\mathbf{t}^{-1})} \cdot K_{\mathbf{v}}(\mathbf{t}^{-1}).$$

The coefficients of t_0^{-m} on the two sides of this equation are equal as Laurent polynomials in $\mathbf{t}$. Hence we can divide both of these coefficients by $\mathbf{D}$ to get

$$\Phi_{m\mathcal{P}}(\mathbf{t}^{-1}) \;=\; \sum_{\mathbf{v}} \mathbf{t}^{-m\mathbf{v}} \cdot \frac{K_{\mathbf{v}}(\mathbf{t}^{-1})}{D_{\mathbf{v}}(\mathbf{t}^{-1})}.$$

Substituting $\mathbf{t}$ for $\mathbf{t}^{-1}$ in the above equation completes the proof. $\qquad \square$

The Ehrhart polynomial of the polytope $\mathcal{P}$ is obtained from the lattice point enumerator $\Phi_{m\mathcal{P}}(\mathbf{t})$, which is a Laurent polynomial, by substituting $\mathbf{t} = \mathbf{1}$. Although this substitution is not possible on the individual terms $\mathbf{t}^{m\mathbf{v}} \cdot \frac{K_{\mathbf{v}}(\mathbf{t})}{D_{\mathbf{v}}(\mathbf{t})}$ in Brion's Formula, because the denominators $D_{\mathbf{v}}(\mathbf{t})$ vanish at $\mathbf{t} = \mathbf{1}$, the substitution can be applied to the sum of rational functions on the right-hand side of Brion's Formula using L'Hôpital's rule. Allowing the values of m in this substitution to vary yields another proof of the polynomiality of the Ehrhart counting function $E_{\mathcal{P}}(m)$, and more.

Corollary 12.16 (Ehrhart reciprocity) *Let $\mathcal{P}$ be a lattice polytope. The function $m \mapsto E_{\mathcal{P}}(m)$ for $m \in \mathbb{N}$ is a polynomial, and moreover, the number of interior lattice points in $m \cdot \mathcal{P}$ for $m \in \mathbb{N}$ equals $(-1)^{\dim(\mathcal{P})} \cdot E_{\mathcal{P}}(-m)$.*

Proof. Let t be a variable, and set $D(t) = \prod_{\mathbf{v}} D_{\mathbf{v}}(t, \ldots, t)$, the product of all the vertex denominators of $\mathcal{P}$ evaluated with $t_i = t$ for $i = 1, \ldots, d$. Factor $D(t)$ as $(t-1)^k C(t)$, where $t-1$ does not divide the polynomial $C(t)$. Writing $|\mathbf{v}| = v_1 + \cdots + v_d$ for each vertex $\mathbf{v} = (v_1, \ldots, v_d)$ of $\mathcal{P}$, the limit

$$\lim_{t \to 1} \left(\frac{1}{(t-1)^k} \cdot \frac{D(t)}{C(t)} \cdot \sum_{\mathbf{v}} t^{m|\mathbf{v}|} \frac{K_{\mathbf{v}}(t, \ldots, t)}{D_{\mathbf{v}}(t, \ldots, t)} \right) \;=\; \Phi_{m\mathcal{P}}(1, \ldots, 1) \quad (12.11)$$

exists for all $m \in \mathbb{Z}$ by Brion's Formula. Define the rational function

$$f(t) \;=\; \frac{D(t)}{C(t)} \cdot \sum_{\mathbf{v}} t^{m|\mathbf{v}|} \frac{K_{\mathbf{v}}(t, \ldots, t)}{D_{\mathbf{v}}(t, \ldots, t)}$$

of t, and set $f_j(t) = \frac{d^j}{dt^j} f(t)$ for $j \in \mathbb{N}$. The existence of the limit in (12.11) guarantees that $\lim_{t \to 1} f_j(t) = 0$ for all derivatives of order $j < k$, by induction on j and L'Hôpital's rule, because the denominator $(t-1)^k$ has j^{th} derivative zero at $t = 1$ for all $j < k$. Furthermore, the limit in (12.11) must (by L'Hôpital's rule again) equal $\frac{1}{k!} \lim_{t \to 1} f_k(t)$.

Since $D_{\mathbf{v}}(t, \ldots, t)$ divides $D(t)$ for all vertices $\mathbf{v} \in \mathcal{P}$, we can write $f(t)$ as a ratio $\frac{C(t)f(t)}{C(t)}$ of rational functions whose denominator is a polynomial not divisible by $t-1$. Although $C(t)f(t)$ need not be a polynomial (the exponent on $t^{m|\mathbf{v}|}$ is allowed to be negative), it is a linear combination of Laurent monomials $t^{m|\mathbf{v}|}$ with polynomial coefficients and therefore can be evaluated at $t = 1$ without taking limits. Consequently, $f_k(t)$ equals a ratio of functions of t whose numerator can be evaluated at 1 and whose polynomial denominator $C(t)^k$ is not divisible by $t - 1$. Furthermore, when $f_k(t)$ is expressed this way, its numerator is a polynomial in m whose coefficients are rational functions of t, since m does not appear in any of the coefficients of $t^{m|\mathbf{v}|}$ in $C(t)f(t)$. Hence $f_k(1)/k!$ is a polynomial in m that agrees with $E_{\mathcal{P}}(m)$ for all $m \in \mathbb{N}$ and counts interior lattice points when $-m \in \mathbb{N}$. $\square$

Example 12.17 For the one-dimensional polytope $\mathcal{P} = [2,3]$ in Example 12.14, substituting $t = 1$ into expression (12.6) yields the Ehrhart polynomial $E_{\mathcal{P}}(m) = m + 1$. Hence

$$(-1)^{\dim(\mathcal{P})} \cdot E_{\mathcal{P}}(-m) \;=\; -((-m) + 1) \;=\; m - 1$$

is the number of interior lattice points in $m \cdot \mathcal{P} = [2m, 3m]$. $\diamond$

12.4 Short rational generating functions

In this section we discuss the computation of Ehrhart polynomials and related computations in combinatorial commutative algebra from the perspective of complexity of algorithms. Here, we assume that the reader is familiar with the language of complexity theory.

A lattice polytope $\mathcal{P}$ in $\mathbb{R}^d$ is presented either by a list of vertices $\mathbf{v} = (v_1, \ldots, v_d)$ or by a list of defining inequalities $\nu_i \cdot \mathcal{P} \geq w_i$ for some vector $\mathbf{w} \in \mathbb{Z}^n$. If we assume that the dimension d is fixed, then the two presentations can be transformed into each other in polynomial time. The space complexity of presenting the polytope $\mathcal{P}$ is measured by the bit size of the integers v_i, ν_{ij}, and w_i. Recall that the bit size of an integer M is $O(\log(|M|))$. The goal of this section is the following complexity result.

Theorem 12.18 (Barvinok's Theorem) *Suppose that the dimension d is fixed and $\mathcal{P}$ is a lattice polytope in $\mathbb{R}^d$. Then the lattice point enumerator*

$$\Phi_{\mathcal{P}}(\mathbf{t}) \;=\; \sum_{\mathbf{v} \in \mathcal{P}} \mathbf{t}^{\mathbf{v}} \cdot \frac{K_{\mathbf{v}}(\mathbf{t})}{D_{\mathbf{v}}(\mathbf{t})}$$

can be computed in polynomial time, in the binary complexity model.

From the rational function on the right-hand side of Theorem 12.18, one can read off the number of lattice points in $\mathcal{P}$. This amounts to substituting $t = (1, \ldots, 1)$ while being careful to avoid the poles. The basic idea is to substitute a numerical vector close to $(1, \ldots, 1)$ and then to round the result to the nearest integer. This can be done by a deterministic algorithm that runs in polynomial time. We recover the Ehrhart polynomial $E_{\mathcal{P}}(m)$ by Lagrange interpolation from its values at $m = 0, 1, 2, \ldots, d$. Each value of the Ehrhart polynomial is computed by the method described earlier.

Corollary 12.19 *The Ehrhart polynomial of a lattice polytope can be computed in polynomial time if the dimension of the polytope is fixed.*

Before presenting the proof of Theorem 12.18, we give a simple example.

Example 12.20 Let $\mathcal{P}$ be the triangle in $\mathbb{R}^2$ with vertices $(0,0)$, $(a,0)$, and (a, a^2), where a is a large positive integer. The Ehrhart polynomial is

$$E_{\mathcal{P}}(m) \;=\; 1 + \left(a + \frac{a^2}{2}\right)m + \frac{a^3}{2}m^2.$$

The binary encoding of both the input $\mathcal{P}$ and output $E_{\mathcal{P}}$ requires $O(\log(a))$ bits. Let us now examine the complexity of the lattice point enumerator

$$\Phi_{\mathcal{P}}(t_1, t_2) \;=\; \sum_{i=0}^{a} \sum_{j=0}^{ia} t_1^i t_2^j.$$

The number of terms is exponential in $O(\log(a))$, so we are not allowed to write down this expanded form. Brion's Formula for this rational function is

$$\Phi_{\mathcal{P}}(t_1, t_2) \;=\; \frac{K_{(0,0)}(t_1, t_2)}{D_{(0,0)}(t_1, t_2)} + t_1^a \cdot \frac{K_{(a,0)}(t_1, t_2)}{D_{(a,0)}(t_1, t_2)} + t_1^a t_2^{a^2} \cdot \frac{K_{(a,a^2)}(t_1, t_2)}{D_{(a,a^2)}(t_1, t_2)}.$$

The last two terms are easily computed in polynomial time:

$$K_{(a,0)} = 1 \quad \text{and} \quad D_{(a,0)} = (1 - t_1^{-1})(1 - t_2),$$
$$K_{(a,a^2)} = 1 \quad \text{and} \quad D_{(a,a^2)} = (1 - t_1^{-1} t_2^{-a})(1 - t_2^{-1}).$$

However, for the first term, we have

$$K_{(0,0)} = 1 + t_1 t_2 + t_1 t_2^2 + \cdots + t_1 t_2^{a-1} \quad \text{and} \quad D_{(0,0)} = (1 - t_1)(1 - t_1 t_2^a).$$

The number of terms in $K_{(0,0)}$ is exponential in $O(\log(a))$, so we are not allowed to write down this K-polynomial in its expanded form. However,

$$K_{(0,0)}(t_1, t_2) \;=\; 1 + \frac{t_1 t_2 - t_1 t_2^a}{1 - t_2}.$$

Assembling the pieces, we arrive at a representation of the rational function $\Phi_{\mathcal{P}}(t_1, t_2)$ that requires only $O(\log(a))$ bits. $\diamond$

Proof of Theorem 12.18. We can triangulate $\mathcal{P}$ in polynomial time since the dimension d is fixed. The lattice point enumerator $\Phi_{\mathcal{P}}(\mathbf{t})$ is an integer sum of the lattice point enumerators of the simplices in the triangulation of $\mathcal{P}$, where the coefficients come from the Möbius function of the face poset of the triangulation. Hence we may assume that $\mathcal{P}$ is a d-simplex. Brion's Formula expresses $\Phi_{\mathcal{P}}(\mathbf{t})$ in terms of the Hilbert series $\frac{K_{\mathbf{v}}(\mathbf{t})}{D_{\mathbf{v}}(\mathbf{t})}$ of the $d+1$ simplicial semigroups $Q_{\mathbf{v}}$, so the claim is reduced to the following lemma. $\square$

Lemma 12.21 *Given linearly independent vectors $\mathbf{a}_1, \ldots, \mathbf{a}_d$ in $\mathbb{Z}^d$ with d fixed, we can compute the Hilbert series of the saturated affine semigroup*

$$Q = \mathbb{R}_{\geq 0}(\mathbf{a}_1, \ldots, \mathbf{a}_d) \cap \mathbb{Z}^d$$

in polynomial time. The output is a "short rational function" in $t_1, \ldots, t_d$.

Proof. Let α denote the absolute value of the determinant of the matrix $(\mathbf{a}_1, \ldots, \mathbf{a}_d)$. The Hilbert series denominator is $\prod_{i=1}^d (1 - \mathbf{t}^{a_i})$. The numerator is a sum of α Laurent monomials, as explained in the proof of Theorem 12.2. If $\alpha = 1$, then the numerator is just 1 and we are done. For general α, however, we are not allowed to write down the numerator in its expanded form because α is exponential in the bit complexity of the input data $(\mathbf{a}_1, \ldots, \mathbf{a}_d)$. For $\alpha \geq 2$, consider the parallelotope

$$\{\lambda_1 \mathbf{a}_1 + \cdots + \lambda_d \mathbf{a}_d \mid -1 \leq \lambda_i \leq +1 \text{ for } i = 1, 2, \ldots, d\}.$$

This is a centrally symmetric convex body of volume $\alpha \cdot 2^d > 2^d$. By Minkowski's Theorem [Gru93, Theorem 4], it has a nonzero lattice point $\mathbf{u} = \mu_1 \mathbf{a}_1 + \cdots + \mu_d \mathbf{a}_d$ in its interior. There exists a positive constant ϵ_d such that $|\mu_i| \leq 1 - \epsilon_d$ for all i. Moreover, the lattice point $\mathbf{u}$ can be found in polynomial time using lattice reduction (the Lenstra–Lenstra–Lovasz (LLL) algorithm [Gru93, Section 6.2]). Now write the Hilbert series of Q as the alternating sum of the Hilbert series of the semigroups

$$Q_i = \mathbb{R}_{\geq 0}(\mathbf{a}_1, \ldots, \mathbf{a}_{i-1}, \mathbf{u}, \mathbf{a}_{i+1}, \ldots, \mathbf{a}_d) \cap \mathbb{Z}^d$$

and their faces. Since d is fixed, this can be done in polynomial time. Now the determinant $\mu_i \alpha$ of the semigroup Q_i is smaller than α by a factor of at least $1 - \epsilon_d$, so we only need to iterate this alternating decomposition at most $O(\log(\alpha))$ times until we get to the base case $\alpha = 1$. This completes the proof of Lemma 12.21 and hence of Theorem 12.18. $\square$

Since every saturated affine semigroup can be decomposed into simplicial semigroups, Lemma 12.21 immediately implies the following result.

Corollary 12.22 *For d fixed, the Hilbert series $H(Q; \mathbf{t})$ of any saturated affine semigroup $Q \subset \mathbb{Z}^d$ can be computed in polynomial time.*

The representation of the Hilbert series $H(Q; \mathbf{t})$ produced by Barvinok's algorithm is called a *short rational function*. This means that its size in the binary encoding is polynomial in the size of the description of the semigroup Q. Short rational functions turn out to be abundant in combinatorial commutative algebra. For example, the Hilbert basis $\mathcal{H}_Q$ of a saturated affine semigroup $Q \subset \mathbb{Z}^d$ can be represented as a Laurent polynomial:

$$\mathcal{H}_Q(\mathbf{t}) \quad = \quad \text{the sum of all monomials } \mathbf{t}^\mathbf{a} \text{ for } \mathbf{a} \in \mathcal{H}_Q.$$

Recent work of Barvinok and Woods implies the following two theorems. Although we will not prove either one, we wish to mention them so as to indicate possible future developments in combinatorial commutative algebra.

Theorem 12.23 *For d fixed, the Hilbert basis $\mathcal{H}_Q$ of any saturated affine semigroup $Q \subset \mathbb{Z}^d$ can be computed in polynomial time.*

The point is that while the size of $\mathcal{H}_Q$ can grow exponentially in the bit complexity of the description of Q, we write the Laurent polynomial $\mathcal{H}_Q$ as a short rational function requiring only polynomially many bits. A simple example comes from the cone generated by $(0, 1)$ and $(a, 1)$ in the plane. Here,

$$\mathcal{H}_Q(t_1, t_2) \quad = \quad t_2 + t_1 t_2 + t_1^2 t_2 + t_1^3 t_2 + \cdots + t_1^a t_2 \quad = \quad \frac{t_2 - t_1^{a+1} t_2}{1 - t_1}.$$

This short rational function encodes the $a+1$ elements in the Hilbert basis.

A similar encoding is available for Gröbner bases of toric ideals. Let $\mathbf{A} = (\mathbf{a}_1, \mathbf{a}_2, \ldots, \mathbf{a}_n)$ be an integer $d \times n$ matrix, L its kernel, and I_L its toric ideal in $S = \mathbb{k}[x_1, \ldots, x_n]$. Represent any finite set of binomials $\mathbf{x}^\mathbf{u} - \mathbf{x}^\mathbf{v}$ in I_L by the sum of the corresponding monomials $\mathbf{x}^\mathbf{u} \mathbf{y}^\mathbf{v}$ in $2n$ unknowns.

Theorem 12.24 *Fix d and n. Let $\mathbf{A}$ be an integer $d \times n$ matrix and L its kernel. Then the following can be computed in polynomial time:*

1. *The $\mathbb{Z}^d$-graded Hilbert series of S/I_L*
2. *A minimal generating set of the toric ideal I_L*
3. *Any reduced Gröbner basis of I_L*
4. *A finite universal Gröbner basis of I_L*

We believe that such short representations of ideals and their Hilbert series, originally introduced for the purpose of computing Ehrhart polynomials, will play an increasingly important role in combinatorial commutative algebra. Here is how such a future toric Gröbner computation will look.

Example 12.25 Fix $n = 4$, set $d = 2$, and let $a \geq 3$ be a large integer. The input is the matrix $\mathbf{A} = \begin{bmatrix} a & a-1 & 1 & 0 \\ 0 & 1 & a-1 & a \end{bmatrix}$. The task is to compute the reduced lexicographic Gröbner basis of I_L for the kernel L of $\mathbf{A}$. Thus I_L is the kernel of the map $\Bbbk[x_1, x_2, x_3, x_4] \to \Bbbk[s, t]$ sending

$$x_1 \mapsto s^a, \qquad x_2 \mapsto s^{a-1}t, \qquad x_3 \mapsto st^{a-1}, \qquad \text{and} \qquad x_4 \mapsto t^a.$$

The output would consist of the rational function

$$G(\mathbf{x}, \mathbf{y}) \;=\; x_1 x_4 y_2 y_3 + x_2 x_4^{a-2} y_3^{a-1} + \frac{x_1 x_3 y_2^2 \big((x_1 y_2)^{a-2} - (x_3 y_4)^{a-2}\big)}{x_1 y_2 - x_3 y_4}.$$

This rational function is a polynomial. Each of the $a+1$ terms in its expansion represents a Gröbner basis element. The cardinality of the Gröbner basis grows exponentially in $\log(a)$, the size of the input data, but the running time for computing $G(\mathbf{x}, \mathbf{y})$ is bounded by a polynomial in $\log(m)$. $\diamond$

Exercises

12.1 Give an example of lattice polytope $\mathcal{P}$ such that the $\Bbbk[Q]$-module $\Bbbk[Q_{\text{sat}}]$ considered in Section 12.1 is not free. Give a general condition implying that $\Bbbk[Q_{\text{sat}}]$ is free over $\Bbbk[Q]$.

12.2 Prove that if $\mathcal{P}$ is a lattice polytope in $\mathbb{R}^d$, then $m\mathcal{P}$ is normal for $m \geq d-1$.

12.3 Let $\mathcal{P}_1, \ldots, \mathcal{P}_s$ be lattice polytopes in $\mathbb{R}^d$ and consider the number of lattice points in the Minkowski sum $m_1\mathcal{P}_1 + \cdots + m_s\mathcal{P}_s$ for any $m_1, \ldots, m_s \in \mathbb{N}$. Show that this function is a polynomial of degree d in the parameters $m_1, \ldots, m_s$.

12.4 For the octahedron $\mathcal{P}$ with vertices $(\pm 1, 0, 0)$, $(0, \pm 1, 0)$, and $(0, 0, \pm 1)$, compute $\Phi_{m\mathcal{P}}(t_1, t_2, t_3)$ using Brion's Formula. (An answer is included after the last exercise in this chapter.) Verify that all roots of $\Phi_{m\mathcal{P}}(t, t, t)$ have real part $-1/2$.

12.5 Prove that the second coefficient (after the leading coefficient) of the Ehrhart polynomial of $\mathcal{P}$ equals $1/2$ times the sum of the volumes of each facet, each normalized with respect to the sublattice in the hyperplane spanned by the facet.

12.6 Complete the derivation of Ehrhart's Theorem from Brion's Formula via Corollary 12.16 by directly calculating the degree of the Ehrhart polynomial.

12.7 A function $q : \mathbb{Z} \to \mathbb{Z}$ is called a **quasi-polynomial** if there are polynomials $p_1, \ldots, p_r$ such that $q(m) = p_i(m)$ whenever $m \equiv i \pmod{r}$. Prove that if $\mathcal{P}$ is a polytope whose vertices have rational coordinates, then the function $E_{\mathcal{P}}(m) = \#\big((m \cdot \mathcal{P}) \cap \mathbb{Z}^d\big)$ counting the integer points in dilations of $\mathcal{P}$ is a quasi-polynomial.

12.8 How would you compute the normal form of $x_1^a x_2^a x_3^a x_4^a$ modulo the reduced Gröbner basis G presented in Example 12.25?

12.9 Let Q be an affine semigroup and $C^\vee$ the cone dual to $C = \mathbb{R}_{\geq 0}Q$. Fix any triangulation of $C^\vee$ into simplicial cones. Each maximal face in that triangulation is dual to a simplicial cone containing C, and hence corresponds to a simplicial semigroup containing Q. Explain how to write the Hilbert series of Q as an alternating sum of the Hilbert series of these simplicial semigroups.

12.10 Are all coefficients of the Ehrhart polynomial nonnegative?

12.11 Compute the $\mathbb{Z}^3$-graded Hilbert series of the semigroup $C \cap \mathbb{Z}^3$, where C is the cone spanned by $(5, 7, 11)$, $(7, 11, 5)$, and $(11, 5, 7)$. Try to use Barvinok's algorithm, which appears in the proof of Lemma 12.21.

12.12 Draw the affine semigroups in $\mathbb{Z}^2$ generated by

  (a) $\{(4, 0), (3, 1), (2, 2), (1, 3), (0, 4)\}$

  (b) $\{(4, 0), (3, 1), (1, 3), (0, 4)\}$

  (c) $\{(4, 0), (3, 1), (2, 2), (0, 4)\}$

In which cohomological degrees do the corresponding dualizing complexes have nonzero cohomology? (See Exercise 13.4 for an explanation.)

12.13 Describe the canonical module of the ring $\Bbbk[t^3, t^4, t^5]$ as the quotient of a Laurent lattice module in $\Bbbk[x^{\pm 1}, y^{\pm 1}, z^{\pm 1}]$ by a lattice action. Is there a relation to Alexander duality?

Answer to Exercise 12.4 Run the following code in Maple.

```
f1:=(1-z^2)/((1-x*z)*(1-x^(-1)*z)*(1-y*z)*(1-y^(-1)*z)): f2:=subs(z=1/z,f1):
f3:=subs({x=y,y=z,z=x},f1): f4 := subs({x=y,y=z,z=x},f2):
f5:=subs({x=z,y=x,z=y},f1): f6 := subs({x=z,y=x,z=y},f2):
ans:=z^(-m)*f1 +z^m*f2 +x^(-m)*f3 +x^m*f4 +y^(-m)*f5 +y^m*f6: print(ans);
```

Notes

Ehrhart developed the theory of polynomiality for lattice point counting functions in multiples of rational and integral polytopes throughout the 1960s, in a long sequence of articles, highlights of which include [Ehr62a, Ehr62b, Ehr67a, Ehr67b, Ehr67c]. Aspects of the reciprocity in Corollary 12.16 are due to Ehrhart in these articles, and to Macdonald [Macd71], who had also been on this track at the time, for example with his article on lattice polytopes [Macd63].

The normalized dualizing complex makes sense for every local or graded ring of geometric interest. The notion is due to Grothendieck and appears in Hartshorne's book on local cohomology [Har66b]. Most of the information on dualizing complexes in this chapter is based on articles by Ishida [Ish80, Ish87]. We will see more about dualizing complexes in Section 13.5.

Brion's Formula originally appeared in [Bri88], where the proof used equivariant K-theory of toric varieties. More recent work of Brion and Vergne [BV97] develops a powerful setting for lattice point counting based on Fourier transforms of measures on piecewise polyhedral regions. The results include volume formulas under continuous parallel translation of the facets of a polytope. We recommend that interested readers start with [Stu95, Ver03].

Barvinok proved Theorem 12.18 in [Bar94]. An excellent survey on various methods for computing Ehrhart polynomials is [BP99]. Some of these are efficiently implemented in the program *LattE* [DH³TY03]. Theorems 12.23 and 12.24 are consequences of general results on short rational functions due to Barvinok and Woods [BW03]. The short encoding of Gröbner bases and its proposed implementation in *LattE* are discussed in [DH³SY03].

A number of the exercises in this chapter, including Exercise 12.4, were suggested by Matthias Beck, who notes that all roots for the generalized octahedra (cross-polytopes) in any dimension have real part $-1/2$ [BCKV00, BDDPS04]. Exercise 12.9 was suggested by Michel Brion; see [BV97, Proposition 3.1].

Chapter 13

Local cohomology

As we have seen in the previous two chapters, $\mathbb{Z}^d$-graded injective modules and resolutions reflect the polyhedral geometry of affine semigroups. In the present chapter, our last on toric algebra, we investigate how this combinatorial structure extends to another construction from homological algebra, namely local cohomology. Roughly speaking, local cohomology modules are defined by starting with an injective resolution, deleting some indecomposable summands, and taking the cohomology of the resulting complex.

Local cohomology provides "derived" information regarding associated primes, analogous to the "derived" data regarding generators and relations provided by higher syzygies. Local cohomology in combinatorial contexts produces modules with interesting Hilbert series, which record homological invariants of simplicial and cellular complexes. In somewhat less combinatorial (but still $\mathbb{Z}^d$-graded) contexts, these modules can be presented in finite data structures relying on polyhedral geometry. This type of presentation is necessary because although local cohomology modules are well-behaved as $\mathbb{Z}^d$-graded vector spaces, they are neither finitely generated nor finitely cogenerated. Finally, local cohomology holds the key to binding together an assortment of criteria all characterizing the ubiquitous Cohen–Macaulay condition, some of which are used in combinatorial applications.

13.1 Equivalent definitions

In earlier parts of this book, we exploited the fact that Betti numbers of a given module M over a polynomial ring can be calculated two ways: either by tensoring a free resolution of M with $\Bbbk$ or by tensoring a Koszul complex with M. Similarly, there is more than one way to calculate local cohomology. Also similarly, we will present the various ways of calculating local cohomology but only sketch the proof of their equivalence, which belongs more properly to homological algebra.

Let us begin with the construction of local cohomology via injective resolutions. This particular choice makes it most clear in what sense local cohomology is derived from taking submodules with given support.

Definition 13.1 For an ideal I in a commutative ring R, and a module M,

$$\Gamma_I M = (0 :_M I^\infty) = \{y \in M \mid I^r y = 0 \text{ for some } r \in \mathbb{N}\}$$

is the submodule **supported on** I. An element in $\Gamma_I M$ is said to have **support on** I. The ith **local cohomology module** of M **with support on** I is the module $H_I^i(M)$ obtained from any (see Exercise 13.8) injective resolution $0 \to E^0 \to E^1 \to \cdots$ of M by taking the ith cohomology of its subcomplex $0 \to \Gamma_I E^0 \to \Gamma_I E^1 \to \cdots$ supported on I.

In more categorical language, H_I^i is the right derived functor of the left-exact functor Γ_I. Definition 13.1 is not as abstract as it may at first seem, as we will see shortly, at least in the case that interests us. This case is where $R = \Bbbk[Q]$ is an affine semigroup ring, I is a monomial ideal, and M is $\mathbb{Z}^d$-graded, as is the injective resolution $E^\cdot$. (Ungraded injectives would suffice, but who prefers those?) We can even assume that I is a radical ideal in $\Bbbk[Q]$, because $\Gamma_I = \Gamma_{\sqrt{I}}$. This means the ideal I has the following form.

Definition 13.2 Suppose Δ is a polyhedral subcomplex of the real cone $\mathbb{R}_{\geq 0} Q$. The **face ideal** I_Δ is generated by all monomials with exponent vectors not lying on faces of Q that are in Δ. Equivalently, the nonzero monomials in the **face ring** $\Bbbk[Q]/I_\Delta$ are precisely those lying on faces of Δ.

Polyhedral face rings and ideals are straightforward generalizations of Stanley–Reisner rings and ideals, which constitute the case $Q = \mathbb{N}^d$. The map from polyhedral subcomplexes of $\mathbb{R}_{\geq 0} Q$ to face ideals in $\Bbbk[Q]$ is an inclusion-reversing bijection.

For the purpose of making Definition 13.1 more concrete, identifying the ideal I as a combinatorial object (a face ideal) is half the battle. The other half comes from the realization that Γ_{I_Δ} is quite a simple operation to carry out on injective modules.

Lemma 13.3 If $E = \bigoplus_{k \in K} \Bbbk\{\mathbf{a}_k + F^k - Q\}$ is an injective module, then

$$\Gamma_{I_\Delta} E = \bigoplus_{F^k \in \Delta} \Bbbk\{\mathbf{a}_k + F^k - Q\}$$

is obtained by taking only those summands whose support faces lie in Δ.

Proof. First make the (easy) check that Γ_I commutes with direct sums and $\mathbb{Z}^d$-graded translation. Then use the fact that $\Gamma_{I_\Delta} \Bbbk\{F - Q\}$ is zero unless $F \in \Delta$, in which case every element is annihilated by some power of I_Δ. $\square$

The last sentence in the above proof is fundamental; the reader should check it carefully. Note how much simpler Γ_I is on injectives than it is on arbitrary modules! We can exploit this to compute examples explicitly.

Example 13.4 Let Q be the subsemigroup of $\mathbb{Z}^3$ from Examples 7.13 and 12.8, generated by $(1,0,0)$, $(1,1,0)$, $(1,1,1)$, and $(1,0,1)$. Its semi-group ring is $\Bbbk[Q] \cong \Bbbk[a,b,c,d]/\langle ac - bd\rangle$. The ideal $\mathfrak{p} = \langle c, d\rangle$ is the face ideal for the 2-dimensional facet corresponding to 'ab', in the xy-plane. Let us compute the local cohomology modules $H_{\mathfrak{p}}^i(\omega_Q)$ of the canonical module using the dualizing complex in Example 12.8, by Theorem 12.11.

Applying $\Gamma_{\mathfrak{p}}$ to the dualizing complex yields a complex

$$
\begin{array}{ccc}
 & & \mathcal{O} \\
a \quad b & & \mathbf{0} \\
\mathbf{0} \quad \mathbf{0} & a\;\mathbf{0}\begin{bmatrix}1\end{bmatrix} \\
ab\;\mathbf{0}\begin{bmatrix}-1 & 1\end{bmatrix} & b\;\mathbf{0}\begin{bmatrix}1\end{bmatrix}
\end{array}
$$

$$
0 \to \Gamma_{\mathfrak{p}}I^1 \xrightarrow{\hspace{2.5cm}} \Gamma_{\mathfrak{p}}I^2 \xrightarrow{\hspace{2cm}} \Gamma_{\mathfrak{p}}I^3 \to 0
$$

$$
\begin{array}{ccc}
\| & \| & \| \\
T_{ab} & T_a \oplus T_b & T_{\mathcal{O}}
\end{array}
$$

by Lemma 13.3. Again using the notation of Example 12.8, consider the contributions of the injective hulls T_{ab}, T_a, T_b, and $T_{\mathcal{O}}$ of the surviving faces to a $\mathbb{Z}^3$-graded degree $\mathbf{b} = (b_x, b_y, b_z)$. If $b_z > 0$, then none of the four injective hulls contribute. The half-space $b_z \le 0$, however, is partitioned into five *sectors*, each of which consists of a collection of degrees where the set of summands contributing a nonzero vector space remains constant. The summands contributing to each sector are listed in Fig. 13.1, which depicts the intersections of the sectors with the plane $b_z = -m$ as the five regions.

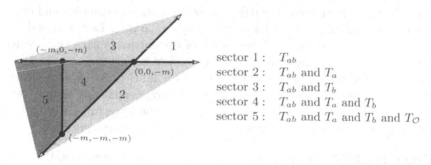

sector 1 : T_{ab}
sector 2 : T_{ab} and T_a
sector 3 : T_{ab} and T_b
sector 4 : T_{ab} and T_a and T_b
sector 5 : T_{ab} and T_a and T_b and $T_{\mathcal{O}}$

Figure 13.1: Intersections of sectors with a horizontal plane

Only in sectors 1 and 4 does $\Gamma_{\mathfrak{p}}I^{\bullet}$ have any cohomology. The cone of integer points in sector 1 and the cohomology of $\Gamma_{\mathfrak{p}}I^{\bullet}$ there are as follows:

$$\text{sector 1}: \quad b_z \le 0 \text{ and } b_x > b_y > 0 \quad \Leftrightarrow \quad H_{\mathfrak{p}}^1(\omega_Q)_{\mathbf{b}} = \Bbbk.$$

For sector 4, we get the cone of integer points and cohomology as follows:

$$\text{sector 4}: \quad 0 \geq b_y \geq b_x > b_z \quad \Leftrightarrow \quad H^2_{\mathfrak{p}}(\omega_Q)_{\mathbf{b}} = \Bbbk.$$

These two modules demonstrate that local cohomology modules of finitely generated modules might be neither finitely generated nor finitely cogenerated (see Example 13.17 for more details about the latter). $\diamond$

Local cohomology of M with support on I can be equivalently characterized by tensoring M with a complex of localizations of R.

Definition 13.5 For elements $m_1, \ldots, m_r$ in a commutative ring R, set $m_\sigma = \prod_{i \in \sigma} m_i$ for $\sigma \subseteq \{1, \ldots, r\}$. The **Čech complex** $\check{C}^{\bullet}(m_1, \ldots, m_r)$ is

$$0 \to R \to \bigoplus_{i=1}^{n} R[m_i^{-1}] \to \cdots \to \bigoplus_{|\sigma|=k} R[m_\sigma^{-1}] \to \cdots \to R[m_{\{1,\ldots,r\}}^{-1}] \to 0.$$

This is to be considered as a cochain complex (upper indices increasing from the copy of R sitting in cohomological degree 0), with the map between the summands $R[m_\sigma^{-1}] \to R[m_{\sigma \cup i}^{-1}]$ in $\check{C}^{\bullet}(m_1, \ldots, m_r)$ being $\mathrm{sign}(i, \sigma \cup i)$ times the canonical localization homomorphism.

The Čech complex would more accurately be called the *stable Koszul complex*, as it sometimes is in the literature. Indeed, another way to define the Čech complex is to formally replace each summand $S[-\sigma]$ in the coKoszul complex $\mathbb{K}^{\bullet}$ from Definition 5.4 by the localization $R[m_i^{-1} \mid i \in \sigma]$. This describes $\check{C}^{\bullet}(m_1, \ldots, m_r)$ as a "cocellular complex of localizations".

Particularly important among Čech complexes is the one on the variables $x_1, \ldots, x_n$ over the polynomial ring $S = \Bbbk[x_1, \ldots, x_n]$. We can use a special variant of monomial matrices to write down this complex and other complexes involving localizations of the polynomial ring. Each row and column has a single vector label, but some of the entries in each such vector are allowed to equal the symbol $*$, which indicates that the corresponding variables have been inverted. Thus the vector label $(3, *, -4, *)$ indicates the localization $S[x_2^{-1}, x_4^{-1}](-3, 0, 4, 0)$ of the polynomial ring in four variables. (We use the symbol $*$ because replacing the zeros in $(-3, 0, 4, 0)$ by arbitrary integers does not change the $\mathbb{Z}^4$-graded isomorphism class.)

Example 13.6 Using the above specialized monomial matrices, the Čech complex $\check{C}^{\bullet}(x_1, x_2, x_3)$ over the polynomial ring S in three variables is

Despite the way we drew the direct sums, elements in them are to be considered as *row* vectors, so the monomial matrices act by multiplication on the right, as is natural for cochain complexes—see Example 1.21.   ◇

Given a module M, its submodule with support on the ideal $I = \langle m_1, \ldots, m_r \rangle$ is the kernel of the homomorphism $M \to \bigoplus_i M \otimes R[m_i^{-1}]$ (Exercise 13.6). Since this homomorphism is just the first map in the complex $M \otimes_R \check{C}^{\bullet}(m_1, \ldots, m_r)$, the next result should at least be plausible.

Theorem 13.7 *The local cohomology of M supported on $I = \langle m_1, \ldots, m_r \rangle$ is the cohomology of the Čech complex tensored with M:*

$$H_I^i(M) \;=\; H^i(M \otimes \check{C}^{\bullet}(m_1, \ldots, m_r)).$$

Sketch of proof. One possibility is to use homological algebra as in Exercise 1.12, where the double complex this time comes from tensoring together an injective resolution $E^{\bullet}$ of M and the Čech complex $\check{C}^{\bullet}$. View $E^{\bullet}$ as going upward and $\check{C}^{\bullet}$ as going to the right, so the arrows in Exercise 1.12 must all be reversed. The flatness of localization implies that the vertical differential makes the columns into a resolution of the complex $M \otimes \check{C}^{\bullet}$. On the other hand, assuming that $R = \Bbbk[Q]$ and everything in sight is $\mathbb{Z}^d$-graded, Exercises 13.6 and 13.7 prove that the horizontal differential makes the rows into a resolution of the complex $\Gamma_I E^{\bullet}$. The same proof actually works in general, without the semigroup ring and $\mathbb{Z}^d$-graded assumptions on R and M, but ungraded injectives are required, for which parts of the arguments are no longer combinatorial [BH98, Chapter 3].   □

The proof of Theorem 13.7 actually yields a more general principle.

Fact 13.8 *Fix an ideal I. Suppose $C^{\bullet}$ is a complex of flat modules such that*

1. *for every module M, the 0th cohomology of $M \otimes C^{\bullet}$ is $\Gamma_I M$; and*

2. *the ith cohomology of $J \otimes C^{\bullet}$ is zero whenever J is injective and $i \geq 1$.*

Then $H_I^i(M)$ is the ith cohomology of $M \otimes C^{\bullet}$ for all modules M.

It is worth bearing in mind that the only requirement for Theorem 13.7 is noetherianity of the base ring R; no gradings or other combinatorial hypotheses are necessary. However, if it is checked that some complex $C^{\bullet}$ satisfies the hypotheses of Fact 13.8 for $\mathbb{Z}^d$-graded injective modules J, then the conclusion only holds a priori for modules M that are $\mathbb{Z}^d$-graded. Observe that the tensor product $M \otimes \check{C}^{\bullet}(m_1, \ldots, m_r)$ happens to be $\mathbb{Z}^d$-graded whenever the ring R, the generators m_i, and the module M are, so the natural $\mathbb{Z}^d$-grading on $H_I^i(M)$ falls out of Theorem 13.7.

There is one more commonly used characterization of local cohomology, namely as a limit of Ext modules, which occasionally arises combinatorially. Since we have not officially seen Ext in this book, let us introduce it now.

Definition 13.9 Given two modules N and M, the module $\underline{\mathrm{Ext}}^i_R(N, M)$ is the ith cohomology of the complex

$$0 \to \underline{\mathrm{Hom}}_R(N, E^0) \to \underline{\mathrm{Hom}}_R(N, E^1) \to \underline{\mathrm{Hom}}_R(N, E^2) \to \cdots$$

for any injective resolution $E^{\boldsymbol{\cdot}}$ of M.

As usual, this definition works verbatim just as well for ungraded rings, modules, and injective resolutions, but in our case, we intend that everything be over a $\mathbb{Z}^d$-graded semigroup ring $R = \Bbbk[Q]$. Part of the power of Ext is that it can also be computed using a free resolution of its first argument.

Proposition 13.10 *The module* $\underline{\mathrm{Ext}}^i_R(N, M)$ *is isomorphic to the ith cohomology of the complex* $\underline{\mathrm{Hom}}_R(\mathcal{F}_{\boldsymbol{\cdot}}, M)$, *for any free resolution* $\mathcal{F}_{\boldsymbol{\cdot}}$ *of* N.

The homological algebra used to prove this fact is the same as in Exercise 1.12, by comparing cohomology to that of the total complex, except that here the tensor product complex is replaced by $\underline{\mathrm{Hom}}_R(\mathcal{F}_{\boldsymbol{\cdot}}, E^{\boldsymbol{\cdot}})$, and the directions of one set of arrows (the horizontal ones, say) must be reversed.

Every homomorphism $N \to N'$ of modules induces a homomorphism $\underline{\mathrm{Hom}}_R(N, E^{\boldsymbol{\cdot}}) \to \underline{\mathrm{Hom}}_R(N', E^{\boldsymbol{\cdot}})$ of complexes and therefore a homomorphism $\underline{\mathrm{Ext}}^i_R(N, M) \to \underline{\mathrm{Ext}}^i_R(N', M)$ for all i. This happens in particular for the surjection $R/I^{t+1} \twoheadrightarrow R/I^t$, where the $\underline{\mathrm{Hom}}$ modules are quite explicit.

Lemma 13.11 $\underline{\mathrm{Hom}}_R(R/I^t, M) = (0 :_M I^t) = \{y \in M \mid I^t y = 0\}$ *is the set of elements in* M *annihilated by* I^t. *Taking direct limits over t yields*

$$\varinjlim_t \underline{\mathrm{Hom}}_R(R/I^t, M) \;=\; \Gamma_I M.$$

Loosely, the union of the homomorphic images of R/I^t inside M for ever-increasing values of t fills up the part of M supported on I. The proof is immediate from the definitions. Lemma 13.11 implies that the corresponding limits of Ext modules have a concrete interpretation.

Theorem 13.12 *Local cohomology with support on I equals the limit*

$$H^i_I(M) \;\cong\; \varinjlim_t \underline{\mathrm{Ext}}^i_R(R/I^t, M).$$

Proof. Apply $\underline{\mathrm{Hom}}(R/I^t, \_)$ to an injective resolution $E^{\boldsymbol{\cdot}}$ of M and take the direct limit as t approaches ∞. By Lemma 13.11 the limit complex is $\Gamma_I E^{\boldsymbol{\cdot}}$. Since taking cohomology commutes with direct limits [Wei94, Theorem 2.6.15], the cohomology of the limit complex $\Gamma_I E^{\boldsymbol{\cdot}}$ is the limit of the finite-level cohomology modules $\underline{\mathrm{Ext}}^{\boldsymbol{\cdot}}_R(R/I^t, M)$. $\qquad\square$

13.2 Hilbert series calculations

Among the first homological objects to be calculated explicitly for square-free monomial ideals I_Δ in polynomial rings $S = \Bbbk[x_1, \ldots, x_n]$ were the local cohomology modules of S/I_Δ with support on the maximal ideal $\mathfrak{m} = \langle x_1, \ldots, x_n \rangle$. Their $\mathbb{Z}^n$-graded Hilbert series $H(H^i_{\mathfrak{m}}(S/I_\Delta); \mathbf{x})$ are expressed in terms of the cohomology of links in the Stanley–Reisner complex Δ.

Theorem 13.13 *The Hilbert series of the ith maximal-support local cohomology module of a Stanley–Reisner ring satisfies*

$$H(H^i_{\mathfrak{m}}(S/I_\Delta); \mathbf{x}) \;=\; \sum_{\sigma \in \Delta} \dim_\Bbbk \widetilde{H}^{i-|\sigma|-1}(\mathrm{link}_\Delta(\sigma); \Bbbk) \prod_{j \in \sigma} \frac{x_j^{-1}}{1 - x_j^{-1}}.$$

Let us parse the statement. The product over $j \in \sigma$ is the sum of all Laurent monomials whose exponent vectors are nonpositive and have support exactly σ. Therefore, the formula for the Hilbert series of $H^i_{\mathfrak{m}}(S/I_\Delta)$ is just like the one for S/I_Δ in the third line of the displayed equation in the proof of Theorem 1.13, except that here we consider monomials with negative exponents and we additionally must take into account the nonnegative coefficients $\dim_\Bbbk \widetilde{H}^{i-|\sigma|-1}(\mathrm{link}_\Delta(\sigma); \Bbbk)$ depending on i and σ.

As one might expect from seeing the similarity of the Čech complex $\check{\mathcal{C}}^\bullet(x_1, \ldots, x_n)$ to the Koszul and coKoszul complexes, the proof of Theorem 13.13 is similar to that of Theorem 1.34, being accomplished (as usual) by checking which simplicial complex has its cochain complex in each $\mathbb{Z}^n$-graded degree. The main complication is in determining what relation localization $S/I_\Delta \otimes S[\mathbf{x}^{-\tau}]$ has to the Stanley–Reisner ring of something.

Proof. Given a vector $\mathbf{b} \in \mathbb{Z}^n$, for the duration of this proof we let $\mathbf{b}^-$ and $\mathbf{b}^+$ denote the subsets of $\{1, \ldots, n\}$, where $\mathbf{b}$ has strictly negative and strictly positive entries, respectively. Having fixed Δ, define the simplicial complex $\Delta(\mathbf{b})$ on the vertex set $\{1, \ldots, n\} \smallsetminus \mathbf{b}^-$ to consist of those faces σ such that $\sigma \cup \mathbf{b}^- \cup \mathbf{b}^+$ is a face of Δ. Note that if $\mathbf{b}^+$ is nonempty, then $\Delta(\mathbf{b})$ is a cone from $\mathbf{b}^+$ and therefore has zero homology.

Now calculate the local cohomology $H^i_{\mathfrak{m}}(S/I_\Delta)$ as the cohomology of the complex $S/I_\Delta \otimes \check{\mathcal{C}}^\bullet$ for $\check{\mathcal{C}}^\bullet = \check{\mathcal{C}}^\bullet(x_1, \ldots, x_n)$. The $\mathbb{Z}^n$-graded degree $\mathbf{b}$ piece of the localization $S/I_\Delta[\mathbf{x}^{-\tau}]$ is nonzero precisely when τ contains $\mathbf{b}^-$ and also $\mathbf{b}^+ \cup \tau$ is a face of Δ. Equivalently, $S/I_\Delta[\mathbf{x}^{-\tau}]_\mathbf{b} \neq 0$ precisely when $\tau \smallsetminus \mathbf{b}^-$ is a face of $\Delta(\mathbf{b})$. The complex $(\check{\mathcal{C}}^\bullet \otimes S/I_\Delta)_\mathbf{b}$ is therefore isomorphic to the cochain complex $\widetilde{\mathcal{C}}^\bullet(\Delta(\mathbf{b}); \Bbbk)$, but cohomologically shifted so that its faces of dimension r lie in cohomological degree $|\mathbf{b}^-|+1+r$. Taking cohomology, we find that $H^i_{\mathfrak{m}}(S/I_\Delta)_\mathbf{b}$ is zero unless every coordinate of $\mathbf{b}$ is nonpositive, in which case $H^i_{\mathfrak{m}}(S/I_\Delta)_\mathbf{b} = \widetilde{H}^{i-|\sigma|-1}(\mathrm{link}_\Delta(\sigma); \Bbbk)$ for $\sigma = \mathbf{b}^- = \mathrm{supp}(\mathbf{b})$. $\square$

The simplicial complex Δ appears in the argument of the local cohomology in Theorem 13.13. In the next result, Δ appears in the support ideal instead, but now the argument is less complicated.

The forthcoming computation works not just for polynomial rings but also for arbitrary normal semigroup rings $\Bbbk[Q]$. Recall from Section 12.2 that $C = \mathbb{R}_{\geq 0} Q$ is the product of its lineality space and the cone over a polytope obtained as a transverse affine-linear section of codimension $\dim(C) + 1$. In this chapter, we write (an arbitrary choice of) this polytope as $\mathcal{P}$ and denote by $\bar{F}$ the face of $\mathcal{P}$ corresponding to the face F of Q. Given $\mathbf{a} \in \mathbb{Z}^d$, we again use the subcomplex $\mathcal{P}_{\mathbf{a}}$ defined in (12.2), whose faces $\bar{F}$ are such that the relative interior of $\mathbb{R}_{\geq 0} F$ lies behind the interior of C as seen from $\mathbf{a}$.

Theorem 13.14 *Fix a saturated affine semigroup Q such that $\dim(\mathcal{P}) = r - 1$. Let Δ be a polyhedral subcomplex of $\mathbb{R}_{\geq 0} Q$ corresponding to a subcomplex $\bar{\Delta} \subseteq \mathcal{P}$. The ith local cohomology of the canonical module ω_Q with support on I_Δ has Hilbert series given by relative homology:*

$$H(H^i_{I_\Delta}(\omega_Q); \mathbf{t}) = \sum_{\mathbf{a} \in \mathbb{Z}^d} H_{r-1-i}(\bar{\Delta}, \mathcal{P}_{\mathbf{a}} \cap \bar{\Delta}; \Bbbk) \cdot \mathbf{t}^{\mathbf{a}}.$$

Proof. By Theorem 12.11, the local cohomology can be calculated using the dualizing complex $\Omega^{\bullet}_Q$. The indecomposable injective summand $\Bbbk\{F - Q\}$ lies inside $\Gamma_{I_\Delta} \Omega^{\bullet}_Q$ if and only if $F \in \Delta$. Now use Lemma 12.12 along with the calculation in (12.3). The homological degree $r - 1 - i$ comes from the fact that the dimension i faces of $\mathcal{P}$ index summands of Ω^{r-1-i}_Q. $\square$

We chose to let $\mathcal{P}$ have dimension $r - 1$ in Theorem 13.14 to respect the most common case, where Q is pointed and $\Bbbk[Q]$ has Krull dimension $r = d$.

Remark 13.15 Although the Hilbert series of a graded $\Bbbk[Q]$-module says much about its gross size, it fails to capture some important details. In general, for instance, it is an open problem to determine which faces of Q correspond to prime ideals associated to $H^i_{I_\Delta}(\omega_Q)$.

When $Q = \mathbb{N}^n$, the Hilbert series in Theorem 13.14 can be expressed in a more "closed" form, quite similar to that in Theorem 13.13. Moreover, in this case the canonical module $\omega_{\mathbb{N}^n}$ is simply $S(-\mathbf{1})$ for $\mathbf{1} = (1, \ldots, 1)$, so we may as well take the local cohomology of S instead of $\omega_{\mathbb{N}^n}$.

Corollary 13.16 *For the polynomial ring $S = \Bbbk[x_1, \ldots, x_n]$ and Δ a simplicial complex on n vertices, the $\mathbb{Z}^n$-graded Hilbert series of $H^i_{I_\Delta}(S)$ is*

$$H(H^i_{I_\Delta}(S); \mathbf{x}) = \sum_{\sigma \in \Delta} \dim_{\Bbbk} \tilde{H}_{n-i-|\sigma|-1}(\mathrm{link}_\Delta(\sigma); \Bbbk) \prod_{j \in \bar{\sigma}} \frac{x_j^{-1}}{1 - x_j^{-1}} \prod_{k \in \sigma} \frac{1}{1 - x_k}.$$

Proof. $\bar{\mathbb{N}}^n_{\mathbf{a}}$ consists precisely of those faces of $\bar{\Delta}$ corresponding to the faces $\sigma \in \Delta$ such that $i \in \sigma$ whenever $a_i \leq 0$. These are all of the faces containing the set $\sigma(\mathbf{a})$ of indices i such that $a_i \leq 0$. Therefore $H_{n-1-i}(\bar{\Delta}, \bar{\mathbb{N}}^n_{\mathbf{a}} \cap \bar{\Delta}; \Bbbk)$ is isomorphic to $\tilde{H}_{n-i-|\sigma|-1}(\mathrm{link}_\Delta(\sigma(\mathbf{a})); \Bbbk)$. The sum of all Laurent monomials with $\sigma(\mathbf{a}) = \sigma$ is $x_1 \cdots x_n$ times the double product in the statement of the corollary. Dividing by $x_1 \cdots x_n$ corresponds to the translation by $\mathbf{1}$. $\square$

We will see in Theorem 13.20 that there is a version of Theorem 13.14 for local cohomology of arbitrary finitely generated modules, although of course the result is less explicit. Theorem 13.14 partitions $\mathbb{Z}^d$ into finitely many equivalence classes of $\mathbb{Z}^d$-graded degrees. On each equivalence class, the vector space structure of $H^i_{I_\Delta}(\omega_Q)$ is constant. Therefore, the formula specifies the vector space structure of local cohomology in a finite data structure. This remains true even though $H^i_{I_\Delta}(\omega_Q)$ need not be presentable using generators and relations, or using cogenerators and "correlations": it will in general have neither finite Betti numbers nor finite Bass numbers.

Example 13.17 (Hartshorne's response to a conjecture of Grothendieck) Resume the notation from Example 13.4. Sector 4 has infinitely many degrees with *cogenerators* in $H^2_{\mathfrak{p}}(\omega_Q)$. These are elements annihilated by the maximal ideal $\mathfrak{m} = \langle a, b, c, d \rangle$, and they occupy all degrees $(0, 0, -t)$ for $t > 0$. This means that the injective hull of $H^2_{\mathfrak{p}}(\omega_Q)$ has infinitely many summands isomorphic to $\mathbb{Z}^3$-graded translates of the injective hull $\Bbbk\{-Q\}$ of $\Bbbk$. Equivalently, the zeroth Bass number of $H^2_{\mathfrak{p}}(\omega_Q)$ is not finite. ◇

The previous example works with the canonical module ω_Q, but in the special case there, ω_Q is isomorphic to a $\mathbb{Z}^3$-graded translate of $\Bbbk[Q]$ itself. (This means by definition that $\Bbbk[Q]$ is *Gorenstein*.) In general, here is an important open question, for arbitrary saturated semigroups Q.

Problem 13.18 *Characterize the face ideals $I_\Delta \subset \Bbbk[Q]$ such that the local cohomology of ω_Q or $\Bbbk[Q]$ supported on I_Δ has infinite-dimensional socle.*

This question remains open in part because there are no known combinatorial descriptions even of the Hilbert series for local cohomology of $\Bbbk[Q]$ with arbitrary support, let alone its module structure. Exploration of this problem requires algorithmic methods. The main issue is how to keep the data structures and computations finite, given that generators and cogenerators have been ruled out. The solution is to mimic the decomposition of $\mathbb{Z}^d$ that we obtained for local cohomology of canonical modules.

Definition 13.19 Suppose H is a $\mathbb{Z}^d$-graded module over an affine semigroup ring $\Bbbk[Q]$. A **sector partition** of H is

1. a finite partition $\mathbb{Z}^d = \bigcup_{S \in \mathcal{S}} S$ of the lattice $\mathbb{Z}^d$ into **sectors**,
2. a finite-dimensional vector space H_S for each $S \in \mathcal{S}$, along with isomorphisms $H_S \cong H_{\mathbf{a}}$ for all $\mathbf{a} \in S$, and
3. vector space homomorphisms $H_S \xrightarrow{\mathbf{x}^{T-S}} H_T$ whenever there exist $\mathbf{a} \in S$ and $\mathbf{b} \in T$ satisfying $\mathbf{b} - \mathbf{a} \in Q$, such that the diagram commutes:

$$\begin{array}{ccc} H_S & \xrightarrow{\mathbf{x}^{T-S}} & H_T \\ \downarrow & & \downarrow \\ H_{\mathbf{a}} & \xrightarrow{\mathbf{x}^{\mathbf{b}-\mathbf{a}}} & H_{\mathbf{b}} \end{array}$$

Write $\mathcal{S} \vdash H$ to indicate the above sector partition.

Sector partitions describe $\mathbb{Z}^d$-graded modules completely; future algorithmic computations of local cohomology will produce them as output and
will be able to calculate associated primes, locations of socle degrees, and
more. For example, Hilbert series simply record the vector space dimensions in each of the finitely many sectors $S \in \mathcal{S}$. The observation we make
here is that sector partitions for local cohomology modules always exist.

Theorem 13.20 *There is a sector partition* $\mathcal{S} \vdash H^i_I(M)$ *of the local coho-*
mology of any finitely generated $\mathbb{Z}^d$-*graded module over any normal semi-*
group ring $\Bbbk[Q]$, *in which each sector in* $\mathcal{S}$ *consists of the lattice points in*
a finite union of convex polyhedra defined as intersections of half-spaces for
hyperplanes that are parallel to the facets of Q.

Proof. Treat the cohomological index i as fixed, and consider the three
terms $E^{i-1} \to E^i \to E^{i+1}$ in a minimal injective resolution of M. The local
cohomology $H^i_I(M)$ is the middle cohomology of the complex $\Gamma_I E^{i-1} \to$
$\Gamma_I E^i \to \Gamma_I E^{i+1}$. The indecomposable injective summands appearing in the
these three terms divide $\mathbb{Z}^d$ into equivalence classes, where $\mathbf{a}$ is equivalent
to $\mathbf{b}$ if the set of summands having nonzero elements of degree $\mathbf{a}$ is precisely
the same at is for $\mathbf{b}$. These equivalence classes are the sectors. That they
are finite unions of convex polyhedra of the desired form follows because
the set of $\mathbb{Z}^d$-graded degrees where an indecomposable injective is nonzero
is a translate of a cone whose facets are parallel to those of Q.

The cohomology of the complex $\Gamma_I E^{\bullet}$ is constant on each sector by construction, proving the second condition for $\mathcal{S} \vdash H^i_I(M)$. The third condition
comes from the natural maps between $\mathbb{Z}^d$-graded degrees of $\Gamma_I E^{\bullet}$. $\square$

We draw the reader's attention at this point back to Example 13.4,
which serves as an instance of Theorem 13.14 as well as Theorem 13.20.

13.3   Toric local cohomology

Even granted the multitude of characterizations of local cohomology over
semigroup rings in Section 13.1, two special cases are so important that
yet more complexes have been found to calculate them. These two cases
are local cohomology with maximal support $\mathfrak{m}$ over an affine semigroup
ring $\Bbbk[Q]$ and with monomial support I_Δ over a polynomial ring S. Their
significance stems from their relation to sheaf cohomology on toric varieties.
We will not make this connection precise in either case (the interested reader
should consult [FM05]) but observe that the former yields sheaf cohomology
on the projective toric variety $\mathrm{Proj}(\Bbbk[Q])$, whereas the latter produces sheaf
cohomology on the toric variety $\mathrm{SpecTor}(S, B)$. There is a general heuristic
here: cohomology over a quotient of an open subset U of a variety V is
obtained from local cohomology over the original variety V with support
on the closed complement $V \smallsetminus U$. Although we will not explicitly compute

any sheaf cohomology, we take it as motivation to study the corresponding local cohomology in more detail.

13.3.1 Maximal support over semigroup rings

All of the computations we have made thus far in this chapter have been in the context of normal semigroup rings. Now we turn to a computation that works for arbitrary semigroups, not just saturated ones. This is surely the most important local cohomology computation for semigroup rings in this chapter. Again recall from before Theorem 13.14 and Section 12.2 the definition of the polytope $\mathcal{P}$ obtained by transverse linear section of $\mathbb{R}_{\geq 0}Q$.

Definition 13.21 The Matlis dual of the dualizing complex is the **Ishida complex** $\mho_Q^{\bullet} = (\Omega_Q^{\bullet})^{\vee}$ of the semigroup Q, or of the semigroup ring $\mathbb{k}[Q]$.

Matlis-dualizing the complex after Definition 12.7 yields $\mho_Q^{\bullet}$:

$$0 \to \mathbb{k}[Q] \to \bigoplus_{\substack{\text{vertices } \bar{v} \\ \text{of } \mathcal{P}}} \mathbb{k}[Q]_v \to \bigoplus_{\substack{\text{2-dim} \\ \text{faces } F \\ \text{of } \mathcal{P}}} \mathbb{k}[Q]_F \to \cdots \to \bigoplus_{\substack{\text{facets } \bar{F} \\ \text{of } \mathcal{P}}} \mathbb{k}[Q]_F \to \mathbb{k}[\mathbb{Z}^d] \to 0,$$

where $\mathbb{k}[Q]_F$ is the localization of $\mathbb{k}[Q]$ inverting all monomials in the face F. The differential of the Ishida complex $\mho_Q^{\bullet}$ is derived from the algebraic cochain complex of the polytope $\mathcal{P}$. The cohomological degrees are set up so that $\mathbb{k}[Q]$ sits in cohomological degree 0 (so this is really the cohomological indexing resulting from choice (ii) after Definition 12.7).

When Q is a pointed semigroup, vertices $\bar{v}$ of $\mathcal{P}$ correspond to rays v of Q, and so on. However, we have adopted notation that works even when Q has nontrivial units, or equivalently, when $\mathbb{R}_{\geq 0}Q$ has positive dimensional lineality. In these cases, vertices $\bar{v}$ correspond to faces of higher dimension, but still these faces are minimal among those not equal to the face of units in Q. In any case, we denote the maximal graded ideal by $\mathfrak{m}$.

Lemma 13.22 $H^0(M \otimes \mho_Q^{\bullet}) = \Gamma_{\mathfrak{m}}M$ *for all* $\mathbb{k}[Q]$*-modules* M.

Proof. An element in M is supported at $\mathfrak{m}$ if and only if its image in every localization M_v for vertices $\bar{v}$ of $\mathcal{P}$ is zero. $\qquad\square$

Before getting to the main result, we need to see what kinds of complexes can result by tensoring the Ishida complex with an injective module.

Proposition 13.23 *Let* F *be a face of an affine semigroup* Q. *The complex* $\mathbb{k}\{F - Q\} \otimes \mho_Q^{\bullet}$ *can only have nonzero cohomology when* $\bar{F} = \varnothing$, *in which case* $H^0(\mathbb{k}\{-Q\} \otimes \mho_Q^{\bullet}) = \mathbb{k}\{-Q\}$ *and all higher cohomology is zero.*

Proof. When $\bar{F} = \varnothing$, so that the prime corresponding to F is $P_F = \mathfrak{m}$, the cohomology is as stated because all localizations of $\mathbb{k}\{-Q\}$ at primes corresponding to nonempty faces of $\mathcal{P}$ are zero. Suppose now that $\bar{F}$ is nonempty. Then $\Bbbk\{F - Q\}_G$ is equal to $\Bbbk\{F - Q\}$ if $G \subseteq F$, and zero otherwise (see Proposition 11.17 if this is not clear). Therefore $\Bbbk\{F-Q\} \otimes \mathring{\mho}_Q$ is just $\Bbbk\{F-Q\}$ tensored over $\Bbbk$ with the reduced cochain complex of $\bar{F}$ over $\Bbbk$, where $\bar{F}$ is considered as a polytope in its own right. That $\bar{F}$ is nonempty means that this reduced cochain complex has zero cohomology. $\square$

Theorem 13.24 *Let $\Bbbk[Q]$ be an affine semigroup ring with multigraded maximal ideal $\mathfrak{m}$. The local cohomology of any $\Bbbk[Q]$-module M supported at $\mathfrak{m}$ is the cohomology of the Ishida complex tensored with M:*

$$H^i_{\mathfrak{m}}(M) \;\cong\; H^i(M \otimes \mathring{\mho}_Q).$$

Proof. Apply Fact 13.8, using Lemma 13.22 and Proposition 13.23. $\square$

Remark 13.25 We only proved Theorem 13.24 for $\mathbb{Z}^d$-graded modules M, because we applied Fact 13.8 using $\mathbb{Z}^d$-graded injectives. However, the fact holds for arbitrarily graded (or ungraded) injectives with the same proof, once one has a handle on their basic properties; see [BH98, Theorem 6.2.5] and its proof. Therefore Theorem 13.24 holds for ungraded modules M.

For an affine semigroup $Q \subseteq \mathbb{Z}^d$, the localization $\Bbbk[Q]_F$ is nonzero in graded degree $\mathbf{b} \in \mathbb{Z}^d$ if and only if $\mathbf{b}$ lies in the localized semigroup $Q - F$ of $\mathbb{Z}^d$ obtained by inverting semigroup elements in the face F. Therefore, the set of faces of $\mathcal{P}$ contributing a nonzero vector space (of dimension 1) to the degree $\mathbf{b}$ piece of the Ishida complex $\mathring{\mho}_Q$ is

$$\nabla_Q(\mathbf{b}) \;=\; \{\text{faces } \bar{F} \text{ of } \mathcal{P} \mid \mathbf{b} \in Q - F\}.$$

This set of faces of the polytope $\mathcal{P}$ is closed under going up, meaning that if $\bar{F} \subseteq \bar{G}$ and $\bar{F} \in \nabla_Q(\mathbf{b})$, then also $\bar{G} \in \nabla_Q(\mathbf{b})$. By definition, this means that $\nabla_Q(\mathbf{b})$ is a *polyhedral cocomplex* inside $\mathcal{P}$. When we write cohomology groups $H^i(\nabla; \Bbbk)$ for such a polyhedral cocomplex ∇, we formally mean that

$$H^i(\nabla) \;=\; H^i(\mathcal{P}, \mathcal{P} \smallsetminus \nabla; \Bbbk)$$

is the relative cohomology with coefficients in $\Bbbk$ of the pair $\mathcal{P} \smallsetminus \nabla \subseteq \mathcal{P}$ of cell complexes inside of $\mathcal{P}$. Here now is a down-to-earth description of the graded pieces of local cohomology of $\Bbbk[Q]$ itself.

Corollary 13.26 *The degree $\mathbf{b}$ part of the local cohomology of the semigroup ring $\Bbbk[Q]$ supported at $\mathfrak{m}$ is isomorphic to the cohomology of $\nabla_Q(\mathbf{b})$:*

$$H^i_{\mathfrak{m}}(\Bbbk[Q])_{\mathbf{b}} \;=\; H^i(\nabla_Q(\mathbf{b}); \Bbbk).$$

Turning the poset of faces in a cocomplex ∇ inside $\mathcal{P}$ upside down yields a corresponding polyhedral cell complex $\nabla^{\vee}$ inside the polar polytope $\mathcal{P}^{\vee}$. The cohomology $H^i(\nabla; \Bbbk)$ of the cocomplex ∇ is canonically isomorphic to the reduced homology $\tilde{H}_{\dim \mathcal{P} - i - 1}(\nabla^{\vee}; \Bbbk)$. This whittles the computation of local cohomology of semigroup rings with maximal support down to computing reduced homology of honest polyhedral cell complexes over $\Bbbk$.

Example 13.27 The semigroup Q generated by the columns of the matrix

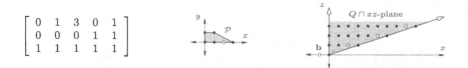

$$\begin{bmatrix} 0 & 1 & 3 & 0 & 1 \\ 0 & 0 & 0 & 1 & 1 \\ 1 & 1 & 1 & 1 & 1 \end{bmatrix}$$

at left consists of almost all of the integer points in the cone over the trapezoid $\mathcal{P}$ (which sits at height $z = 1$ over the xy-plane). Missing are the lattice points in the cone $\mathbb{R}_{\geq 0} Q$ on the real line through the two white points $(2, 0, 1)$ and $\mathbf{b} = (-1, 0, 0)$. The localized semigroup $Q - F$ equals the corresponding localization $Q_{\mathrm{sat}} - F$ of the saturation of Q unless either $F = \varnothing$, in which case $Q - F = Q$, or F is the ray generated by $(3, 0, 1)$, in which case all of the lattice points on the line are still missing.

The cocomplex $\nabla_{Q_{\mathrm{sat}}}(\mathbf{b})$ for the saturation of Q in degree $\mathbf{b} = (-1, 0, 0)$ consists of all faces of $\mathcal{P}$ not lying in the yz-plane. The polar complex $\nabla_{Q_{\mathrm{sat}}}(\mathbf{b})^{\vee}$ is two line segments joined at a point, which has zero reduced homology, so $H^i_{\mathfrak{m}}(\Bbbk[Q_{\mathrm{sat}}])_{\mathbf{b}} = 0$ for all i. In contrast, $\nabla_Q(\mathbf{b})$ is $\nabla_{Q_{\mathrm{sat}}}(\mathbf{b})$ minus the vertex $(3, 0, 1)$ of $\mathcal{P}$, so its polar complex $\nabla_Q(\mathbf{b})^{\vee}$ is a line segment and a disjoint point. The cohomology $H^i_{\mathfrak{m}}(\Bbbk[Q])_{\mathbf{b}} = H^i(\nabla_Q(\mathbf{b}); \Bbbk) = \widetilde{H}_{1-i}(\nabla_Q(\mathbf{b})^{\vee}; \Bbbk)$ is therefore $\Bbbk$ if $i = 1$ and zero otherwise. $\diamond$

13.3.2 Monomial support over polynomial rings

Recall the special $*$ notation from Section 13.1 in monomial matrices for complexes of localizations of free modules over the $\mathbb{Z}^n$-graded polynomial ring $S = \Bbbk[x_1, \ldots, x_n]$. Next we identify a class of such complexes that can be used in place of Čech complexes in computing local cohomology.

Definition 13.28 Suppose that $\mathcal{F}_{\bullet}$ is a free resolution of S/I_{Δ} that has monomial matrices in which every row and column label is squarefree. The **generalized Čech complex** $\check{C}^{\bullet}_{\mathcal{F}}$ is the complex of localizations of S obtained by replacing every 1 in every row and column label with the symbol $*$. This complex is to be considered as a cohomological complex as in Example 13.6. When $\mathcal{F}_{\bullet}$ is minimal, $\check{C}^{\bullet}_{\mathcal{F}}$ is called the **canonical Čech complex** of I_{Δ} and we use $\check{C}^{\bullet}_{\Delta}$ to denote it.

Example 13.29 Start with the triangular, square, and pentagonal minimal cellular resolutions of

$$S/\langle a, b, c \rangle, \quad S/\langle ab, bc, cd, ad \rangle, \quad \text{and} \quad S/\langle abc, bcd, cde, ade, abe \rangle$$

for appropriate S in Example 4.12. The associated canonical Čech complexes have monomial matrices filled with the coboundary complexes of the following labeled cell complexes:

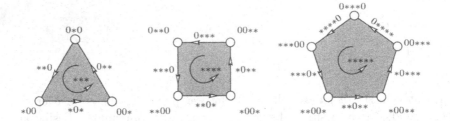

The empty set is labeled $0 \cdots 0$ in all three pictures. The triangle here gives monomial matrices for the usual Čech complex in Example 13.6, whereas the triangle in Example 4.12 is the Koszul complex in Example 1.27 (both with different sign conventions). This example works more generally for irrelevant ideals of smooth (or simplicial) projective toric varieties.   ◇

The usual Čech complex is a generalized Čech complex.

Proposition 13.30 *Suppose that I_Δ is generated by squarefree monomials $m_1, \ldots, m_r$. If $\mathcal{F}_\bullet$ is the Taylor resolution on these generators, then $\check{C}^\bullet_{\mathcal{F}} = \check{C}^\bullet(m_1, \ldots, m_r)$ is the usual Čech complex.*

This is a key point, and it follows immediately from the definitions. Now we come to the main result on generalized Čech complexes.

Theorem 13.31 *The local cohomology of M supported on I_Δ is the cohomology of any generalized Čech complex tensored with M:*

$$H^i_{I_\Delta}(M) = H^i(M \otimes \check{C}^\bullet_{\mathcal{F}}).$$

The proof, at the end of this section, relies on a construction that extends the construction in Definition 13.28 to arbitrary $\mathbb{Z}^n$-graded modules.

Definition 13.32 The **Čech hull** of a $\mathbb{Z}^n$-graded module M is the $\mathbb{Z}^n$-graded module $\check{C}M$ whose degree $\mathbf{b}$ piece is

$$(\check{C}M)_{\mathbf{b}} = M_{\mathbf{b}_+} \quad \text{where} \quad \mathbf{b}_+ = \sum_{b_i \geq 0} b_i \mathbf{e}_i$$

and $\mathbf{e}_i$ is the ith standard basis vector of $\mathbb{Z}^n$. Equivalently,

$$\check{C}M = \bigoplus_{\mathbf{b} \in \mathbb{N}^n} M_{\mathbf{b}} \otimes_{\Bbbk} \Bbbk[x_i^{-1} \mid b_i = 0].$$

The action of multiplication by x_i is

$$\cdot x_i : (\check{C}M)_{\mathbf{b}} \to (\check{C}M)_{\mathbf{e}_i + \mathbf{b}} = \begin{cases} \text{identity} & \text{if } b_i < 0 \\ \cdot x_i : M_{\mathbf{b}_+} \to M_{\mathbf{e}_i + \mathbf{b}_+} & \text{if } b_i \geq 0. \end{cases}$$

Note that $\mathbf{e}_i + \mathbf{b}_+ = (\mathbf{e}_i + \mathbf{b})_+$ whenever $b_i \geq 0$.

The staircase diagram of $\check{C}I$ for any ideal I (not necessarily square-free) is obtained by pushing to negative infinity any point on the staircase diagram for I that touches the boundary of the positive orthant:

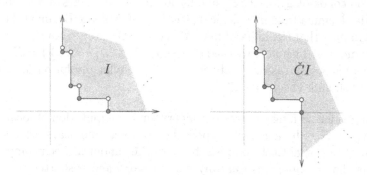

Heuristically, the first description of $\check{C}M$ in the definition says that if you want to know what $\check{C}M$ looks like in degree $\mathbf{b} \in \mathbb{Z}^n$, then check what M looks like in the nonnegative degree closest to $\mathbf{b}$; the second description says that the vector space $M_{\mathbf{a}}$ for $\mathbf{a} \in \mathbb{N}^n$ is copied into all degrees $\mathbf{b}$ such that $\mathbf{b}_+ = \mathbf{a}$. The Čech hull "forgets" everything about the original module that occurred in degrees outside $\mathbb{N}^n$.

The Čech hull can be applied to a homogeneous map of degree $\mathbf{0}$ between two modules, by copying the maps in the $\mathbb{N}^n$-graded degrees as prescribed. Checking $\mathbb{Z}^n$-degree by $\mathbb{Z}^n$-degree yields the following simple result.

Lemma 13.33 *The Čech hull takes exact sequences to exact sequences.* □

Next we need to see how to recover the construction in Definition 13.28 using the Čech hull. Set $\mathbf{1} = (1, \ldots, 1)$ and write $\omega_S = S(-\mathbf{1})$, the free module generated in degree $\mathbf{1}$.

Proposition 13.34 *If $\mathcal{F}_{\bullet}$ is a free resolution of S/I_Δ with squarefree row and column labels, then the generalized Čech complex can be expressed as*

$$\check{C}^{\bullet}_{\mathcal{F}} = (\check{C}\mathcal{F}^{\bullet})(\mathbf{1}),$$

the $\mathbb{Z}^n$-graded translate down by $\mathbf{1}$ of the Čech hull of $\mathcal{F}^{\bullet} = \underline{\mathrm{Hom}}(\mathcal{F}_{\bullet}, \omega_S)$.

Proof. Every summand $S(-\sigma)$ in $\mathcal{F}_{\bullet}$ becomes a summand $S(-\bar{\sigma})$ with generator of degree $\bar{\sigma} = \mathbf{1} - \sigma$ in $\mathcal{F}^{\bullet}$. It is straightforward to check that $\check{C}(S(-\bar{\sigma})) = \Bbbk\{\mathbf{x}^{\mathbf{b}} \mid \mathbf{b}_+ \succeq \bar{\sigma}\} = S[\mathbf{x}^{-\sigma}](-\bar{\sigma})$. Consequently, the summand $\check{C}(S(-\bar{\sigma}))(\mathbf{1})$ of $\check{C}(\mathcal{F}^{\bullet})(\mathbf{1})$ is the localization whose vector label has a $*$ precisely where σ has a 1. □

Proof of Theorem 13.31. Every squarefree resolution $\mathcal{F}_{\bullet}$ of S/I_Δ contains a minimal free resolution. Applying $\underline{\mathrm{Hom}}(-, \omega_S)$ produces a surjection from $\mathcal{F}^{\bullet}$ to the dual of the minimal free resolution, and this surjection induces an isomorphism on cohomology (which is $\underline{\mathrm{Ext}}^{\bullet}(S/I_{\Delta}, \omega_S)$ in both cases). By Proposition 13.34, taking Čech hulls and subsequently translating by $\mathbf{1}$ yields a map $\check{\mathcal{C}}^{\bullet}_{\mathcal{F}} \to \check{\mathcal{C}}^{\bullet}_{\Delta}$, and Lemma 13.33 implies that it induces an isomorphism on cohomology. Since $\check{\mathcal{C}}^{\bullet}_{\Delta}$ and $\check{\mathcal{C}}^{\bullet}_{\mathcal{F}}$ are both complexes of flat modules, a standard lemma from homological algebra (see [Mil00b, Lemma 6.11] for a proof) implies that the induced map $M \otimes \check{\mathcal{C}}^{\bullet}_{\mathcal{F}} \to M \otimes \check{\mathcal{C}}^{\bullet}_{\Delta}$ is an isomorphism on cohomology. Therefore we need only show that $H^i(M \otimes \check{\mathcal{C}}^{\bullet}_{\Delta}) = H^i_{I_{\Delta}}(M)$. But this follows by taking $\mathcal{F}_{\bullet}$ above to be the Taylor resolution, by Proposition 13.30 and Theorem 13.7.   □

The reader wishing to carry out algorithmic computation of local cohomology over S with monomial support should use a canonical Čech complex instead of the usual Čech complex, because the canonical Čech complex always has fewer summands—usually many fewer—and is shorter.

13.4 Cohen–Macaulay conditions

The importance of a commutative ring or module being Cohen–Macaulay cannot be overstated. We have already seen the Cohen–Macaulay condition in the context of Alexander duality for resolutions (Section 5.5) and for generic monomial ideals (Section 6.2).

In general, there are numerous equivalent ways to detect the Cohen–Macaulay condition for a module, and many of these fit nicely into the realm of combinatorial commutative algebra. Unfortunately, the equivalences of many of these criteria require homological methods from general—that is, not really combinatorial—commutative algebra, so it would take us too far astray to present a self-contained proof of them all. That being said, the Cohen–Macaulay condition is so robust, comes up so often, and is so useful in combinatorics that we would be remiss were we not to at least present some of the equivalent conditions. This we shall do, with references to where missing parts of the proofs can be found. Afterward, we give some examples of how the criteria can be applied in combinatorial situations.

A few of the Cohen–Macaulay criteria involve notions from commutative algebra that we have not yet seen in this book.

Definition 13.35 Fix a positive multigrading of $S = \Bbbk[x_1, \ldots, x_n]$ by $\mathbb{Z}^d$, and a graded ideal I. A sequence $\mathbf{y} = y_1, \ldots, y_r$ of $\mathbb{Z}^d$-graded homogeneous elements in the graded maximal ideal of S/I is called a

- **system of parameters** for a graded module M if M has Krull dimension r and $M/\mathbf{y}M$ has dimension 0.

- **regular sequence** of **length** r on a graded module M if $M/\mathbf{y}M \neq 0$ and y_i is a nonzerodivisor on $M/\langle y_1, \ldots, y_{i-1}\rangle M$ for each $i = 1, \ldots, r$.

Note that $y_1, \ldots, y_r$ are algebraically independent over $\Bbbk$ in either case, so $\Bbbk[\mathbf{y}]$ is a polynomial subring of dimension r inside of S/I.

Although we stated the above definitions in the presence of a positive $\mathbb{Z}^d$-grading, the Cohen–Macaulay conditions that refer to them require not just an arbitrary positive grading by $\mathbb{Z}^d$, but a positive grading by $\mathbb{N}$. Therefore, in the statement of Theorem 13.37, we fix a *coarsening* of the positive $\mathbb{Z}^d$-grading to a positive $\mathbb{N}$-grading, by which we mean a linear map $\mathbb{Z}^d \to \mathbb{Z}$ such that $\deg(x_i) \in \mathbb{Z}^d$ maps to a strictly positive integer $\deg_{\mathbb{N}}(x_i) \in \mathbb{N}$ for all $i = 1, \ldots, n$. The maximal $\mathbb{Z}^d$-graded ideal $\mathfrak{m} = \langle x_1, \ldots, x_n \rangle$ of S is also the unique maximal $\mathbb{N}$-graded ideal.

Under a positive $\mathbb{N}$-grading, every finitely generated module M admits a system of parameters. (Proof: It suffices by induction on $\dim(M)$ to find the first element in the system; now use prime avoidance [Eis95, Lemma 3.3] to pick an element in $\mathfrak{m}$ but outside any remaining prime ideals associated to M.) In contrast, even for positive gradings by $\mathbb{Z}^2$, multigraded modules need not admit $\mathbb{Z}^2$-graded systems of parameters (Exercise 13.9).

The following module will be used in Criteria 10–13 of Theorem 13.37.

Definition 13.36 Let S be any multigraded polynomial ring. If the variables x_i have degrees $\mathbf{a}_i = \deg(x_i)$, write $\omega_S = S(-\mathbf{a}_1 - \cdots - \mathbf{a}_n)$. For any S-module M of dimension r, define the **canonical module** of M to be

$$\omega_M = \underline{\mathrm{Ext}}_S^{n-r}(M, \omega_S),$$

so $\omega_M = H^{n-r} \underline{\mathrm{Hom}}(\mathcal{F}_{\bullet}, \omega_S)$ for any resolution $\mathcal{F}_{\bullet}$ of M by free S-modules.

Theorem 13.37 *Let M be a finitely generated graded module of dimension r over a positively $\mathbb{Z}^d$-graded polynomial ring S with maximal ideal $\mathfrak{m}$, and fix a coarsening to a positive $\mathbb{N}$-grading. The following are equivalent.*

1. *M is Cohen–Macaulay.*
2. *Every minimal resolution $\mathcal{F}_{\bullet}$ of M by free S-modules has length $n - r$.*
3. *There is an $\mathbb{N}$-graded regular sequence of length r on M.*
4. *Every $\mathbb{N}$-graded system of parameters for M is regular on M.*
5. *M is a free module over the subalgebra $\Bbbk[\mathbf{y}] \subseteq S$ for some (and hence every) $\mathbb{N}$-graded system of parameters $\mathbf{y}$ for M.*
6. *For some (and hence every) $\mathbb{N}$-graded system of parameters $\mathbf{y}$ for M, $H(M/\mathbf{y}M; t) = H(M; t) \cdot \prod_{i=1}^{r}(1 - t^{b_i})$, where $\deg_{\mathbb{N}}(y_i) = b_i$.*
7. *The smallest index i for which $\underline{\mathrm{Ext}}_S^i(\Bbbk, M)$ is nonzero is $i = r$.*
8. *The smallest index i for which $H_{\mathfrak{m}}^i(M)$ is nonzero is $i = r$.*
9. *The only index i for which $H_{\mathfrak{m}}^i(M)$ is nonzero is $i = r$.*
10. *$H_{\mathfrak{m}}^i(M)$ is zero unless $i = r$, and $H_{\mathfrak{m}}^r(M) \cong \omega_M^{\vee}$ is Matlis dual to ω_M.*
11. *$\underline{\mathrm{Hom}}_S(\mathcal{F}_{\bullet}, \omega_S)$ is a minimal free resolution of ω_M as an S-module.*

If $M = R = S/I$, then the following are equivalent to the above conditions.

12. *$\underline{\mathrm{Ext}}_R^i(\Bbbk, \omega_R) = 0$ unless $i = r$, and $\underline{\mathrm{Ext}}_R^r(\Bbbk, \omega_R) = \Bbbk$.*
13. *ω_R has a resolution of finite length by graded-injective R-modules.*

Proof. 2 ⇔ 1 by Definition 5.52, which works in any positive multigrading.

3 ⇔ 2 by the Auslander–Buchsbaum formula [BH98, Theorem 1.3.3].

4 ⇔ 3: For the ⇐ direction, use the fact that every finitely generated module has a homogeneous system of parameters. For ⇒, we can safely replace S with the quotient $R = S/\mathrm{ann}(M)$ by the annihilator of M, and $\mathfrak{m}$ with $\mathfrak{m}R$. The maximal length of a regular sequence on M in any ideal $I \subset R$ only depends only on the radical of I, not on I itself; this follows from part (e) of [BH98, Proposition 1.2.10]. Taking $I = \langle \mathbf{y} \rangle$, whose radical is $\mathfrak{m}R$, the maximal length of a regular sequence in $\langle \mathbf{y} \rangle$ is therefore r. On the other hand, the ideal $\langle \mathbf{y} \rangle$ in R contains a regular sequence of length r if and only if $\mathbf{y}$ itself is a regular sequence; this is proved in [BH98, Corollary 1.6.19] using Koszul complexes.

4 ⇒ 5: Since we already proved 3 ⇔ 4, the desired result—including the "some (and therefore every)" clause—will follow once we show that M is free over $\Bbbk[\mathbf{y}]$ for every regular sequence $\mathbf{y}$ of length r on M. Requiring that $\mathbf{y} = y_1, \ldots, y_r$ be a system of parameters for M is equivalent to requiring that M be finitely generated as a module over the subalgebra $\Bbbk[\mathbf{y}]$ of S (this uses the fact that M is graded and has dimension r as an S-module, along with the slightly nonstandard version [Eis95, Exercise 4.6a] of Nakayama's Lemma). Now repeatedly apply Lemma 8.27, first to M with $y = y_1$, then to $M/y_1 M$ with $y = y_2$, and so on, to deduce that $\mathrm{Tor}_i^{\Bbbk[\mathbf{y}]}(\Bbbk, M) = 0$ for all $i \geq 1$. Use [Eis95, Exercise 6.1] to conclude that M is a graded flat module over the ℕ-graded ring $\Bbbk[\mathbf{y}]$, whence M is free over $\Bbbk[\mathbf{y}]$ by [Eis95, Exercise 6.2].

5 ⇒ 3 because $\mathbf{y}$ is a regular sequence on every free $\Bbbk[\mathbf{y}]$-module.

4 ⇒ 6. It is enough to show that if y is a nonzerodivisor of degree b on M, then $H(M/\mathbf{y}M; t) = H(M; t) \cdot (1 - t^b)$. This follows by additivity of Hilbert series on the short exact sequence $0 \to M(-b) \xrightarrow{\cdot y} M \to M/yM \to 0$.

6 ⇒ 3: It suffices to show that any system of parameters satisfying the Hilbert series condition must be a regular sequence. In fact, a stronger statement can be derived by repeatedly applying the following.

Claim 13.38 *Let S be positively multigraded by $\mathbb{Z}^d$. Fix a homogeneous polynomial $y \in S$ of degree $\mathbf{b}$ and a finitely generated graded S-module N.*

1. *For every degree $\mathbf{a} \in \mathbb{Z}^d$, the coefficient on $\mathbf{t}^{\mathbf{a}}$ in the Hilbert series $H(N/yN; \mathbf{t})$ is at least the coefficient on $\mathbf{t}^{\mathbf{a}}$ in $(1 - \mathbf{t}^{\mathbf{b}})H(N; \mathbf{t})$.*

2. *If y is a zerodivisor on N, then the coefficient on $\mathbf{t}^{\mathbf{a}}$ in $H(N/yN; \mathbf{t})$ is greater than the coefficient on $\mathbf{t}^{\mathbf{a}}$ in $(1 - \mathbf{t}^{\mathbf{b}})H(N; \mathbf{t})$ for some $\mathbf{a} \in \mathbb{Z}^d$.*

To prove both parts simultaneously, use the additivity of Hilbert series on the exact sequence $0 \to K \to N(-\mathbf{b}) \to N \to N/yN \to 0$, where $K = (0 :_N y)(-\mathbf{b})$ is the submodule killed by y. This yields $H(N/yN; \mathbf{t}) = (1 - \mathbf{t}^{\mathbf{b}})H(N; \mathbf{t}) + H(K; \mathbf{t})$. More details can be found in [Sta78, Section 3].

7 ⇔ 3 is the content of [BH98, Theorem 1.2.8].

$8 \Leftrightarrow 7$ because we claim that, in general, both numbers equal the smallest cohomological degree i for which the Bass number $\mu^i(\mathfrak{m}, M)$ of M as an S-module is nonzero. Call this smallest cohomological degree i_0. For condition 7, our claim follows from [BH98, Proposition 3.2.9]. For condition 8, let $E^{\cdot}$ be a minimal resolution of M by injective S-modules, and use the definition of $H_{\mathfrak{m}}^i(\_)$ as the cohomology of $\Gamma_{\mathfrak{m}} E^{\cdot}$. The complex $\Gamma_{\mathfrak{m}} E^{\cdot}$ has no terms in cohomological degrees less than i_0, so clearly $H_{\mathfrak{m}}^i(M) = 0$ for $i < i_0$. On the other hand, for each i, the map $\Gamma_{\mathfrak{m}} E^i \to \Gamma_{\mathfrak{m}} E^{i+1}$ on elements with maximal support induces the zero map on the socle of E^i, by definition of minimality for injective resolutions. Hence for $i = i_0$ the socle of E^i gives rise to nonzero cohomology of $\Gamma_{\mathfrak{m}} E^{\cdot}$.

$9 \Leftrightarrow 8$ because, in general, the local cohomology $H_{\mathfrak{m}}^i(M)$ can be calculated as the cohomology of the length r complex obtained by tensoring M with the Čech complex on a system of parameters for M.

$10 \Leftrightarrow 9$: The direction $\Rightarrow$ is trivial. That the sole nonzero local cohomology module is Matlis dual to ω_M is an immediate consequence of local duality [BH98, Corollary 3.5.9].

$11 \Rightarrow 2$: The minimality of $\mathcal{F}_{\cdot}$ implies that the last nonzero map in $\underline{\mathrm{Hom}}_S(\mathcal{F}_{\cdot}, \omega_S)$ cannot be a surjection. Consequently, there must be cohomology at the last nonzero cohomological degree. We are done because the only nonzero cohomology, namely ω_M, sits in cohomological degree $n - r$.

$3 \Rightarrow 11$: The cohomology of the complex in Criterion 11 is $\underline{\mathrm{Ext}}_S^i(M, \omega_S)$ by definition. Hence we need that $\underline{\mathrm{Ext}}_S^i(M, \omega_S) = 0$ for $i \neq n - r$. This fact is part (b)(i) of [BH98, Proposition 3.3.3], given that [BH98] defines the Cohen–Macaulay condition in terms of Criterion 3 [BH98, Definition 2.1.1].

$12 \Leftrightarrow 11$ by a general theorem on dualizing complexes [Har66b, Proposition 3.4]; the complex $\underline{\mathrm{Hom}}_S(\mathcal{F}_{\cdot}, \omega_S)$ is a dualizing complex by [Har66b, Proposition 2.4], so the one-term complex ω_R is, too, by Criterion 11.

$12 \Rightarrow 13$: The dimension of $\underline{\mathrm{Ext}}_R^i(\Bbbk, \omega_R)$ as a vector space over $\Bbbk$ is the ith Bass number of ω_R by [BH98, Proposition 3.2.9]. An infinite injective resolution would have injective hulls of $\Bbbk$ appearing in all sufficiently large cohomological degrees.

$13 \Rightarrow 3$: Given that ω_R has a finite injective resolution, [BH98, Theorem 3.1.17] says that the maximal length of a regular sequence on R is bounded from below by the dimension of ω_R. Now apply [BH98, Theorem 8.1.1], which says that the module ω_R has dimension at least r. $\square$

Remark 13.39 The *depth* of a graded module M is the maximal length of a regular sequence on M. Criterion 3 says that M is Cohen–Macaulay precisely when $\mathrm{depth}(M) = \dim(M)$. In many sources, such as [BH98], the Cohen–Macaulay condition is defined by this particular condition.

Remark 13.40 The $\mathbb{N}$-grading fixed in Theorem 13.37 is arbitrary. Therefore, instead of fixing the coarsening, we could have stated those criteria involving $\mathbb{N}$-gradings using the phrase "for every $\mathbb{N}$-graded coarsening".

Remark 13.41 Two of the conditions in Theorem 13.37 do not make reference to the polynomial ring S, namely Criteria 12 and 13. As it turns out, most of the others—in fact all of them except for 1, 2, and 11—do not really depend on S, either. To be more precise, suppose that M in Theorem 13.37 is a module over a graded $\Bbbk$-algebra R' (which may differ from both S and R) with maximal ideal $\mathfrak{m}'$ such that $R'/\mathfrak{m}' = \Bbbk$. Then Criteria 3–10 are still equivalent to each other (as well as to Criteria 12 and 13) when $(S, \mathfrak{m})$ is replaced by $(R', \mathfrak{m}')$. Hence Criteria 2 and 11 are the only two that depend on S being a polynomial ring.

Observe that Criteria 12 and 13, which involve the canonical module ω_R of the ring R (which we conclude in the theorem is Cohen–Macaulay), seem to implicitly use the fact that R is a quotient of S. In Exercise 13.12 we define *Gorenstein* rings. In general, existence of a canonical R-module with the properties of ω_R in Theorem 13.37 is equivalent to R being a quotient of some Gorenstein ring [BH98, Theorem 3.3.6]. Thus there is nothing so special about the presentation of R as a quotient of S rather than, say, as a finitely generated module over some other polynomial ring.

13.5   Examples of Cohen–Macaulay rings

The list of conditions in Theorem 13.37 may appear daunting, but nearly every one of them is useful for some combinatorial purpose. For example, the Hilbert series Criterion 6 implies that the h-polynomials of standard graded Cohen–Macaulay rings are nonnegative, which is crucial to Stanley's proof of the Upper Bound Theorem [Sta96, Corollary II.4.5]. For another example, the local (as opposed to graded) version of Criterion 5 plays a role in the proof of Haiman's $n!$ and $(n + 1)^{n-1}$ Theorems [Hai01, Hai02] (though our presentation of these results in Chapter 18 does not go far enough to include this application).

The remainder of this chapter presents some important combinatorial consequences of Theorem 13.37. One of them, Reisner's criterion, has already been used in Chapter 5. Another, that shellable simplicial complexes are Cohen–Macaulay, will find uses in later chapters; see Theorem 16.43 and its consequences, including Corollary 16.44 and Theorem 17.23.

13.5.1   Normal semigroup rings

The injective resolution in Criterion 13 can be made combinatorially explicit when $R = \Bbbk[Q]$ is a Cohen–Macaulay affine semigroup ring.

Theorem 13.42 *Let $\Bbbk[Q]$ be an affine semigroup ring, and express it as a quotient $\Bbbk[Q] \cong S/I_L$, as in Theorem 7.3. Then $\Bbbk[Q]$ is Cohen–Macaulay if and only if its dualizing complex from Definition 12.7 is a $\mathbb{Z}^d$-graded injective resolution of the canonical module $\omega_{\Bbbk[Q]}$ from Definition 13.36.*

Proof. Theorem 13.24 plus Criterion 9 of Theorem 13.37 together imply that $M = \Bbbk[Q]$ is Cohen–Macaulay precisely when the Ishida complex $\mho_Q^{\bullet}$ has cohomology only in the latest possible place. This occurs if and only if the Matlis dual of $\mho_Q^{\bullet}$, namely the dualizing complex $\Omega_Q^{\bullet}$, is an injective resolution of some module. Criterion 10 implies that this module is $\omega_{\Bbbk[Q]}$. $\square$

Given Theorem 12.11 for normal semigroup rings, Theorem 13.42 specializes to an amazing fact, equating the abstract, homologically defined module $\omega_{\Bbbk[Q]}$ with the concrete, combinatorially defined module ω_Q. Here, as elsewhere, Theorem 13.37 shows itself to be a powerful tool for identifying deep homological significance in combinatorial constructions.

Corollary 13.43 *If $\Bbbk[Q]$ is a normal affine semigroup ring, then $\Bbbk[Q]$ is Cohen–Macaulay and the module $\omega_{\Bbbk[Q]}$ in Theorem 13.42 equals the module ω_Q from Definition 12.9 (which we already called the canonical module).*

13.5.2 Reisner's criterion

Our next combinatorial application of Theorem 13.37 is the proof of Reisner's criterion, Theorem 5.53, as promised in Chapter 5. To recap, it says that a Stanley–Reisner ring S/I_Δ is Cohen–Macaulay if and only if $\widetilde{H}^i(\mathrm{link}_\Delta(\sigma); \Bbbk) = 0$ for all $i \neq \dim(\Delta) - |\sigma|$ and all faces $\sigma \in \Delta$.

Proof of Theorem 5.53. Using Theorem 13.13 and Criterion 9 of Theorem 13.37, we find the Stanley–Reisner ring S/I_Δ to be Cohen–Macaulay if and only if $\widetilde{H}^{i-|\sigma|-1}(\mathrm{link}_\Delta(\sigma); \Bbbk)$ is nonzero only for $i = r$, where r is the Krull dimension of S/I_Δ. Now simply note that $r = \dim(\Delta) + 1$. $\square$

13.5.3 Shellable simplicial complexes

One of the simplest (and most ubiquitous) criteria for verifying that a Stanley–Reisner ring is Cohen–Macaulay is to check that the corresponding simplicial complex is shellable. For utmost clarity, given a face F of a simplicial complex Δ, denote by $\hat{F}$ the (closed) simplex in Δ generated by F, so that $\hat{F}$ consists of all faces of F.

Definition 13.44 A **shelling** of Δ is an ordered list $F_1, F_2, \ldots, F_m$ of its facets such that $\bigcup_{j<i} \hat{F}_j \cap \hat{F}_i$ is a subcomplex generated by codimension 1 faces of F_i for all $i \leq m$. If Δ is pure and has a shelling then it is **shellable**.

Theorem 13.45 *The Stanley–Reisner ring of a shellable simplicial complex is Cohen–Macaulay.*

Proof. There are many proofs of this result, but given the definition of shellable, it seems they all have no choice but to proceed by induction on the number of facets. Suppose $F_1, \ldots, F_{m-1}, F_m$ is a shelling of Δ, and let Δ' be the subcomplex generated by $F_1, \ldots, F_{m-1}$. We will show that Δ is Cohen–Macaulay using the fact Δ' is Cohen–Macaulay.

Renumbering the variables if necessary, assume that the vertices of F_m are $1, \ldots, r$ (so Δ has dimension $r - 1$). Renumbering the first r variables $x_1, \ldots, x_r$ again if necessary, assume that the facets of $F_m \cap \Delta'$ are precisely those simplices obtained by deleting one element of the set $\{1, \ldots, q\}$ from F_m, where $q \leq r$. Then the simplex $\{1, \ldots, q\}$ is the unique minimal face in $\Delta \smallsetminus \Delta'$. It follows that $\Bbbk[\Delta'] = \Bbbk[\Delta]/\langle x_1 \cdots x_q\rangle\Bbbk[\Delta]$.

On the other hand, $x_j \cdot x_1 \cdots x_q$ is zero in $\Bbbk[\Delta]$ whenever $j > r$, because then $\{j\} \cup \{1, \ldots, q\}$ is not a face of Δ. Therefore the principal ideal generated by $\langle x_1 \cdots x_q\rangle$ is a free module over the polynomial subring $\Bbbk[x_1, \ldots, x_r]$ of $\Bbbk[\Delta]$. In particular, this ideal is a Cohen–Macaulay module of Krull dimension r.

Now we have an exact sequence

$$0 \longrightarrow \langle x_1 \cdots x_q\rangle\Bbbk[x_1, \ldots, x_r] \longrightarrow \Bbbk[\Delta] \longrightarrow \Bbbk[\Delta'] \longrightarrow 0$$

in which the first and third nonzero modules are Cohen–Macaulay of dimension r. Apply $\underline{\mathrm{Ext}}\,_S^*(\Bbbk, \_)$ to the above short exact sequence. Denoting the first nonzero module in the above sequence by $N = \langle x_1 \cdots x_q\rangle\Bbbk[x_1, \ldots, x_r]$, the long exact sequence for this application of $\underline{\mathrm{Ext}}$ then has segments

$$\cdots \longrightarrow \underline{\mathrm{Ext}}\,_S^i(\Bbbk, N) \longrightarrow \underline{\mathrm{Ext}}\,_S^i(\Bbbk, \Bbbk[\Delta]) \longrightarrow \underline{\mathrm{Ext}}\,_S^i(\Bbbk, \Bbbk[\Delta']) \longrightarrow \cdots$$

for all $i \geq 0$. Criterion 7 says that the left and right modules above are zero when $i < r$ and nonzero when $i = r$. Therefore the same holds for the middle module above, and the result follows from the same criterion. $\square$

Exercises

13.1 Compute the associated primes of the local cohomology modules in Example 13.4. Locate the degrees of the elements whose annihilators are these primes. Which of the submodules annihilated by these primes are finitely generated?

13.2 Use a Čech complex to recompute the local cohomology in Example 13.4. Was this easier or harder than the calculation in Example 13.4? What if you use Theorem 13.14 instead?

13.3 Compute all of the local cohomology modules with support at $\mathfrak{m}$, and their Hilbert series, of the semigroup rings $\Bbbk[Q]$ and $\Bbbk[Q_{\mathrm{sat}}]$ from Example 13.27.

13.4 The affine semigroup in part (a) of Exercise 12.12 is the saturation of the semigroups in parts (b) and (c). Show that the semigroup ring for part (b) is regular in codimension one but not Cohen–Macaulay. Conversely, show that the semigroup ring for part (c) is Cohen–Macaulay but not regular in codimension one.

13.5 Use part (b) of the previous exercise and Criterion 4 to get a quick solution to Exercise 7.6. (One sentence will do; see the end of the exercises in this chapter.)

13.6 Prove directly, using Definition 13.1, that $H_I^0(M) = \Gamma_I M$. Do the same using the characterization of local cohomology in Theorem 13.7.

13.7 Assume that $m_1, \ldots, m_r$ are monomials in an affine semigroup ring $\Bbbk[Q]$. Prove directly that $\Bbbk\{F - Q\} \otimes \check{C}^\bullet(m_1, \ldots, m_r)$ has no i^{th} cohomology for $i \geq 1$.

13.8 Verify that $H_I^i(M)$ does not depend on the injective resolution of M chosen in Definition 13.1. Hint: Exercise 11.6.

13.9 Show that $\Bbbk[x, y]/\langle xy \rangle$, with the usual grading by $\mathbb{N}^2 \subset \mathbb{Z}^2$, does not admit a $\mathbb{Z}^2$-graded system of parameters.

13.10 Given $\mathbf{a} \in \mathbb{N}^n$ and a $\mathbb{Z}^n$-graded module M over $S = \Bbbk[\mathbb{N}^n]$, write $M_{\preceq \mathbf{a}}$ for the quotient module $\bigoplus_{\mathbf{b} \preceq \mathbf{a}} M_{\mathbf{b}}$. Prove that Alexander duality for ideals can be expressed using Matlis duality and the Čech hull as $I^{[\mathbf{a}]} = (\check{C}(S/I)_{\preceq \mathbf{a}})^\vee(-\mathbf{a})$. Use this to define an Alexander duality functor on $\mathbb{N}^n$-graded S-modules and prove that it is exact. Verify that Exercise 11.2 is an instance of this exact functor.

13.11 Fix a saturated semigroup Q, and let Δ be a union of codimension 1 faces. Prove that $\Bbbk[Q]/I_\Delta$ is Cohen–Macaulay if and only if the face ideal I_Δ, as a module over $\Bbbk[Q]$, is Cohen–Macaulay.

13.12 A positively $\mathbb{Z}^d$-graded ring R is **Gorenstein** if R is Cohen–Macaulay, and the Matlis dual $\omega_R := H_{\mathfrak{m}}^{\dim(R)}(R)^\vee$ of the top local cohomology is isomorphic to a $\mathbb{Z}^d$-graded translate of R. Show that the Stanley–Reisner ring of every simplicial sphere is Gorenstein. More generally, if $Q = Q_{\text{sat}}$ and Δ is a subcomplex of $\mathbb{R}_{\geq 0}Q$ such that $\bar{\Delta} \subseteq \mathcal{P}$ is a sphere, show that the face ring $\Bbbk[Q]/I_\Delta$ is Gorenstein.

13.13 Formulate what it means for a polyhedral (not necessarily simplicial) cell complex to be shellable. Prove that if Q is saturated and $\bar{\Delta} \subseteq \mathcal{P}$ is a shellable subcomplex of the transverse slice $\mathcal{P}$, then the face ring $\Bbbk[Q]/I_\Delta$ is Cohen–Macaulay.

Answer to Exercise 13.5 Criterion 4 implies, by part (b) of Exercise 13.4, that the maximal ideal $\mathfrak{m}$ of $\Bbbk[Q']$ consists of zerodivisors modulo every nonunit principal monomial ideal of $\Bbbk[Q']$, which is equivalent to saying that $\mathfrak{m}$ is associated to every nonunit principal monomial ideal.

Notes

The presentation in Section 13.1 is standard, apart from its focus on $\mathbb{Z}^d$-gradings. These gradings have their origins in papers such as [Hoc77], [GW78], and [TH86]. Further reading on local cohomology in general contexts can be found in [BH98] and [BrS98]. Weibel's book [Wei94] is a good reference on homological algebra, although a more leisurely introduction would be Mac Lane's classic [MacL95].

The Hilbert series in Theorem 13.13 was an unpublished result of Hochster until it finally found its way into [Sta96, Theorem II.4.1]. The module structure of that local cohomology was described explicitly by Gräbe [Grä84]. Corollary 13.16 is due to Terai [Ter99b], inspired by Theorem 13.13 and its proof. Independently and simultaneously, Mustaţă computed the module structure [Mus00], motivated by sheaf cohomology on toric varieties [EMS00, Mus02]. The similarity of the Hilbert series in Theorem 13.13 and Corollary 13.16, for local cohomology of S/I_Δ supported at $\mathfrak{m}$ and of S supported on I_Δ, can be explained as in [Mil00a] using the strong form of Alexander duality in Exercise 13.10. Yanagawa proved Theorem 13.14 for canonical modules [Yan01] in response to Corollary 13.16.

The result of Example 13.17 is due to Hartshorne [Har70]; it provided the first counterexample to Grothendieck's conjecture that local cohomology always has finite Bass numbers. Hartshorne's method of calculating was different than the one presented here, which has been used by Helm and Miller to generalize Hartshorne's result to arbitrary nonsimplicial semigroups [HM03].

The sector partitions in Theorem 13.20 can be calculated algorithmically [HM04]. The idea is to compute finitely many stages of an injective resolution of M using irreducible resolutions and then compute representatives for the maps required in part 3 of Definition 13.19. Complications arise for unsaturated semigroups, and it is an open problem to produce an algorithm in that case. Even in the saturated case, it remains to find algorithms for associated primes and socle degrees (or more generally, Bass numbers) of local cohomology.

The Ishida complex appears in work of Ishida [Ish80, Ish87], though similar constructions were made by Goto and Watanabe [GW78], Trung and Hoa [TH86], and Schäfer and Schenzel [SS90]. Ishida began by proving that the dualizing complex in Definition 12.7 really is one, according to the general theory in [Har66b], and then he concluded using local duality that its Matlis dual computes local cohomology. Schafer and Schenzel used their combinatorially defined dualizing complex in [SS90] to prove useful results on Serre's conditions S_k for semigroup rings (Cohen–Macaulay being "S_k for all k").

The Čech hull was defined in [Mil98] as a constituent of Alexander duality theory for monomial ideals. Its relation to local cohomology was realized [Mil00a] after Corollary 13.16 appeared in [Ter99b, Mus00]. The motivation for Theorem 13.31 was to prove *local duality with monomial support* [Mil00a]. This duality is a special case of *Greenlees–May duality*, which generalizes the usual local duality theorem [GM92] (see [Mil02a] for an introduction from a combinatorial point of view). This is where one really needs the fact that the canonical Čech complex has minimal length. Canonical Čech complexes have been further developed for semigroup rings by Yanagawa [Yan02, Section 6].

The long list of equivalent Cohen–Macaulay criteria in Theorem 13.37 is intended as an aid to the working combinatorialist. Our presentation of these criteria is an honest reflection of which ones are derived from others, starting from scratch. Thus, since our line of proof mainly follows [BH98], citations to that book essentially proceed forward as the proof progresses. We confined our mention of the Gorenstein condition to only one exercise because there is too much material from which to choose. The interested reader can begin with [BH98] and [Sta96].

The Cohen–Macaulayness of normal semigroup rings in Corollary 13.43 is due to Hochster [Hoc72]. Theorem 13.42, on the other hand, is a special case of a very general theorem of Grothendieck [Har66b], given Ishida's result that the dualizing complex really is one. As we mentioned in the Notes to Chapter 5, Reisner's criterion originated in [Rei76]. That shellability implies Cohen–Macaulayness was first shown by Kind and Kleinschmidt [KK79].

Part III

Determinants

Chapter 14

Plücker coordinates

Homogeneous coordinates, by which we mean lists $(\theta_1, \ldots, \theta_n) \in \mathbb{k}^n$ up to scale and with not all θ_i equaling zero, correspond to points in projective space $\mathbb{P}^{n-1}$. Equivalently, such lists correspond to lines in the vector space $\mathbb{k}^n$. The notion of *Grassmannian*, a variety whose points are the vector subspaces of a fixed dimension in $\mathbb{k}^n$, therefore encompasses projective space as a special case. More generally yet, the *flag variety* consists of all flags of vector subspaces, each one contained in the next. Grassmannians and flag varieties appear in many branches of mathematics and its applications. Like projective spaces, these varieties come equipped with their own versions of homogeneous coordinates. This chapter gives an introduction from several perspectives within commutative algebra and combinatorics. The central result, Theorem 14.11, says that the homogeneous coordinates on flag varieties, called the *Plücker coordinates*, form a *sagbi basis*.

14.1 The complete flag variety

Assume throughout this chapter that $\mathbb{k}$ is an algebraically closed field. A *complete flag* in the vector space $\mathbb{k}^n$ is a chain

$$V_\bullet : \quad V_0 \subset V_1 \subset \cdots \subset V_{n-1} \subset V_n$$

of vector subspaces of $\mathbb{k}^n$ such that $\dim_{\mathbb{k}}(V_d) = d$. The *flag variety* $\mathcal{F}\ell_n$ is the set of all complete flags in $\mathbb{k}^n$. We will be concerned primarily with the commutative algebra of a certain ring associated to the flag variety. It may not be clear from the definition that the set $\mathcal{F}\ell_n$ is an algebraic variety, but we will derive it using the properties of this ring. Our main purpose in this section is to motivate Definition 14.5.

Before tackling complete flags, let us begin with individual subspaces. Every d-dimensional subspace of $\mathbb{k}^n$ can be expressed as the row span of some $d \times n$ matrix Θ with entries θ_{ij} in $\mathbb{k}$. Such a matrix Θ must have rank d, because its d rows span a vector space of dimension d. Hence there are d columns of Θ forming a square matrix with nonzero determinant.

Definition 14.1 The determinant $\det(\Xi)$ of a square $r \times r$ submatrix Ξ inside of Θ is called a **minor** of size r. The r-minor $\det(\Xi)$ is **maximal** if $r = d$ is as large as possible.

Maximal minors Ξ in a $d \times n$ matrix Θ with $d < n$ depend, up to sign, only on a choice of d column indices from $[n] = \{1, \ldots, n\}$. We denote the submatrix with column indices $\sigma \subseteq [n]$ by Θ_σ, so the corresponding maximal minor is $\det(\Theta_\sigma)$.

Proposition 14.2 *The list* $(\det(\Theta_\sigma) \mid \sigma \subseteq [n]$ *and* $|\sigma| = d)$ *of maximal minors up to scale identifies the row span of* Θ *uniquely. More precisely, a matrix* Θ' *has the same row span as* Θ *if and only if there exists a nonzero scalar* $\gamma \in \Bbbk$ *such that*

$$\det(\Theta_\sigma) = \gamma \det(\Theta'_\sigma) \quad \text{for all} \quad \sigma \subseteq [n] \text{ and } |\sigma| = d.$$

Proof. If the matrices Θ' and Θ have the same row span, then Θ' equals $\Gamma\Theta$ for some invertible $d \times d$ matrix Γ over $\Bbbk$. It follows that $\det(\Theta'_\sigma) = \det(\Gamma)\det(\Theta_\sigma)$ for all σ, so we take $\gamma = \det(\Gamma)^{-1}$.

Conversely, suppose that the desired identity holds for some γ. Since Θ and Θ' have rank d, there exists $\sigma \subseteq [n]$ such that $\det(\Theta_\sigma) = \gamma \cdot \det(\Theta'_\sigma)$ is nonzero. Replace Θ by $\Theta_\sigma^{-1} \cdot \Theta$, and replace Θ' by $(\Theta'_\sigma)^{-1} \cdot \Theta'$. This leaves their row spans unchanged; moreover, Θ and Θ' now contain a unit matrix in columns from σ and have identical lists of maximal minors. Each entry of the matrix Θ can be expressed as a maximal minor $\det(\Theta_\tau)$ in which τ differs from σ by one element (cf. Exercise 14.1). Since the same holds for Θ', we conclude that $\Theta = \Theta'$, so they have the same row span. $\square$

Given a subspace $V_d \subset \Bbbk^n$, the maximal minors of any $d \times n$ matrix Θ having row span V_d are called the *Plücker coordinates* for V_d. Proposition 14.2 says that the list of Plücker coordinates for a fixed subspace is well-defined up to a global scalar, just like homogeneous coordinates for a point in projective space. In coordinate-free language, the subspace $V_d \subseteq \Bbbk^n$ determines a point (unique up to scale) inside the dth exterior power $\bigwedge^d(\Bbbk^n)$. Choosing a basis $e_1, \ldots, e_n$ in which to write the rows $\theta_1, \ldots, \theta_d$ of Θ yields automatically the basis $\{e_{\sigma_1} \wedge \cdots \wedge e_{\sigma_d}\}_{|\sigma|=d}$ for the dth exterior power of $\Bbbk^n$. The point in $\bigwedge^d \Bbbk^n$ corresponding to V_d is the wedge product $\theta_1 \wedge \cdots \wedge \theta_d$, which the Plücker coordinates express in the basis $\{e_{\sigma_1} \wedge \cdots \wedge e_{\sigma_d}\}_{|\sigma|=d}$. Returning now to the case of the complete flags, we have to deal with subspaces of arbitrary dimension inside $\Bbbk^n$ and therefore with Plücker coordinates that are minors of various sizes.

Definition 14.3 For any subset $\sigma \subseteq [n]$ and any $n \times n$ matrix Θ, let Θ_σ be the submatrix with rows $1, \ldots, d$ and columns $\sigma_1, \ldots, \sigma_d$, where $d = |\sigma|$.

The **Plücker coordinates** of an invertible $n \times n$ matrix Θ are the minors $\det(\Theta_\sigma)$ for subsets $\sigma \subseteq [n]$.

The list of all $2^n - 1$ nonunit Plücker coordinates of an invertible $n \times n$ matrix Θ should be parsed by separating out first the n minors of size 1, then the $\binom{n}{2}$ minors of size 2, and so on. Each list of $\binom{n}{d}$ minors of size d represents a subspace $V_d \subset \Bbbk^n$ of dimension d, and it follows from the above discussion that $0 \subset V_1 \subset V_2 \subset \cdots \subset V_{n-1} \subset \Bbbk^n$ is a complete flag of vector subspaces. Taken together, therefore, the Plücker coordinates can be thought of as functions taking each invertible $n \times n$ matrix to a list of homogeneous coordinates for its associated flag.

Definition 14.4 The subvariety $G_{d,n}$ of the projective space $\mathbb{P}^{\binom{n}{d}-1}$ consisting of all (Plücker coordinate vectors representing) d-dimensional subspaces of $\Bbbk^n$ is called the **Grassmannian**. Likewise, the **flag variety** $\mathcal{F}\ell_n$ is parametrically given by the Plücker coordinate vectors representing complete flags in $\Bbbk^n$. It is a subvariety of the product of projective spaces

$$\mathbb{P}^{n-1} \times \mathbb{P}^{\binom{n}{2}-1} \times \mathbb{P}^{\binom{n}{3}-1} \times \cdots \times \mathbb{P}^{\binom{n}{n-2}-1} \times \mathbb{P}^{\binom{n}{n-1}-1}.$$

Returning to commutative algebra, we think of the Plücker coordinate indexed by $\sigma = \sigma_1 < \cdots < \sigma_d$ as a function on matrices, or as the *generic minor* $\det(\mathbf{x}_\sigma)$ of the $n \times n$ matrix $\mathbf{x} = (x_{ij})$ of variables. As in Definition 14.3, the $d \times d$ submatrix $\mathbf{x}_\sigma$ has row indices $1, \ldots, d$ and column indices $\sigma_1, \ldots, \sigma_d$. The Plücker coordinates are therefore elements inside the polynomial ring $\Bbbk[\mathbf{x}]$ in the n^2 variables x_{ij} for $i, j = 1, \ldots, n$. Here is the central algebraic object of this chapter.

Definition 14.5 The subalgebra of $\Bbbk[\mathbf{x}]$ generated by the 2^n Plücker coordinates $\det(\mathbf{x}_\sigma)$ is called the **Plücker algebra**.

The remaining sections in this chapter explore combinatorial aspects of Plücker coordinates and the algebraic relations they satisfy, as they pertain to Gröbner bases, subalgebra bases, and the resulting semigroup rings. For a construction of $\mathcal{F}\ell_n$ from the Plücker algebra, see Exercise 14.16.

14.2 Quadratic Plücker relations

As in the previous section, let $\mathbf{x} = (x_{ij})$ be an $n \times n$ matrix of indeterminates and let $\Bbbk[\mathbf{x}]$ denote the polynomial ring over a field $\Bbbk$ generated by these indeterminates. Define a second polynomial ring $\Bbbk[\mathbf{p}]$ by introducing a variable p_σ for each subset of $[n] = \{1, \ldots, n\}$. Thus $\Bbbk[\mathbf{x}]$ and $\Bbbk[\mathbf{p}]$ are polynomial rings of dimensions n^2 and 2^n, respectively, and the indexing on the variables in $\Bbbk[\mathbf{p}]$ suggests that we define the ring homomorphism

$$\phi_n : \Bbbk[\mathbf{p}] \to \Bbbk[\mathbf{x}] \quad \text{sending} \quad p_\sigma \mapsto \det(\mathbf{x}_\sigma).$$

The map ϕ_n gives a presentation for the Plücker algebra as a quotient of $\Bbbk[\mathbf{p}]$. For convenience, we identify each subset $\sigma \subseteq [n]$ with the ordered string of its elements. Then we can allow arbitrary substrings as indices of the variables in $\Bbbk[\mathbf{p}]$, subject to the usual sign conventions for permutations. For instance,

$$p_{275} = -p_{257} \quad \text{and} \quad p_{725} = p_{257} \quad \text{and} \quad p_{272} = 0 \quad \text{in } \Bbbk[\mathbf{p}].$$

The same convention governs the change in sign on $\det(\mathbf{x}_\sigma)$ after permuting the columns of $\mathbf{x}_\sigma$ or choosing some column twice, so this sign convention respects the map ϕ_n.

Our object of study is the homogeneous prime ideal $I_n = \ker(\phi_n)$ of *Plücker relations*. For $n = 1$ and $n = 2$, this ideal is zero. The first interesting case is $n = 3$. The ideal I_3 is principal and generated by the quadric $p_{23}p_1 - p_{13}p_2 + p_{12}p_3$. For $n = 4$, the ideal I_n is minimally generated by the following 10 quadrics:

$$\begin{array}{ll}
\underline{p_{23}p_1} - p_{13}p_2 + p_{12}p_3, & \underline{p_{24}p_1} - p_{14}p_2 + p_{12}p_4, \\
\underline{p_{34}p_1} - p_{14}p_3 + p_{13}p_4, & \underline{p_{34}p_2} - p_{24}p_3 + p_{23}p_4, \\
\underline{p_{14}p_{23}} - p_{13}p_{24} + p_{12}p_{34}, & \underline{p_{234}p_1} - p_{134}p_2 + p_{124}p_3 - p_{123}p_4, \\
\underline{p_{134}p_{12}} - p_{124}p_{13} + p_{123}p_{14}, & \underline{p_{234}p_{12}} - p_{124}p_{23} + p_{123}p_{24}, \\
\underline{p_{234}p_{13}} - p_{134}p_{23} + p_{123}p_{34}, & \underline{p_{234}p_{14}} - p_{134}p_{24} + p_{124}p_{34}.
\end{array}$$

These 10 quadrics form a Gröbner basis for the ideal I_4 with respect to any term order that selects the underlined initial terms. We will generalize this quadratic Gröbner basis to arbitrary n.

First, introduce a poset $\mathcal{P}$ whose underlying set consists of the variables $\mathbf{p} = \{p_\sigma \mid \sigma \subseteq [n]\}$. When $\sigma = \{\sigma_1 < \cdots < \sigma_s\}$ and $\tau = \{\tau_1 < \cdots < \tau_t\}$ are two subsets of $[n]$, set $p_\sigma \le p_\tau$ in the poset $\mathcal{P}$ if $s \ge t$ and $\sigma_i \le \tau_i$ for all $i = 1, \ldots, t$. (A weak chain $p_{\sigma_1} \le \cdots \le p_{\sigma_\ell}$ is thus a semistandard tableau of length ℓ; see Definition 14.12.) Here is the Hasse diagram of $\mathcal{P}$ for $n = 4$:

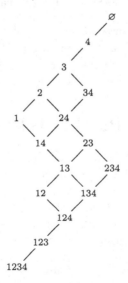

The 10 incomparable pairs in this poset are precisely the underlined initial terms in our Gröbner basis for I_4. This fact generalizes.

Totally order the variables in $\Bbbk[\mathbf{p}]$ by setting $p_\sigma \prec p_\tau$ if $|\sigma| > |\tau|$ or $[|\sigma| = |\tau|$ and σ comes before τ in the lexicographic order]. This total order is a linear extension of the poset $\mathcal{P}$. Let $\prec$ also denote the reverse lexicographic term order on $\Bbbk[\mathbf{p}]$ induced by the variable ordering $\prec$.

Theorem 14.6 *The ideal I_n of Plücker relations has a Gröbner basis under $\prec$ consisting of homogeneous quadrics. More precisely, the products $p_\sigma p_\tau$ of incomparable pairs of the poset $\mathcal{P}$ generate the initial ideal* $\mathrm{in}_\prec(I_n)$.

Proof. We first show that each incomparable product $p_\sigma p_\tau$ lies inside $\mathrm{in}_\prec(I_n)$. Fix such a product. We may assume that $p_\sigma \prec p_\tau$ and hence $s = |\sigma| \geq |\tau| = t$. Since p_σ and p_τ are incomparable, there exists an index $i \in \{1, \ldots, t\}$ such that $\tau_i < \sigma_i$, and we take i to be the smallest index with this property. Consider the strictly increasing $(s+1)$-chain of indices

$$\tau_1 < \cdots < \tau_i < \sigma_i < \cdots < \sigma_s. \tag{14.1}$$

For any permutation π of these $s+1$ indices, let $\pi(\tau)$ be defined by $\pi(\tau)_j = \pi(\tau_j)$ if $j \leq i$, and $\pi(\tau)_j = \tau_j$ otherwise. Similarly define $\pi(\sigma)$. Use all of the $(s+1)!$ quadratic monomials $p_{\pi(\sigma)}p_{\pi(\tau)}$ to form the alternating sum

$$\sum_\pi \mathrm{sign}(\pi) \cdot p_{\pi(\sigma)} \cdot p_{\pi(\tau)}, \tag{14.2}$$

and divide by a constant so that all of the terms have coefficient $+1$ or -1. By summing only over *shuffles* of the sequence (14.1), which by definition result in sequences that increase in the first i slots and (separately) increase in the remaining $s+1-i$ slots, this division can be avoided, so the construction is also valid in positive characteristic. The result is a homogeneous quadratic polynomial.

We claim that the monomial $p_\sigma \cdot p_\tau$ is the initial term of (14.2) with respect to the reverse lexicographic term order $\prec$. This follows from the fact that, for any nonidentity permutation π of (14.1),

$$p_{\pi(\sigma)} \prec p_\sigma \prec p_\tau \prec p_{\pi(\tau)}$$

whenever these variables are nonzero. We next claim that

$$\sum_\pi \mathrm{sign}(\pi) \cdot \det(\mathbf{x}_{\pi(\sigma)}) \cdot \det(\mathbf{x}_{\pi(\tau)}) \;=\; 0 \tag{14.3}$$

is a valid algebraic relation among the Plücker coordinates of a generic matrix $\mathbf{x}$. Let $\mathbf{x}_{s \times n}$ denote the $s \times n$ matrix consisting of the top s rows of $\mathbf{x}$. The left-hand side of (14.3) is multilinear and alternating as a function of the $s+1$ columns of $\mathbf{x}_{s \times n}$ indexed by (14.1). As the columns of an $s \times n$ matrix over $\Bbbk$ span a vector space of dimension at most s, the polynomial on the left-hand side of (14.3) vanishes whenever values in $\Bbbk$ are chosen for the variables in $\mathbf{x}_{s \times n}$. Hence this polynomial is identically zero, proving (14.3).

It follows from (14.3) that the quadratic polynomial (14.2) lies in the ideal $I_n = \ker(\phi_n)$, so $p_\sigma p_\tau$ lies inside $\mathrm{in}_\prec(I_n)$ whenever p_σ and p_τ are incomparable in the poset $\mathcal{P}$. It remains to be shown that $\mathrm{in}_\prec(I_n)$ has no other monomial generators.

Introduce a term order on the other polynomial ring $\Bbbk[\mathbf{x}]$, namely the purely lexicographic term order on $\Bbbk[\mathbf{x}]$ induced by the row-wise order

$$x_{11} > x_{12} > \cdots > x_{1n} > x_{21} > \cdots > x_{2n} > \cdots > x_{n1} > \cdots > x_{nn}$$

on the n^2 variables. If $\sigma = \{\sigma_1 < \cdots < \sigma_s\}$ then the initial term of the minor $\phi_n(p_\sigma) = \det(\mathbf{x}_\sigma)$ is its *diagonal term*

$$\mathrm{in}(\phi_n(p_\sigma)) \;=\; x_{1\sigma_1} x_{2\sigma_2} \cdots x_{s\sigma_s}.$$

Every monomial in $\Bbbk[\mathbf{x}]$ can be written uniquely as an ordered product

$$x_{1i_{11}} x_{1i_{12}} \cdots x_{1i_{1\ell_1}} x_{2i_{21}} x_{2i_{22}} \cdots x_{2i_{2\ell_2}} \cdots x_{ni_{n1}} x_{ni_{n2}} \cdots x_{ni_{n\ell_n}} \qquad (14.4)$$

of variables, with repetition allowed. Here, $i_{j,k} \leq i_{j,k+1}$ holds for all j and k. The monomial (14.4) is the initial term of $\phi_n(\mathbf{p}^{\mathbf{a}})$ for some monomial $\mathbf{p}^{\mathbf{a}}$ in $\Bbbk[\mathbf{p}]$ if and only if

$$\ell_1 \geq \ell_2 \geq \cdots \geq \ell_n \quad \text{and} \quad i_{j,k} < i_{j+1,k} \quad \text{for all } j,k. \qquad (14.5)$$

In fact, if condition (14.5) is satisfied then there exists a unique monomial $\mathbf{p}^{\mathbf{a}}$ in $\Bbbk[\mathbf{p}]$ such that the support of $\mathbf{p}^{\mathbf{a}}$ is a chain in $\mathcal{P}$ and $\phi_n(\mathbf{p}^{\mathbf{a}})$ has the initial term (14.4). This monomial equals

$$\mathbf{p}^{\mathbf{a}} \;=\; p_{i_{11}i_{21}\ldots} \cdot p_{i_{12}i_{22}\ldots} \cdot p_{i_{13}i_{23}\ldots} \cdots . \qquad (14.6)$$

In summary, we have shown that the $\mathbf{x}$-monomials $\mathrm{in}(\phi_n(\mathbf{p}^{\mathbf{a}}))$ are all distinct as $\mathbf{p}^{\mathbf{a}}$ runs over $\mathbf{p}$-monomials that are supported on chains in the poset $\mathcal{P}$.

To complete the proof of the theorem, it remains to show that $\mathrm{in}_\prec(I_n)$ is contained in the ideal generated by the incomparable products. Suppose this is not the case. Then there exists a nonzero polynomial f in the ideal I_n whose initial term under $\prec$ is not a multiple of any incomparable product. Thus the variables p_σ appearing in this initial term form a chain in the poset $\mathcal{P}$. Assuming that the polynomial f is minimal with respect to the term order $\prec$, we can actually write

$$f \;=\; \sum c_{\mathbf{a}} \cdot \mathbf{p}^{\mathbf{a}},$$

where every monomial $\mathbf{p}^{\mathbf{a}}$ appearing in f is supported on a chain in $\mathcal{P}$. Consider the identity

$$\sum c_{\mathbf{a}} \cdot \phi_n(\mathbf{p}^{\mathbf{a}}) \;=\; 0 \quad \text{in } \Bbbk[\mathbf{x}].$$

Let m denote the highest monomial appearing in any of the expressions $\phi_n(\mathbf{p^a})$ appearing in this identity. There exist at least two distinct terms $\mathbf{p^a}$ and $\mathbf{p^b}$ in f such that $\operatorname{in}(\phi_n(\mathbf{p^a})) = \operatorname{in}(\phi_n(\mathbf{p^b})) = m$. Since both $\mathbf{p^a}$ and $\mathbf{p^b}$ are supported on a chain of $\mathcal{P}$, this contradicts the conclusion at the end of the last paragraph and completes the proof of Theorem 14.6. $\square$

Corollary 14.7 *The initial ideal* $\operatorname{in}_\prec(I_n)$ *is the Stanley–Reisner ideal of the simplicial complex of chains in the poset* $\mathcal{P}$ *(the* **order complex** *of* $\mathcal{P}$*).*

Example 14.8 We illustrate how the standard $\mathbf{p}$-monomial in (14.6) is reconstructed from the initial $\mathbf{x}$-term (14.4) in its expansion under ϕ_n. Let $n = 4$. The following $\mathbf{x}$-monomial satisfies condition (14.5):

$$m = x_{11}^2 x_{12}^3 x_{13} x_{14} x_{22} x_{23}^3 x_{24} x_{33} x_{34}.$$

The corresponding standard $\mathbf{p}$-monomial (14.6) is found to be

$$\mathbf{p^a} = p_{123} p_{134} p_{23}^2 p_{24} p_3 p_4.$$

Indeed, this monomial lies outside of $\operatorname{in}_\prec(I_4)$, and $\operatorname{in}(\phi_4(\mathbf{p^a})) = m$. Note that there are other $\mathbf{p}$-monomials $\mathbf{p^b}$ with $\operatorname{in}(\phi_4(\mathbf{p^b})) = m$, for instance

$$\mathbf{p^b} = p_{123} p_{234} p_{13} p_{23} p_{34} p_2 p_4,$$

but these monomials necessarily lie inside $\operatorname{in}_\prec(I_4)$. $\diamond$

A monomial $\mathbf{p^a}$ in $\Bbbk[\mathbf{p}]$ is called *semistandard* if its support is a chain in the poset $\mathcal{P}$. We have proved that $\mathbf{p^a}$ is semistandard if and only if $\mathbf{p^a}$ is standard—i.e., not in the initial ideal $\operatorname{in}_\prec(I_n)$. Hence we get the following.

Corollary 14.9 *The set of semistandard monomials* $\mathbf{p^a}$ *constitutes a basis for the Plücker algebra as a vector space over* $\Bbbk$.

14.3 Minors form sagbi bases

This section concerns objects of the following type consisting of minors.

Definition 14.10 A set $\{f_1, \dots, f_r\}$ of polynomials in a polynomial ring is a **sagbi basis** with respect to a given term order if every polynomial f in the subalgebra $\Bbbk[f_1, \dots, f_r]$ has the following property: the initial term $\operatorname{in}(f)$ is a monomial $\operatorname{in}(f_1)^{i_1} \cdots \operatorname{in}(f_r)^{i_r}$ in the initial terms $\operatorname{in}(f_1), \dots, \operatorname{in}(f_r)$.

The term "sagbi" is an acronym for "subalgebra analogue of Gröbner bases for ideals". In contrast to the situation for ideals, the initial algebra of a finitely generated subalgebra of a polynomial ring need not be finitely generated. The existence of a finite sagbi basis is a special property for a subalgebra. It turns out that our Plücker algebra enjoys this property.

Theorem 14.11 *The 2^n Plücker coordinates of the $n \times n$ generic matrix $\mathbf{x}$ form a sagbi basis under any **diagonal** term order (meaning that the initial term of each minor $\det(\mathbf{x}_\sigma)$ is its diagonal term $x_{1\sigma_1} x_{2\sigma_2} \cdots x_{s\sigma_s}$) or any **antidiagonal** term order (meaning that the initial term of each minor $\det(\mathbf{x}_\sigma)$ is its antidiagonal term $x_{1\sigma_s} x_{2\sigma_{s-1}} \cdots x_{s\sigma_1}$).*

Before getting to the proof, we need a definition and a lemma.

Definition 14.12 A **tableau** with n rows is an array

$$
m \;=\; \begin{bmatrix}
i_{11} & i_{12} & i_{13} & \cdots & i_{1\ell_1} \\
i_{21} & i_{22} & i_{23} & i_{24} & \cdots & i_{2\ell_2} \\
\vdots & \vdots & & & \\
i_{n1} & i_{n2} & \cdots & i_{n\ell_n}
\end{bmatrix}.
$$

of nonnegative integers in which the rows need not have equal lengths $\ell_1, \ldots, \ell_n$. A tableau is **semistandard** if

- the rows get shorter as they go down (that is, $\ell_1 \geq \ell_2 \geq \cdots \geq \ell_n$),
- the rows are weakly increasing (that is, $i_{j,k} \leq i_{j,k+1}$ for all j, k), and
- the columns are strictly increasing (that is, $i_{j,k} < i_{j+1,k}$ for all j, k).

The point is that indices coming from a general monomial m as in (14.4) can be written in tableau form. For example, the semistandard tableau

1	1	3	4	4	4
2	4	5	7
5	5	6
6

corresponds to $\quad x_{11}^2 x_{13} x_{14}^3 x_{22} x_{24} x_{25} x_{27} x_{35}^2 x_{36} x_{46}.$

Using this form, we have the following.

Lemma 14.13 *A monomial m in $\Bbbk[\mathbf{x}]$ is the initial term of a polynomial in the Plücker algebra $\mathrm{image}(\phi_n)$ if and only if the tableau corresponding to m is semistandard.*

Proof. We have seen in (14.4)–(14.6) that every semistandard tableau is associated to the initial term of $\phi_n(\mathbf{p}^{\mathbf{a}})$ for some monomial $\mathbf{p}^{\mathbf{a}}$ supported on a chain in $\mathcal{P}$. On the other hand, Corollary 14.9 implies that every polynomial $f(\mathbf{x})$ in the Plücker algebra $\mathrm{image}(\phi_n)$ is a $\Bbbk$-linear combination of the images of such $\mathbf{p}$-monomials. Hence the tableau corresponding to the initial term $\mathrm{in}(f)$ is semistandard. $\qquad\square$

Proof of Theorem 14.11. By symmetry, we need only prove the diagonal term order case. Lemma 14.13 says that the initial algebra of the Plücker algebra is the vector space over $\Bbbk$ spanned by all semistandard tableaux. Each monomial corresponding to a semistandard tableau is the product of

the monomials corresponding to its columns, so the initial algebra is the
k-algebra generated by the semistandard tableaux with only one column.
These one-column tableaux are precisely the diagonal term monomials

$$\begin{bmatrix} \sigma_1 \\ \sigma_2 \\ \vdots \\ \sigma_s \end{bmatrix} \;\; = \;\; x_{1\sigma_1} x_{2\sigma_2} \cdots x_{s\sigma_s} \;\; = \;\; \mathrm{in}(\det(\mathbf{x}_\sigma)).$$

Hence the minors $\det(\mathbf{x}_\sigma)$ form a sagbi basis. □

If R is any subalgebra of a polynomial ring that possesses a finite sagbi
basis, then this sagbi basis defines a *flat degeneration* from R to its initial
algebra $\mathrm{in}(R)$. The initial algebra is generated by monomials, so it corresponds to a toric variety. Hence, geometrically, a finite sagbi basis provides
a flat family connecting the given variety $\mathrm{Spec}(R)$ to the affine toric variety
$\mathrm{Spec}(\mathrm{in}(R))$. Of course, we can replace "Spec" by "Proj" in the presence
of a $\mathbb{Z}$-grading (or even SpecTor in the presence of a multigrading; see Exercise 14.16). Hence Theorem 14.11 states that the flag variety and the
Grassmannian can be sagbi-degenerated to toric varieties. In what follows,
we make this degeneration explicit at the level of presentation ideals.

Example 14.14 Consider the special case of the Grassmannian $G_{2,4}$. Its
homogeneous coordinate ring is generated by the six 2×2 minors of a 2×4
matrix of indeterminates, and its presentation ideal is

$$\mathrm{ker}(\phi_4) \quad - \quad \langle p_{14}p_{23} - p_{13}p_{24} + p_{12}p_{34} \rangle.$$

The presentation ideal of the sagbi degeneration of $G_{2,4}$ is

$$\mathrm{ker}(\psi_4) \quad = \quad \langle p_{14}p_{23} - p_{13}p_{24} \rangle,$$

with ψ_n as defined below. This is the ideal of algebraic relations on the
initial terms $x_{11}x_{22}, x_{11}x_{23}, \ldots, x_{13}x_{24}$ of the 2×2 minors. ◇

Let us examine the toric variety corresponding to the initial algebra of
the Plücker algebra in general. Consider the monomial map

$$\psi_n : \mathbb{k}[\mathbf{p}] \to \mathbb{k}[\mathbf{x}] \quad \text{sending} \quad p_\sigma \mapsto \mathrm{in}(\det(\mathbf{x}_\sigma)).$$

Our toric variety is the zero set of the toric ideal $J_n = \mathrm{ker}(\psi_n)$. We will
prove that the quadratic Gröbner basis for I_n in Theorem 14.6 factors
through a Gröbner basis for J_n, the latter being obtained by setting to zero
all but the *two* highest terms in the quadrics from Theorem 14.6.

To express this Gröbner basis in the most succinct form, first observe
that $\mathcal{P}$ is a *distributive lattice*. The lattice operations *meet* $\wedge$ and *join* $\vee$

are defined as follows: If $\sigma = \{\sigma_1 < \cdots < \sigma_s\}$ and $\tau = \{\tau_1 < \cdots < \tau_t\}$ with $s \geq t$, then

$$\sigma \wedge \tau = \{\alpha_1, \ldots, \alpha_s\} \quad \text{and} \quad \sigma \vee \tau = \{\beta_1, \ldots, \beta_t\},$$

where $\alpha_i = \min\{\sigma_i, \tau_i\}$ and $\beta_i = \max\{\sigma_i, \tau_i\}$ for $i = 1, \ldots, t$, and we set $\alpha_i = \sigma_i$ for $i = t+1, \ldots, s$. The lattice being distributive means that

$$\rho \wedge (\sigma \vee \tau) = (\rho \wedge \sigma) \vee (\rho \wedge \tau),$$
$$\rho \vee (\sigma \wedge \tau) = (\rho \vee \sigma) \wedge (\rho \vee \tau).$$

The diagonal term order on $\Bbbk[\mathbf{x}]$ induces a partial term order $\leq$ on $\Bbbk[\mathbf{p}]$. This partial term order is defined as follows:

$$\mathbf{p}^{\mathbf{a}} \leq \mathbf{p}^{\mathbf{b}} \quad \text{if and only if} \quad \text{in}(\phi_n(\mathbf{p}^{\mathbf{a}})) \leq \text{in}(\phi_n(\mathbf{p}^{\mathbf{b}})). \tag{14.7}$$

We note that the reverse lexicographic order $\prec$ is not a refinement of the partial order $\leq$. It is this fact that makes our next theorem subtle.

Example 14.15 Consider the 10 quadratic monomials on $G_{3,6}$ that involve all 6 indices. In the reverse lexicographic order $\prec$ used in Theorem 14.6, they are ordered

$$p_{123}p_{456} \prec p_{124}p_{356} \prec \cdots \prec p_{134}p_{256} \prec \cdots \prec p_{146}p_{235} \prec p_{156}p_{234}.$$

In the partial term order (14.7), we have

$$p_{123}p_{456} < p_{124}p_{356} < \cdots < p_{134}p_{256} = p_{156}p_{234} < p_{146}p_{235}.$$

The first order is not a refinement of the second order. $\diamond$

Theorem 14.16 *The toric ideal $J_n = \ker(\psi_n)$ equals the initial ideal for the ideal I_n of Plücker relations with respect to the partial term order $\leq$. The reduced Gröbner basis of J_n under the reverse lexicographic term order on $\Bbbk[\mathbf{p}]$ defined above consists of all nonzero binomials*

$$p_\sigma p_\tau - p_{\sigma \vee \tau} p_{\sigma \wedge \tau}.$$

Proof. The Gröbner basis constructed in the proof of Theorem 14.6 is *minimal* (meaning that no element in the Gröbner basis can be omitted) but not reduced. For the following argument we replace it by the reduced Gröbner basis. Consider any quadratic polynomial in the reduced Gröbner basis of I_n with respect to the reverse lexicographic order $\prec$. It has the form

$$g = p_\sigma p_\tau + \text{semistandard tableaux strictly lower in } \prec.$$

We claim that each term $p_\rho p_\pi$ appearing in g is also less than or equal to $p_\sigma p_\tau$ in the order $\leq$. At most one such semistandard tableau equals $p_\sigma p_\tau$

in the partial order $\leq$. That term is $p_{\sigma \vee \tau} p_{\sigma \wedge \tau}$, and since diagonal initial terms cancel in $\phi_n(g) = 0$, the term $p_{\sigma \vee \tau} p_{\sigma \wedge \tau}$ must have coefficient -1 in g. All other tableaux $p_\rho p_\pi$ in g are semistandard; hence the corresponding **x**-monomials $\text{in}(\phi_n(p_\rho p_\pi))$ are all distinct. Since $\phi_n(g) = 0$, the initial terms with respect to $\leq$ must be attained twice, and hence all tableaux $p_\rho p_\pi$ lie strictly below $p_\sigma p_\tau$ in the partial order $\leq$ as well.

We conclude that the initial form of the quadric g with respect to the partial term order $\leq$ is precisely the desired binomial $p_\sigma p_\tau - p_{\sigma \vee \tau} p_{\sigma \wedge \tau}$:

$$g \;=\; p_\sigma p_\tau - p_{\sigma \vee \tau} p_{\sigma \wedge \tau} \;+\; \text{strictly lower terms in } \leq.$$

Now let K_n be the ideal generated by all the binomials $p_\sigma p_\tau - p_{\sigma \vee \tau} p_{\sigma \wedge \tau}$. Then K_n is contained in J_n by the definition of ψ_n. The definition of the partial term order $\leq$ implies that J_n is contained inside $\text{in}_\leq(I_n)$. Hence

$$K_n \;\subseteq\; J_n \;\subseteq\; \text{in}_\leq(I_n). \tag{14.8}$$

The initial monomial ideal of $\text{in}_\leq(I_n)$ with respect to the reverse lexicographic term order $\prec$ is generated by the incomparable products $p_\sigma p_\tau$. But these products lie in the initial monomial ideal of K_n, by the observations in the previous paragraph. We conclude that all three ideals in (14.8) have the same initial monomial ideal, and hence they are equal. This implies both assertions in the statement of the theorem. $\qquad\square$

Example 14.17 In the above proof, it was essential that we used the reduced Gröbner basis of I_n instead of the Gröbner basis of Theorem 14.6. We illustrate the distinction for $n = 8$. For the ideal of the Grassmannian $G_{4,8}$, both Gröbner bases consist of 721 quadrics in 70 unknowns p_{ijkl}. A typical element in the Gröbner basis of Theorem 14.6 looks like

$$\underline{p_{1278}p_{3456}} + \underline{p_{1258}p_{3467}} - p_{1257}p_{3468} - \underline{p_{1248}p_{3567}} + p_{1247}p_{3568}$$
$$+ \, p_{1245}p_{3678} - \underline{p_{1238}p_{4576}} - p_{1237}p_{4568} - p_{1235}p_{4678} + p_{1234}p_{5678}.$$

This quadric is not in the reduced Gröbner basis since the four underlined monomials are not semistandard. The element of the reduced Gröbner basis with the same initial term is

$$\underline{p_{1278}p_{3456}} - p_{1256}p_{3478} + p_{1246}p_{3578} - p_{1245}p_{3678}$$
$$- \, p_{1236}p_{4578} + p_{1235}p_{4678} - p_{1234}p_{5678}.$$

Not every coefficient in the reduced Gröbner basis of I_8 is $+1$ or -1. The following quadric lies the reduced Gröbner basis and has a coefficient $+2$:

$$\underline{p_{1567}p_{2348}} - p_{1347}p_{2568} + p_{1346}p_{2578} - p_{1345}p_{2678} + p_{1247}p_{3568} - p_{1246}p_{3578}$$
$$+ \, p_{1245}p_{3678} - p_{1237}p_{4568} + p_{1236}p_{4578} - p_{1235}p_{4678} + 2p_{1234}p_{5678}.$$

Note that the first two terms constitute a binomial in the toric ideal J_8. $\diamond$

14.4　Gelfand–Tsetlin semigroups

Theorem 14.11 degenerates the Plücker algebra of Definition 14.5 to the affine semigroup ring generated by the diagonal or antidiagonal terms of the $2^n - 1$ Plücker coordinates. In this section, we consider the corresponding *antidiagonal semigroup* $\mathcal{A}_n$, whose Hilbert basis consists of the exponent matrices of the antidiagonal terms of the $2^n - 1$ nonunit Plücker coordinates. We denote this Hilbert basis by $\mathcal{H}_n$.

Example 14.18 When $n = 3$, the $2^3 - 1 = 7$ Hilbert basis elements lie in $\mathbb{Z}^{3 \times 3} = \mathbb{Z}^9$, and we draw vectors as 3×3 square grids of integers. Thus

When $n = 4$, the $2^4 - 1 = 15$ Hilbert basis elements lie in $\mathbb{Z}^{4 \times 4} = \mathbb{Z}^{16}$, so

In these square grids, the empty boxes denote entries equal to zero.　　◇

The semigroup $\mathcal{A}_n$ turns out to be isomorphic (although not equal) to another semigroup, the integer points in the cone of so-called Gelfand–Tsetlin patterns. The importance of this cone and its integer points stem from their connections to representation theory and symplectic geometry.

Definition 14.19 An array $\Lambda = (\lambda_{i,j})_{i,j=1}^n$ of real numbers is a **Gelfand–Tsetlin pattern** if $\lambda_{i,j} \geq \lambda_{i,j+1} \geq \lambda_{i+1,j} \geq 0$ for $i,j = 1, \ldots, n$, and $\lambda_{i,j} = 0$ whenever $i + j > n + 1$ (so $\lambda_{i,j}$ lies strictly below the main antidiagonal). Denote the semigroup of integer Gelfand–Tsetlin patterns by $\mathcal{GT}_n$.

Equivalently, the entries in Gelfand–Tsetlin patterns Λ are nonnegative, decrease in the directions indicated by the arrows in diagram

$$
\begin{array}{l}
\lambda_{1,1} \to \lambda_{1,2} \to \lambda_{1,3} \to \cdots \\
\quad \downarrow \;\swarrow\; \downarrow \;\swarrow \\
\lambda_{2,1} \to \lambda_{2,2} \to \cdots \\
\quad \downarrow \;\swarrow \\
\lambda_{3,1} \to \cdots \\
\quad \downarrow \\
\quad \vdots
\end{array}
\tag{14.9}
$$

and vanish outside the upper left triangle. As suggested by the diagram, the array should be thought of as triangular rather than square. Nonetheless, for convenience, we consider $\mathcal{GT}_n$ as a semigroup inside $\mathbb{Z}^{n \times n}$.

In the language of Part I, integer Gelfand–Tsetlin patterns correspond to certain special kinds of monomial ideals in three variables. Indeed, stacking $\lambda_{i,j}$ three-dimensional blocks on the square (i,j) yields the staircase of standard monomials for an ideal because of the rightward and downward pointing arrows in (14.9). The staircase decreases "diagonally" from the x-axis to the y-axis. There is another way to biject integer Gelfand–Tsetlin patterns with a class of monomial ideals in three variables (Exercise 14.11).

To identify the Hilbert basis of $\mathcal{GT}_n$, we need to introduce partitions.

Definition 14.20 A **partition** is a sequence $\lambda = (\lambda_1, \lambda_2, \dots)$ of weakly decreasing nonnegative integers λ_i, called the **parts** of λ.

Partitions can be drawn in a number of ways using *Ferrers diagrams* or *shapes*. Each of these is a collection of boxes lined up in rows or columns whose lengths correspond to the parts of λ. Here, we use the "English" style, where the ith row from the top has λ_i boxes, justified at the left. Note that the parts are distinct if and only if the rows get strictly shorter.

Example 14.21 The partitions having distinct parts of size at most 3 are

$$\mathcal{H}_3' = \left\{ \square \quad \square\square \quad \square\square\square \quad \text{⊟} \quad \text{⊞} \quad \text{⊞} \quad \text{⊞} \right\},$$

while those having distinct parts of size at most 4 are

$$\mathcal{H}_4' = \left\{ \square \quad \square\square \quad \square\square\square \quad \square\square\square\square \quad \dots \right\}.$$

Compare these to the antidiagonal Hilbert bases in Example 14.18. ◇

A partition λ with distinct parts of size at most n can be viewed as a matrix in $\mathbb{Z}^{n \times n}$, by placing a 1 in each box of the shape of λ and setting the other entries equal to zero.

Proposition 14.22 *The semigroup $\mathcal{GT}_n$ has Hilbert basis $\mathcal{H}_n'$ consisting of partitions with distinct parts of size at most n.*

Proof. Each such partition clearly lies inside $\mathcal{GT}_n$. Furthermore, given a Gelfand–Tsetlin pattern $(\lambda_{i,j})_{i+j<n}$, drawn in the manner of (14.9), its set λ of nonzero entries lies in $\mathcal{H}_n'$. Subtracting the partition λ from the Gelfand–Tsetlin pattern $(\lambda_{i,j})_{i+j<n}$ yields another Gelfand–Tsetlin pattern $(\lambda_{i,j} - 1)_{i+j<n}$ by definition. Now we can continue subtracting partitions taken from $\mathcal{H}_n'$ until we get the zero Gelfand–Tsetlin pattern. □

Theorem 14.23 *There is an automorphism of $\mathbb{Z}^{n \times n}$ taking the antidiagonal semigroup to the Gelfand–Tsetlin semigroup. In particular, $\mathcal{A}_n \cong \mathcal{GT}_n$.*

Proof. It is enough to demonstrate an automorphism of $\mathbb{Z}^{n \times n}$ that induces a bijection between the Hilbert bases of $\mathcal{A}_n$ and $\mathcal{GT}_n$ (Exercise 7.2). In our case, the automorphism of $\mathbb{Z}^{n \times n}$ that takes an array $(\lambda_{i,j})_{i,j=1}^n$ to the array $(\lambda_{i,j} - \lambda_{i,j+1})_{i,j=1}^n$ induces a bijection from $\mathcal{H}'_n$ to $\mathcal{H}_n$. To check that this automorphism really is invertible, note either that it is unitriangular with respect to an ordering of the basis elements of $\mathbb{Z}^{n \times n}$ starting at the northwest corner and snaking its way to the southeast, or that the homomorphism taking the array $(\lambda_{i,j})$ to the array $(\sum_{j' \geq j} \lambda_{ij'})$ is its inverse. □

Corollary 14.24 *The initial algebra of the Plücker algebra resulting from either term order in Theorem 14.11 is isomorphic (as a semigroup ring) to the Gelfand–Tsetlin semigroup ring.*

Theorem 14.16 specifies a flat (sagbi) degeneration from the Plücker algebra $\Bbbk[\mathbf{p}]/I_n$ to the Gelfand–Tsetlin semigroup ring $\Bbbk[\mathcal{GT}_n] = \Bbbk[\mathbf{p}]/J_n$. The latter algebra is a normal affine semigroup ring: $\mathcal{GT}_n$ is saturated by definition, although it can be proved via $\mathcal{A}_n$ (Exercise 14.14), and it also holds by [Stu96, Chapter 13] because J_n has the squarefree initial monomial ideal given by Theorem 14.16. Hochster's result in Corollary 13.43 tells us that $\Bbbk[\mathcal{GT}_n]$ is therefore Cohen–Macaulay. From Corollary 8.31 we conclude that the same result holds for the Plücker algebra.

Corollary 14.25 *The Plücker algebra $\Bbbk[\mathbf{p}]/I_n$ is Cohen–Macaulay.*

Proving Cohen–Macaulayness by Gröbner degeneration is a staple of combinatorial commutative algebra. This technique will be applied again, to Schubert determinantal rings, in Corollary 16.44. Flat degenerations also provide quantitative information such as multidegrees, K-polynomials, and Hilbert series. For the Plücker algebra, these invariants can be determined from the Gelfand–Tsetlin semigroup or the distributive lattice $\mathcal{P}$, and we invite the reader to carry out such a calculation in Exercise 14.12. Even if the multidegree is known, Gröbner degeneration can provide a new combinatorial formula, as we will see for Schubert polynomials in Corollary 16.30.

Exercises

14.1 Let X be a $d \times (n-d)$ matrix, and let $\mathbf{1}$ denote the $d \times d$ unit matrix, and concatenate them to form the $d \times n$ matrix $\Theta = [X \ \mathbf{1}]$. Show that every minor (of any size) of the matrix X equals $\det(\Theta_\sigma)$ for some $\sigma \subseteq [n]$ with $|\sigma| = d$.

14.2 Let $n = 8$, $\sigma = \{1, 5, 6, 7\}$, and $\tau = \{2, 3, 4, 8\}$. Write the quadric (14.2) and compare it to the quadric with the same initial term in Example 14.17.

14.3 Find an $(s+t) \times (s+t)$ matrix such that its nonzero entries are from the top s rows of $\mathbf{x}$, its determinant equals the left side of (14.3), and it has a submatrix of size $(s+t) \times (s+1)$ with only s distinct rows. Conclude again that (14.3) holds.

14.4 List the minimal generators of the ideal I_5.

14.5 The squarefree monomial ideal $\mathrm{in}_{\prec}(I_5)$ determines a simplicial complex on 31 vertices. Find the f-vector of this simplicial complex.

14.6 Is there a nice formula for the number of minimal generators of the ideal I_n? What about the number of first syzygies?

14.7 Express the monomials $p_{167}p_{258}p_{349}$ and $p_{14589}p_{2367}$ as linear combinations of semistandard tableaux modulo I_9.

14.8 Let J be the Alexander dual of the squarefree monomial ideal $\mathrm{in}_{\prec}(I_n)$. How many minimal generators does J have? Is $\Bbbk[\mathbf{p}]/J$ a Cohen–Macaulay ring?

14.9 The ideal of the Grassmannian $G_{3,6}$ is homogeneous with respect to a natural $\mathbb{Z}^6$-grading. Determine the multidegree of this ideal.

14.10 There exist term orders such that the 3×3 minors of a generic 3×6 matrix do not constitute a sagbi basis. Find such a term order.

14.11 Construct a bijection from integer Gelfand–Tsetlin patterns to monomial ideals in three variables that are symmetric under switching x and y. Hint: Shear every $\vec{\nearrow}$ triangle into a $\searrow$ triangle.

14.12 Determine the K-polynomial of the Gelfand–Tsetlin semigroup $\mathcal{GT}_5$.

14.13 Describe the singular locus of the projective toric variety defined by the toric ideal J_n. Hint: Apply the results in [Wag96] to the distributive lattice $\mathcal{P}$.

14.14 Prove directly that the antidiagonal semigroup $\mathcal{A}_n$ is saturated.

14.15 For Corollary 14.25 we were able to apply Corollary 8.31 because we realized the sagbi degeneration of Theorem 14.11 as a Gröbner degeneration in Theorem 14.16. Show that this always works: State and prove an upper-semicontinuity result that is analogous to Theorem 8.29 but holds for sagbi degenerations.

14.16 In this exercise we describe how to get the flag variety $\mathcal{F}\ell_n$ directly from the Plücker algebra. Consider the $\mathbb{Z}^n$-grading on $\Bbbk[\mathbf{x}]$ in which the variables from row i have degree $\mathbf{e}_i$, the ith standard basis vector in $\mathbb{Z}^n$.

(a) Prove that there is an induced $\mathbb{Z}^n$-grading on $\Bbbk[\mathbf{p}]$ in which $\deg(p_\sigma) = \mathbf{e}_1 + \cdots + \mathbf{e}_d$ whenever $|\sigma| = d$.

(b) Let $\mathfrak{m}_i$ be the ideal in the Plücker algebra that is generated by the elements of degree $\mathbf{e}_1 + \cdots + \mathbf{e}_i$, and set $B = \mathfrak{m}_1 \cap \cdots \cap \mathfrak{m}_n$. Explain why B is the image in the Plücker algebra under ϕ_n of a squarefree monomial ideal in $\Bbbk[\mathbf{p}]$.

(c) Prove that the flag variety $\mathcal{F}\ell_n$ is isomorphic to the spector of the Plücker algebra with the irrelevant ideal B (Definition 10.25).

(d) How does this construction of $\mathcal{F}\ell_n$ as a spector reflect the embedding of $\mathcal{F}\ell_n$ into a product of projective spaces?

Notes

The multigraded commutative algebra of determinants, the theme of Part III, is partly motivated by representation theory. The occurring ideals and their quotients are representations of the general linear group GL_n or other semisimple algebraic groups. The occurring varieties are orbits and homogeneous spaces, such as the Grassmannian and flag variety. Whereas Hilbert series in Part II can be interpreted as generating functions for weight spaces of torus representations, Hilbert series in Part III, under larger group actions, become character formulas. Furthermore, Gröbner and sagbi bases yield decompositions of representations into weight spaces indexed by classical combinatorial objects, as in Corollary 14.9.

The Gröbner basis in Section 14.2 is known classically as the *straightening law* for minors of a matrix. The proof presented here follows Doubilet, Rota, and Stein [DRS74]. The interpretation of the straightening law as a statement about Gröbner bases was given by Sturmfels and White [SW89]. Similar results appear in greater generality in the *standard monomial theory* of Lakshmibai, Littelmann, Seshadri, and others (see [Ses95, Lak03, Mus03] for some accounts). Hibi introduced the binomial ideal in Theorem 14.16 for an arbitrary finite lattice and showed that they generate a prime ideal if and only if the lattice is distributive [Hib87]. Hence the ring $\Bbbk[\mathbf{p}]/J_n$ is also known as the *Hibi ring* of the lattice $\mathcal{P}$.

Sagbi bases for subrings of a polynomial ring were introduced by Robbiano and Sweedler [RS90]. These two authors also coined the acronym "sagbi". The sagbi basis analogous to Section 14.3 for Grassmannians was published in [Stu93, Theorem 3.2.9], whereas the case here for flag varieties appeared implicitly in work of Gonciulea and Lakshmibai [GL96]. In the meantime, generalizations, refinements, and variations on this construction have been produced by numerous authors, including Chirivì [Chi00], Caldero [Cal02], and Kogan and Miller [KoM04] (this last reference contains an explicit sagbi statement for flag varieties). A very general result to the effect that every spherical variety degenerates flatly to a toric variety was recently proved by Alexeev and Brion [AB04].

Gelfand–Tsetlin patterns were originally constructed in [GT50] to elucidate polyhedral structures in representations of the general linear group GL_n. From there, Gelfand–Tsetlin patterns lead to symplectic geometry via the Borel–Weil Theorem [Bot57] (see [DK00] for an exposition) and moment maps [GS83]. Recent work by Kogan and others has explored more intricate geometric and combinatorial aspects of Gelfand–Tsetlin patterns that relate to the *Schubert polynomials* we will study in Chapters 15 and 16 [Kog00, KoM04]. In particular, certain faces of the Gelfand–Tsetlin cone correspond to the *reduced pipe dreams* in Definition 16.2. There are non-type A analogues of Gelfand–Tsetlin patterns [Lit98a, AB04].

Of all the combinatorial objects appearing in this chapter, partitions are the most ubiquitous across the mathematical sciences. Among their appearances that relate to Plücker coordinates, the most prominent include the fact that partitions index Schubert varieties in Grassmannians as well as irreducible representations of S_n and GL_n. The tableaux in Definition 14.12 are often called semistandard *Young* tableaux, after Alfred Young [You77]. The multiplication of Young tableaux, codified in the *Littlewood–Richardson rule*, is fundamental for the cohomology ring of Grassmannians and for the multiplicities of irreducibles in tensor products of representations [Ful97]. Connections to *quiver polynomials*, via the Buch–Fulton formula in [BF99], will be discussed in the Notes to Chapter 17.

The "French" style of drawing Ferrers shapes is obtained from the English style by reflecting through a horizontal line. The "Russian" style is obtained by rotating the English style counterclockwise through $135°$; the upward corners on the top edge of the shape can be interpreted as a function on the half-integers $\frac{1}{2}\mathbb{Z}$.

In the literature on flag manifolds, the spector construction in Exercise 14.16 has sometimes been called the "multiple Proj" of the Plücker algebra.

Chapter 15

Matrix Schubert varieties

In the previous chapter, we saw how Plücker coordinates parametrize flags in vector spaces. We found that the list of Plücker coordinates is, up to scale, invariant on the set of invertible matrices mapping to a single flag. Here, we focus on larger sets of matrices, called *matrix Schubert varieties*, that map not to single points in the flag variety, but to subvarieties called *Schubert varieties*. Matrix Schubert varieties are sets of rectangular matrices satisfying certain constraints on the ranks of their submatrices. Commutative algebra enters the picture through their defining ideals, which are generated by minors in the generic rectangular matrix of variables.

This chapter and the two after it offer a self-contained introduction to determinantal ideals. Our presentation complements the existing extensive literature (see the Notes to Chapter 16) concerning both quantitative and qualitative attributes, such as dimension, degree, primality, and Cohen–Macaulayness, of varieties of matrices with rank constraints. We consider the finest possible multigrading, which demands the refined toolkit of a new generation of combinatorialists. Besides primality and Cohen–Macaulayness, the main results are that the essential minors form a Gröbner basis and that the multidegree equals a *double Schubert polynomial*.

Harvesting combinatorics from algebraic fields of study requires sowing combinatorial seeds. In our case, the seed is a partial permutation matrix from which the submatrix ranks are determined. Partial permutations lead us naturally in this chapter to the *Bruhat* and *weak orders* on the symmetric group. Part of this story is the notion of *length* for partial permutations, which is characterized in our algebraic context in terms of operations interchanging pairs of rows in matrices. These combinatorial considerations are fertilized by the geometry of Borel group orbits, on which the rank conditions are fixed. This geometry under row exchanges allows us to reap our reward: the multidegrees of matrix Schubert varieties satisfy the *divided difference* recurrence, characterizing them as double Schubert polynomials.

15.1   Schubert determinantal ideals

Throughout this chapter, $M_{k\ell}$ will denote the vector space of matrices with k rows and ℓ columns over the field $\Bbbk$, which we assume is algebraically closed, for convenience. Denote the coordinate ring of $M_{k\ell}$ by $\Bbbk[\mathbf{x}]$, where

$$\mathbf{x} = (x_{\alpha\beta} \mid \alpha = 1, \ldots, k \text{ and } \beta = 1, \ldots, \ell)$$

is a set of variables filling the *generic $k \times \ell$ matrix*. Our interests in this chapter lie with the following loci inside $\Bbbk^{m \times \ell}$.

Definition 15.1 Let $w \in M_{k\ell}$ be a **partial permutation**, meaning that w is a $k \times \ell$ matrix having all entries equal to 0 except for at most one entry equal to 1 in each row and column. The **matrix Schubert variety** $\overline{X}_w$ inside $M_{k\ell}$ is the subvariety

$$\overline{X}_w = \{Z \in M_{k\ell} \mid \operatorname{rank}(Z_{p \times q}) \leq \operatorname{rank}(w_{p \times q}) \text{ for all } p \text{ and } q\},$$

where $Z_{p \times q}$ is the upper left $p \times q$ rectangular submatrix of Z. Let $r(w)$ be the $k \times \ell$ **rank array** whose entry at (p, q) is $r_{pq}(w) = \operatorname{rank}(w_{p \times q})$.

Example 15.2 The **classical determinantal variety** is the set of all $k \times \ell$ matrices over $\Bbbk$ of rank at most r. This variety is the matrix Schubert variety $\overline{X}_w$ for the partial permutation matrix w with r nonzero entries

$$w_{11} = w_{22} = \cdots = w_{rr} = 1$$

along the diagonal, and all other entries $w_{\alpha\beta}$ equal to zero. The **classical determinantal ideal**, generated by the set of all $(r + 1) \times (r + 1)$ minors of the $k \times \ell$ matrix of variables, vanishes on this variety. In Definition 15.5 this ideal will be called the *Schubert determinantal ideal I_w* for the special partial permutation w above. We will see in Corollary 16.29 that in fact I_w is the prime ideal of $\overline{X}_w$. In Example 15.39 we show that the multidegree of this classical determinantal ideal is a *Schur polynomial*. Our results in Chapter 16 imply that the set of all $(r + 1) \times (r + 1)$ minors is a Gröbner basis and its determinantal variety is Cohen–Macaulay.

Some readers may wonder whether the machinery developed below is really the right way to prove the Gröbner basis property in this classical case. Our answer to this question is "yes": the weak order transitions in Sections 15.3–15.5 provide the steps for an elementary and self-contained proof by an induction involving *all* matrix Schubert varieties (starting from the case of a coordinate subspace in Example 15.3), and it is this induction that captures the combinatorics of determinantal ideals so richly. The combinatorics is inherent in the multigrading that is universal among those making all minors homogeneous (see Exercise 15.1), and the induction relies crucially on having a multigrading beyond the standard $\mathbb{Z}$-grading.

For the classical determinantal ideal of *maximal* minors, where $r = \min(k, \ell)$, the induction is particularly simple and explicit, as is the combinatorial multidegree formula; see Exercises 15.4, 15.5, and 15.12.   ◇

Partial permutation matrices w are sometimes called *rook placements*, because rooks placed on the 1 entries in w are not attacking one another. The number $\mathrm{rank}(w_{p\times q})$ appearing in Definition 15.1 is simply the number of 1 entries (rooks) in the northwest $p\times q$ submatrix of w. A partial permutation can be viewed as a correspondence that takes some of the integers $1,\ldots,k$ to distinct integers from $\{1,\ldots,\ell\}$. We write $w(i) = j$ if the partial permutation w has a 1 in row i and column j. This convention results from viewing matrices in $M_{k\ell}$ as acting on row vectors from the right; it is therefore transposed from the more common convention for writing permutation matrices using columns.

When w is an honest square permutation matrix of size n, so $k = n = \ell$ and there are exactly n entries of w equal to 1, then we can express w in one-line notation: the permutation $w = w_1 \ldots w_n$ of $\{1,\ldots,n\}$ sends $i \mapsto w_i$. This is not to be confused with cycle notation, where (for instance) the permutation $\sigma_i = (i, i+1)$ is the *adjacent transposition* switching i and $i+1$. The number $\mathrm{rank}(w_{p\times q})$ can alternatively be expressed as

$$r_{pq}(w) \;=\; \mathrm{rank}(w_{p\times q}) \;=\; \#\{(i,j) \le (p,q) \mid w(i) = j\}$$

for permutations w. The *symmetric group* of permutations of $\{1,\ldots,n\}$ is denoted by S_n. There is a special permutation $w_0 = n\ldots 321$ called the *long word* inside S_n, which reverses the order of $1,\ldots,n$.

Example 15.3 The variety $\overline{X}_{w_0}$ inside M_{nn} for the long word $w_0 \in S_n$ is just the linear subspace of lower-right-triangular matrices; its ideal is $\langle x_{ij} \mid i+j \le n\rangle$. As we will see in Section 15.3, this is the smallest matrix Schubert variety indexed by an honest permutation in S_n. ◇

Example 15.4 Five of the six 3×3 matrix Schubert varieties for honest permutations are linear subspaces:

$$
\begin{aligned}
I_{123} &= 0 & \overline{X}_{123} &= M_{33} \\
I_{213} &= \langle x_{11}\rangle & \overline{X}_{213} &= \{Z \in M_{33} \mid x_{11} = 0\} \\
I_{231} &= \langle x_{11}, x_{12}\rangle & \overline{X}_{231} &= \{Z \in M_{33} \mid x_{11} = x_{12} = 0\} \\
I_{231} &= \langle x_{11}, x_{21}\rangle & \overline{X}_{312} &= \{Z \in M_{33} \mid x_{11} = x_{21} = 0\} \\
I_{321} &= \langle x_{11}, x_{12}, x_{21}\rangle & \overline{X}_{321} &= \{Z \in M_{33} \mid x_{11} = x_{12} = x_{21} = 0\}
\end{aligned}
$$

The remaining permutation, $w = 132$, has matrix ⊞, so that

$$I_{132} \;=\; \langle x_{11}x_{22} - x_{12}x_{21}\rangle \qquad \overline{X}_{132} \;=\; \{Z \in M_{33} \mid \mathrm{rank}(Z_{2\times 2}) \le 1\}.$$

Thus $\overline{X}_{132}$ is the set of matrices whose upper left 2×2 block is singular. ◇

Since a matrix has rank at most r if and only if its minors of size $r + 1$ all vanish, the matrix Schubert variety $\overline{X}_w$ is the (reduced) subvariety of $\Bbbk^{k\times\ell}$ cut out by the ideal I_w defined as follows.

Definition 15.5 Let $w \in M_{k\ell}$ be a partial permutation. The **Schubert determinantal ideal** $I_w \subset \mathbb{k}[\mathbf{x}]$ is generated by all minors in $\mathbf{x}_{p \times q}$ of size $1 + r_{pq}(w)$ for all p and q, where $\mathbf{x} = (x_{\alpha\beta})$ is the $k \times \ell$ matrix of variables.

It is a nontrivial fact that Schubert determinantal ideals are prime, but we will not need it in this chapter, where we work exclusively with the zero set $\overline{X}_w$ of I_w. We therefore write $I(\overline{X}_w)$ instead of I_w when we mean the radical of I_w. Chapter 16 gives a combinatorial algebraic primality proof.

Example 15.6 Let $w = 13865742$, so that the matrix for w is given by replacing each $\times$ by 1 in the left matrix below.

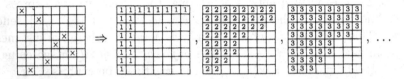

Each 8×8 matrix in $\overline{X}_w$ has the property that every rectangular submatrix contained in the region filled with 1's has rank ≤ 1, and every rectangular submatrix contained in the region filled with 2's has rank ≤ 2, and so on. The ideal I_w contains the 21 minors of size 2×2 in the first region and the 144 minors of size 3×3 in the second region. These 165 minors in fact generate I_w; see Theorem 15.15. $\Diamond$

Example 15.7 Let w be the 3×3 partial permutation matrix $\begin{smallmatrix} & 1 & \\ 1 & & \\ & & \end{smallmatrix}$. The matrix Schubert variety $\overline{X}_w$ is the set of 3×3 matrices whose upper left entry is 0, and whose determinant vanishes. The ideal I_w is

$$\left\langle x_{11}, \ \det \begin{bmatrix} x_{11} & x_{12} & x_{13} \\ x_{21} & x_{22} & x_{23} \\ x_{31} & x_{32} & x_{33} \end{bmatrix} \right\rangle.$$

The generators of I_w are the same as those of the ideal I_{2143} for the permutation in S_4 sending $1 \mapsto 2, 2 \mapsto 1, 3 \mapsto 4$ and $4 \mapsto 3$. $\Diamond$

It might seem a bit unhelpful of us to have ignored partial permutations until the very last example above, but in fact there is a general principle illustrated by Example 15.7 that gets us off the hook. Let us say that a partial permutation $\widetilde{w}$ *extends* w if the matrix $\widetilde{w}$ has northwest corner w.

Proposition 15.8 *Every partial permutation matrix w can be extended canonically to a square permutation matrix $\widetilde{w}$ whose Schubert determinantal ideal $I_{\widetilde{w}}$ has the same minimal generating minors as I_w.*

Proof. Suppose that w is not already a permutation and that by symmetry there is a row (as opposed to a column) of w that has no 1 entries. Define w'

by adding a new column and placing a 1 entry in its highest possible row. Define $\widetilde{w}$ by continuing until there is a 1 entry in every row and column.

The Schubert determinantal ideal of any partial permutation matrix extending w contains the generators of I_w by definition. Therefore it is enough—by induction on the number of rows and columns added to get $\widetilde{w}$— to show that the Schubert determinantal ideal $I_{w'}$ is contained inside I_w. The only generators of $I_{w'}$ that are not obviously in I_w are the minors of size $1 + \mathrm{rank}(w'_{p \times (\ell+1)})$ in the generic matrix $\mathbf{x}_{p \times (\ell+1)}$ for $p = 1, \ldots, k$. Now use the next lemma. $\square$

Lemma 15.9 *The ideal generated by all minors of size r in $\mathbf{x}_{p \times q}$ contains every minor of size $r + 1$ in $\mathbf{x}_{p \times (q+1)}$ or in $\mathbf{x}_{(p+1) \times q}$.*

Proof. Laplace expand each minor of size $r + 1$ along its rightmost column or bottom row, respectively. $\square$

The canonical extension $\widetilde{w}$ in Proposition 15.8 has the properties that

- no 1 entry in columns $\ell + 1, \ell + 2, \ldots$ of $\widetilde{w}$ is northeast of another, and
- no 1 entry in rows $k + 1, k + 2, \ldots$ of $\widetilde{w}$ is northeast of another.

Although $\widetilde{w}$ has size n for some fixed n, any size $n + n'$ permutation matrix extending w with these properties is a matrix for $\widetilde{w}$, viewed as lying in $S_{n+n'}$. Thus this size $n + n'$ permutation matrix is $\widetilde{w}$ plus some extra 1 entries on the main diagonal in the southeast corner. Fortunately, Schubert determinantal ideals are insensitive to the choice of n'.

Proposition 15.10 *If $w \in S_n$ and $\widetilde{w}$ extends w to an element of $S_{n+n'}$ fixing $n + 1, \ldots, n + n'$, then I_w and $I_{\widetilde{w}}$ have the same minimal generators.*

Proof. Add an extra column to w containing no 1 entries to get a partial permutation matrix w'. Since every row of w contains a 1, the "new" minors generating $I_{w'}$ are of size $1 + p$ inside $\mathbf{x}_{p \times (n+1)}$ for each p. These minors are all zero, because they do not fit inside $\mathbf{x}_{p \times (n+1)}$. (If this last statement is unconvincing, think rankwise: the rank conditions coming from the last column of w' say that the first p rows of matrices in $\overline{X}_{w'}$ have rank at most p; but this is a vacuous condition, since it is always satisfied.) Now apply Proposition 15.8 to w' and repeat n' times to get $\widetilde{w}$. $\square$

Remark 15.11 Geometrically, Propositions 15.8 and 15.10 both say that $\overline{X}_{\widetilde{w}} = \overline{X}_w \times \mathbb{k}^m$, where m is the area of the matrix $\widetilde{w}$ minus the area of w.

As a final note on definitions, let us say what matrix Schubert varieties have to do with flag varieties. Recall from Section 14.1 that every invertible matrix $\Theta \in GL_n$ determines a flag in $\mathbb{k}^n$ by its Plücker coordinates.

Definition 15.12 Let $w \in S_n$ be a permutation. The **Schubert variety** X_w in the flag variety $\mathcal{F}\ell_n$ consists of the flags determined by *invertible* matrices lying in the matrix Schubert variety $\overline{X}_w$.

15.2   Essential sets

As we have seen in Example 15.2, some of the most classical ideals in commutative algebra are certain types of Schubert determinantal ideals. To identify other special types and to reduce the number of generating minors from the set given in Definition 15.5, we use the following tools.

Definition 15.13 The **diagram** $D(w)$ of a partial permutation matrix w consists of all locations (called "boxes") in the $k \times \ell$ grid neither due south nor due east of a nonzero entry in w. The **length** of w is the cardinality $l(w)$ of its diagram $D(w)$. The **essential set** $\mathcal{E}ss(w)$ consists of the boxes (p, q) in $D(w)$ such that neither $(p, q+1)$ nor $(p+1, q)$ lies in $D(w)$.

   The diagram of w determines w up to extension as in Propositions 15.8 and 15.10. The length of w is a fundamental combinatorial invariant that will be used repeatedly, starting in the next section. The diagram can be described more graphically by crossing out all the locations due south and east of nonzero entries in w; this leaves precisely the diagram $D(w)$ remaining. The essential set $\mathcal{E}ss(w)$ consists of the "southeast corners" in $D(w)$.

Example 15.14 Consider the 8×8 square matrix for the permutation $w = 48627315$, whose 1 entries are indicated by × in the following array:

$$
\begin{array}{c|cccccccc|}
4 & \Box & \Box & \Box & \times & \cdot & \cdot & \cdot & \cdot \\
8 & \Box & \Box & \Box & \cdot & \Box & \Box & \boxed{1} & \times \\
6 & \Box & \Box & \boxed{0} & \cdot & \boxed{1} & \times & \cdot & \cdot \\
2 & \Box & \times & \cdot & \cdot & \cdot & \cdot & \cdot & \cdot \\
7 & \Box & \cdot & \boxed{1} & \cdot & \boxed{2} & \cdot & \times & \cdot \\
3 & \boxed{0} & \cdot & \times & \cdot & \cdot & \cdot & \cdot & \cdot \\
1 & \times & \cdot & \cdot & \cdot & \cdot & \cdot & \cdot & \cdot \\
5 & \cdot & \cdot & \cdot & \cdot & \times & \cdot & \cdot & \cdot \\
\end{array}
$$

Locations in the diagram of w are indicated by boxes, and its essential set consists of the subset of boxes with numbers in them. The number in the box $(p, q) \in \mathcal{E}ss(w)$ is $\mathrm{rank}(w_{p \times q})$. $\diamond$

Theorem 15.15 *The Schubert determinantal ideal $I_w \subset \Bbbk[\mathbf{x}]$ is generated by minors coming from ranks in the essential set of w:*

$$
I_w = \langle \text{minors of size } 1 + \mathrm{rank}(w_{p \times q}) \text{ in } \mathbf{x}_{p \times q} \mid (p, q) \in \mathcal{E}ss(w) \rangle.
$$

Proof. Suppose (p, q) does not lie in $\mathcal{E}ss(w)$. Either (p, q) lies outside $D(w)$, or one of the two locations $(p, q+1)$ and $(p+1, q)$ lies in $D(w)$.

   In the former case, we demonstrate that the ideal generated by the minors of size $1 + r_{pq}(w)$ in $\mathbf{x}_{p \times q}$ is contained either in the ideal generated by the minors in I_w from $\mathbf{x}_{(p-1) \times q}$ or in the ideal generated by the minors in I_w from $\mathbf{x}_{p \times (q-1)}$. Suppose by symmetry that a nonzero entry of w lies due north of (p, q). Using Lemma 15.9, the minors of size $1 + r_{pq}(w)$ in $\mathbf{x}_{p \times q}$ are stipulated by the rank condition at $(p, q-1)$. Continuing in this way, we can move north and/or west until we get to a box in $D(w)$, or to a location outside the matrix. The first possibility reduces to the case $(p, q) \in D(w)$. For the other possibility, we find that $r_{pq}(w) = \min\{p, q\}$, so there are no minors of size $1 + r_{pq}(w)$ in $\mathbf{x}_{p \times q}$.

Now we treat the case $(p, q) \in D(w)$, where we assume by symmetry that $D(w)$ has a box at $(p, q + 1)$. The rank at (p, q) equals the rank at $(p, q+1)$ in this case, so the minors of size $1 + \mathrm{rank}(w_{p \times q})$ in $\mathbf{x}_{p \times q}$ are contained (as a set) inside the set of minors of size $1 + \mathrm{rank}(w_{p \times (q+1)})$ in $\mathbf{x}_{p \times (q+1)}$. Now continue east and/or south until a box in $\mathcal{E}ss(w)$ is reached. □

Example 15.16 Let w be the partial permutation of Example 15.2, so $\overline{X}_w$ is the variety of all matrices of rank $\leq r$. Then the essential set $\mathcal{E}ss(w)$ is the singleton $\{(k, \ell)\}$ consisting of a box in the southeast corner of w. ◇

Example 15.17 Suppose the (partial) permutation w has essential set

$$\mathcal{E}ss(w) \quad = \quad \{(\alpha_1, \beta_1), \ldots, (\alpha_m, \beta_m)\}$$

in which the α's weakly decrease and the β's weakly increase:

$$\alpha_1 \geq \cdots \geq \alpha_m \quad \text{and} \quad \beta_1 \leq \cdots \leq \beta_m.$$

Thus $\mathcal{E}ss(w)$ lies along a path snaking its way east and north. Such (partial) permutations are called **vexillary**. Various classes of vexillary Schubert determinantal ideals have been objects of study in recent years, under the name "ladder determinantal ideals" (usually with prepended adjectives). ◇

15.3   Bruhat and weak orders

Given a determinantal ideal I, one can analyze I by examining the relations between its generating minors. On the other hand, a main theme of this chapter is that one can also learn a great deal by examining the relations between ideals in a combinatorially structured family containing I as a member. In our case, the set of matrix Schubert varieties inside $M_{k\ell}$ forms a poset under inclusion. The first goal of this section is to analyze this poset, to get a criterion on a permutation v and a partial permutation w for when $\overline{X}_v$ contains $\overline{X}_w$ (Proposition 15.23). Then we define a related but weaker order on the partial permutations in $M_{k\ell}$ (Definition 15.24). These partial orders will result in our being able (in Section 15.4) to calculate the dimensions of matrix Schubert varieties and to show that they are *irreducible*, meaning that the radical of I_w is a prime ideal.

Definition 15.18 For $k \times \ell$ partial permutations v and w, write $v \leq w$ and say that v precedes w in **Bruhat order** if $\overline{X}_v$ contains $\overline{X}_w$.

Thus the Bruhat partial order reverses the partial order by containment.

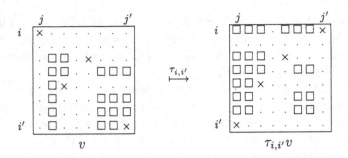

Figure 15.1: Change in length under switching rows

Lemma 15.19 *In Bruhat order, $v \leq w$ if and only if w lies in $\overline{X}_v$. In other words, $v \leq w$ if and only if $\mathrm{rank}(v_{p \times q}) \geq \mathrm{rank}(w_{p \times q})$ for all p, q.*

Proof. Clearly $v \leq w$ implies $w \in \overline{X}_v$. For the other direction, note that $w \in \overline{X}_v$ implies $\mathrm{rank}(Z_{p \times q}) \leq \mathrm{rank}(w_{p \times q}) \leq \mathrm{rank}(v_{p \times q})$ for all $Z \in \overline{X}_w$. $\square$

Remark 15.20 Lemma 15.19 can be taken as motivation to consider the difference $r(v) - r(w)$ of the rank arrays of v and w: the entry of $r(v) - r(w)$ at (p, q) is the integer $r_{pq}(v) - r_{pq}(w)$, which is nonnegative precisely if $v \leq w$. Some simple observations about $r(v) - r(w)$ will form the basis of the proof of Proposition 15.23, using the next two lemmas.

Let $\tau_{i,i'}$ be the operator switching rows i and i' in partial permutations.

Lemma 15.21 *Fix a $k \times \ell$ partial permutation matrix v with nonzero entries $v(i) = j$ and $v(i') = j'$. If $(i, j) < (i', j')$, then the following hold.*

1. *$l(\tau_{i,i'}v) = l(v) + 1 +$ twice the number of nonzero entries of v strictly inside the rectangle enclosed by (i, j) and (i', j').*
2. *$r_{pq}(\tau_{i,i'}v) = r_{pq}(v)$ unless (p, q) lies inside the rectangle enclosed by (i, j) and $(i' - 1, j' - 1)$, in which case $r_{pq}(\tau_{i,i'}v) = r_{pq}(v) - 1$.*

Proof. Outside the mentioned rectangle, the diagram stays the same after the row switch. Inside the rectangle, nothing changes except that before the switch, no boxes in the diagram lie across the top edge or down the left edge, whereas after the switch, no boxes in the diagram lie on the bottom or right edges. In the process, new boxes appear at the upper left corner, as well as above the top of and to the left of every nonzero entry inside the rectangle. Boxes already along the bottom row or right column of the rectangle before the switch move instead to the top or left. See Fig. 15.1 for an illustrative example, where the notation follows that of Example 15.14.
The claim concerning $r_{pq}(v)$ is easier and is left as an exercise. $\square$

Lemma 15.22 *Fix a $k \times \ell$ partial permutation matrix w with a nonzero entry $w(i) = j$ and a zero row i'. If $i < i'$ then the following hold.*

1. $l(\tau_{i,i'}w) = l(w) + 1 + 2e + f$, *where* e *is the number of nonzero entries of* w *inside the rectangle enclosed by* $(i+1, j+1)$ *and* $(i'-1, \ell)$, *whereas* f *is the number of zero rows of* w *indexed by* $i + 1, \ldots, i' - 1$.

2. $r_{pq}(\tau_{i,i'}w) = r_{pq}(w)$ *unless* (p, q) *lies inside the rectangle enclosed by* (i, j) *and* $(i' - 1, \ell)$, *in which case* $r_{pq}(\tau_{i,i'}w) = r_{pq}(w) - 1$.

Proof. Essentially the same as that of Lemma 15.21, except that the zero rows have no boxes in column $\ell + 1$ to move back to column j. $\square$

The following characterization of Bruhat order will be applied in the proof of Theorem 15.31.

Proposition 15.23 *Fix an* $n \times n$ *permutation* v *and an* $n \times n$ *partial permutation* w. *If* $v < w$ *in Bruhat order, then there is a partial permutation* w' *such that* $v \leq w' < w$ *and such that one of the following is satisfied:*

1. $w' = \tau_{i,i'}$ *for some transposition* $\tau_{i,i'}$.

2. w' *is obtained by changing a zero entry of* w *into a nonzero entry.*

Proof. Suppose that v agrees with w in rows $1, \ldots, i-1$ but not in row i. Set $v_i = v(i)$. Either row i of w is zero, in which case set $j = \ell+1$, or else let $j = w(i)$. By comparing rank conditions along row i, we find in both cases that $v_i < j$. Let $\mathcal{R}$ be the rectangle whose northwest corner is $(i+1, v_i)$ and whose southeast corner is $(n, j-1)$. There are two possibilities: (a) the array $r(v) - r(w)$ has an entry $r_{pq}(v) - r_{pq}(w) = 0$ in the rectangle $\mathcal{R}$, or (b) the entries of $r(v) - r(w)$ in $\mathcal{R}$ are all strictly positive.

Case (a): Among all of the zero entries of $r(v) - r(w)$ in $\mathcal{R}$, pick the one in the topmost possible row (with smallest row index p) and as far left in that row of $\mathcal{R}$ as possible (with smallest column index q). Then $\mathcal{R}$ must contain a nonzero entry of w due north of (p, q). In particular, $\mathcal{R}$ contains a nonzero entry of w in the rectangle with southwest and northeast corners (p, q) and $(i, j - 1)$. Let i' be the row index of any maximally northeast nonzero entry of w in this rectangle, and set $w' = \tau_{i,i'}w$. By Lemmas 15.21 and 15.22, $r(w') - r(w)$ is zero except in a rectangle $\mathcal{R}'$, which is filled with 1's. The corresponding rectangle $\mathcal{R}'$ is strictly positive in $r(v) - r(w)$ by construction. Hence $r(v) - r(w')$ is nonnegative, so $v \leq w' < w$.

Case (b): If $\mathcal{R}$ contains a nonzero entry of w, then choose one maximally northeast within $\mathcal{R}$ and argue as in the previous paragraph. Thus we assume that w is zero in the rectangle $\mathcal{R}$. There are two possibilities: either $j = \ell + 1$ or $j \leq \ell$. In the case $j = \ell + 1$, define w' by changing the zero in w at (i, v_i) to a nonzero entry; this creates $r(v) - r(w')$ by subtracting 1 from the entire part of $r(v) - r(w)$ that is southeast of (i, v_i). The array $r(v) - r(w')$ is nonnegative because the entire part of $r(v) - r(w)$ southeast of (and including) the entry at (i, v_i) is strictly positive.

In the case $j \leq \ell$, there must be a nonzero entry of v that is due south of (i, j), say at (p, j), since v agrees with w in rows $< i$ and v is a permutation.

If v has a nonzero entry in $\mathcal{R}$ northwest of (p, j), then let i' be the row of any maximally northwest such entry; otherwise, set $i' = p$. Lemmas 15.21 and 15.22 imply that $v < \tau_{i,i'}v \leq w$. Now argue with $\tau_{i,i'}v$ in place of v. This process of replacing v with τv must halt—either when $v = w'$ and $\tau v = w$, or when w' is found as above—because $l(\tau v) > l(v)$.   □

Proposition 15.23 says that if v is a permutation and $v < w$ in Bruhat order, then w can be obtained from v by some mixture of performing length-increasing transpositions and setting nonzero entries to zero.

The Bruhat order is important in combinatorics, geometry, and representation theory. Usually, it is applied only when both partial permutations are honest square permutations. Restricting to that case would have made the characterization of Bruhat order in Proposition 15.23 substantially simpler, as only the possibility $w' = \tau_{i,i'}w$ could occur. However, the extra generality will come in handy in the process of calculating the dimensions of matrix Schubert varieties, particularly in Lemma 15.29.

That being said, in Section 15.5 and Chapter 16 we will be even more concerned with an equally important partial order that has fewer relations than Bruhat order.

Definition 15.24 Let v and w be $k \times \ell$ partial permutations. The **adjacent transposition** σ_i for $i < k$ takes w to the result $\sigma_i w$ of switching rows i and $i+1$ of w. If $l(v) = l(w) - 1$ and either $v = \sigma_i w$ or v differs from w only in row k, then v is covered by w in **weak order**.

More generally, if $v = v_0 < v_1 < \cdots < v_{r-1} < v_r = w$ is a sequence of covers, then v precedes w in weak order; in particular, $l(v) = l(w) - r$. The operator σ_i should be thought of as simply the $k \times k$ permutation matrix for σ_i, whose only off-diagonal unit entries are at $(i, i+1)$ and $(i+1, i)$. As such, σ_i acts on all of $M_{k\ell}$, as well as on its coordinate ring $\Bbbk[\mathbf{x}] = \Bbbk[M_{k\ell}]$. Lemmas 15.21 and 15.22 imply that $l(\sigma_i w)$ must equal either $l(w) + 1$ or $l(w) - 1$, as long as rows i and $i+1$ of w are not both zero (in which case $\sigma_i w = w$). When rows i and $i+1$ of w are both nonzero, the hypothesis $\sigma_i w < w$ means that w and $\sigma_i w$ look heuristically like

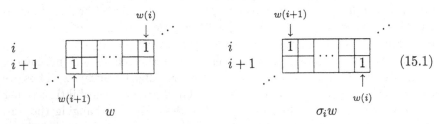

$$(15.1)$$

between columns $w(i+1)$ and $w(i)$ in rows i and $i+1$.

Remark 15.25 Weak order is usually considered only as a partial order on honest $n \times n$ permutation matrices, where the only covers are given by transposition of adjacent rows. Indeed, distinct permutation matrices must differ in at least two rows.

The other kind of cover, that of altering only row k, also has a concrete (although longer) description, through operators written with suggestive symbols $\overleftarrow{\sigma}_k$ and $\overrightarrow{\sigma}_k$. Roughly, these move nonzero entries in row k to the left and right, respectively, as little as possible. More precisely, they act on $k \times \ell$ partial permutations w with the convention that

- if w has zero last row, then $\overrightarrow{\sigma}_k$ has no effect, while $\overleftarrow{\sigma}_k$ adds a nonzero entry in the last zero column (if there is one); otherwise,
- $\overrightarrow{\sigma}_k$ moves the nonzero entry in row k of w to the next column (to the right) in which w has a zero column, unless there is no such next column, in which case $\overrightarrow{\sigma}_k$ sets the last row to zero; and
- $\overleftarrow{\sigma}_k$ moves the nonzero entry of w in its last row to the bottom of the previous zero column (if it has one).

When parenthesized "if" clauses are not satisfied, $\overleftarrow{\sigma}_k$ and $\overrightarrow{\sigma}_k$ have no effect. Note that σ_i for $i < k$ can also have no effect if rows i and $i+1$ are zero.

Example 15.26 The operators $\overrightarrow{\sigma}_7$ and $\overleftarrow{\sigma}_7$ move the bottom $\times$ as follows:

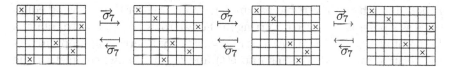

If we had chosen a tall and thin ambient rectangle, then it would be possible to have a blank last row and no blank columns (this would fail to satisfy the first of the two parenthesized "if" clauses). To understand why these operators must work the way they have been defined, draw the diagrams of these partial permutations and compare the numbers of boxes in them. ◇

We have written $v < w$ for covers in weak order because these are also relations in Bruhat order; this follows from Lemmas 15.21 and 15.22 along with an easy calculation for row k covers (by $\overleftarrow{\sigma}_k$ or $\overrightarrow{\sigma}_k$).

15.4 Borel group orbits

Matrix Schubert varieties are clearly stable under separate rescaling of each row or column. Moreover, since they only impose rank conditions on submatrices that are as far north and west as possible, any operation that adds a multiple of some row to a row below it ("sweeping downward") or that adds a multiple of some column to another column to its right ("sweeping to the right") preserves every matrix Schubert variety.

In terms of group theory, let B denote the Borel group of invertible *lower* triangular $k \times k$ matrices and B_+ the invertible upper triangular $\ell \times \ell$ matrices. (Borel groups appeared briefly in Chapter 2.) The previous paragraph says exactly that matrix Schubert varieties $\overline{X}_w$ are preserved by the action of $B \times B_+$ on $M_{k\ell}$ in which $(b, b_+) \cdot Z = bZb_+^{-1}$. This is a left group action, in the sense that $(b, b_+) \cdot ((b', b'_+) \cdot Z)$ equals $((b, b_+) \cdot (b', b'_+)) \cdot Z$ instead of $((b', b'_+) \cdot (b, b_+)) \cdot Z$, even though—in fact, because—the b_+ acts via its inverse on the *right*. We will get a lot of mileage out of the following fact, which implies that $B \times B_+$ has finitely many orbits on $M_{k\ell}$.

Proposition 15.27 *In each orbit of $B \times B_+$ on $M_{k\ell}$ lies a unique partial permutation w, and the orbit $\mathcal{O}_w$ through w is contained inside $\overline{X}_w$.*

Proof. Row and column operations that sweep down and to the right can get us from an arbitrary matrix Z to a partial permutation matrix w. Such sweeping preserves the ranks of northwest $p \times q$ submatrices. This proves uniqueness of the partial permutation w in its orbit and also shows that the minors cutting out $\overline{X}_w$ vanish on $\mathcal{O}_w$.   □

Lemma 15.28 *Set $w' = \tau_{i,i'} w$, where $i < i'$. If $w < w'$, then the closure $\overline{\mathcal{O}}_{w'}$ of the orbit through w' in $M_{k\ell}$ is properly contained inside $\overline{\mathcal{O}}_w$. The same proper inclusion of orbit closures holds for weak order covers $w < w'$.*

Proof. Let t be an invertible parameter. View each of the following equations as a possible scenario occurring in the two rows $\{i, i'\}$ and two columns $\{j, j'\}$ of an equation $b(t) \cdot w \cdot b_+(t)^{-1} = w(t)$, by inserting appropriate extra identity rows and columns into $b(t)$ and $b_+(t)^{-1}$ while completing the middle matrices to w:

$$\begin{bmatrix} 1 & 0 \\ t^{-1} & 1 \end{bmatrix} \cdot \begin{bmatrix} 1 & 0 \\ 0 & 0 \end{bmatrix} \cdot \begin{bmatrix} t & 0 \\ 0 & 1 \end{bmatrix} = \begin{bmatrix} t & 0 \\ 1 & 0 \end{bmatrix},$$

$$\begin{bmatrix} 1 & 0 \\ t^{-1} & 1 \end{bmatrix} \cdot \begin{bmatrix} 1 & 0 \\ 0 & 1 \end{bmatrix} \cdot \begin{bmatrix} t & 1 \\ 0 & -t^{-1} \end{bmatrix} = \begin{bmatrix} t & 1 \\ 1 & 0 \end{bmatrix}.$$

Each equation yields a 1-parameter family of matrices in $\mathcal{O}_w = BwB_+$. The limit at $t = 0$ is w' in both cases, as seen from the right-hand sides.

For weak order covers $w < w' = \vec{\sigma}_k w$ moving a nonzero entry in column j of row k to column j', simply "sweep w to the right" by first adding column j to column j' and then multiplying column j by t. Again, taking the limit at $t = 0$ yields w'.

By the previous paragraphs, w' lies in $\overline{\mathcal{O}}_w$. Hence we get $\overline{\mathcal{O}}_{w'} \subset \overline{\mathcal{O}}_w$ because $\overline{\mathcal{O}}_w$ is stable under the action of $B \times B_+$ and closed inside $M_{k\ell}$. The containment $\overline{\mathcal{O}}_{w'} \subset \overline{\mathcal{O}}_w$ is proper because $r_{pq}(w') < r_{pq}(w)$ for some pair (p, q), so the corresponding minors vanish on $\overline{\mathcal{O}}_{w'}$ but not on $\overline{\mathcal{O}}_w$.   □

Lemma 15.29 *Given a $k \times \ell$ partial permutation w, there exists a chain $v_0 < v_1 < v_2 < \cdots < v_{k\ell-1} < v_{k\ell}$ of covers in the weak order, in which $l(v_i) = i$ for all i, and $v_{l(w)} = w$.*

Proof. It is enough to show that if $0 < l(w)$ then there exists a cover $v < w$, and if $l(w) < k\ell$ then there exists a cover $w < v$. In the former case, choose

$v = \sigma_i w$ for the row index i on any box in the essential set $\mathcal{E}ss(w)$, using $v = \overleftarrow{\sigma}_k w$ if $i = k$. In the latter case we have $w \neq 0$; either $w(i) < w(i+1)$ for some $i < k$, in which case take $v = \sigma_i w$, or choose $v = \sigma_i w$ for the row index i on the lowest nonzero entry of w, using $v = \overrightarrow{\sigma}_k w$ if $i = k$. $\square$

Proposition 15.30 *If w is a $k \times \ell$ partial permutation, then the orbit closure $\overline{\mathcal{O}}_w$ is an irreducible variety of dimension* $\dim(\overline{\mathcal{O}}_w) = k\ell - l(w)$.

Proof. The map $B \times B_+ \to \mathcal{O}_w$ that expresses $\mathcal{O}_w$ as an orbit of $B \times B_+$ takes $(b, b_+) \mapsto bwb_+^{-1}$. This map of varieties induces a homomorphism $\Bbbk[M_{k\ell}] \to \Bbbk[B \times B_+]$ in which the target is a domain and the kernel is the ideal of $\overline{\mathcal{O}}_w$. Hence the ideal of $\overline{\mathcal{O}}_w$ is prime so $\overline{\mathcal{O}}_w$ is irreducible.

A weak order chain as in Lemma 15.29 gives a corresponding chain of prime ideals of orbits, properly containing one another, as in the second sentence of Lemma 15.28. Since the polynomial ring $\Bbbk[M_{k\ell}]$ itself has dimension $k\ell$, the part of this chain of primes consisting of those containing $I(\overline{\mathcal{O}}_w)$ must have maximal length among all chains of primes containing $I(\overline{\mathcal{O}}_w)$. Lemma 15.29 therefore implies that $\overline{\mathcal{O}}_w$ has Krull dimension $k\ell - l(w)$. $\square$

Next comes the result toward which we have been building for the last two sections. To say that w is a *smooth point* means that localizing at

$$\mathfrak{m}_w = \langle x_{\alpha\beta} \mid w(\alpha) \neq \beta \rangle + \langle x_{\alpha\beta} - 1 \mid w(\alpha) = \beta \rangle, \qquad (15.2)$$

its maximal ideal, yields a regular local ring [Eis95, Section 10.3].

Theorem 15.31 *Let w be a $k \times \ell$ partial permutation. The matrix Schubert variety $\overline{X}_w$ is the closure $\overline{\mathcal{O}}_w$ of the $B \times B_+$ orbit through $w \in M_{k\ell}$ and is irreducible of dimension $k\ell - l(w)$. The matrix w is a smooth point of $\overline{X}_w$.*

Proof. Every point on an orbit $\mathcal{O}$ of an algebraic group is a smooth point of $\mathcal{O}$, because $\mathcal{O}$ has a smooth point [Har77, Theorem I.5.3], and the group action is transitive on $\mathcal{O}$. Hence the smoothness at w follows from the rest.

Let $\widetilde{w}$ be the extension of w to a permutation as in Proposition 15.8. If the theorem holds for $\widetilde{w}$, then the irreducibility and dimension count hold for $\overline{X}_w$ by Remark 15.11; moreover, as the orbit closure $\overline{\mathcal{O}}_w$ is closed and has dimension $k\ell - l(w)$ by Proposition 15.30, the containment in Proposition 15.27 implies the whole theorem for w. Hence we assume $w = \widetilde{w}$.

The irreducibility and dimension count follow from Proposition 15.30 as soon as we show that $\overline{X}_w = \overline{\mathcal{O}}_w$. For this, Proposition 15.27 plus the stability of $\overline{X}_w$ under $B \times B_+$ imply that $\overline{X}_w$ is the union of the orbits $\mathcal{O}_{w'}$ through the partial permutations $w' \in \overline{X}_w$, so we need only show that these partial permutations all lie in $\overline{\mathcal{O}}_w$. Indeed, then we can conclude that $\overline{X}_w$ is contained in $\overline{\mathcal{O}}_w$, whence it equals $\overline{\mathcal{O}}_w$ because $\mathcal{O}_w \subseteq \overline{X}_w \subseteq \overline{\mathcal{O}}_w$.

Since $w' \in \overline{X}_w$ if and only if $w \leq w'$ by Lemma 15.19, we must show that $w \leq w'$ implies $w' \in \overline{\mathcal{O}}_w$. Proposition 15.23 says that w' is obtained from w by sequentially applying some length-increasing transpositions and setting entries to zero. The transpositions stay inside $\overline{\mathcal{O}}_w$ by Lemma 15.28. Setting entries of partial permutation matrices to zero stays inside $\overline{\mathcal{O}}_w$ because $\overline{\mathcal{O}}_w$ is stable under independent scaling of each row or column.   □

Corollary 15.32 *If $v < w$ in Bruhat order, then $l(v) < l(w)$.*

This section concludes with some results about *boundary components* of matrix Schubert varieties $\overline{X}_v$, meaning components of $\overline{X}_v \smallsetminus \mathcal{O}_v$, to be used in the proof of Proposition 15.37 (which is crucial for Theorem 15.40). Once we see that $\overline{X}_{\sigma_i w}$ for $\sigma_i w < w$ is fixed under multiplication by the permutation matrix σ_i, the subsequent lemmas say that all of its boundary components except $\overline{X}_w$ are fixed by σ_i and that the variable $x_{i+1,w(i+1)}$ maps to a generator of the maximal ideal in the local ring of $\sigma_i w$ in $\overline{X}_{\sigma_i w}$.

Corollary 15.33 *Let w be a $k \times \ell$ partial permutation and fix $i < k$. If $\sigma_i w < w$, then $\sigma_i(\overline{X}_{\sigma_i w}) = \overline{X}_{\sigma_i w}$.*

Proof. Let $B \times B_+$ act on $\sigma_i(\overline{X}_{\sigma_i w})$ and take the closure. As $\sigma_i(\overline{X}_{\sigma_i w})$ is irreducible, its image under the morphism $\mu : B \times B_+ \times \sigma_i(\overline{X}_{\sigma_i w}) \to M_{k\ell}$ is irreducible. Since $B \times B_+$ has only finitely many orbits in $M_{k\ell}$ and the image of μ is stable under the $B \times B_+$ action, the closure of the image of μ is a matrix Schubert variety. By Theorem 15.31 we need only check that its codimension is $l(\sigma_i w)$, because $\sigma_i w \in \sigma_i \overline{X}_{\sigma_i w}$ (apply σ_i to $w \in \overline{X}_{\sigma_i w}$).

Every partial permutation $w' \in \overline{X}_{\sigma_i w}$ other than $\sigma_i w$ has length greater than $l(\sigma_i w)$. Since $l(\sigma_i w')$ is at least $l(w')-1$, and $l(\sigma_i \sigma_i w) = l(w) > l(\sigma_i w)$, every partial permutation in $\sigma_i \overline{X}_{\sigma_i w}$ has length at least $l(\sigma_i w)$.   □

Remark 15.34 Corollary 15.33 is really a combinatorial statement about weak order, as the second paragraph of its proof indicates. It is equivalent to the following statement: If $w < \sigma_i w$, then $w < v$ if and only if $w < \sigma_i v$.

Lemma 15.35 *Let v be an $n \times n$ partial permutation and w an $n \times n$ permutation with $\sigma_i w < w$ and $\sigma_i w < v$. If $l(v) = l(w)$, then $\overline{X}_v$ has codimension 1 inside $\overline{X}_{\sigma_i w}$, and $\overline{X}_v$ is mapped to itself by σ_i unless $v = w$.*

Proof. The codimension statement comes from Theorem 15.31. Using Proposition 15.23, we find that v is obtained from $\sigma_i w$ either by switching a pair of rows or deleting a single nonzero entry from $\sigma_i w$.

Any 1 that we delete from $\sigma_i w$ must have no 1's southeast of it, else the length increases by more than one. Thus the 1 in row i of $\sigma_i w$ cannot be deleted, by (15.1), leaving us in the situation of Corollary 15.33 with $v = w$ and completing the case where an entry of $\sigma_i w$ has been set to zero.

Suppose now that v is obtained by switching rows p and p' of $\sigma_i w$, and assume that $\sigma_i(\overline{X}_v) \neq \overline{X}_v$. Then $v(i) > v(i+1)$ by Corollary 15.33. At least one of p and p' must lie in $\{i, i+1\}$ because moving neither row p nor row p' of $\sigma_i w$ leaves $v(i) < v(i+1)$. On the other hand, it is impossible for exactly one of p and p' to lie in $\{i, i+1\}$; indeed, since switching rows p and p' increases length, either the 1 at $(i, w(i+1))$ or the 1 at $(i+1, w(i))$ would lie in the rectangle formed by the switched 1's, making $l(v)$ too big by Lemma 15.21. Thus $\{p, p'\} = \{i, i+1\}$ and $v = w$, completing the proof. $\square$

Lemma 15.36 *Let w be an $n \times n$ permutation with $\sigma_i w < w$. If $\mathfrak{m} = \mathfrak{m}_{\sigma_i w}$ is the maximal ideal of $\sigma_i w \in \overline{X}_{\sigma_i w}$, then the variable $x_{i+1, w(i+1)}$ maps to $\mathfrak{m} \smallsetminus \mathfrak{m}^2$ under the natural map $\Bbbk[\mathbf{x}] \to (\Bbbk[\mathbf{x}]/I(\overline{X}_{\sigma_i w}))_{\mathfrak{m}}$.*

Proof. Let v be the permutation $\sigma_i w$, and consider the map $B \times B_+ \to M_{nn}$ sending $(b, b^+) \mapsto bvb^+$. The image of this map is the orbit $\mathcal{O}_v \subset \overline{X}_v$, and the identity $\mathrm{id} := (\mathrm{id}_B, \mathrm{id}_{B_+})$ maps to v. The induced map of local rings the other way thus takes $\mathfrak{m}_v$ to the maximal ideal

$$\mathfrak{m}_{\mathrm{id}} := \langle b_{ii} - 1, b_{ii}^+ - 1 \mid 1 \le i \le n \rangle + \langle b_{ij}, b_{ji}^+ \mid i > j \rangle$$

in the local ring at the identity $\mathrm{id} \in B \times B_+$. It is enough to demonstrate that the image of $x_{i+1, w(i+1)}$ lies in $\mathfrak{m}_{\mathrm{id}} \smallsetminus \mathfrak{m}_{\mathrm{id}}^2$.

Direct calculation shows that $x_{i+1, w(i+1)}$ maps to the entry

$$b_{i+1, i} b_{w(i+1), w(i+1)}^+ + \sum_{p \in P} b_{i+1, p} b_{p, w(i+1)}^+$$

at $(i+1, w(i+1))$ in bvb_+, where $P = \{p < i \mid w(p) < w(i+1)\}$ consists of the row indices of 1's in $\sigma_i w$ northwest of $(i, w(i+1))$. In particular, all of the summands $b_{i+1, p} b_{p, w(i+1)}^+$ lie in $\mathfrak{m}_{\mathrm{id}}^2$. On the other hand, $b_{w(i+1), w(i+1)}^+$ is a unit in the local ring at id, so $b_{i+1, i} b_{w(i+1), w(i+1)}^+$ lies in $\mathfrak{m}_{\mathrm{id}} \smallsetminus \mathfrak{m}_{\mathrm{id}}^2$. $\square$

Certain functions on matrix Schubert varieties are obviously nonzero. For instance, if v has its nonzero entries in rows $i_1, \ldots, i_r$ and columns $j_1, \ldots, j_r$, then the minor Δ of the generic matrix $\mathbf{x}$ using those rows and columns is nowhere zero on $\mathcal{O}_v$. Therefore the zero set of Δ inside $\overline{X}_v$ is a union of its boundary components, although the multiplicities may be more than 1. The transposition σ_i acts on the coordinate ring $\Bbbk[\mathbf{x}]$ by switching rows i and $i+1$. Therefore, if Δ uses row i, then $\sigma_i \Delta$ uses row $i+1$ instead.

Proposition 15.37 *Assume $\sigma_i w < w$, set $j = w(i) - 1$, and define Δ as the minor in $\mathbf{x}$ using all rows and columns in which $(\sigma_i w)_{i \times j}$ is nonzero. The images of Δ and $\sigma_i \Delta$ in $\Bbbk[\mathbf{x}]/I(\overline{X}_{\sigma_i w})$ have equal multiplicity along every boundary component of $\overline{X}_{\sigma_i w}$ other than $\overline{X}_w$, and Δ has multiplicity 1 along the component $\overline{X}_w$. In particular, $\sigma_i \Delta$ is not the zero function on $\overline{X}_w$.*

Proof. Lemma 15.35 says that σ_i induces an automorphism of the local ring at the prime ideal of $\overline{X}_v$ inside $\overline{X}_{\sigma_i w}$, for every boundary component $\overline{X}_v$ of $\overline{X}_{\sigma_i w}$ other than $\overline{X}_w$. This automorphism takes Δ to $\sigma_i \Delta$, so these two functions have the same multiplicity along $\overline{X}_v$. The only remaining codimension 1 boundary component of $\overline{X}_{\sigma_i w}$ is $\overline{X}_w$, and we shall now verify that Δ has multiplicity 1 there.

By Theorem 15.31, the local ring of $\sigma_i w$ in $\overline{X}_{\sigma_i w}$ is regular. Since σ_i is an automorphism of $\overline{X}_{\sigma_i w}$ (Corollary 15.33), we find that the localization of $\Bbbk[\mathbf{x}]/I(\overline{X}_{\sigma_i w})$ at the maximal ideal $\mathfrak{m}_w$ (15.2) of w is also regular. In this localization, the variables $x_{\alpha\beta}$ corresponding to the locations of nonzero entries in $w_{i \times j}$ are units. This implies that the coefficient of $x_{i,w(i+1)}$ in Δ is a unit in the local ring of $w \in \overline{X}_{\sigma_i w}$. On the other hand, the variables in spots where w has zeros generate $\mathfrak{m}_w$. Therefore, all terms of Δ lie in the square of $\mathfrak{m}_w$ in the localization, except for the unit times $x_{i,w(i+1)}$ term produced earlier. Hence, to prove multiplicity 1, it is enough to prove that $x_{i,w(i+1)}$ itself lies in $\mathfrak{m}_w \smallsetminus \mathfrak{m}_w^2$, or equivalently (after applying σ_i), that $x_{i+1,w(i+1)}$ lies in $\mathfrak{m}_{\sigma_i w} \smallsetminus \mathfrak{m}_{\sigma_i w}^2$. This is Lemma 15.36. □

15.5 Schubert polynomials

Having proved that matrix Schubert varieties are reduced and irreducible, let us begin to unravel their homologically hidden combinatorics. Working with multigradings here instead of the usual $\mathbb{Z}$-grading means that the homological invariants we seek possess algebraic structure themselves: they are polynomials, as opposed to the integers resulting in the $\mathbb{Z}$-graded case. The forthcoming definition will let us mine this algebraic structure to compare the multidegrees of all of the different matrix Schubert varieties by downward induction on weak order.

Definition 15.38 Let R be a commutative ring, and $\mathbf{t} = t_1, t_2, \ldots$ an infinite set of independent variables. The ith **divided difference** operator ∂_i takes each polynomial $f \in R[\mathbf{t}]$ to

$$\partial_i f(t_1, t_2, \ldots) = \frac{f(t_1, t_2, \ldots,) - f(t_1, \ldots, t_{i-1}, t_{i+1}, t_i, t_{i+2}, \ldots)}{t_i - t_{i+1}}.$$

Letting $\mathbf{s}$ be another set of variables and $R = \mathbb{Z}[\mathbf{s}]$, the **double Schubert polynomial** for a permutation matrix w is defined recursively by

$$\mathfrak{S}_{\sigma_i w}(\mathbf{t} - \mathbf{s}) = \partial_i \mathfrak{S}_w(\mathbf{t} - \mathbf{s})$$

whenever $\sigma_i w < w$, and the initial conditions

$$\mathfrak{S}_{w_0}(\mathbf{t} - \mathbf{s}) = \prod_{i+j \leq n} (t_i - s_j)$$

for all n, where $w_0 = n \cdots 321$ is the long word in S_n. The **(ordinary) Schubert polynomial** $\mathfrak{S}_w(\mathbf{t})$ is defined by setting $\mathbf{s} = \mathbf{0}$ everywhere. For partial permutations w, define $\mathfrak{S}_w = \mathfrak{S}_{\widetilde{w}}$ as the Schubert polynomial for the minimal extension of w to a permutation (Proposition 15.8).

Example 15.39 Let w be the partial permutation matrix in Example 15.2 with $k \leq \ell$. In this classical case, the double Schubert polynomial $\mathfrak{S}_w$ is the *Schur polynomial* associated to the partition with rectangular Ferrers shape $(k-r) \times (\ell-r)$. The *Jacobi–Trudi formula* expresses $\mathfrak{S}_w(\mathbf{t}-\mathbf{s})$ as the determinant of a Hankel matrix of size $(k-r) \times (k-r)$. The (α, β)-entry in this matrix is the coefficient of $q^{\ell-r+\beta-\alpha}$ in the generating function

$$\frac{\prod_{j=1}^{\ell}(1 - s_j q)}{\prod_{i=1}^{k}(1 - t_i q)}. \tag{15.3}$$

This formula appears in any book on symmetric functions, e.g. [Macd95]. $\diamond$

In the definition of $\mathfrak{S}_w(\mathbf{t}-\mathbf{s})$, the operator ∂_i acts only on the $\mathbf{t}$ variables and not on the $\mathbf{s}$ variables. Checking monomial by monomial verifies that $t_i - t_{i+1}$ divides the numerator of $\partial_i(f)$, so $\partial_i(f)$ is again a polynomial, homogeneous of degree $d - 1$ if f is homogeneous of degree d. Note that only finitely many variables from $\mathbf{t}$ and $\mathbf{s}$ are ever used at once. Also, setting all $\mathbf{s}$ variables to zero commutes with divided differences.

In the literature, double Schubert polynomials are usually written with $\mathbf{x}$ and $\mathbf{y}$ instead of $\mathbf{t}$ and $\mathbf{s}$; but we have used $\mathbf{x}$ throughout this book to mean coordinates on affine space, whereas $\mathbf{t}$ has been used for multidegrees.

Every $n \times n$ permutation matrix w can be expressed as a product $w = \sigma_{i_r} \cdots \sigma_{i_1} w_0$ of matrices, where the $n \times n$ matrix w_0 is the long word in S_n and $l(w_0) - l(w) = r$. The condition $l(w_0) - l(w) = r$ implies by definition that r is minimal, so $w w_0 = \sigma_{i_r} \cdots \sigma_{i_1}$ is what is known as a *reduced expression* for the permutation matrix $w w_0$. The recursion for both single and double Schubert polynomials can be summarized as $\mathfrak{S}_w = \partial_{i_r} \cdots \partial_{i_1} \mathfrak{S}_{w_0}$. More generally, if $w = \sigma_{i_r} \cdots \sigma_{i_1} v$ and $l(w) = l(v) - r$, then it holds that

$$\mathfrak{S}_w = \partial_{i_r} \cdots \partial_{i_1} \mathfrak{S}_v. \tag{15.4}$$

Indeed, this reduces to the case where $v = w_0$ by writing $\mathfrak{S}_v = \partial_{j_s} \cdots \partial_{j_1} w_0$.

It is not immediately obvious from Definition 15.38 that $\mathfrak{S}_w$ is well-defined, because we could have used any downward chain of covers in weak order to define $\mathfrak{S}_w$ from $\mathfrak{S}_{w_0}$. However, the well-definedness will follow from our main theorem in this chapter, Theorem 15.40. It is also a consequence of the fact that divided differences satisfy the braid relations in Exercise 15.3, which the reader is encouraged to check directly.

We are interested in a multigrading of $\Bbbk[\mathbf{x}]$ by $\mathbb{Z}^{k+\ell}$, which we take to have basis $\mathbf{t} \cup \mathbf{s}$, where $\mathbf{t} = t_1, \ldots, t_k$ and $\mathbf{s} = s_1, \ldots, s_\ell$.

Theorem 15.40 *If w is a $k \times \ell$ partial permutation and $\Bbbk[\mathbf{x}]$ is $\mathbb{Z}^{k+\ell}$-graded with $\deg(x_{ij}) = t_i - s_j$, then the matrix Schubert variety $\overline{X}_w$ has multidegree*

$$\mathcal{C}(\overline{X}_w; \mathbf{t}, \mathbf{s}) = \mathfrak{S}_w(\mathbf{t} - \mathbf{s})$$

equal to the double Schubert polynomial for w.

Example 15.41 The multidegree of the classical determinantal variety $\overline{X}_w$ in Example 15.2 equals the Schur polynomial in Example 15.39. Replacing every t_i by t and every s_j by 0 yields the classical degree of that projective variety. This substitution replaces (15.3) by $1/(1-tq)^k$, and the (α, β)-entry of the Jacobi matrix specializes to $\binom{k+\ell-r+\beta-\alpha-1}{k-1}t^{\ell-r+\beta-\alpha}$. The determinant of this matrix (and hence the classical degree of $\overline{X}_w$) equals the number of semistandard Young tableaux of rectangular shape $(k-r) \times (\ell-r)$. This statement holds more generally for the matrix Schubert varieties associated with Grassmannians; see Exercise 16.9.   ◊

   The proof of Theorem 15.40 will compare the zero sets of two functions on $\overline{X}_{\sigma_i w} \times \Bbbk$ with equal degrees. The zeros of the first function consist of $\overline{X}_w \times \Bbbk$ plus some boundary components, whereas the second function has zeros $(\sigma_i \overline{X}_w \times \Bbbk) \cup (\overline{X}_{\sigma_i w} \times \{0\})$ plus the *same* boundary components. When the (equal) multidegrees of the zero sets of our two functions are decomposed by additivity and compared, the extra components cancel.

Proof of Theorem 15.40. As the matrix Schubert varieties for w and its minimal completion to a permutation have equal multidegrees by Proposition 15.8, we assume that w is a permutation. The result for $\mathfrak{S}_{w_0}$ follows immediately from Proposition 8.49 and Example 15.3. For other permutations w we shall use downward induction on weak order.
   Consider the polynomials Δ and $\sigma_i \Delta$ from Proposition 15.37 not as elements in $\Bbbk[\mathbf{x}]$, but as elements in the polynomial ring $\Bbbk[\mathbf{x}, y]$ with $k\ell + 1$ variables. Setting the degree of the new variable y equal to $\deg(y) = t_i - t_{i+1}$ makes Δ and the product $y\sigma_i \Delta$ in $\Bbbk[\mathbf{x}, y]$ have the *same* degree $\delta \in \mathbb{Z}^{k+\ell}$. Since the affine coordinate ring $\Bbbk[\mathbf{x}]/I(\overline{X}_{\sigma_i w})$ of $\overline{X}_{\sigma_i w}$ is a domain, neither Δ nor $\sigma_i \Delta$ vanishes on $\overline{X}_{\sigma_i w}$, so we get two short exact sequences

$$0 \to \Bbbk[\mathbf{x}, y](-\delta)/I(\overline{X}_{\sigma_i w}) \xrightarrow{\Theta} \Bbbk[\mathbf{x}, y]/I(\overline{X}_{\sigma_i w}) \longrightarrow Q(\Theta) \to 0,$$

in which Θ equals either Δ or $y\sigma_i \Delta$. The quotients $Q(\Delta)$ and $Q(y\sigma_i \Delta)$ have equal $\mathbb{Z}^{k+\ell}$-graded K-polynomials and hence equal multidegrees.
   Note that $\Bbbk[\mathbf{x}, y]$ is the coordinate ring of $M_{k\ell} \times \Bbbk$. The minimal primes of $Q(\Delta)$ all correspond to varieties $\overline{X}_v \times \Bbbk$ for boundary components $\overline{X}_v$ of $\overline{X}_{\sigma_i w}$. Similarly, almost all minimal primes of $Q(y\sigma_i \Delta)$ correspond by Proposition 15.37 to varieties $\overline{X}_v \times \Bbbk$. The only exceptions are $\overline{X}_{\sigma_i w} \times \{0\}$, because of the factor y, and the image $\sigma_i \overline{X}_w \times \Bbbk$ of $\overline{X}_w \times \Bbbk$ under the automorphism σ_i. As a consequence of Proposition 15.37, the multiplicity of $y\sigma_i \Delta$ along $\sigma_i \overline{X}_w \times \Bbbk$ equals 1, just as Δ has multiplicity 1 along $\overline{X}_w \times \Bbbk$.
   Now break the multidegrees of $Q(\Delta)$ and $Q(y\sigma_i \Delta)$ into sums over top-dimension components by additivity (Theorem 8.53). Proposition 15.37 implies that almost all terms in the equation $\mathcal{C}(Q(\Delta); \mathbf{t}, \mathbf{s}) = \mathcal{C}(Q(y\sigma_i \Delta); \mathbf{t}, \mathbf{s})$ cancel; the only terms that remain yield the equation

$$\mathcal{C}(\overline{X}_w \times \Bbbk; \mathbf{t}, \mathbf{s}) = \mathcal{C}(\sigma_i \overline{X}_w \times \Bbbk; \mathbf{t}, \mathbf{s}) + \mathcal{C}(\overline{X}_{\sigma_i w} \times \{0\}; \mathbf{t}, \mathbf{s}) \quad (15.5)$$

on multidegrees. Since the equations in $\Bbbk[\mathbf{x}, y]$ for $\overline{X}_w \times \Bbbk$ are the same as those for $\overline{X}_w$ in $\Bbbk[\mathbf{x}]$, the K-polynomials of $\overline{X}_w$ and $\overline{X}_w \times \Bbbk$ agree. Hence the multidegree on the left-hand side of (15.5) equals $\mathcal{C}(\overline{X}_w; \mathbf{t}, \mathbf{s})$. For the same reason, the first multidegree on the right-hand side of (15.5) equals the result $\sigma_i \mathcal{C}(\overline{X}_w; \mathbf{t}, \mathbf{s})$ of switching t_i and t_{i+1} in the multidegree of $\overline{X}_w$.

The equations defining $\overline{X}_{\sigma_i w} \times \{0\}$, on the other hand, are those defining $\overline{X}_{\sigma_i w}$ along with the equation $y = 0$. The K-polynomial of $\overline{X}_{\sigma_i w} \times \{0\}$ therefore equals $(t_i/t_{i+1})\mathcal{K}(\overline{X}_{\sigma_i w}; \mathbf{t}, \mathbf{s})$, which is the "exponential weight" t_i/t_{i+1} of y times the K-polynomial of $\overline{X}_{\sigma_i w}$. Therefore the second multidegree on the right-hand side of (15.5) equals $(t_i - t_{i+1})\mathcal{C}(\overline{X}_{\sigma_i w}; \mathbf{t}, \mathbf{s})$.

Substituting these multidegree calculations into (15.5), we find that

$$\mathcal{C}(\overline{X}_w; \mathbf{t}, \mathbf{s}) = \sigma_i \mathcal{C}(\overline{X}_w; \mathbf{t}, \mathbf{s}) + (t_i - t_{i+1})\mathcal{C}(\overline{X}_{\sigma_i w}; \mathbf{t}, \mathbf{s})$$

as polynomials in $\mathbf{t}$ and $\mathbf{s}$. Subtracting $\sigma_i \mathcal{C}(\overline{X}_w; \mathbf{t}, \mathbf{s})$ from both sides and dividing through by $t_i - t_{i+1}$ yields $\partial_i \mathcal{C}(\overline{X}_w; \mathbf{t}, \mathbf{s}) = \mathcal{C}(\overline{X}_{\sigma_i w}; \mathbf{t}, \mathbf{s})$. $\square$

Example 15.42 The first five of the six 3×3 matrix Schubert varieties in Example 15.4 have $\mathbb{Z}^{3+3}$-graded multidegrees that are products of expressions having the form $t_i - s_j$ by Proposition 8.49. They are, in the order they appear in Example 15.4: $1, t_1 - s_1, (t_1 - s_1)(t_1 - s_2), (t_1 - s_1)(t_2 - s_1)$, and $(t_1 - s_1)(t_1 - s_2)(t_2 - s_1)$. This last one is $\mathcal{C}(\overline{X}_{321}; \mathbf{t}, \mathbf{s})$, and applying $\partial_2 \partial_1$ to it yields the multidegree

$$\mathcal{C}(\overline{X}_{132}; \mathbf{t}, \mathbf{s}) = t_1 + t_2 - s_1 - s_2$$

of $\overline{X}_{132}$, as the reader should check. $\diamond$

Example 15.43 The ideal I_{2143} from Example 15.7 equals $I(\overline{X}_{2143})$, since it has a squarefree initial ideal $\langle x_{11}, x_{13}x_{22}x_{31}\rangle$ and is therefore a radical ideal. The multidegree of $\overline{X}_{2143}$ is the double Schubert polynomial

$$\begin{aligned}
&\mathfrak{S}_{2143}(\mathbf{t} - \mathbf{s}) \\
&= \partial_2\partial_1\partial_3\partial_2\big((t_1 - s_3)(t_1 - s_2)(t_1 - s_1)(t_2 - s_2)(t_2 - s_1)(t_3 - s_1)\big) \\
&= \partial_2\partial_1\partial_3\big((t_1 - s_3)(t_1 - s_2)(t_1 - s_1)(t_2 - s_1)(t_3 - s_1)\big) \\
&= \partial_2\partial_1\big((t_1 - s_3)(t_1 - s_2)(t_1 - s_1)(t_2 - s_1)\big) \\
&= \partial_2\big((t_1 - s_1)(t_2 - s_1)(t_1 + t_2 - s_2 - s_3)\big) \\
&= (t_1 - s_1)(t_1 + t_2 + t_3 - s_1 - s_2 - s_3).
\end{aligned}$$

Compare this to the multidegree of $\Bbbk[\mathbf{x}_{4\times 4}]/\langle x_{11}, x_{13}x_{22}x_{31}\rangle$. $\diamond$

Setting $\mathbf{s} = \mathbf{0}$ in Theorem 15.40 yields the "ordinary" version.

Corollary 15.44 *If w is a $k \times \ell$ partial permutation and $\Bbbk[\mathbf{x}]$ is $\mathbb{Z}^k$-graded with $\deg(x_{ij}) = t_i$, then the multidegree of the matrix Schubert variety $\overline{X}_w$ equals the ordinary Schubert polynomial for w: $\mathcal{C}(\overline{X}_w; \mathbf{t}) = \mathfrak{S}_w(\mathbf{t})$.*

Remark 15.45 The K-polynomials of matrix Schubert varieties satisfy similarly nice recursions under the so-called *isobaric divided differences* (or *Demazure operators*) $f \mapsto -\partial_i(t_{i+1}f)$; see the Notes to this chapter.

Exercises

15.1 Prove that the unique finest multigrading on $\Bbbk[\mathbf{x}]$ in which all Schubert determinantal ideals I_w are homogeneous is the $\mathbb{Z}^{k+l}$-grading here. Prove that this multigrading is also universal for the set of classical determinantal ideals.

15.2 Express the ideal I in Exercise 8.5 as an ideal of the form $I(\overline{X}_w)$. Compute the multidegree of $\Bbbk[\mathbf{x}]/I$ for the $\mathbb{Z}^{4+4}$-grading $\deg(x_{ij}) = t_i - s_j$, and show that it specializes to the $\mathbb{Z}^4$-graded multidegrees you computed in Exercise 8.5.

15.3 Verify that divided difference operators ∂_i satisfy the relations $\partial_i \partial_j = \partial_j \partial_i$ for $|i - j| \geq 2$ and the **braid relations**, which say that $\partial_i \partial_{i+1} \partial_i = \partial_{i+1} \partial_i \partial_{i+1}$.

15.4 Using cycle notation, let $v = (n \cdots 321)$ be the permutation cycling $n, \ldots, 1$.

(a) Write down generators for the Schubert determinantal ideal I_v.

(b) Calculate that $\mathfrak{S}_v(\mathbf{t}) = t_1^{n-1}$. Hint: Don't use divided differences.

15.5 Let I be the ideal of maximal minors in the generic $k \times l$ matrix, where $k \leq l$, and let X be the zero set of I in M_{kl}, so X consists of the singular $k \times l$ matrices.

(a) Prove that I has the same minimal generators as I_w for the $(l+1) \times (l+1)$ permutation $w = \sigma_k \cdots \sigma_2 \sigma_1 v$, for v as in Exercise 15.4 with $n = l + 1$.

(b) Deduce using Eq. (15.4) that $\mathcal{C}(X; \mathbf{t}) = h_{l+1-k}(t_1, \ldots, t_k)$ is the complete homogeneous symmetric function of degree $l + 1 - k$ in k variables.

(c) Conclude that X has ordinary $\mathbb{Z}$-graded degree $\binom{l}{k-1}$.

15.6 An $n \times n$ permutation w is **Grassmannian** if it has at most one **descent**— that is, if $w(k) > w(k+1)$ for at most one value of $k < n$. Show that a permutation is Grassmannian with descent at k if and only if its essential set lies along row k. Describe the Schubert determinantal ideals for Grassmannian permutations.

15.7 Consider positive integers $i_1 < \cdots < i_m \leq k$ and $j_1 < \cdots < j_m \leq l$, and let $\mathbf{x}$ be the $k \times l$ matrix of variables. Find a partial permutation w such that I_w is generated by the size $m + 1$ minors of $\mathbf{x}$ along with the union over $r = 1, \ldots, m$ of the minors of size r in the top $i_r - 1$ rows of $\mathbf{x}$ and the minors of size r in the left $j_r - 1$ columns of $\mathbf{x}$. Compute the extension of w to a permutation in S_{k+l}.

15.8 Prove that the Bruhat poset is a graded poset, with rank function $w \mapsto l(w)$.

15.9 Write down explicitly the degree δ in the proof of Theorem 15.40.

15.10 Let w be a permutation matrix. Show that $\mathfrak{S}_{w^{-1}}(\mathbf{s} - \mathbf{t})$ can be expressed as $(-1)^{l(w)} \mathfrak{S}_w((-\mathbf{t}) - (-\mathbf{s}))$. In other words, $\mathfrak{S}_{w^{-1}}(\mathbf{s} - \mathbf{t})$ is obtained by substituting each variable with its negative in the argument of $(-1)^{l(w)} \mathfrak{S}_w(\mathbf{t} - \mathbf{s})$. Hint: Consider the rank conditions transpose to those determined by w.

15.11 Consider divided difference operators ∂_i' that act only on $\mathbf{s}$ variables instead of on $\mathbf{t}$ variables. Deduce from Theorem 15.40 applied to the transpose of w that $\mathfrak{S}_w(\mathbf{t} - \mathbf{s})$ can be obtained (with a global sign factor of $(-1)^{l(w)}$) from $\mathfrak{S}_{w_0}(\mathbf{t} - \mathbf{s})$ by using the divided differences ∂_i' in the $\mathbf{s}$ variables.

15.12 As in Exercise 15.5, let X be the variety of singular $k \times l$ matrices, where we assume $k \leq l$. This time, though, use the multigrading of $\Bbbk[\mathbf{x}]$ by $\mathbb{Z}^l$ in which $\deg(x_{ij}) = s_j$. Prove that $\mathcal{C}(X; \mathbf{s}) = e_{l+1-k}(s_1, \ldots, s_l)$ is an elementary symmetric function, and conclude again that X has $\mathbb{Z}$-graded degree $\binom{l}{k-1}$.

15.13 Let f and g be polynomials in $R(t_1, \ldots, t_n)$ over a commutative ring R.

(a) Prove that if f is symmetric in t_i and t_{i+1}, then $\partial_i(fg) = f\partial_i(g)$.

(b) Deduce that f is symmetric in t_i and t_{i+1} if and only if $\partial_i f = 0$.

(c) Show that $\partial_i f$ is symmetric in t_i and t_{i+1}.

(d) Conclude that $\partial_i^2 = \partial_i \circ \partial_i$ is the zero operator, so $\partial_i^2 f = 0$ for all f.

15.14 For a permutation w, let $m+w$ be the result of letting w act in the obvious way on $m+1, m+2, m+3, \ldots$ instead of $1, 2, 3, \ldots$, so $m+w$ fixes $1, \ldots, m$. Show that $\mathfrak{S}_{m+w}(\mathbf{t} - \mathbf{s})$ is symmetric in $t_1, \ldots, t_m$ as well as (separately) in $s_1, \ldots, s_m$.

Notes

The class of determinantal ideals in Definition 15.1 was identified by Fulton in [Ful92], which is also where the essential set, Example 15.14, and the characterization of vexillary permutations in Example 15.17 come from. A permutation is vexillary precisely when it is "2143-avoiding". Treatments of various aspects of vexillary (a.k.a. ladder determinantal) ideals include [Mul89, HT92, Ful92, Con95, MS96, CH97, GL97, KP99, BL00, GL00, GM00], and much more can be found by looking at the articles cited in the references to these.

Proposition 15.23, applied in the case where both v and w are permutations, is a characterization of Bruhat order on the symmetric group. As in Remark 15.25, our weak order on partial permutations restricts to the standard definition of weak order on the symmetric group. For readers wishing to see the various characterizations of Bruhat and weak order, their generalizations to other Coxeter groups, and further areas where they arise, we suggest [Hum90] and [BB04].

Schubert polynomials were invented by Lascoux and Schützenberger [LS82a], based on general notions of divided differences developed by Bernstein–Gelfand–Gelfand [BGG73] and Demazure [Dem74]. Their purpose was to isolate representatives for the cohomology classes of Schubert varieties (Definition 15.12) that are polynomials with desirable algebraic and combinatorial properties, some of which we will see in Chapter 16. Our indexing of Schubert polynomials is standard, but paradoxically, it is common practice to index Schubert *varieties* backward from Definition 15.12, replacing w with w_0w. For an introduction (beyond Chapter 16) to the algebra, combinatorics, and geometry of Schubert polynomials related to flag varieties, we recommend [Man01]. Other sources include [Macd91] for a more algebraic perspective and [FP98] for a more global geometric perspective.

The characterization of Schubert polynomials as multidegrees of matrix Schubert varieties in Theorem 15.40 is due to Knutson and Miller [KnM04b, Theorem A]. The original motivation was to geometrically explain the desirable algebraic and combinatorial properties of Schubert polynomials. Theorem 15.40 can be viewed as a statement in the equivariant Chow group of $M_{k\ell}$ [Tot99, EG98]. It is essentially equivalent to the main theorem of [Ful92] expressing double Schubert polynomials as classes of certain degeneracy loci for vector bundle morphisms.

Remark 15.45 means that the K-polynomials of matrix Schubert varieties are the *Grothendieck polynomials* of Lascoux and Schützenberger [LS82b]. The proof of this statement in [KnM04b] does not rely on theory more general than what appears in Chapters 15 and 16, though it does require more intricate combinatorics. Viewing the K-polynomial statement as taking place in the equivariant K-theory of $M_{k\ell}$, it is essentially equivalent to a theorem of Buch [Buc02, Theorem 2.1].

The $\mathbb{Z}$-graded result of Exercises 15.5 and 15.12, which follows from work of Giambelli [Gia04], is the most classical of all. The method of starting from Exercise 15.4 and using divided differences to prove Exercise 15.5 by induction on k demonstrates the utility of replacing the integer $\mathbb{Z}$-graded degree with a polynomial multidegree. The more finely graded statements in both of these two exercises are special cases of Exercise 16.9 in Chapter 16. The determinantal ideals described in Exercise 15.7 constitute the class of ideals *cogenerated by a minor* discussed in [HT92].

We have more references and comments to make on Schubert polynomials and determinantal ideals, but we postpone them until the Notes to Chapter 16.

Chapter 16

Antidiagonal initial ideals

Schubert polynomials have integer coefficients. This, at least, is clear from the algebraic recursion via divided differences in Section 15.5, where we also saw the geometric expression of Schubert polynomials as multidegrees. In contrast, this chapter explores the combinatorial properties of Schubert polynomials, particularly why their integer coefficients are positive.

One of our main goals is to illustrate the combinatorial importance of Gröbner bases and their geometric interpretation. Suppose a polynomial is expressed as the multidegree of some variety. Gröbner degeneration of that variety yields pieces whose multidegrees add up to the given polynomial. This process can provide geometric explanations for positive combinatorial formulas. The example pervading this chapter comes from Theorem 15.40:

Corollary 16.1 *Schubert polynomials have nonnegative coefficients.*

Proof. Write $\mathfrak{S}_w(\mathbf{t}) = \mathcal{C}(\overline{X}_w; \mathbf{t})$ as in Theorem 15.40. Choosing a term order on $\Bbbk[\mathbf{x}]$, Corollary 8.47 implies that $\mathfrak{S}_w(\mathbf{t}) = \mathcal{C}(\Bbbk[\mathbf{x}]/\mathrm{in}(I(\overline{X}_w)); \mathbf{t})$. Now use Theorem 8.53 to write $\mathfrak{S}_w(\mathbf{t})$ as a positive sum of multidegrees of quotients $\Bbbk[\mathbf{x}]/\langle x_{i_1,j_1}, \ldots, x_{i_r,j_r} \rangle$ by monomial primes. Proposition 8.49 says that the multidegree of this quotient of $\Bbbk[\mathbf{x}]$ is the monomial $t_{i_1} \cdots t_{i_r}$. $\square$

The *existence* of a Gröbner basis proves the positivity in Corollary 16.1. After choosing an especially nice term order, our efforts in this chapter will identify the prime components of the initial ideal explicitly as combinatorial diagrams called *reduced pipe dreams*. Hence adding up monomials corresponding to reduced pipe dreams yields Schubert polynomials.

This positive formula will be our motivation for a combinatorial study of reduced pipe dreams, in terms of reduced expressions in the permutation group S_n. Applications include the primality of Schubert determinantal ideals, and the fact that matrix Schubert varieties are Cohen–Macaulay.

16.1   Pipe dreams

Combinatorics of Schubert polynomials—and as it will turn out in Section 16.4, of Schubert determinantal ideals—is governed by certain "drawings" of (partial) permutations. Consider a $k \times \ell$ grid of squares, with the box in row i and column j labeled (i, j), as in a $k \times \ell$ matrix. If each box in the grid is covered with a square tile containing either $+$ or $\text{-}\!\!\!\text{r}$, then the tiled grid looks like a network of pipes. Each such tiling corresponds to a subset of the $k \times \ell$ rectangle, namely the set of its crossing tiles:

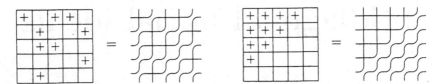

We omit the square tile boundaries in the right-hand versions.

Definition 16.2 A $k \times \ell$ **pipe dream** is a tiling of the $k \times \ell$ rectangle by **crosses** $+$ and **elbow joints** $\text{-}\!\!\!\text{r}$. A pipe dream is **reduced** if each pair of pipes crosses at most once. The set $\mathcal{RP}(w)$ of reduced pipe dreams for a $k \times \ell$ partial permutation w consists of those pipe dreams D with $l(w)$ crossing tiles such that the pipe entering row i exits from column $w(i)$.

Example 16.3 The long permutation $w_0 = n \ldots 321$ in S_n has a unique $n \times n$ reduced pipe dream D_0, whose $+$ tiles fill the region strictly above the main antidiagonal, in spots (i, j) with $i + j \leq n$. The right-hand pipe dream displayed before Definition 16.2 is D_0 for $n = 5$. $\diamond$

Example 16.4 The permutation $w = 2143$ has three reduced pipe dreams:

$$\mathcal{RP}(2143) \;=\; \left\{ \begin{matrix} & \end{matrix} \right\}.$$

The permutation is written down the left edge of each pipe dream; thus each row is labeled with the destination of its pipe. Reduced pipe dreams for permutations are contained in D_0 (Exercise 16.1), so the crossing tiles only occur strictly above the main antidiagonal. Therefore we omit the wavy "sea" of elbow pipes below the main antidiagonal. $\diamond$

If $\widetilde{w}$ is the minimal-length extension of a $k \times \ell$ partial permutation w to an $n \times n$ permutation, then $\mathcal{RP}(w)$ is the set of $k \times \ell$ pipe dreams to which adding elbow tiles in the region $(n \times n) \smallsetminus (k \times \ell)$ yields a reduced pipe dream for $\widetilde{w}$. In other words, $\mathcal{RP}(w) = \mathcal{RP}(\widetilde{w})_{k \times \ell}$ consists of the restrictions to the northwest $k \times \ell$ rectangle of reduced pipe dreams for $\widetilde{w}$ (Exercise 16.2). Pipes can exit out of the east side of a pipe dream $D \in \mathcal{RP}(w)$, rather than out the top; when w is zero in row i, for example, this holds for the pipe entering row i.

Although we always draw crossing tiles as some sort of cross (either "+" or "┼", the former with square tile boundary and the latter without), we often leave elbow tiles blank or denote them by dots, to minimize clutter.

Here is an easy criterion, to be used in Theorem 16.11, for when removing a ┼ from a pipe dream $D \in \mathcal{RP}(w)$ leaves a pipe dream in $\mathcal{RP}(\sigma_i w)$.

Lemma 16.5 *Suppose that $D \in \mathcal{RP}(w)$, and let j be a fixed column index with $(i+1, j) \notin D$, but $(i, p) \in D$ for all $p \leq j$, and $(i+1, p) \in D$ for all $p < j$. Then $l(\sigma_i w) < l(w)$, and if $D' = D \smallsetminus (i, j)$, then $D' \in \mathcal{RP}(\sigma_i w)$.*

The hypotheses of the lemma say precisely that D looks like

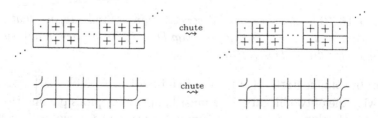

at the left end of rows i and $i+1$ in D, and the ┼ to be deleted sits at (i, j).

Proof. Removing (i, j) only switches the exit points of the two pipes starting in rows i and $i+1$. Thus the pipe starting in row p of D' exits out of column $\sigma_i w(p)$ for every row index p. No pair of pipes can cross twice in D' because there are $l(\sigma_i w) = l(w) - 1$ crossings.   □

Definition 16.6 A **chutable rectangle** is a $2 \times r$ block C of ┼ and ╶╮ tiles such that $r \geq 2$, and the only elbows in C are its northwest, southwest, and southeast corners. Applying a **chute move** to a pipe dream D is accomplished by placing a ┼ in the southwest corner of a chutable rectangle $C \subseteq D$ and removing the ┼ from the northeast corner of the same C.

Heuristically, chuting looks like the following:

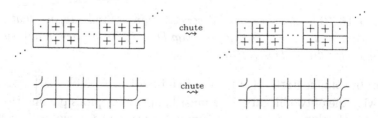

Lemma 16.7 *Chuting $D \in \mathcal{RP}(w)$ yields another reduced pipe dream for w.*

Proof. If two pipes intersect at the ┼ in the northeast corner of a chutable rectangle C, then chuting that ┼ only relocates the crossing point of those two pipes to the southwest corner of C. No other pipes are affected.   □

The rest of this section is devoted to a procedure generating all reduced pipe dreams. Given a $k \times \ell$ pipe dream D, row i of D is filled solidly with $+$ tiles until the first elbow tile (or until the end of the row). In what follows, we need a notation for the column index of this first elbow:

$$\text{start}_i(D) \quad = \quad \min\left(\{j \mid (i,j) \text{ is an elbow tile in } D\} \cup \{k+1\}\right).$$

Definition 16.8 Let D be a pipe dream, and fix a row index i. Suppose there is a smallest column index j such that $(i+1,j)$ is an elbow tile but (i,p) is a $+$ tile in D for all $p \leq j$. Construct the mth **offspring** of D by

1. removing (i,j), and then
2. performing $m-1$ chute moves west of $\text{start}_i(D)$ from row i to row $i+1$.

The ith **mitosis** operator sends a pipe dream $D \in \mathcal{RP}(w)$ to the set $\text{mitosis}_i(D)$ of its offspring. Write $\text{mitosis}_i(\mathcal{P}) = \bigcup_{D \in \mathcal{P}} \text{mitosis}_i(D)$ whenever $\mathcal{P}$ is a set of pipe dreams.

The total number of offspring is the number of ⊞ configurations in rows i and $i+1$ that are west of $\text{start}_i(D)$. This number, which is allowed to equal zero (so D is "barren"), equals 3 in the next example.

Example 16.9 The pipe dream D at left is a reduced pipe dream for $w = 13865742$. Applying mitosis_3 yields the indicated set of pipe dreams:

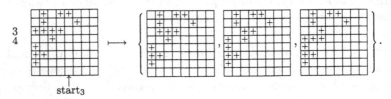

The three offspring on the right are listed in the order they are produced by successive chute moves.   ◇

Mitosis can be reversed. Equivalently, "Parentage can be determined."

Lemma 16.10 *Fix a $k \times \ell$ partial permutation w, and suppose that $i < k$ satisfies $\sigma_i w < w$. Then every pipe dream $D' \in \mathcal{RP}(\sigma_i w)$ lies in $\text{mitosis}_i(D)$ for some pipe dream $D \in \mathcal{RP}(w)$.*

Proof. In column $\text{start}_{i+1}(D')$, rows i and $i+1$ in D' look like ⊟, because otherwise one of two illegal things must happen: the pipes passing through the row i of column start_{i+1} in D' intersect again at the closest ⊟ column to the left in rows i and $i+1$ of D', or the pipe entering row $i+1$ of D' crosses the pipe entering row i of D'. This latter occurrence is illegal because $\sigma_i w < w$, so $\sigma_i w$ has no descent at i.

Consequently, we can perform a sequence of inverse chute moves on D', the first one with its northeast corner at $(i, \text{start}_{i+1}(D'))$, and the last with its west end immediately east of the solid $\boxplus$ part of rows i and $i+1$. These chute moves preserve the property of being in $\mathcal{RP}(\sigma_i w)$ by Lemma 16.7. Now adding the $+$ into row i of the last vacated column yields a pipe dream D whose pipes go to the correct destinations to be in $\mathcal{RP}(w)$ (see Lemma 16.5). That D is reduced follows because it has $l(w)$ crossing tiles. That $D' \in \mathrm{mitosis}_i(D)$ is by construction. $\square$

Theorem 16.11 *If w is a $k \times \ell$ partial permutation and $i < k$ is a row index that satisfies $\sigma_i w < w$, then the set of reduced pipe dreams for $\sigma_i w$ is the disjoint union $\bigcup_{D \in \mathcal{RP}(w)} \mathrm{mitosis}_i(D)$.*

Proof. Lemmas 16.5 and 16.7 imply that $\mathrm{mitosis}_i(D) \subseteq \mathcal{RP}(\sigma_i w)$ whenever $D \in \mathcal{RP}(w)$, and Lemma 16.10 gives the reverse containment. That the union is disjoint (i.e., that $\mathrm{mitosis}_i(D) \cap \mathrm{mitosis}_i(D') = \varnothing$ if $D \neq D'$ are reduced pipe dreams for w) is easy to deduce directly from Definition 16.8. $\square$

Corollary 16.12 *Let w be an $n \times n$ permutation. If $w = \sigma_{i_1} \cdots \sigma_{i_m} w_0$ with $m = l(w_0) - l(w)$, then $\mathcal{RP}(w) = \mathrm{mitosis}_{i_m} \cdots \mathrm{mitosis}_{i_1}(D_0)$.*

The previous corollary says that mitosis (irredundantly) generates all reduced pipe dreams for honest permutations. By replacing a partial permutation w with an extension $\widetilde{w}$ to a permutation, this implies that mitosis generates all reduced pipe dreams for w, with no restriction on w.

16.2 A combinatorial formula

The manner in which mitosis generates reduced pipe dreams has substantial algebraic structure, to be exploited in this section. In particular, we shall prove the following positive combinatorial formula for Schubert polynomials. (The corresponding formula for double Schubert polynomials will appear in Corollary 16.30.) Recall that $k \times \ell$ pipe dreams are identified with their sets of $+$ tiles in the $k \times \ell$ grid.

Theorem 16.13 $\mathfrak{S}_w(\mathbf{t}) = \displaystyle\sum_{D \in \mathcal{RP}(w)} \mathbf{t}^D$, *where* $\mathbf{t}^D = \displaystyle\prod_{(i,j) \in D} t_i$.

The proof, at the end of this section, comes down to an attempt at calculating $\partial_i(\mathbf{t}^D)$ directly. Fixing the loose ends in this method requires the involution in Proposition 16.16, to gather terms together in pairs. The involution is defined by first partitioning rows i and $i+1$.

Definition 16.14 Let D be a pipe dream and i a fixed row index. Order the tiles in rows i and $i+1$ of D as in the following diagram:

	1	2	3	4	$\cdots$
i	1	3	5	7	
$i+1$	2	4	6	8	$\cdots$

An **intron** in these two adjacent rows is a height 2 rectangle C such that the following two conditions hold:

1. The first and last tiles in C (the northwest and southeast corners) are elbow tiles.

2. No elbow tile in C is strictly northeast or strictly southwest of another elbow (so due north, due south, due east, or due west are all okay).

Ignoring all ⊞ columns in rows i and $i+1$, an intron is thus just a sequence of ⊟ columns in rows i and $i+1$, followed by a sequence of ⊟ columns, possibly with one ⊟ column in between. (Columns ⊞ with two crosses can be ignored for the purpose of the proof of the next result.)

An intron C is **maximal** if it satisfies the following extra condition:

3. The elbow with largest index before C (if there is one) lies in row $i+1$, and the elbow with smallest index after C (if there is one) lies in row i.

Lemma 16.15 *Let C be an intron in a reduced pipe dream. There is a unique intron $\tau(C)$ satisfying the following two conditions.*

1. *The sets of ⊞ columns are the same in C and $\tau(C)$.*

2. *The number c_i of $+$ tiles in row i of C equals the number of $+$ tiles in row $i+1$ of $\tau(C)$, and the same holds with i and $i+1$ switched.*

*The involution τ, called **intron mutation**, can always be accomplished by a sequence of chute moves or inverse chute moves.*

Proof. First assume $c_i > c_{i+1}$ and work by induction on $c = c_i - c_{i+1}$. If $c = 0$, then $\tau(C) = C$ and the lemma is obvious. If $c > 0$, then consider the leftmost ⊞ column. Moving to the left from this column, there must be a column not equal to ⊞, since the northwest entry of C is an elbow. The rightmost such column must be ⊟, because its row i entry is an elbow (by construction) and its row $i+1$ entry cannot be a $+$ (for then the pipes crossing there would also cross in the ⊞ column). This means that we can chute the $+$ in ⊞ into the ⊟ column and proceed by induction.

Flip the argument $180°$ if $c_i < c_{i+1}$, so the chute move becomes an inverse chute move. □

For example, here is an intron mutation accomplished by chuting the crossing tiles in columns $4, 6$, and then 7 of row i. The zigzag shapes formed by the dots in these introns are typical.

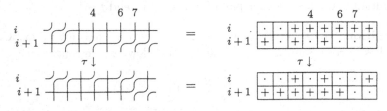

Proposition 16.16 *For each i there is an involution $\tau_i : \mathcal{RP}(w) \to \mathcal{RP}(w)$ such that $\tau_i^2 = 1$, and for all $D \in \mathcal{RP}(w)$, the following hold:*

1. *$\tau_i D$ agrees with D outside rows i and $i+1$.*
2. *$\mathrm{start}_i(\tau_i D) = \mathrm{start}_i(D)$, and $\tau_i D$ equals D strictly west of this column.*
3. *The number of $+$ tiles in $\tau_i D$ from row i in columns $\geq \mathrm{start}_i(\tau_i D)$ equals the number of $+$ tiles in D from row $i+1$ in these columns.*

Proof. Let $D \in \mathcal{RP}(w)$. Consider the union of all columns in rows i and $i+1$ of D that are east of or coincide with column $\mathrm{start}_i(D)$. Since the first and last tiles in this region (numbered as in Definition 16.14) are elbows, this region breaks uniquely into a disjoint union of height 2 rectangles, each of which is either a maximal intron or completely filled with $+$ tiles. Indeed, this follows from the definition of start_i and Definition 16.14. Applying intron mutation to each maximal intron therein leaves a pipe dream that breaks up uniquely into maximal introns and solid regions of $+$ tiles in the same way. Therefore the proposition comes down to verifying that intron mutation preserves the property of being in $\mathcal{RP}(w)$. This is an immediate consequence of Lemmas 16.7 and 16.15.   □

Proof of Theorem 16.13. It suffices to prove the result for honest permutations, so we use downward induction on weak order in S_n. The result for the long permutation $w = w_0$ holds because $\mathcal{RP}(w_0) = \{D_0\}$ (Example 16.3).

Fix $D \in \mathcal{RP}(w)$, write $\mathbf{t}^D = \prod_{(i,j) \in D} t_i$, and let $m = |\mathrm{mitosis}_i(D)|$ be the number of mitosis offspring of D. This number m equals the number of $+$ tiles in $\boxplus$ configurations located west of $\mathrm{start}_i(D)$ in rows i and $i+1$ of D. Let D' be the pipe dream (*not* reduced) that results after deleting these $+$ tiles from D. The monomial $\mathbf{t}^D$ is then the product $t_i^m \mathbf{t}^{D'}$. Definition 16.8 immediately implies that

$$\sum_{E \in \mathrm{mitosis}_i(D)} \mathbf{t}^E \;=\; \sum_{d=1}^{m} t_i^{m-d} t_{i+1}^{d-1} \cdot \mathbf{t}^{D'} \;=\; \partial_i(t_i^m) \cdot \mathbf{t}^{D'}. \qquad (16.1)$$

If $\tau_i D = D$, then $\mathbf{t}^{D'}$ is symmetric in t_i and t_{i+1} by Proposition 16.16, so

$$\partial_i(t_i^m) \cdot \mathbf{t}^{D'} \;=\; \partial_i(t_i^m \cdot \mathbf{t}^{D'}) \;=\; \partial_i(\mathbf{t}^D)$$

in this case (see Exercise 15.13). On the other hand, if $\tau_i D \neq D$, then letting the transposition σ_i act on polynomials by switching t_i and t_{i+1}, Proposition 16.16 implies that adding the sums in (16.1) for D and $\tau_i D$ yields

$$\partial_i(t_i^m) \cdot (\mathbf{t}^{D'} + \sigma_i \mathbf{t}^{D'}) \;=\; \partial_i\big(t_i^m(\mathbf{t}^{D'} + \sigma_i \mathbf{t}^{D'})\big) \;=\; \partial_i(\mathbf{t}^D + \mathbf{t}^{\tau_i D}).$$

Pairing off the elements of $\mathcal{RP}(w)$ not fixed by τ_i, we conclude that

$$\partial_i\left(\sum_{D \in \mathcal{RP}(w)} \mathbf{t}^D \right) \;=\; \sum_{E \in \mathrm{mitosis}_i(\mathcal{RP}(w))} \mathbf{t}^E.$$

The left-hand side is $\mathfrak{S}_{\sigma_i w}(\mathbf{t})$ by induction and the recursion for $\mathfrak{S}_w(\mathbf{t})$ in Definition 15.38, while the right side is $\sum_{E \in \mathcal{RP}(\sigma_i w)} \mathbf{t}^E$ by Theorem 16.11. $\square$

16.3   Antidiagonal simplicial complexes

It should come as no surprise that we wish to reduce questions about determinantal ideals to computations with monomials, since this is a major theme in combinatorial commutative algebra. The rest of this chapter is devoted to deriving facts about sets of minors defined by the rank conditions $r_{pq}(w) = \text{rank}(w_{p \times q})$ by exploring (and exploiting) the combinatorics of their antidiagonal terms.

As in the previous chapter, let $Z_{p \times q}$ be the northwest $p \times q$ subarray of any rectangular array Z (such as a matrix or a pipe dream).

Definition 16.17 Let $\mathbf{x} = (x_{\alpha\beta})$ be the $k \times \ell$ matrix of variables. An **antidiagonal** of size r in $\Bbbk[\mathbf{x}]$ is the antidiagonal term of a minor of size r, i.e., the product of entries along the antidiagonal of an $r \times r$ submatrix of $\mathbf{x}$. For a $k \times \ell$ partial permutation w, the **antidiagonal ideal** $J_w \subset \Bbbk[\mathbf{x}]$ is generated by all antidiagonals in $\mathbf{x}_{p \times q}$ of size $1 + r_{pq}(w)$ for all p and q. Write $\mathcal{L}_w$ for the **antidiagonal complex**, the Stanley–Reisner complex of J_w.

Observe that J_w is indeed a squarefree monomial ideal. This section is essentially a complicated verification that two Stanley–Reisner ideals are Alexander dual. These ideals are the antidiagonal ideal J_w and the ideal whose generators are the monomials $\mathbf{x}^D$ for reduced pipe dreams $D \in \mathcal{RP}(w)$. Here is an equivalent, more geometric statement.

Theorem 16.18 *The facets of the antidiagonal complex $\mathcal{L}_w$ are the complements of the reduced pipe dreams for w, yielding the prime decomposition*

$$J_w = \bigcap_{D \in \mathcal{RP}(w)} \langle x_{ij} \mid (i,j) \text{ is a crossing tile in } D \rangle.$$

It is convenient to identify each antidiagonal $\underline{a} \in \Bbbk[\mathbf{x}]$ with the subset of the $k \times \ell$ array of variables dividing $\underline{a}$, just as we identify pipe dreams with their sets of $+$ tiles. Then Theorem 16.18 can be equivalently rephrased as saying that a pipe dream D meets every antidiagonal in J_w and is minimal with this property if and only if D lies in $\mathcal{RP}(w)$. This is the statement that we will actually be thinking of in our proofs.

Example 16.19 The antidiagonal ideal J_{2143} for the 4×4 permutation 2143 equals $\langle x_{11}, x_{13}x_{22}x_{31} \rangle$. The antidiagonal complex $\mathcal{L}_{2143}$ is the union of three coordinate subspaces $L_{11,13}$, $L_{11,22}$, and $L_{11,31}$, with ideals

$$I(L_{11,13}) = \langle x_{11}, x_{13} \rangle, \quad I(L_{11,22}) = \langle x_{11}, x_{22} \rangle, \quad \text{and} \quad I(L_{11,31}) = \langle x_{11}, x_{31} \rangle$$

whose intersection yields the prime decomposition of J_{2143}. Pictorially, represent the subspaces $L_{11,13}$, $L_{11,22}$, and $L_{11,31}$ by pipe dreams

$$D_{L_{11,13}} = \begin{array}{|c|c|c|c|}\hline + & & + & \\ \hline & & & \\ \hline & & & \\ \hline & & & \\ \hline\end{array}, \quad D_{L_{11,22}} = \begin{array}{|c|c|c|c|}\hline + & & & \\ \hline & + & & \\ \hline & & & \\ \hline & & & \\ \hline\end{array}, \quad \text{and} \quad D_{L_{11,31}} = \begin{array}{|c|c|c|c|}\hline + & & & \\ \hline & & & \\ \hline + & & & \\ \hline & & & \\ \hline\end{array}$$

inside the 4×4 grid that have $+$ entries wherever the corresponding subspace is required to be zero. These three pipe dreams coincide with the reduced pipe dreams in $\mathcal{RP}(2143)$ from Example 16.4. $\diamond$

The next lemma is a key combinatorial observation. Its proof (in each case, check that each rank condition is still satisfied) is omitted.

Lemma 16.20 *If $\underline{a}$ and $\underline{a}'$ are antidiagonals in $\Bbbk[\mathbf{x}]$, with $\underline{a} \in J_w$, then $\underline{a}'$ also lies in J_w if it is obtained from $\underline{a}$ by one of the following operations:*

   (W) *moving west one or more of the variables in $\underline{a}$*
   (E) *moving east any variable except the northeast one in $\underline{a}$*
   (N) *moving north one or more of the variables in $\underline{a}$*
   (S) *moving south any variable except the southwest one in $\underline{a}$*

For each subset L of the $k \times \ell$ grid, let D_L be its complement. Thus the subspace of the $k \times \ell$ matrices corresponding to L is $\langle x_{ij} \mid (i,j) \in D_L \rangle$.

Lemma 16.21 *The set of complements D_L of faces $L \in \mathcal{L}_w$ is closed under chute moves and inverse chute moves.*

Proof. A pipe dream D is equal to D_L for some face $L \in \mathcal{L}_w$ if and only if D meets every antidiagonal in J_w. Suppose that C is a chutable rectangle in D_L for $L \in \mathcal{L}_w$. For chutes, it is enough to show that the intersection $\underline{a} \cap D_L$ of any antidiagonal $\underline{a} \in J_w$ with D_L does not consist entirely of the single $+$ in the northeast corner of C, unless $\underline{a}$ also contains the southwest corner of C. So assume that $\underline{a}$ contains the $+$ in the northeast corner (p,q) of C, but not the $+$ in the southwest corner, and split into cases:

   (i) $\underline{a}$ does not continue south of row p.
   (ii) $\underline{a}$ continues south of row p but skips row $p + 1$.
   (iii) $\underline{a}$ intersects row $p + 1$, but strictly east of the southwest corner of C.
   (iv) $\underline{a}$ intersects row $p + 1$, but strictly west of the southwest corner of C.

Letting $(p + 1, t)$ be the southwest corner of C, construct antidiagonals $\underline{a}'$ that are in J_w, and hence intersect D_L, by moving the $+$ at (p,q) in $\underline{a}$ to:

   (i) (p, t), using Lemma 16.20(W);
   (ii) $(p + 1, q)$, using Lemma 16.20(S);
   (iii) (p, q), so $\underline{a} = \underline{a}'$ trivially; or
   (iv) (p, t), using Lemma 16.20(W).

Observe that in case (iii), $\underline{a}$ already shares a box in row $p + 1$ where D_L has a $+$. Each of the other antidiagonals $\underline{a}'$ intersects *both* $\underline{a}$ and D_L in some box that is not (p, q), since the location of $\underline{a}' \smallsetminus \underline{a}$ has been constructed not to be a crossing tile in D_L.

The proof for inverse chutes is just as easy and is left to the reader. $\square$

Call a pipe dream *top-justified* if no + tile is due south of a ⌐ tile—in other words, if the configuration ⊟ does not occur.

Lemma 16.22 *Given a face* $L \in \mathcal{L}_w$, *there is a sequence* $L_0, \ldots, L_m$ *of faces of* $\mathcal{L}_w$ *in which* $L_0 = L$, *the complement* D_{L_m} *is top-justified, and* $D_{L_{e+1}}$ *is obtained from* D_{L_e} *by either deleting a* + *or performing an inverse chute.*

Proof. Suppose that D_L for some face $L \in \mathcal{L}_w$ is not top-justified, and has no inverse-chutable rectangles. Consider a configuration ⊟ in the most eastern column containing one. To the east of this configuration, the first 2×1 configuration that is not ⊞ must be ⊟ because of the absence of inverse-chutable rectangles. The union of the original ⊟ configuration along with this ⊟ configuration and all intervening ⊞ configurations is a chutable rectangle with an extra crossing tile in its southwest corner. Reasoning exactly as in the proof of Lemma 16.21 shows that we may delete the crossing tile at the northeast corner of this rectangle.   □

Given a $k \times \ell$ pipe dream D, denote by L_D the coordinate subspace inside the $k \times \ell$ matrices whose ideal in $\Bbbk[\mathbf{x}]$ is $\langle x_{ij} \mid (i,j)$ is a + tile in $D \rangle$.

Proposition 16.23 *Let* D *and* E *be pipe dreams, with* E *obtained from* D *by a chute move. Then* L_D *is a facet of* $\mathcal{L}_w$ *if and only if* L_E *is.*

Proof. Suppose that L_D is not a facet. This means that deleting from D some +, let us call it ⊞, yields a pipe dream D' whose subspace $L_{D'}$ is still a face of $\mathcal{L}_w$. We will show that some + may be deleted from E to yield a pipe dream whose face still lies in $\mathcal{L}_w$. Let C be the chutable rectangle on which the chute move acts.

If ⊞ lies outside of the rectangle C, then deleting it from E yields the result E' of chuting C in D'; that $L_{E'}$ still lies in $\mathcal{L}_w$ is by Lemma 16.21. If ⊞ is the northeast corner of C, then deleting the southwest corner of C from E again yields D'. Thus we may assume ⊞ lies in C, and not at either end. Every antidiagonal $\underline{a} \in J_w$ contains a + in D in some row other than that of ⊞. If ⊞ lies in the top row of C, then this other + lies in E, as well; hence deleting ⊞ from E has the desired effect. Finally, if ⊞ lies in the bottom row of C, then let ⊞′ be the crossing tile immediately due north of it in D. Chuting ⊞′ and subsequently the northeast corner of C in $D \smallsetminus ⊞$ yields $D' \smallsetminus ⊞'$, and $L_{D' \smallsetminus ⊞'}$ lies in $\mathcal{L}_w$ by Lemma 16.21 again.

We have shown that L_D is a facet if L_E is; the converse is similar.   □

The previous result implies that the facets of $\mathcal{L}_w$ constitute the nodes of a graph whose edges connect pairs of facets related by chute moves. The main hurdle to jump before the proof of Theorem 16.18 is the connectedness of this graph. By Lemma 16.22, this amounts to the uniqueness of a top-justified facet complement for $\mathcal{L}_w$, which we will show in Proposition 16.26. In general, top-justified pipe dreams enjoy some desirable properties.

Proposition 16.24 *Every top-justified pipe dream is reduced, and* $\mathcal{RP}(w)$ *contains a unique one, called the* **top reduced pipe dream** $\mathrm{top}(w)$. *Every pipe dream* $D \in \mathcal{RP}(w)$ *can be reached by a sequence of chutes from* $\mathrm{top}(w)$.

Proof. Replacing w with $\tilde{w}$ if necessary, it suffices to consider honest permutations, as usual. Next we show that reduced pipe dreams that are not top-justified always admit inverse chute moves. Consider a configuration ⊟ in the most eastern column containing one. To the east of this configuration, the first 2×1 configuration that is not ⊞ must either be ⊟ or ⊡. The former is impossible because the pipes passing through the $+$ in ⊟ would also intersect at the $+$ in ⊡. Hence the union of the original ⊟ along with this ⊡ and all intervening ⊞ configurations is an inverse-chutable rectangle.

Now simply count: there are $n!$ top pipe dreams contained in the long permutation pipe dream D_0 for S_n, and we have just finished showing that there must be at least $n!$ distinct top-justified reduced pipe dreams inside D_0. The result follows immediately. $\qquad\square$

Let us say that a rank condition $r_{pq} \leq r$ *causes* an antidiagonal $\underline{a}$ of the generic matrix $\mathbf{x}$ if $\mathbf{x}_{p \times q}$ contains $\underline{a}$ and $\underline{a}$ has size at least $r + 1$. For instance, when the rank condition comes from $r(w)$, the antidiagonals it causes include those antidiagonals $\underline{a} \in J_w$ that are contained in $\mathbf{x}_{p \times q}$ but not in any smaller northwest rectangular submatrix of $\mathbf{x}$.

Lemma 16.25 *Antidiagonals in* $J_w \smallsetminus J_{\sigma_i w}$ *are contained in* $\mathbf{x}_{i \times w(i)-1}$ *and intersect row* i.

Proof. If an antidiagonal in J_w is either contained in $\mathbf{x}_{i-1 \times w(i)}$ or not contained in $\mathbf{x}_{i \times w(i)}$, then some rank condition causing it is in both $r(w)$ and $r(\sigma_i w)$. Indeed, it is easy to check that the rank matrices $r(\sigma_i w)$ and $r(w)$ differ only in row i between columns $w(i+1)$ and $w(i) - 1$, inclusive. $\quad\square$

Proposition 16.26 *For each partial permutation* w, *there is a unique facet* $L \in \mathcal{L}_w$ *whose complementary pipe dream* D_L *is top-justified, and in fact* $D_L = \mathrm{top}(w)$ *is the top reduced pipe dream for* w.

Proof. This is clearly true for $w = w_0$. Assuming it for all $n \times n$ permutations of length at least l, we prove it for $n \times n$ permutations of length $l - 1$.

Let $v \in S_n \smallsetminus w_0$ be a permutation. Then v has an ascent, $v(i) < v(i + 1)$. Choose i minimal with this property, and set $j = v(i)$. Then let $w = \sigma_i v$, so that $v = \sigma_i w < w$, as usual. Since i is minimal, the northwest $i + 1 \times j$ rectangle $r(\sigma_i w)_{i+1 \times j}$ of the rank matrix of $\sigma_i w$ is zero except at (i, j) and $(i + 1, j)$, where $r(\sigma_i w)$ takes the value 1. Every variable x_{pq} sitting on one of these zero entries is an antidiagonal of size 1 in $J_{\sigma_i w}$. Therefore *every* pipe dream D_L that is the complement of a facet $L \in \mathcal{L}_{\sigma_i w}$ has $+$ tiles in these locations. See the following figure for a medium-sized example.

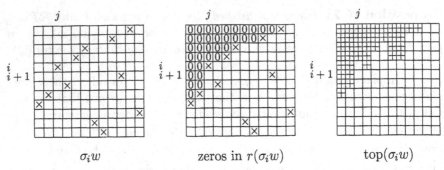

$$\sigma_i w \qquad\qquad \text{zeros in } r(\sigma_i w) \qquad\qquad \text{top}(\sigma_i w)$$

The same statements hold with w in place of $\sigma_i w$, except that the only nonzero entry of $r(w)_{i+1 \times j}$ is $r_{i+1,j}(w) = 1$. We will need the consequence $J_w \supseteq J_{\sigma_i w}$ of the componentwise inequality $r(w) \le r(\sigma_i w)$.

Let D be a top-justified facet complement for $\mathcal{L}_{\sigma_i w}$. These exist by Lemma 16.22. If $\underline{a}$ is an antidiagonal in $J_w \smallsetminus J_{\sigma_i w}$, then either $\underline{a}$ intersects D or $\underline{a}$ contains x_{ij}. Indeed, suppose that $\underline{a}$ misses D and that also x_{ij} does not lie in a. Since $\underline{a}$ lies in $J_w \smallsetminus J_{\sigma_i w}$, Lemma 16.25 implies that $\underline{a}$ is caused by a row i rank condition $r_{iq}(w)$ in some column q between $w(i+1)$ and $w(i) - 1$. This rank matrix entry is one less than the entry due south of it by (15.1): $r_{i+1,q}(w) = r_{iq}(w) + 1$. Hence $x_{i+1,j}\underline{a}$ is an antidiagonal in $J_{\sigma_i w}$ that misses D, which is impossible. Therefore adding a $+$ tile at (i,j) to D yields a pipe dream D' whose complement is a face $L \in \mathcal{L}_w$.

It remains to show that L is a facet and that $D = D' \smallsetminus (i,j)$. Indeed, using the former we deduce that D' must be the top reduced pipe dream for w by induction, and using the latter we conclude that $D = \text{top}(\sigma_i w)$ by Lemma 16.5 and Proposition 16.24. First we show that L is a facet.

Every variable in $\mathbf{x}_{i+1 \times j}$ except for $x_{i+1,j}$ itself is actually an antidiagonal in J_w, so no $+$ tile in $D'_{i+1 \times j}$ can be deleted. Suppose $\boxplus$ is one of the remaining $+$ tiles in D'. Then $\boxplus$ equals the unique intersection of some antidiagonal $\underline{a} \in J_{\sigma_i w}$ with D. If $\underline{a}$ misses x_{ij}, then $\{\boxplus\} = \underline{a} \cap D'$, so $\boxplus$ cannot be deleted from D'.

On the other hand, suppose $\underline{a}$ contains x_{ij}. If $\underline{a}$ continues southwest of x_{ij}, then $\underline{a}$ skips row $i+1$ because we assumed $\boxplus$ does not lie in $D'_{i+1 \times j}$. Hence we can replace $\underline{a}$ with $(x_{i+1,j}/x_{ij})\underline{a}$, which misses x_{ij}, using Lemma 16.20(S). Finally, if $\underline{a}$ has its southwest end at (i,j), suppose the northeast end of $\underline{a}$ lies in column q. Using (15.1) as a guide, calculate that $r_{i-1,q}(\sigma_i w) = r_{iq}(\sigma_i w) - 1$. Thus $\underline{a}/x_{ij}$ lies in $J_{\sigma_i w}$ and misses x_{ij}.

Finally, note that the arguments in the previous two paragraphs can also be used to show that deleting (i,j) from D' yields a face complement for $\mathcal{L}_{\sigma_i w}$. Hence $D = D' \smallsetminus (i,j)$, as required. $\qquad\square$

Proof of Theorem 16.18. The set $\mathcal{RP}(w)$ of reduced pipe dreams for w is characterized by Proposition 16.24 as the set of pipe dreams obtained from top(w) by applying chute moves. Lemma 16.22, Proposition 16.23, and Proposition 16.26 imply that the set of complements D_L of facets $L \in \mathcal{L}_w$ is characterized by the same property. $\qquad\square$

16.4 Minors form Gröbner bases

Theorem 16.18 immediately implies some useful statements about Schubert determinantal ideals. As we will see, the next result will be enough to conclude that the minors generating Schubert determinantal ideals I_w form Gröbner bases and therefore that the ideals I_w are prime.

Corollary 16.27 *The antidiagonal simplicial complex $\mathcal{L}_w$ is pure. In the multigrading with $\deg(x_{ij}) = t_i$, it has multidegree $C(\Bbbk[\mathbf{x}]/J_w; \mathbf{t}) = \mathfrak{S}_w(\mathbf{t})$.*

Proof. Purity of $\mathcal{L}_w$ is immediate from Theorem 16.18 and the fact that all reduced pipe dreams for w have the same number of $+$ tiles. Using purity, Theorem 8.53 and Proposition 8.49 together imply that $C(\Bbbk[\mathbf{x}]/J_w; \mathbf{t})$ is the sum of monomials $\mathbf{t}^{D_L}$ for complements D_L of facets $L \in \mathcal{L}_w$. As these facet complements are precisely the reduced pipe dreams for w by Theorem 16.18, the result follows from the formula in Theorem 16.13. □

A term order on $\Bbbk[\mathbf{x}]$ is called *antidiagonal* if the initial term of every minor of $\mathbf{x}$ is its antidiagonal term. Thus, if $[i_1 \cdots i_r | j_1 \cdots j_r]$ is the determinant of the square submatrix of the generic matrix $\mathbf{x}$ whose rows and columns are indexed by $i_1 < \cdots < i_r$ and $j_1 < \cdots < j_r$, respectively, then

$$\mathsf{in}([i_1 \cdots i_r | j_1 \cdots j_r]) = x_{i_r j_1} x_{i_{r-1} j_2} \cdots x_{i_2 j_{r-1}} x_{i_1 j_r}.$$

There are numerous antidiagonal term orders (Exercise 16.11).

Theorem 16.28 *The minors inside the Schubert determinantal ideal I_w constitute a Gröbner basis under any antidiagonal term order:*

$$\mathsf{in}(I_w) \quad = \quad J_w.$$

Proof. The multidegree $C(\Bbbk[\mathbf{x}]/\mathsf{in}(I_w); \mathbf{t})$ equals the Schubert polynomial $\mathfrak{S}_w(\mathbf{t})$ by Corollary 15.44 and Corollary 8.47. As J_w is obviously contained inside the initial ideal $\mathsf{in}(I_w)$ under any antidiagonal term order, and moreover $\mathsf{in}(I_w) \subseteq \mathsf{in}(I(\overline{X}_w))$, we can apply Exercise 8.13 with $I = \mathsf{in}(I(\overline{X}_w))$ and $J = J_w$, by Corollary 16.27. Hence $J_w = \mathsf{in}(I_w) = \mathsf{in}(I(\overline{X}_w))$. □

Geometrically, Theorem 16.28 exhibits a Gröbner degeneration of each matrix Schubert variety: the fiber at 1 is $\overline{X}_w$, and the fiber at 0 is $\mathcal{L}_w$ (realized as a union of coordinate subspaces). Actually, Gröbner degenerations are only defined once suitable weight vectors are chosen; see Definition 8.25 and do Exercise 16.11.

Here is the first important consequence of the Gröbner basis statement.

Corollary 16.29 *Schubert determinantal ideals I_w are prime.*

Proof. The zero set of I_w is the matrix Schubert variety $\overline{X}_w$, which is irreducible by Theorem 15.31. Hence the radical of I_w is prime. However, Theorem 16.28 says that I_w has a squarefree initial ideal, which automatically implies that I_w is a radical ideal. □

The primality of Schubert determinantal ideals means that I_w equals the radical ideal $I(\overline{X}_w)$ of polynomials vanishing on the matrix Schubert variety for w. Therefore the multidegree calculation for matrix Schubert varieties in Theorem 15.40 holds for $\Bbbk[\mathbf{x}]/I_w$. This enables us to deduce the double version of Theorem 16.13.

Corollary 16.30 *The double Schubert polynomial for w satisfies*

$$\mathfrak{S}_w(\mathbf{t} - \mathbf{s}) = \sum_{D \in \mathcal{RP}(w)} (\mathbf{t} - \mathbf{s})^D, \quad where \quad (\mathbf{t} - \mathbf{s})^D = \prod_{(i,j) \in D} (t_i - s_j).$$

Proof. The multidegree $\mathcal{C}(\Bbbk[\mathbf{x}]/J_w; \mathbf{t}, \mathbf{s})$ equals the double Schubert polynomial by Theorem 15.40, Corollary 8.47, and Theorem 16.28, using the fact that $I_w = I(\overline{X}_w)$ (Corollary 16.29). Now apply additivity of multidegrees on components (Theorem 8.53) and the explicit calculation of multidegrees for coordinate subspaces (Proposition 8.49), using Theorem 16.18 to get the sum to be over reduced pipe dreams.   □

Generally speaking, the minors generating I_w fail to be Gröbner bases for other term orders, although these can still be used to get formulas for double Schubert polynomials.

Example 16.31 Consider the Schubert determinantal ideal I_{2143} for the 4×4 permutation 2143. This ideal has the same generators as the ideal I_w in Example 15.7, although in a bigger polynomial ring. We discussed the antidiagonal ideal $J_w = \text{in}(I_w)$ in Example 16.19. Note that the two minors generating I_{2143} never form a Gröbner basis for a *diagonal* term order, because x_{11} divides the diagonal term $x_{11}x_{22}x_{33}$.

In the multigrading where $\deg(x_{ij}) = t_i - s_j$, the multidegree of $L_{i_1 j_1, i_2 j_2}$ equals $(t_{i_1} - s_{j_1})(t_{i_2} - s_{j_2})$. The formula in Corollary 16.30 says that

$$\mathfrak{S}_{2143}(\mathbf{t} - \mathbf{s}) \quad = \quad (t_1 - s_1)(t_1 - s_3) + (t_1 - s_1)(t_2 - s_2) + (t_1 - s_1)(t_3 - s_1),$$

which agrees with the calculation of this double Schubert polynomial in Example 15.43. On the other hand, there is a diagonal term order under which $\langle x_{11}, x_{13}x_{21}x_{32} \rangle = \langle x_{11}, x_{13} \rangle \cap \langle x_{11}, x_{21} \rangle \cap \langle x_{11}, x_{32} \rangle$ is the initial ideal of I_{2143}. Thus we can also calculate

$$\mathfrak{S}_{2143}(\mathbf{t} - \mathbf{s}) \quad = \quad (t_1 - s_1)(t_1 - s_3) + (t_1 - s_1)(t_2 - s_1) + (t_1 - s_1)(t_3 - s_2),$$

using additivity and the explicit calculation for subspaces.   ◇

The title of this section alludes to that of Section 14.3, where antidiagonals are initial terms of Plücker coordinates. The differences are that Theorem 14.11 works in a sagbi (subalgebra) context and speaks only of top-justified minors, whereas Theorem 16.28 works in a Gröbner basis (ideal) context and allows certain more general collections of minors.

16.5 Subword complexes

Our goal in this section is to prove that Schubert determinantal rings $\Bbbk[\mathbf{x}]/I_w$ are Cohen–Macaulay. We shall in fact show that antidiagonal complexes are Cohen-Macaulay. The argument involves some satisfying combinatorics of reduced expressions in symmetric groups.

Every $n \times n$ permutation matrix can be expressed as a product of elements in the set $\{\sigma_1, \ldots, \sigma_{n-1}\}$ of *simple $n \times n$ reflection* matrices—that is, permutation matrices for adjacent transpositions (see Definition 15.24 and the paragraph after it). Simple reflections σ_i are allowed to appear more than once in such an expression.

Definition 16.32 A **reduced expression** for a permutation matrix w is an expression $w = \sigma_{i_m} \cdots \sigma_{i_1}$ as a product of $m = l(w)$ simple reflections.

Lemma 16.33 *The minimal number of matrices required to express a permutation matrix w as a product of simple reflections is $l(w)$.*

Proof. For the identity matrix this is obvious, since it has length zero. Multiplying an arbitrary permutation matrix on the left by a simple reflection either increases length by 1 or decreases it by 1; this is a special case of Lemma 15.21. Ascending in weak order from the identity to a permutation matrix w therefore requires at least $l(w)$ many simple reflections. □

It is easy to see that reduced expressions exist—in other words, that the minimum $l(w)$ in Lemma 16.33 is actually attained. In fact, we are about to produce a number of reduced expressions explicitly, using pipe dreams. For notation, let us say that a $+$ tile at (p, q) in a pipe dream D sits on the ith *antidiagonal* if $p + q - 1 = i$.

Let $Q(D)$ be the ordered sequence of simple reflections σ_i corresponding to the antidiagonals on which the $+$ tiles of D sit, starting from the southwest corner of D and reading left to right in each row, snaking up to the northeast corner. For a random example, the pipe dream

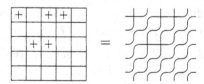

yields the ordered sequence $\sigma_4\sigma_5\sigma_1\sigma_3\sigma_4$. We should mention that compared to conventions in the literature (see the Notes), this convention looks like the sequence is read backward; but this is only because our permutation matrices here have corresponding abstract permutations obtained by reading the column indices of the nonzero entries instead of the rows. Transposing matrices inverts the permutations and reverses the reduced expressions.

Example 16.34 The unique pipe dream D_0 for the $n \times n$ long permutation (antidiagonal matrix) w_0 corresponds to the ordered sequence

$$Q(D_0) \;=\; \underbrace{\sigma_{n-1}}\underbrace{\sigma_{n-2}\sigma_{n-1}} \;\cdots\cdots\; \underbrace{\sigma_2\sigma_3\cdots\sigma_{n-1}}\underbrace{\sigma_1\sigma_2\cdots\sigma_{n-1}},$$

the *reverse triangular* reduced expression for w_0. The part of $Q(D_0)$ arising from each row of D_0 has its own underbrace. When $n = 4$, the above expression simplifies to $Q_0 = \sigma_3\sigma_2\sigma_3\sigma_1\sigma_2\sigma_3$. $\diamond$

Example 16.35 The ordered sequence constructed from the pipe dream whose crossing tiles entirely fill the $n \times n$ grid is the *reverse square word*

$$Q_{n\times n} \;=\; \underbrace{\sigma_n\sigma_{n+1}\cdots\sigma_{2n-2}\sigma_{2n-1}}_{\text{bottom row}} \;\cdots\; \underbrace{\sigma_2\sigma_3\cdots\sigma_n\sigma_{n+1}}_{\text{second row}}\;\underbrace{\sigma_1\sigma_2\cdots\sigma_{n-1}\sigma_n}_{\text{top row}}.$$

This sequence necessarily involves reflections $\sigma_1, \ldots, \sigma_{2n-1}$, which lie in S_{2n}, even though reduced expressions for permutation matrices $w \in S_n$ never involve reflections σ_i with $i \geq n$. $\diamond$

Lemma 16.36 *Suppose that the pipe entering row i of an $n \times n$ pipe dream D exits column $w(i)$ for some $n \times n$ permutation w. Multiplying the reflections in $Q(D)$ yields the permutation matrix w. Thus $Q(D)$ is a reduced expression for w if and only if $D \in \mathcal{RP}(w)$.*

Proof. Use induction on the number of crossing tiles: adding a $+$ in the ith antidiagonal at the start of the list switches the destinations of the pipes entering through rows i and $i + 1$. $\square$

In other words, pipe dreams in the $n \times n$ grid are naturally "subwords" of the reverse square word, while reduced pipe dreams are naturally *reduced* subwords. This explains the adjective "reduced" for pipe dreams.

Definition 16.37 A **word** of size m is a sequence $Q = (\sigma_{i_m} \ldots, \sigma_{i_1})$ of simple reflections. An ordered subsequence P of Q is a **subword** of Q.

1. P **represents** an $n \times n$ permutation matrix w if the ordered product of the simple reflections in P is a reduced expression for w.
2. P **contains** w if some subsequence of P represents w.

The **subword complex** $\Delta(Q, w)$ is the set of subwords whose complements contain w: $\Delta(Q, w) = \{Q \smallsetminus P \mid P \text{ contains } w\}$.

In other words, deleting a face of $\Delta(Q, w)$ from Q leaves a reduced expression for w as a subword of what remains. If $Q \smallsetminus D$ is a facet of the subword complex $\Delta(Q, w)$, then the reflections in D constitute a reduced expression for w. Note that subwords of Q come with their embeddings into Q, so two subwords P and P' involving reflections at different positions in Q are unequal, even if the sequences of reflections in P and P' are equal.

Usually we write Q as a string without parentheses or commas, and we abuse notation by saying that Q is a word in S_n, without explicit reference to the set of simple reflections. Note that Q need not itself be a reduced expression. The following lemma is immediate from the definitions and the fact that all reduced expressions for $w \in S_n$ have the same length.

Lemma 16.38 $\Delta(Q, w)$ *is a pure simplicial complex whose facets are the subwords* $Q \smallsetminus P$ *such that* $P \subseteq Q$ *represents* w. $\qquad\square$

Example 16.39 Consider the subword complex $\Delta = \Delta(\sigma_3\sigma_2\sigma_3\sigma_2\sigma_3, 1432)$ for the 4×4 permutation $w = 1432$. This permutation has two reduced expressions, namely $\sigma_3\sigma_2\sigma_3$ and $\sigma_2\sigma_3\sigma_2$. Labeling the vertices of a pentagon with the reflections in $Q = \sigma_3\sigma_2\sigma_3\sigma_2\sigma_3$ (in cyclic order), the facets of Δ are the pairs of adjacent vertices. Thus Δ is the boundary of the pentagon. $\diamond$

Proposition 16.40 *Antidiagonal complexes* $\mathcal{L}_w$ *are subword complexes.*

Proof. When w is a permutation matrix, the fact that

$$\mathcal{L}_w = \Delta(Q_{n \times n}, w)$$

is a subword complex for the $n \times n$ reverse square word is immediate from Theorem 16.18 and Lemma 16.33. When w is an arbitrary $k \times \ell$ partial permutation, simply replace w by a minimal extension to a permutation $\widetilde{w}$, and replace $Q_{n \times n}$ by the word corresponding to tiles in a $k \times \ell$ rectangle. $\square$

We will show that subword complexes are Cohen–Macaulay via Theorem 13.45 by proving that they are shellable. In fact, we shall verify a substantially stronger, but less widely known criterion. Recall from Definition 1.38 the notion of the *link* of a face in a simplicial complex.

Definition 16.41 Let Δ be a simplicial complex and $F \in \Delta$ a face.

1. The **deletion** of F from Δ is $\mathrm{del}_\Delta(F) = \{G \in \Delta \mid F \cap G = \varnothing\}$.
2. The simplicial complex Δ is **vertex-decomposable** if Δ is pure and either (i) $\Delta = \{\varnothing\}$, or (ii) for some vertex $v \in \Delta$, both $\mathrm{del}_\Delta(v)$ and $\mathrm{link}_\Delta(v)$ are vertex-decomposable.

The definition of vertex-decomposability is not circular; rather, it is inductive on the number of vertices in Δ. Here is a typical example of how this inductive structure can be mined.

Proposition 16.42 *Vertex-decomposable complexes are shellable.*

Proof. Use induction on the number of vertices by first shelling $\mathrm{del}_\Delta(v)$ and then shelling the cone from v over $\mathrm{link}_\Delta(v)$ to get a shelling of Δ. $\square$

Theorem 16.43 *Antidiagonal complexes are shellable and hence Cohen–Macaulay. More generally, subword complexes are vertex-decomposable.*

Proof. By Proposition 16.40 and Proposition 16.42, it is enough to prove the second sentence. With $Q = (\sigma_{i_m}, \sigma_{i_{m-1}}, \ldots, \sigma_{i_1})$, it suffices by induction on the number of vertices to demonstrate that both the link and the deletion of σ_{i_m} from $\Delta(Q, w)$ are subword complexes. By definition, both consist of subwords of $Q' = (\sigma_{i_{m-1}}, \ldots, \sigma_{i_1})$. The link is naturally identified with the subword complex $\Delta(Q', w)$. For the deletion, there are two cases. If $\sigma_{i_m} w$ is longer than w, then the deletion of σ_{i_m} equals its link because no reduced expression for w has σ_{i_m} at its left end. On the other hand, when $\sigma_{i_m} w$ is shorter than w, the deletion is $\Delta(Q', \sigma_{i_m} w)$.   □

Corollary 16.44 *Schubert determinantal rings are Cohen–Macaulay.*

Proof. Apply Theorems 16.43 and 8.31 to Theorem 16.28.   □

Exercises

16.1 Use the length condition in Definition 16.2 to show that crossing tiles in reduced pipe dreams for permutations all occur strictly above the main antidiagonal.

16.2 Suppose that a $k \times \ell$ partial permutation matrix w is given, and that $\widetilde{w}$ is a permutation matrix extending w. Prove that the crossing tiles in every reduced pipe dream for $\widetilde{w}$ all fit inside the northwest $k \times \ell$ rectangle.

16.3 What permutation in S_4 has the most reduced pipe dreams? In S_5? In S_n?

16.4 Prove directly, using the algebra of antidiagonals and without using Theorem 16.11 or Theorem 16.18, that if L is a facet of $\mathcal{L}_w$ and $\sigma_i w < w$, then mitosis$_i(D_L)$ consists of pipe dreams $D_{L'}$ for facets L' of $\mathcal{L}_{\sigma_i w}$.

16.5 Change each box in the diagram of w (Definition 15.13) into a $+$ tile and then push all of these tiles due north as far as possible. Show that the resulting top-justified pipe dream is top(w).

16.6 Given a $+$ tile in the top reduced pipe dream top(w), construct an explicit antidiagonal in J_w whose intersection with top(w) is precisely the given $+$ tile. Hint: Consider the $\diagdown\!\!\diagup$ tiles along the pipe passing vertically through the $+$ tile.

16.7 Show that each partial permutation has a unique **bottom** reduced pipe dream in which no $\diagdown\!\!\diagup$ is due west of a $+$ in the same row.

16.8 Prove that the bottom reduced pipe dream for a Grassmannian permutation (Exercise 15.6) forms the Ferrers shape (in "French" position) of a partition. Verify that the resulting map from the set of $n \times n$ Grassmannian permutations with descent at k to the set of partitions that fit into the $k \times (n - k)$ grid is bijective.

16.9 A semistandard tableau T (Definition 14.12) determines a monomial $\mathbf{t}^T$ whose degree in t_i is the number of entries of T equal to i. Given a Grassmannian permutation w, let $\lambda(w)$ be the partition from Exercise 16.8. Exhibit a *monomial-preserving* bijection from reduced pipe dreams for w to semistandard Young tableaux of shape $\lambda(w)$, meaning that $\mathbf{t}^D = \mathbf{t}^T$ when $D \mapsto T$. Conclude that Schubert polynomials for Grassmannian permutations are Schur polynomials.

16.10 Under the bijection of the previous exercise, characterize intron mutation on Grassmannian reduced pipe dreams directly in terms of semistandard Young tableaux. (Note: Readers who know tableaux will recognize that intron mutation thus specializes to the *Bender–Knuth involution* on semistandard tableaux.)

16.11 Find a total ordering of the variables in the $k \times \ell$ array $\mathbf{x} = (x_{ij})$ whose reverse lexicographic term order is antidiagonal. Do the same for lexicographic order. Find an explicit weight vector inducing an antidiagonal partial term order.

16.12 Let P be obtained from a pipe dream in $\mathcal{RP}(w)$ by adding a single extra $+$ tile. Explain why there is at most one other $+$ tile that can be deleted from P to get a reduced pipe dream for w.

16.13 By a general theorem, shellable complexes whose ridges (codimension 1 faces) each lie in at most 2 facets is a ball or sphere [BLSWZ99, Proposition 4.7.22]. Use Exercise 16.12 to deduce that antidiagonal complexes are balls or spheres.

16.14 (For those who know about Coxeter groups) Define subword complexes for arbitrary Coxeter groups. Show they are pure and vertex-decomposable. Prove that if Δ is a subword complex for a Coxeter group, then no ridge is contained in more than two facets. Conclude that Δ is homeomorphic to a ball or sphere.

16.15 What conditions on a word Q and an element w guarantee that $\Delta(Q, w)$ is (i) Gorenstein or (ii) spherical?

16.16 Recall the notation from Exercise 15.14. If w is a permutation of length $l(w) \leq m$, prove that the coefficient on $t_1 t_2 \cdots t_m$ in the Schubert polynomial $\mathfrak{S}_{m+w}(\mathbf{t})$ equals the number of reduced expressions for w.

Notes

There are many important ways of extracting combinatorics from determinantal ideals other than via pipe dreams. For example, there are vast literatures on this topic concerned with *straightening laws* [DRS74, DEP82, Hib86] and the *Robinson–Schensted–Knuth correspondence* [Stu90, HT92, BC01]. The former is treated in [BV88], as well as more briefly in [BH98, Chapter 7] and [Hib92, Part III], and the state of the art in RSK methods is explained in the excellent expository article [BC03].

Reduced pipe dreams are special cases of the curve diagrams invented by Fomin and Kirillov [FK96]. Our notation follows Bergeron and Billey [BB93], who called them *rc-graphs*; the corresponding objects in [FK96] are rotated by 135°. The definition of chute move comes from [BB93], as does the characterization of reduced pipe dreams in Proposition 16.24, which is [BB93, Theorem 3.7].

The mitosis recursion in Theorem 16.11 is [KnM04b, Theorem C]. Our proof here is approximately the one in [Mil03a], although Lemma 16.10 is new, as is the resulting argument proving the formula in Theorem 16.13. This result was first proved by Billey, Jockusch, and Stanley [BJS93], although independently(!) and almost simultaneously, Fomin and Stanley gave a shorter, more elegant combinatorial proof [FS94]. Corollary 16.30 is due to Fomin and Kirillov [FK96].

Theorem 16.18 and Theorem 16.28 together with the shellability of antidiagonal simplicial complexes is due to Knutson and Miller [KnM04b, Theorem B]. Our proof here is much simplified, because we avoid proving any K-polynomial statements along the way, focusing instead on multidegrees. Mitosis is a shadow of the combinatorial transitions in the weak order on S_n that govern the standard monomials of Schubert determinantal ideals and antidiagonal ideals, which are necessary for proving Hilbert series formulas, as in Remark 15.45. The use of Exercise 8.13 in Theorem 16.28 is the same as in [KnM04b] and similar to [Mar03]. Martin's applications are to *picture spaces* [Mar03], which parametrize drawings of graphs in the projective plane, and particularly to *slope varieties*, which record the edge slopes.

The appearances of antidiagonal initial terms in Theorem 14.11 and Theorem 16.28 are not coincidentally similar: there is a direct geometric connection [KoM04], in which each reduced pipe dream subspace from the Gröbner degeneration maps to a face of the Gelfand–Tsetlin toric variety.

Primality and Cohen–Macaulayness of Schubert determinantal ideals is due to Fulton [Ful92], who originally defined them. His elegant proof relied on related statements for Schubert varieties in flag varieties [Ram85] that use positive characteristic methods and vanishing theorems for sheaf cohomology. These Schubert variety statements follow from Corollary 16.29 and Corollary 16.44.

Subword complexes were introduced in [KnM04b] for the same purpose as they appear in this chapter; their vertex-decomposability is Theorem E of that article. The notion of vertex-decomposability was introduced by Billera and Provan, who proved that it implies shellability [BP79]. Further treatment of subword complexes, in the Coxeter group generality of Exercise 16.14, can be found in [KnM04a]. Included there are Hilbert series calculations and explicit characterizations of when balls and spheres occur. In addition, several down-to-earth open problems on combinatorics of reduced expressions in Coxeter groups appear there.

Exercise 16.3 is inspired by a computation due to Woo [Woo04a, Woo04b]. We learned the bijection in Exercise 16.9 from Kogan [Kog00]. The last sentence of Exercise 16.9 is essentially the statement that the Schubert classes on Grassmannians are represented by the Schur polynomials. The result of Exercise 16.9 holds more generally for double Schubert polynomials and supersymmetric Schur polynomials. Explicit weight orders as in Exercise 16.11 are crucial in some applications of Schubert determinantal ideals and the closely related quiver ideals of Chapter 17 [KoM04, KMS04]. Exercise 16.16 leads into the theory of *stable Schubert polynomials*, which are also known as *Stanley symmetric functions*. This important part of the theory surrounding Schubert polynomials is due to Stanley [Sta84], along with subsequent positivity results and connections to combinatorics contributed by [LS85, EG87, LS89, Hai92, RS95], among others. See [BB04] for an introductory account of this story.

Chapter 17

Minors in matrix products

Chapters 14–16 dealt with minors inside a single matrix. In this chapter, we consider *quiver ideals*, which are generated by minors in products of matrices. The zero sets of these ideals are called *quiver loci*. Surprisingly, we can reduce questions about quiver ideals and quiver loci to questions about Schubert determinantal ideals, by using the *Zelevinsky map*, which embeds a sequence of matrices as blocks in a single larger matrix. As a consequence, we deduce that quiver ideals are prime and that quiver loci are Cohen–Macaulay. In addition, we get a glimpse of how the combinatorics of quivers, pipe dreams, and Schubert polynomials are reflected in formulas for *quiver polynomials*, which are the multidegrees of quiver loci.

17.1 Quiver ideals and quiver loci

The questions we asked about ideals generated by minors in a matrix of variables—concerning primality, Cohen–Macaulayness, and explicit formulas for multidegrees—also make sense for ideals generated by minors in products of two or more such matrices. Whereas the former correspond geometrically to varieties of linear maps with specified ranks between two fixed vector spaces, the latter correspond to varieties of *sequences* of linear maps. Naturally, if we hope to get combinatorics out of this situation, then we should first isolate the combinatorics that goes into it: what kinds of conditions on the ranks of composite maps is it reasonable for us to request?

Example 17.1 The sequence of three matrices in Fig. 17.1 constitutes an element in the vector space $M_{23} \times M_{34} \times M_{43}$ of sequences of linear maps $\Bbbk^2 \to \Bbbk^3 \to \Bbbk^4 \to \Bbbk^3$ (so $\Bbbk^r$ consists of row vectors for each r). Note that these matrices—call them w_1, w_2, and w_3—can be multiplied in the order they are given. Since they are all partial permutations (Definition 15.1), we can represent the sequence $\mathbf{w} = (w_1, w_2, w_3)$ by the graph above it, called its *lacing diagram*. When w_i has a 1 entry in row α and column β, the

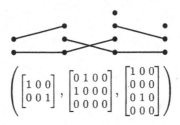

Figure 17.1: A sequence of partial permutations and its lacing diagram

lacing diagram has a segment directly above w_i joining the dot at height α to the dot in the next column to the right at height β.

Let $\Bbbk[\mathbf{f}]$ denote the coordinate ring of $M_{23} \times M_{34} \times M_{43}$. Thus $\Bbbk[\mathbf{f}]$ is a polynomial ring $6 + 12 + 12 = 30$ variables $\mathbf{f} = \{f_{\alpha\beta}^i\}$, arranged in three rectangular *generic matrices* $\Phi_1 = (f_{\alpha\beta}^1)$, $\Phi_2 = (f_{\alpha\beta}^2)$, and $\Phi_3 = (f_{\alpha\beta}^3)$. We would like to think of the sequence $\mathbf{w}$ as lying in the zero set of an ideal generated by minors in the products of these generic matrices Φ_i.

What size minors should we take? The ranks of the three maps w_1, w_2, and w_3 are all 2, as this is the number of nonzero entries in each matrix. Hence $\mathbf{w}$ lies in the zero set of the ideal generated by all 3×3 minors Φ_1, Φ_2, and Φ_3. However, $\mathbf{w}$ satisfies additional conditions: the composite maps $\Bbbk^2 \xrightarrow{w_1 w_2} \Bbbk^4$ and $\Bbbk^3 \xrightarrow{w_2 w_3} \Bbbk^3$ both have rank 1. One way to see this without multiplying the matrices is to count the number of length 2 laces (one each) spanning the first three or the last three columns of dots. Therefore $\mathbf{w}$ also lies in the ideal generated by the 2×2 minors of $\Phi_1\Phi_2$ and $\Phi_2\Phi_3$. Finally, the composite map $\Bbbk^2 \xrightarrow{w_1 w_2 w_3} \Bbbk^3$ is zero, since no laces span all of the columns, so $\mathbf{w}$ lies in the zero set of the entries of the product $\Phi_1\Phi_2\Phi_3$. Hence the rank conditions that best describe $\mathbf{w}$ are the bounds determined by $\mathbf{w}$ on the ranks of the $6 = 3+2+1$ consecutive products of generic matrices Φ_i. $\diamond$

Sequences of partial permutations given by lacing diagrams are in many ways fundamental. In particular, our goal in this section is to show in Proposition 17.9 that they are the *only examples* of matrix lists, up to changes of basis. Therefore, let us formalize the notion of lacing diagram.

Definition 17.2 Fix $r_0, \ldots, r_n \in \mathbb{N}$. Let $\mathbf{w} = (w_1, \ldots, w_n)$ be a list of partial permutations, with w_i of size $r_{i-1} \times r_i$. The **lacing diagram** of $\mathbf{w}$ is a graph having r_i vertices in column i for $i = 0, \ldots, n$, and an edge from the αth dot in column $i - 1$ to the βth dot in column i whenever $w_i(\alpha) = \beta$. We identify $\mathbf{w}$ with its lacing diagram and call its connected components **laces**.

To describe the general framework for ideals in products of matrices, fix nonnegative integers $r_0, \ldots, r_n$. Denote by $Mat = M_{r_0 r_1} \times \cdots \times M_{r_{n-1} r_n}$ the variety of *quiver representations* over the field $\Bbbk$ with *dimension vector* $(r_0, \ldots, r_n)$; that is, Mat equals the vector space of sequences

$$\phi: \quad \Bbbk^{r_0} \xrightarrow{\phi_1} \Bbbk^{r_1} \xrightarrow{\phi_2} \cdots \xrightarrow{\phi_{n-1}} \Bbbk^{r_{n-1}} \xrightarrow{\phi_n} \Bbbk^{r_n} .$$

of linear transformations. By convention, set $\phi_0 = 0 = \phi_{n+1}$. As the above notation suggests, we have fixed a basis for each of the vector spaces $\Bbbk^{r_i}$, and we express elements of $\Bbbk^{r_i}$ as row vectors of length r_i. Each map ϕ_i in the quiver representation ϕ therefore becomes identified with a matrix over $\Bbbk$ of size $r_{i-1} \times r_i$. The coordinate ring of *Mat* is a polynomial ring $\Bbbk[\mathbf{f}]$ in variables $\mathbf{f} = \{f^i_{\alpha\beta}\} = (f^1_{\alpha\beta}), \ldots, (f^n_{\alpha\beta})$, where the ith index β and the $(i+1)^{\text{st}}$ index α run from 1 to r_i. Let Φ be the *generic* quiver representation, in which the entries in the matrices Φ_i are the variables $f^i_{\alpha\beta}$.

Definition 17.3 For an array $\mathbf{r} = (r_{ij})_{0 \leq i \leq j \leq n}$ of nonnegative integers with $r_{ii} = r_i$, the **quiver ideal** $I_\mathbf{r} \subseteq \Bbbk[\mathbf{f}]$ is generated by the union over $i < j$ of the size $1 + r_{ij}$ minors in the product $\Phi_{i+1} \cdots \Phi_j$ of generic matrices:

$$I_\mathbf{r} = \langle \text{minors of size } 1 + r_{ij} \text{ in } \Phi_{i+1} \cdots \Phi_j \text{ for } i < j \rangle.$$

The **quiver locus** $\Omega_\mathbf{r} \subseteq Mat$ is the zero set of the quiver ideal $I_\mathbf{r}$.

The quiver locus $\Omega_\mathbf{r}$ consists exactly of those ϕ satisfying $r_{ij}(\phi) \leq r_{ij}$ for all $i < j$, where $r_{ij}(\phi)$ is the rank of the composite map $\Bbbk^{r_i} \to \Bbbk^{r_j}$:

$$r_{ij}(\phi) = \text{rank}(\phi_{i+1} \cdots \phi_j) \quad \text{for } i < j. \tag{17.1}$$

Example 17.4 Whenever $0 \leq k \leq \ell \leq n$, there is a quiver representation

$$\mathbf{w}(k, \ell): \quad 0 \to \cdots \to 0 \to \underset{k}{\Bbbk} = \cdots = \underset{\ell}{\Bbbk} \to 0 \to \cdots \to 0$$

having copies of the field $\Bbbk$ in spots between k and ℓ, with identity maps between them and zeros elsewhere. The array $\mathbf{r} = \mathbf{r}(\mathbf{w}(k, \ell))$ in this case has entry $r_{ij} = 1$ if $k \leq i \leq j \leq \ell$, and $r_{ij} = 0$ otherwise. Quiver representations of this form are called **indecomposable**. The matrices in $\mathbf{w}(k, \ell)$ are all 1×1, filled with either 0 or 1, so $\mathbf{w}(k, \ell)$ is a lacing diagram with one lace stretching from the dot in column k to the dot in column ℓ. $\diamond$

It was particularly simple to determine the array $\mathbf{r}$ for the lacing diagrams $\mathbf{w}(k, \ell)$. As it turns out, it is not much harder to do so for arbitrary lacing diagrams. The (easy) proof of the following is left to Exercise 17.2.

Lemma 17.5 *If $\mathbf{w} \in Mat$ is a lacing diagram with precisely $q_{k\ell}$ laces beginning in column k and ending in column ℓ, for each $k \leq \ell$, then $r_{ij}(\mathbf{w})$ equals the number of laces passing through both column i and column j:*

$$r_{ij}(\mathbf{w}) = \sum_{k=0}^{i} \sum_{\ell=j}^{n} q_{k\ell}.$$

Definition 17.6 Let $\mathbf{q} = (q_{ij})$ be a **lace array** filled with arbitrary nonnegative integers for $0 \leq i \leq j \leq n$. The associated **rank array** is the nonnegative integer array $\mathbf{r} = (r_{ij})$ for $0 \leq i \leq j \leq n$ defined by Lemma 17.5. The **rectangle array** of $\mathbf{r}$ (or of $\mathbf{q}$) is the array $R = (R_{ij})$ of rectangles for $0 \leq i < j \leq n$ such that R_{ij} has height $r_{i,j-1} - r_{ij}$ and width $r_{i+1,j} - r_{ij}$.

The point is that for a lacing diagram, we could just as well specify the ranks $\mathbf{r}$ by giving the lace array $\mathbf{q}$. We defined the rectangle array here because it fits naturally with $\mathbf{r}$ and $\mathbf{q}$, but we will not use it until Lemma 17.13.

Example 17.7 The lacing diagram from Example 17.1 has rank array $\mathbf{r} = (r_{ij})$, lace array $\mathbf{q} = (q_{ij})$, and rectangle array $\boldsymbol{R} = (R_{ij})$ as follows:

$\mathbf{r} = $

3	2	1	0	i/j
		2		0
	3	2		1
4	2	1		2
3	2	1	0	3

$\mathbf{q} = $

3	2	1	0	i/j
		0		0
	0	1		1
1	0	1		2
1	1	1	0	3

$\boldsymbol{R} = $

3	2	1	0	i/j
				0
		—		1
	▭	□		2
⊟	□	□		3

Lemma 17.5 says that each entry of $\mathbf{r}$ is the sum of the entries in $\mathbf{q}$ weakly southeast of the corresponding location. The height of R_{ij} is obtained by subtracting the entry r_{ij} from the one above it, whereas the width of R_{ij} is obtained by subtracting the entry r_{ij} from the one to its left. The reason for writing the arrays in this orientation will come from the Zelevinsky map; compare $\mathbf{q}$ and $\boldsymbol{R}$ here to the illustration in Example 17.14. $\diamond$

The reason why the ranks of lacing diagrams decompose as sums is because the lacing diagrams themselves decompose into sums. In general, if ϕ and ψ are two quiver representations with dimension vectors $(r_0, \ldots, r_n)$ and $(r'_0, \ldots, r'_n)$, then the *direct sum* of ϕ and ψ is the quiver representation $\phi \oplus \psi = (\phi_1 \oplus \psi_1, \ldots, \phi_n \oplus \psi_n)$, whose i^{th} vector space is $\Bbbk^{r_i} \oplus \Bbbk^{r'_i}$. Every direct sum of indecomposables is represented by a sequence of partial permutations and hence is a lacing diagram; but not every lacing diagram is equal to a such a direct sum (try the lacing diagram in Example 17.1). On the other hand, with the right notion of isomorphism, every lacing diagram is isomorphic to such a direct sum, after permuting the dots (basis vectors) in each column. To make a precise statement, two quiver representations ϕ and ψ are called *isomorphic* if there are invertible $r_i \times r_i$ matrices η_i for $i = 0, \ldots, n$ such that $\phi_i \eta_i = \eta_{i-1} \psi_i$. In other words, η gives invertible maps $\Bbbk^{r_i} \to \Bbbk^{r_i}$ making every square in the diagram $\phi \xrightarrow{\eta} \psi$ commute.

Lemma 17.8 *Every lacing diagram $\mathbf{w} \in \mathrm{Mat}$ is isomorphic to the direct sum of the indecomposable lacing diagrams corresponding to its laces. Two lacing diagrams are isomorphic if and only if they have the same lace array.*

The (easy) proof is left to Exercise 17.2; note that the second sentence is a consequence of the first. The lemma brings us to the main result of the section. It is the sequences-of-maps analogue of the fact that every linear map between two vector spaces can be written as a diagonal matrix with only zeros and ones, after changing bases in both the source and target.

Proposition 17.9 *Every quiver representation $\phi \in \mathrm{Mat}$ is isomorphic to a lacing diagram $\mathbf{w}$, whose lace array $\mathbf{q}$ is independent of the choice of $\mathbf{w}$.*

Proof. It suffices by Lemma 17.8 to show that ϕ is isomorphic to a direct sum of indecomposables. We may as well assume that $r_0 \neq 0$ and let j be the largest index for which the composite $\mathbf{k}^{r_0} \to \mathbf{k}^{r_j}$ is nonzero. Choose a linearly independent set $B_0 \subset \mathbf{k}^{r_0}$ whose span maps isomorphically to the image of $\mathbf{k}^{r_0}$ in $\mathbf{k}^{r_j}$ under the composite $\mathbf{k}^{r_0} \to \mathbf{k}^{r_j}$. The image $B_i \subset \mathbf{k}^{r_i}$ of B_0 under $\mathbf{k}^{r_0} \to \mathbf{k}^{r_i}$ for $i \leq j$ is independent. Setting $B_i = \varnothing$ for $i > j$, let ψ be the induced quiver representation on $(\mathrm{span}(B_0), \ldots, \mathrm{span}(B_n))$.

Choose a splitting $\mathbf{k}^{r_j} = V'_j \oplus \mathrm{span}(B_j)$. This induces, for each $i \leq j$, a splitting $\mathbf{k}^{r_i} = V'_i \oplus \mathrm{span}(B_i)$, where V'_i is the preimage of V'_j under the composite map $\mathbf{k}^{r_i} \to \mathbf{k}^{r_j}$. Set $V'_i = \mathbf{k}^{r_i}$ for $i > j$, so we get a quiver representation ϕ' on $V' = (V'_0, \ldots, V'_n)$ by restriction of ϕ.

By construction, $\phi = \psi \oplus \phi'$. Since ψ is isomorphic to a direct sum of $\#B_0$ copies of $\mathbf{w}(0, j)$, induction on $r_0 + \cdots + r_n$ completes the proof. $\square$

Definition 17.3 made no assumptions about the array $\mathbf{r}$ of nonnegative integers, and the ranks r_{ij} there are only upper bounds. Unless every rank r_{ij} equals zero, there will always be matrix lists $\phi \in \Omega_{\mathbf{r}}$ whose composite maps have strictly smaller rank than $\mathbf{r}$; and if some rank r_{ij} for $i < j$ is very big, then *all* matrix lists $\phi \in \Omega_{\mathbf{r}}$ will have strictly smaller rank $r_{ij}(\phi)$. The point is that only certain arrays $\mathbf{r}$ can actually occur as ranks of quiver representations: nontrivial restrictions on the array $\mathbf{r}(\phi)$ are imposed by Proposition 17.9. In more detail, after choosing an isomorphism $\phi \cong \mathbf{w}$ with a lacing diagram $\mathbf{w}$, inverting Lemma 17.5 yields

$$q_{ij} \;=\; r_{ij} - r_{i-1,j} - r_{i,j+1} + r_{i-1,j+1}$$

for $i \leq j$, where $r_{ij} = 0$ if i and j do not both lie between 0 and n. Therefore the array $\mathbf{r}$ can occur as in (17.1) if and only if $r_{ii} = r_i$ for $i = 0, \ldots, n$, and

$$r_{ij} - r_{i-1,j} - r_{i,j+1} + r_{i-1,j+1} \;\geq\; 0 \tag{17.2}$$

for $i \leq j$, since the left-hand side is simply q_{ij}. Here, finally, is the answer to what kinds of rank conditions can we reasonably request.

Convention 17.10 Starting in Section 17.2, *we consider only rank arrays $\mathbf{r}$ that occur as in (17.1)*, and we call these rank arrays **irreducible** if we need to emphasize this point. Thus we can interchangeably use a lace array $\mathbf{q}$ or its corresponding rank array $\mathbf{r}$ to specify a quiver ideal or locus.

We close this section with an example to demonstrate how the irreducibility of a rank array $\mathbf{r}$ can detect good properties of the quiver ideal $I_{\mathbf{r}}$.

Example 17.11 (Minors of fixed size in a product of two matrices) Consider two matrices of variables, Φ_1 and Φ_2, where Φ_1 has size $r_0 \times r_1$ and Φ_2 has size $r_1 \times r_2$. We are interested in the ideal I generated by all of the minors of size $\rho + 1$ in the product $\Phi_1 \Phi_2$, so the quiver locus Ω consists of the pairs (ϕ_1, ϕ_2) such that $\phi_1 \phi_2$ has rank at most ρ. This rank condition is automatically satisfied unless $\rho < \min\{r_0, r_1, r_2\}$, so we assume this inequality. The question is whether I is prime. Suppose that $I = I_{\mathbf{r}}$ for some rank array $\mathbf{r}$. In order for the only equations generating I to be the minors in $\Phi_1 \Phi_2$, there must be no rank conditions on Φ_1 individually, and also none on Φ_2 individually. In other words, we must stipulate that $r_{01} = \min(r_0, r_1)$ and $r_{12} = \min(r_1, r_2)$ are as large as possible. Suppose this is the case, and consider a quiver representation $\phi \in \Omega_{\mathbf{r}}$.

In terms of elementary linear algebra, ϕ_1 and ϕ_2 are matrices of maximal rank, and we want $\mathrm{rank}(\phi_1 \phi_2) \le \rho$. However, if the middle vector space in $\Bbbk^{r_0} \xrightarrow{\phi_1} \Bbbk^{r_1} \xrightarrow{\phi_2} \Bbbk^{r_2}$ is the smallest of the three, so $r_0 \ge r_1$ and $r_1 \le r_2$, then ϕ_1 is surjective and ϕ_2 is injective. Hence $\phi_1 \phi_2$ has rank precisely r_1 in this case, and we require that $\rho < r_1$. We conclude that r_1 cannot be too small.

How large must r_1 be? Answering this question from first principles is possible, but with lacing diagrams, it becomes easy. So suppose our ϕ is actually a lacing diagram $\mathbf{w}$. Then $\mathbf{w}$ has ρ laces spanning all three columns of dots because $r_{02} = \rho$, so $q_{02} = \rho$. Next, to make w_1 of maximal rank r_{01}, we must have $r_{01} - \rho$ laces from column 0 to column 1; that is, $q_{01} = r_{01} - \rho$, where we recall that $r_{01} = \min(r_0, r_1)$. Similarly, $q_{12} = r_{12} - \rho$. Graph-theoretically, we must find a matching on the set of dots above height ρ that saturates the two outside columns of dots, because no endpoint of the q_{01} laces can be shared with one of the q_{12} laces. In the diagrams

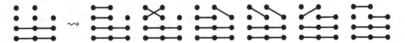   (17.3)

$(r_0, r_1, r_2) = (4, 5, 3)$ and $\rho = 2$. We conclude that the array $\mathbf{r}$ is irreducible if and only if $r_1 \ge \rho + q_{01} + q_{12} = r_{01} + r_{12} - \rho$. We will see in Theorem 17.23 that in this case $\Omega_{\mathbf{r}}$ is an irreducible variety and in fact $I_{\mathbf{r}}$ is prime.

What happens if $\mathbf{r}$ is not irreducible? Take the lacing diagrams below:

Here, $(r_0, r_1, r_2) = (4, 4, 3)$ and $\rho = 2$. In contrast to (17.3), the six choices of matchings (edges of length 1) in this case give rise to lacing diagrams with *two different rank arrays*. It follows that $\Omega_{\mathbf{r}}$ contains the quiver loci for both, so $\Omega_{\mathbf{r}}$ is reducible as a variety. Therefore $I_{\mathbf{r}}$ is not prime. $\diamond$

17.2   Zelevinsky map

The rank conditions given by arrays in Definition 17.3 are essentially forced upon us by naturality: they are the only rank conditions that are invariant under arbitrary changes of basis, by Proposition 17.9. In this section we will see how the irreducible rank conditions in Convention 17.10 can be transformed into the rank conditions for a Schubert determinantal ideal.

Given a rank array $\mathbf{r}$, or equivalently, a lace array $\mathbf{q}$, we shall construct a permutation $v(\mathbf{r})$ in the symmetric group S_d, where $d = r_0 + \cdots + r_n$. In general, any matrix in the space M_d of $d \times d$ matrices comes with a decomposition into block rows of heights $r_0, \ldots, r_n$ (block rows listed from top to bottom) and block columns of widths $r_n, \ldots, r_0$ (block columns listed from left to right). Note that our indexing convention may be unexpected, with the square blocks lying along the main block *anti*diagonal rather than on the diagonal as usual. With these conventions, the ith block column refers to the block column of width r_i, which sits i blocks from the *right*.

Draw the matrix for each permutation $v \in S_d$ by placing a symbol $\times$ (instead of a 1) at each position $(k, v(k))$, and zeros elsewhere.

Proposition–Definition 17.12 *Given a rank array $\mathbf{r}$, there is a unique* **Zelevinsky permutation** $v(\Omega_{\mathbf{r}}) = v(\mathbf{r})$ *in S_d, satisfying the following conditions. Consider the block in the ith column and j^{th} row.*

1. *If $i \leq j$ (that is, the block sits on or below the main block antidiagonal), then the number of $\times$ entries in that block equals q_{ij}.*
2. *If $i = j+1$ (that is, the block sits on the main block superantidiagonal), then the number of $\times$ entries in that block equals $r_{j,j+1}$.*
3. *If $i \geq j + 2$ (that is, the block lies strictly above the main block superantidiagonal), then there are no $\times$ entries in that block.*
4. *Within every block row or block column, the $\times$ entries proceed from northwest to southeast; that is, no $\times$ entry is northeast of another.*

Proof. We need the number of $\times$ entries in any block row, as dictated by conditions 1–3, to equal the height of that block row (and transposed for columns), since condition 4 then stipulates uniquely how to arrange the $\times$ entries within each block. In other words, the height $r_j = r_{jj}$ of the jth block row must equal the number $r_{j,j+1}$ of $\times$ entries in the superantidiagonal block in that block row, plus the sum $\sum_{i \leq j} q_{ij}$ of the number of $\times$ entries in the rest of the blocks in that block row (and a similar statement must hold for block columns). These statements follow from Lemma 17.5. $\square$

The diagram (Definition 15.13) of a Zelevinsky permutation refines the data contained in the rectangle array $\mathbf{R}$ for the corresponding ranks (Definition 17.6). The next lemma is a straightforward consequence of the definition of Zelevinsky permutation, and we leave it to Exercise 17.2.

Lemma 17.13 *In each block of the diagram of a Zelevinsky permutation $v(\mathbf{r})$ that is on or below the superantidiagonal, the boxes form a rectangle justified in the southeast corner of the block. Moreover, the rectangle in the ith block column and j^{th} block row is the rectangle $R_{i-1,j+1}$ in the array $\mathbf{R}$.*

Example 17.14 Let $\mathbf{r}$, $\mathbf{q}$, and $\mathbf{R}$ be as in Example 17.7. The Zelevinsky permutation for this data is

$$v(\mathbf{r}) = \begin{bmatrix} 1 & 2 & 3 & 4 & 5 & 6 & 7 & 8 & 9 & 10 & 11 & 12 \\ 8 & 9 & 4 & 5 & 11 & 1 & 2 & 6 & 12 & 3 & 7 & 10 \end{bmatrix},$$

whose permutation matrix is indicated by the × entries in the array

```
 8 | *  *  * | *  *  *  * | ×  .  . | .  .
 9 | *  *  * | *  *  *  * | .  ×  . | .  .
 4 | *  *  * | ×  .  .  . | .  .  . | .  .
 5 | *  *  * | .  ×  .  . | .  .  . | .  .
11 | *  *  * | .  .  □  □ | .  .  □ | ×  .
 1 | ×  .  . | .  .  .  . | .  .  . | .  .
 2 | .  ×  . | .  .  .  . | .  .  . | .  .
 6 | .  .  □ | .  .  ×  . | .  .  . | .  .
12 | .  .  □ | .  .  .  □ | .  .  □ | .  ×
 3 | .  .  × | .  .  .  . | .  .  . | .  .
 7 | .  .  . | .  .  ×  . | .  .  . | .  .
10 | .  .  . | .  .  .  . | .  .  × | .  .
```

and whose diagram $D(v(\mathbf{r}))$ is indicated by the set of all $*$ and $\square$ entries. ◇

The locations in the diagram of $v(\mathbf{r})$ strictly above the block superantidiagonal are drawn as $*$ entries instead of boxes because they are contained in the diagram of the Zelevinsky permutation $v(\mathbf{r})$ for every rank array $\mathbf{r}$ with fixed dimension vector $(r_0, \ldots, r_n)$. In fact, the $*$ entries form the diagram of the Zelevinsky permutation $v(Mat)$ corresponding to the quiver locus that equals the entire quiver space Mat. We henceforth denote this unique Zelevinsky permutation of minimal length by $v_* = v(Mat)$.

It is clear from the combinatorics of Zelevinsky permutations that we can read off the rank array $\mathbf{r}$ and the lace array $\mathbf{q}$ from $v(\mathbf{r})$. We next demonstrate that the combinatorial encoding of $\mathbf{r}$ by its Zelevinsky permutation reflects a simple geometric map that translates between quiver loci and matrix Schubert varieties.

Definition 17.15 The **Zelevinsky map** $\mathcal{Z} : Mat \to M_d$ takes

$$
(\phi_1, \phi_2, \ldots, \phi_n) \xmapsto{\ \mathcal{Z}\ }
\begin{bmatrix}
0 & & 0 & \phi_1 & \mathbf{1} \\
0 & & \phi_2 & \mathbf{1} & 0 \\
0 & \cdot^{\cdot^{\cdot}} & \mathbf{1} & 0 & 0 \\
\phi_n & \cdot^{\cdot^{\cdot}} & 0 & 0 & 0 \\
\mathbf{1} & & 0 & 0 & 0
\end{bmatrix}, \qquad (17.4)
$$

so $\mathcal{Z}(\phi)$ is a block matrix of total size $d \times d$. If $\Bbbk[\mathbf{x}]$ denotes the coordinate ring of M_d, then denote the kernel of the induced map $\Bbbk[\mathbf{x}] \twoheadrightarrow \Bbbk[\mathbf{f}]$ by $\mathfrak{m}_{\mathbf{f}}$.

Indexing for the $d \times d$ matrix $\mathbf{x}$ of variables in $\Bbbk[\mathbf{x}] = \Bbbk[M_d]$ does not arise in this section, so it will be introduced later, as necessary. The ideal $\mathfrak{m}_{\mathbf{f}}$ is generated by equations setting the appropriate variables in $\Bbbk[\mathbf{x}]$ to 0 or 1. To be more precise, $\mathfrak{m}_{\mathbf{f}}$ contains every $\mathbf{x}$ variable except

- those in superantidiagonal blocks, as they correspond to the coordinates $\mathbf{f}$ on Mat and map isomorphically to their images in $\Bbbk[\mathbf{f}]$, and
- those on the diagonals of the antidiagonal blocks; for each such variable x, the ideal $\mathfrak{m}_{\mathbf{f}}$ contains $x - 1$ instead.

The proof of Theorem 17.17 will use the following handy general lemma.

Lemma 17.16 *Let $\Gamma\Phi$ be the product of two matrices with entries in a commutative ring R. If Γ is square and $\det(\Gamma)$ is a unit, then for each fixed $u \in \mathbb{N}$, the ideals generated by the size u minors in Φ and in $\Gamma\Phi$ coincide.*

Proof. The result is easy when $u = 1$. The case of arbitrary u reduces to the case $u = 1$ by noting that the minors of size u in a matrix for a map $R^k \to R^\ell$ of free modules are simply the entries in a particular choice of matrix for the associated map $\bigwedge^u R^k \to \bigwedge^u R^\ell$ between u^{th} exterior powers. $\square$

Here now is our comparison connecting the algebra of quiver ideals, which are generated by minors of fixed size in products of generic matrices, to that of Schubert determinantal ideals, which are generated by minors of varying sizes in a single generic matrix. We remark that it does *not* imply that the generators in Definition 17.3 form a Gröbner basis; see the Notes.

Theorem 17.17 *Let $\mathbf{r}$ be a rank array and $v(\mathbf{r})$ its Zelevinsky permutation. Under the map $\Bbbk[\mathbf{x}] \twoheadrightarrow \Bbbk[\mathbf{f}]$, the image of the Schubert determinantal ideal $I_{v(\mathbf{r})}$ equals the quiver ideal $I_{\mathbf{r}}$. Equivalently, $\Bbbk[\mathbf{f}]/I_{\mathbf{r}} \cong \Bbbk[\mathbf{x}]/(I_{v(\mathbf{r})} + \mathfrak{m}_{\mathbf{f}})$.*

Example 17.18 For a generic 4×5 matrix Φ_1 and 5×3 matrix Φ_2, let I be the ideal of 3×3 minors in $\Phi_1\Phi_2$. Thus $I = I_{\mathbf{r}}$ for the rank array $\mathbf{r}$ of the lacing diagrams on the right-hand side of (17.3). The essential set of $v(\mathbf{r})$ consists of the two boxes at $(9,8)$ and $(4,3)$, so $I_{v(\mathbf{r})}$ is generated by the 7×7 minors in $\mathbf{x}_{9\times 8}$ and the entries of $\mathbf{x}_{4\times 3}$ (Exercise 17.4). Using the generators of $\mathfrak{m}_{\mathbf{f}}$ to set $\mathbf{x}$ variables equal to 0 or 1 yields the block 2×2 matrix

$$
\Phi \quad - \quad
\begin{bmatrix}
\begin{smallmatrix} 0\ 0\ 0 \\ 0\ 0\ 0 \\ 0\ 0\ 0 \\ 0\ 0\ 0 \end{smallmatrix} & \Phi_1 \\
\Phi_2 & \begin{smallmatrix} 1 \\ 1 \\ 1 \\ 1 \end{smallmatrix}
\end{bmatrix}
$$

in the northwest 9×8 corner. It follows from the Binet–Cauchy formula for the minors of $\Phi_1\Phi_2$ as sums of products of minors in Φ_1 and in Φ_2 that the ideal generated by all of the 7×7 minors in the block matrix Φ equals I. Compare the above block matrix with the general version in (17.6). $\Diamond$

Proof of Theorem 17.17. By Lemma 17.13 the essential set $\mathcal{E}ss(v(\mathbf{r}))$ consists of boxes (k,ℓ) at the southeast corners of blocks. Therefore, by Theorem 15.15, $I_{v(\mathbf{r})}$ is generated by the minors of size $1 + \mathrm{rank}(v(\mathbf{r})_{k\times\ell})$ in $\mathbf{x}_{k\times\ell}$ for the southeast corners (k,ℓ) of blocks on or below the superantidiagonal, along with all variables strictly above the block superantidiagonal.

Consider a box (k,ℓ) at the southeast corner of $B_{i+1,j-1}$, the intersection of block column $i + 1$ and block row $j - 1$, so that

$$
\mathrm{rank}(v(\mathbf{r})_{k\times\ell}) \;=\; \sum_{\substack{\alpha>i \\ \beta<j}} q_{\alpha\beta} + \sum_{m=i+1}^{j} r_{m-1,m} \tag{17.5}
$$

by definition of Zelevinsky permutation. We pause to prove the following.

Lemma 17.19 *The number* $\mathrm{rank}(v(\mathbf{r})_{k\times\ell})$ *in (17.5) is* $r_{ij} + \sum_{m=i+1}^{j-1} r_m$.

Proof. The coefficient on $q_{\alpha\beta}$ in $r_{ij} + \sum_{m=i+1}^{j-1} r_m$ is the number of elements in $\{r_{ij}\} \cup \{r_{i+1,i+1},\ldots,r_{j-1,j-1}\}$ that are weakly northwest of $r_{\alpha\beta}$ in the rank array $\mathbf{r}$ (when the array $\mathbf{r}$ is oriented so that its southeast corner is r_{0n}). This number equals the number of elements in $\{r_{i,i+1},\ldots,r_{j-1,j}\}$ that are weakly northwest of $r_{\alpha\beta}$, unless $r_{\alpha\beta}$ happens to lie strictly north and strictly west of r_{ij}, in which case we get one fewer. This one fewer is exactly made up by the sum of entries from $\mathbf{q}$ in (17.5). □

Resuming the proof of Theorem 17.17, consider the minors in $I_{v(\mathbf{r})}$ coming from the northwest $k \times \ell$ submatrix $\mathbf{x}_{k\times\ell}$, for (k,ℓ) in the southeast corner of $B_{i+1,j-1}$. Taking their images in $\Bbbk[\mathbf{f}]$ has the effect of setting the appropriate $\mathbf{x}$ variables to 0 or 1 and then changing the block superantidiagonal $\mathbf{x}$ variables into the corresponding $\mathbf{f}$ variables. Therefore the minors in $I_{v(\mathbf{r})}$ become minors of the matrix in (17.4) if each ϕ_i is replaced by the generic matrix Φ_i of $\mathbf{f}$ variables. In particular, using Lemma 17.19, the equations in $\Bbbk[\mathbf{f}]$ from $\mathbf{x}_{k\times\ell}$ are the minors of size $1 + u + r_{ij}$ in the generic $(u+r_i)\times(u+r_i)$ block matrix

$$
\begin{bmatrix}
0 & 0 & & 0 & \Phi_{i+1} \\
0 & 0 & & \Phi_{i+2} & 1 \\
0 & 0 & \cdot^{\cdot} & 1 & 0 \\
0 & \Phi_{j-1} & \cdot^{\cdot} & 0 & 0 \\
\Phi_j & 1 & & 0 & 0
\end{bmatrix},
\tag{17.6}
$$

where $u = \sum_{m=i+1}^{j-1} r_m$ is the sum of the ranks of the subantidiagonal identity blocks. The ideal generated by these minors of size $1 + u + r_{ij}$ is preserved under multiplication of (17.6) by any determinant 1 matrix with entries in $\Bbbk[\mathbf{f}]$, by Lemma 17.16. Now multiply (17.6) on the left by

$$
\begin{bmatrix}
1 & -\Phi_{i+1} & \Phi_{i+1}\Phi_{i+2} & \cdots & \pm\Phi_{i+1,j-2} & \mp\Phi_{i+1,j-1} \\
 & 1 & & & & \\
 & & 1 & & & \\
 & & & \cdot^{\cdot\cdot} & & \\
 & & & & 1 & \\
 & & & & & 1
\end{bmatrix},
$$

where $\Phi_{i+1,m} = \Phi_{i+1}\cdots\Phi_m$ for $i+1 \leq m$. The result agrees with (17.6) except in its top block row, which has left block $(-1)^{j-1-i}\Phi_{i+1}\cdots\Phi_j$ and all other blocks zero. Crossing out the top block row and the left block column leaves a block upper-left-triangular matrix that is square of size u, so the size $1+u+r_{ij}$ minors in (17.6) generate the same ideal in $\Bbbk[\mathbf{f}]$ as the size $1 + r_{ij}$ minors in $\Phi_{i+1}\cdots\Phi_j$. This holds for all $i \leq j$, completing the proof. □

Corollary 17.20 *The Zelevinsky map takes the quiver locus* $\Omega_{\mathbf{r}} \subseteq$ Mat *isomorphically to the intersection* $\mathcal{Z}(\Omega_{\mathbf{r}}) = \overline{X}_{v(\mathbf{r})} \cap \mathcal{Z}(\mathrm{Mat})$ *of the matrix Schubert variety* $\overline{X}_{v(\mathbf{r})}$ *with the affine space* $\mathcal{Z}(\mathrm{Mat})$ *inside of* M_d.

17.3 Primality and Cohen–Macaulayness

Theorem 17.17 shows how to get the equations for $I_{\mathbf{r}}$ directly from those for $I_{v(\mathbf{r})}$: set the appropriate $\mathbf{x}$ variables to 0 or 1. Its proof never needed that Schubert determinantal ideals are prime or Cohen–Macaulay (Corollaries 16.29 and 16.44). Our next goal is to put these assertions to good use, to reach the same conclusions for quiver ideals. This involves a more detailed study of the group theory surrounding the geometry in Corollary 17.20.

Let P be the *parabolic subgroup* of block lower triangular matrices in GL_d, where the diagonal blocks have sizes $r_0, \ldots, r_n$ (proceeding from left to right). The quotient $P \backslash GL_d$ of the general linear group GL_d by the parabolic subgroup P is called a *partial flag variety*. By definition, the Schubert variety $X_{v(\mathbf{r})}$ is the image of $\overline{X}_{v(\mathbf{r})} \cap GL_d$ in $P \backslash GL_d$. The Zelevinsky image $\mathcal{Z}(\Omega_{\mathbf{r}})$ of the quiver locus maps isomorphically to its image inside of $X_{v(\mathbf{r})}$, and this image is often called the *opposite Schubert cell* in $X_{v(\mathbf{r})}$, even though $\mathcal{Z}(\Omega_{\mathbf{r}})$ is usually not isomorphic to an affine space (i.e., $\Omega_{\mathbf{r}}$ is not a cell).

Note that the block structure on P is block-column reversed from the one considered earlier in this chapter. The coordinate ring $\Bbbk[P]$ is obtained from the polynomial ring $\Bbbk[\mathbf{p}]$ in the variables from the block lower triangle by inverting the determinant polynomial. In particular, the square blocks $\mathbf{p}^{00}, \ldots, \mathbf{p}^{nn}$ in (17.7) have inverses filled with regular functions on P.

Lemma 17.21 *The multiplication map* $P \times \mathrm{Mat} \to P \cdot \mathcal{Z}(\mathrm{Mat})$ *that sends* (π, ϕ) *to the product* $\pi \mathcal{Z}(\phi)$ *of matrices in* M_d *is an isomorphism of varieties that takes* $P \times \Omega_{\mathbf{r}}$ *isomorphically to* $P \cdot \mathcal{Z}(\Omega_{\mathbf{r}})$ *for each rank array* $\mathbf{r}$.

Proof. It is enough to treat the case where $\Omega_{\mathbf{r}} = \mathrm{Mat}$. Denote by Φ the generic matrix obtained from (17.4) after replacing its blocks ϕ_i by Φ_i, and let $\mathbf{x}_{v_*}$ be the block matrix of coordinate variables on $\overline{X}_{v_*}$, in the left-hand matrix below.

$$
\begin{bmatrix}
0 & 0 & 0 & \mathbf{x}^{01} & \mathbf{x}^{00} \\
0 & 0 & \mathbf{x}^{12} & \mathbf{x}^{11} & \mathbf{x}^{10} \\
0 & \cdot\cdot & \mathbf{x}^{22} & \mathbf{x}^{21} & \mathbf{x}^{20} \\
\mathbf{x}^{n-1,n} & \vdots & \vdots & \vdots & \vdots \\
\mathbf{x}^{nn} & \cdots & \mathbf{x}^{n2} & \mathbf{x}^{n1} & \mathbf{x}^{n0}
\end{bmatrix}
\mapsto
\begin{bmatrix}
\mathbf{p}^{00} & 0 & 0 & 0 & 0 \\
\mathbf{p}^{10} & \mathbf{p}^{11} & 0 & 0 & 0 \\
\mathbf{p}^{20} & \mathbf{p}^{21} & \mathbf{p}^{22} & 0 & 0 \\
\vdots & \vdots & \vdots & \ddots & 0 \\
\mathbf{p}^{n0} & \mathbf{p}^{n1} & \mathbf{p}^{n2} & \cdots & \mathbf{p}^{nn}
\end{bmatrix}
\cdot \Phi \quad (17.7)
$$

Direct calculation show that $P \cdot \mathcal{Z}(\mathrm{Mat}) \subseteq \overline{X}_{v_*}$. Therefore, the morphism $P \times \mathrm{Mat} \to P \cdot \mathcal{Z}(\mathrm{Mat})$ is determined by (and is basically equivalent to—see

Exercise 17.12) the map (17.7) of algebras $\Bbbk[\mathbf{x}_{v_*}] \to \Bbbk[\mathbf{p}, \mathbf{f}]$, which sends

$$
\begin{aligned}
\mathbf{x}^{i,i+1} &\mapsto \mathbf{p}^{ii}\Phi_{i+1} & \text{for } i = 0, \ldots, n-1, \\
\mathbf{x}^{j0} &\mapsto \mathbf{p}^{j0} & \text{for } j = 0, \ldots, n, \\
\mathbf{x}^{ji} &\mapsto \mathbf{p}^{ji} + \mathbf{p}^{j,i-1}\Phi_i & \text{for } 1 \le i \le j \le n.
\end{aligned}
\tag{17.8}
$$

Observe that the inverse of $\mathbf{x}^{00} = \mathbf{p}^{00}$ is regular on $P \cdot \mathcal{Z}(Mat)$, so we can recover $\Phi_1 = (\mathbf{x}^{00})^{-1}\mathbf{x}^{01}$. Then we can recover the first column $\mathbf{p}^{j1}$ of $\mathbf{p}$ by subtracting (the zeroth column of $\mathbf{p}$) $\cdot \Phi_1$ from (the penultimate column of $\mathbf{x}_{v_*}$). Continuing similarly, we can produce Φ and $\mathbf{p}$ as regular functions on $P \cdot \mathcal{Z}(Mat)$ to construct the inverse map $\Bbbk[P \times Mat] \to \Bbbk[P \cdot \mathcal{Z}(Mat)]$. $\square$

Proposition 17.22 *Multiplication by P on the left preserves the matrix Schubert variety $\overline{X}_{v(\mathbf{r})}$. In fact, $\overline{X}_{v(\mathbf{r})}$ is the closure in M_d of $P \cdot \mathcal{Z}(\Omega_{\mathbf{r}})$.*

Proof. Definition 17.12.4 and Corollary 15.33 imply that the matrix Schubert variety $\overline{X}_{v(\mathbf{r})}$ is stable under the action of $S_{r_0} \times \cdots \times S_{r_n}$, the block diagonal permutation matrices whose blocks have sizes $r_0, \ldots, r_n$ acting on the left. This finite group and the lower triangular Borel group B together generate P. Combining this with the stability of $\overline{X}_{v(\mathbf{r})}$ under the left action of B in Theorem 15.31 completes the proof of the first statement.

Theorem 17.17 states that $\Bbbk[\mathbf{f}]/I_{\mathbf{r}} \cong \Bbbk[\mathbf{x}]/(I_{v(\mathbf{r})} + \mathfrak{m}_{\mathbf{f}})$. But $I_{v(\mathbf{r})}$ already contains the variables in $\mathfrak{m}_{\mathbf{f}}$ above the block antidiagonal, and only $\dim(P)$ many generators of $\mathfrak{m}_{\mathbf{f}}$ remain. Thus the codimension of $\mathcal{Z}(\Omega_{\mathbf{r}})$ inside $\overline{X}_{v(\mathbf{r})}$ is at most $\dim(P)$. However, $\overline{X}_{v(\mathbf{r})}$ contains $P \cdot \mathcal{Z}(\Omega_{\mathbf{r}})$ by stability of $\overline{X}_{v(\mathbf{r})}$ under P, and $\dim(P \cdot \mathcal{Z}(\Omega_{\mathbf{r}})) = \dim(\mathcal{Z}(\Omega_{\mathbf{r}})) + \dim(P)$ by Lemma 17.21. Thus the codimension of $\mathcal{Z}(\Omega_{\mathbf{r}})$ inside $\overline{X}_{v(\mathbf{r})}$ is at least $\dim(P)$. We conclude that $\mathcal{Z}(\Omega_{\mathbf{r}})$ has codimension exactly $\dim(P)$ inside $\overline{X}_{v(\mathbf{r})}$, so that $\dim(P \cdot \mathcal{Z}(\Omega_{\mathbf{r}})) = \dim(\overline{X}_{v(\mathbf{r})})$. Since $\overline{X}_{v(\mathbf{r})}$ is an irreducible variety, it follows that $P \cdot \mathcal{Z}(\Omega_{\mathbf{r}})$ is Zariski dense in $\overline{X}_{v(\mathbf{r})}$, proving the second statement. $\square$

Theorem 17.23 *Given an irreducible rank array $\mathbf{r}$, the quiver ideal $I_{\mathbf{r}}$ inside $\Bbbk[\mathbf{f}]$ is prime and the quiver locus $\Omega_{\mathbf{r}}$ is Cohen–Macaulay.*

Proof. The matrix Schubert variety $\overline{X}_{v(\mathbf{r})}$ is Cohen–Macaulay by Corollary 16.44, and $P \cdot \mathcal{Z}(\Omega_{\mathbf{r}})$ is a dense subvariety of $\overline{X}_{v(\mathbf{r})}$ by Proposition 17.22. Since being Cohen–Macaulay is a local property [BH98, Definition 2.1.1], we conclude that $P \cdot \mathcal{Z}(\Omega_{\mathbf{r}})$ is Cohen–Macaulay. By Lemma 17.21, the coordinate ring $\Bbbk[P \cdot \mathcal{Z}(\Omega_{\mathbf{r}})]$, which we have just seen is Cohen–Macaulay, is isomorphic to the localization by $\det(\mathbf{p})$ of the polynomial ring $\Bbbk[\Omega_{\mathbf{r}}][\mathbf{p}]$ over the coordinate ring of $\Omega_{\mathbf{r}}$. This localization is Cohen–Macaulay if and only if $\Bbbk[\Omega_{\mathbf{r}}][\mathbf{p}]$ is; see Exercise 17.16. As the equations setting the off-diagonal $\mathbf{p}$ variables to 0 and the diagonal $\mathbf{p}$ variables to 1 constitute a regular sequence on $\Bbbk[\Omega_{\mathbf{r}}][\mathbf{p}]$, we conclude by Criterion 2 of Theorem 13.37 and repeated application of Lemma 8.27 that $\Bbbk[\Omega_{\mathbf{r}}]$ is Cohen–Macaulay.

The variety $\Omega_{\mathbf{r}}$ is irreducible by Lemma 17.21 and Proposition 17.22, because $\overline{X}_{v(\mathbf{r})}$ is, so the radical of $I_{\mathbf{r}}$ is prime; but it still remains to prove that $I_\mathbf{r}$ is itself prime. By Theorem 17.17, we need that the image of $\mathfrak{m}_\mathbf{f}$ in $\Bbbk[\mathbf{x}]/I_{v(\mathbf{r})}$ is prime. As the homomorphism $\Bbbk[\mathbf{x}]/I_{v(\mathbf{r})} \to \Bbbk[P \cdot \mathcal{Z}(\Omega_\mathbf{r})]$ is injective by Proposition 17.22, we only need the image of $\mathfrak{m}_\mathbf{f}$ in $\Bbbk[P \cdot \mathcal{Z}(\Omega_\mathbf{r})]$ to generate a prime ideal. To that end, we identify the ideal generated by the image of $\mathfrak{m}_\mathbf{f}$ in $\Bbbk[P \times \Omega_\mathbf{r}]$ under the isomorphism with $\Bbbk[P \cdot \mathcal{Z}(\Omega_\mathbf{r})]$ in Lemma 17.21, given by (17.7) and (17.8). The generators of $\mathfrak{m}_\mathbf{f}$ set $\mathbf{x}^{ji} = 0$ in (17.8) for $i < j$ and $\mathbf{x}^{ii} = 1$. By induction on i, the images in $\Bbbk[\mathbf{p}, \mathbf{f}]$ of these equations imply the equations setting $\mathbf{p}^{ii} = 1$ and $\mathbf{p}^{ji} = 0$ for $i < j$. Hence $\mathbf{p}^{ii}\Phi_{i+1} = \Phi_{i+1}$ modulo $\mathfrak{m}_\mathbf{f}$, so the image of $\mathfrak{m}_\mathbf{f}$ generates the kernel of the homomorphism $\Bbbk[P \times \Omega_\mathbf{r}] \to \Bbbk[\Omega_\mathbf{r}]$ coming from the inclusion $\Omega_\mathbf{r} \cong \mathrm{id} \times \Omega_\mathbf{r} \hookrightarrow P \times \Omega_\mathbf{r}$. The result follows because $\Bbbk[\Omega_\mathbf{r}]$ is a domain. $\qquad\square$

17.4 Quiver polynomials

Having exploited the algebra and geometry of matrix Schubert varieties to deduce qualitative statements about quiver ideals and loci, we now turn to more quantitative data, namely multidegrees and K-polynomials. For this, we (finally) need full details on the indexing of all the variables involved.

Again setting $d = r_0 + \cdots + r_n$, the coordinate ring $\Bbbk[\mathbf{f}]$ of *Mat* is graded by $\mathbb{Z}^d$. To describe this grading efficiently, write

$$\mathbb{Z}^d = \mathbb{Z}^{r_0} \oplus \cdots \oplus \mathbb{Z}^{r_n}, \quad \text{where} \quad \mathbb{Z}^{r_i} = \mathbb{Z} \cdot \{\mathbf{e}_1^i, \ldots \mathbf{e}_{r_i}^i\}.$$

Thus the basis of $\mathbb{Z}^d$ splits into a sequence of $n+1$ subsets $\mathbf{e}^0, \ldots, \mathbf{e}^n$ of sizes $r_0, \ldots, r_n$. We declare the variable $f_{\alpha\beta}^i \in \Bbbk[\mathbf{f}]$ to have degree

$$\deg(f_{\alpha\beta}^i) = \mathbf{e}_\alpha^{i-1} - \mathbf{e}_\beta^i \tag{17.9}$$

in $\mathbb{Z}^d$ for each $i = 1, \ldots, n$. Under this multigrading, the quiver ideal $I_\mathbf{r}$ is homogeneous. Indeed, the entries in products $\Phi_{i+1} \cdots \Phi_j$ of consecutive matrices are $\mathbb{Z}^d$-graded (check this!), so minors in such products are, too. When we write multidegrees and K-polynomials for this $\mathbb{Z}^d$-grading, we similarly split the list $\mathbf{t}$ of d variables into a sequence of $n+1$ alphabets $\mathbf{t}^0, \ldots, \mathbf{t}^n$ of sizes $r_0, \ldots, r_n$, so that the ith alphabet is $\mathbf{t}^i = t_1^i, \ldots, t_{r_i}^i$.

Definition 17.24 Under the above $\mathbb{Z}^d$-grading on $\Bbbk[\mathbf{f}]$, the multidegree

$$\mathcal{Q}_\mathbf{r}(\mathbf{t} - \mathring{\mathbf{t}}) = \mathcal{C}(\Omega_\mathbf{r}; \mathbf{t})$$

of $\Bbbk[\mathbf{f}]/I_\mathbf{r}$ is the (**ordinary**) **quiver polynomial** for the rank array $\mathbf{r}$.

For the moment, the argument $\mathbf{t} - \mathring{\mathbf{t}}$ of $\mathcal{Q}_\mathbf{r}$ can be regarded as a formal symbol, denoting that $n+1$ alphabets $\mathbf{t} = \mathbf{t}^0, \ldots, \mathbf{t}^n$ are required as input. Later in this section we will define "double quiver polynomials", with arguments $\mathbf{t} - \mathring{\mathbf{s}}$ indicating two sequences of alphabets as input. Then, in Theorem 17.34, the symbol $\mathring{\mathbf{t}}$ will take on additional meaning as the reversed sequence $\mathbf{t}^n, \ldots, \mathbf{t}^0$ of alphabets constructed from $\mathbf{t}$.

Example 17.25 The quiver ideal $I_{\mathbf{r}}$ for the rank array determined by the lacing diagram is a complete intersection of codimension 2. It is generated by the two entries in the product $\Phi_1\Phi_2\Phi_3$ of the generic 2×3, 3×3, and 3×1 matrices. These two entries have degree $\mathbf{e}_1^0 - \mathbf{e}_1^3$ and $\mathbf{e}_2^0 - \mathbf{e}_1^3$, so the multidegree of $\Bbbk[\mathbf{f}]/I_{\mathbf{r}}$ is $\mathcal{Q}_{\mathbf{r}}(\mathbf{t} - \overset{\circ}{\mathbf{t}}) = (t_1^0 - t_1^3)(t_2^0 - t_1^3)$. $\diamond$

Example 17.26 Consider a sequence of $2n$ vector spaces with dimensions $1, 2, 3 \ldots, n-1, n, n, n-1, \ldots, 3, 2, 1$. For a size $n + 1$ permutation matrix w, let $\mathbf{q}_w$ be the lace array whose entries are zero outside of the southeast $n \times n$ corner, which is filled with 1's and 0's by rotating $w_{n \times n}$ around $180°$. Exercise 17.14 explores the combinatorics of the arrays $\mathbf{q}_w$. Quiver polynomials for the associated rank arrays $\mathbf{r}_w$ are called **Fulton polynomials**. $\diamond$

The algebraic connection from quiver ideals to Schubert determinantal ideals will allow us to compute quiver polynomials in terms of double Schubert polynomials. For this, the coordinate ring $\Bbbk[\mathbf{x}] = \Bbbk[M_d]$ of the $d \times d$ matrices is multigraded by the group $\mathbb{Z}^{2d} = (\mathbb{Z}^{r_0} \oplus \cdots \oplus \mathbb{Z}^{r_n})^2$, which we take to have basis $\{\mathbf{e}_\alpha^i, \dot{\mathbf{e}}_\alpha^i \mid i = 0, \ldots, n \text{ and } \alpha = 1, \ldots, r_i\}$; note the dot over the second $\mathbf{e}_\alpha^i$. In our context, it is most natural to index the variables

$$\mathbf{x} = (x_{\alpha\beta}^{ji} \mid i, j = 0, \ldots, n \text{ and } \alpha = 1, \ldots, r_j \text{ and } \beta = 1, \ldots, r_i)$$

in the generic $d \times d$ matrix in a slightly unusual manner: $x_{\alpha\beta}^{ji} \in \Bbbk[\mathbf{x}]$ occupies the spot in row α and column β within the rectangle at the intersection of the ith block column and the jth block row, where we label block columns starting from the *right*. Declare the variable $x_{\alpha\beta}^{ji}$ to have degree

$$\deg(x_{\alpha\beta}^{ji}) = \mathbf{e}_\alpha^j - \dot{\mathbf{e}}_\beta^i. \qquad (17.10)$$

To write multidegrees and K-polynomials we use two sets of $n + 1$ alphabets

$$\mathbf{t} = \mathbf{t}^0, \ldots, \mathbf{t}^n \quad \text{and} \quad \overset{\circ}{\mathbf{s}} = \mathbf{s}^n, \ldots, \mathbf{s}^0, \qquad (17.11)$$

where

$$\mathbf{t}^j = t_1^j, \ldots, t_{r_j}^j \quad \text{and} \quad \mathbf{s}^j = s_1^j, \ldots, s_{r_j}^j. \qquad (17.12)$$

We rarely see the degree (17.10) directly; more often, we see the polynomial $t_\alpha^j - s_\beta^i = \mathcal{C}(\Bbbk[\mathbf{x}]/\langle x_{\alpha\beta}^{ji}\rangle; \mathbf{t}, \mathbf{s})$, which we call the *ordinary weight* of $x_{\alpha\beta}^{ji}$. Ordinary weights are the building blocks for multidegrees because of Theorem 8.44. (The analogous building block for K-polynomials, namely the ratio $t_\alpha^j/s_\beta^i = \mathcal{K}(\Bbbk[\mathbf{x}]/\langle x_{\alpha\beta}^{ji}\rangle; \mathbf{t}, \mathbf{s})$, is called the *exponential weight* of $x_{\alpha\beta}^{ji}$.) Pictorially, label the rows of the $d \times d$ grid with the $\mathbf{t}$ variables in the order they are given, from top to bottom, and similarly label the columns with $\overset{\circ}{\mathbf{s}} = \mathbf{s}^n, \ldots, \mathbf{s}^0$, from left to right. The ordinary weight of the variable $x_{\alpha\beta}^{ji}$ is then its row $\mathbf{t}$-label minus its column $\mathbf{s}$-label.

For notational clarity in examples, it is convenient to use alphabets $\mathbf{t}^0 = \mathbf{a}$ and $\mathbf{t}^1 = \mathbf{b}$ and $\mathbf{t}^2 = \mathbf{c}$, and so on, rather than upper indices, where

$$\mathbf{a} = a_1, a_2, a_3, \ldots, \quad \mathbf{b} = b_1, b_2, b_3, \ldots, \quad \text{and} \quad \mathbf{c} = c_1, c_2, c_3, \ldots.$$

The quiver polynomial in Example 17.25 is $(a_1 - d_1)(a_2 - d_1)$ in this notation. For the $\mathbf{s}$ alphabets, we use $\mathbf{s}^0 = \dot{\mathbf{a}}$ and $\mathbf{s}^1 = \dot{\mathbf{b}}$ and $\mathbf{s}^2 = \dot{\mathbf{c}}$, and so on, where

$$\dot{\mathbf{a}} = \dot{a}_1, \dot{a}_2, \dot{a}_3, \ldots, \quad \dot{\mathbf{b}} = \dot{b}_1, \dot{b}_2, \dot{b}_3, \ldots, \quad \text{and} \quad \dot{\mathbf{c}} = \dot{c}_1, \dot{c}_2, \dot{c}_3, \ldots$$

are the same as $\mathbf{a}, \mathbf{b}, \mathbf{c}, \ldots$ but with dots on top. All of the above notation should be made clearer by the following example.

Example 17.27 If $(r_0, r_1, r_2) = (2, 3, 1)$ then $\Bbbk[\mathbf{x}]$ has variables $x_{\alpha\beta}^{ji}$ as they appear in the following matrices (the x variables are the same in both):

	s_1^2	s_1^1	s_2^1	s_3^1	s_1^0	s_2^0
t_1^0	x_{11}^{02}	x_{11}^{01}	x_{12}^{01}	x_{13}^{01}	x_{11}^{00}	x_{12}^{00}
t_2^0	x_{21}^{02}	x_{21}^{01}	x_{22}^{01}	x_{23}^{01}	x_{21}^{00}	x_{22}^{00}
t_1^1	x_{11}^{12}	x_{11}^{11}	x_{12}^{11}	x_{13}^{11}	x_{11}^{10}	x_{12}^{10}
t_2^1	x_{21}^{12}	x_{21}^{11}	x_{22}^{11}	x_{23}^{11}	x_{21}^{10}	x_{22}^{10}
t_3^1	x_{31}^{12}	x_{31}^{11}	x_{32}^{11}	x_{33}^{11}	x_{31}^{10}	x_{32}^{10}
t_1^2	x_{11}^{22}	x_{11}^{21}	x_{12}^{21}	x_{13}^{21}	x_{11}^{20}	x_{12}^{20}

$=$

	$\dot{c}_1$	$\dot{b}_1$	$\dot{b}_2$	$\dot{b}_3$	$\dot{a}_1$	$\dot{a}_2$
a_1	x_{11}^{02}	x_{11}^{01}	x_{12}^{01}	x_{13}^{01}	x_{11}^{00}	x_{12}^{00}
a_2	x_{21}^{02}	x_{21}^{01}	x_{22}^{01}	x_{23}^{01}	x_{21}^{00}	x_{22}^{00}
b_1	x_{11}^{12}	x_{11}^{11}	x_{12}^{11}	x_{13}^{11}	x_{11}^{10}	x_{12}^{10}
b_2	x_{21}^{12}	x_{21}^{11}	x_{22}^{11}	x_{23}^{11}	x_{21}^{10}	x_{22}^{10}
b_3	x_{31}^{12}	x_{31}^{11}	x_{32}^{11}	x_{33}^{11}	x_{31}^{10}	x_{32}^{10}
c_1	x_{11}^{22}	x_{11}^{21}	x_{12}^{21}	x_{13}^{21}	x_{11}^{20}	x_{12}^{20}

The ordinary weight of each x variable equals its row label minus its column label. For example, the variable x_{23}^{01} has ordinary weight $t_2^0 - s_3^1 = a_2 - b_3$. ◇

The coordinate ring $\Bbbk[\mathcal{Z}(Mat)] = \Bbbk[\mathbf{x}]/\mathfrak{m}_{\mathbf{f}}$ of the image of the Zelevinsky map is not naturally multigraded by all of $\mathbb{Z}^{2d}$, but only by $\mathbb{Z}^d$, with the variable $x_{\alpha\beta}^{ji} \in \Bbbk[\mathcal{Z}(Mat)]$ having ordinary weight $t_\alpha^j - t_\beta^i$. This convention is consistent with the multigrading on $\Bbbk[\mathbf{f}]$ in (17.9) under the isomorphism to $\Bbbk[\mathbf{x}]/\mathfrak{m}_{\mathbf{f}}$ induced by the Zelevinsky map. Indeed, the $\mathbf{x}$ variable $x_{\alpha\beta}^{i-1,i} \in \Bbbk[\mathbf{x}]/\mathfrak{m}_{\mathbf{f}}$ maps to $f_{\alpha\beta}^i \in \Bbbk[\mathbf{f}]$, and their ordinary weights $t_\alpha^{i-1} - t_\beta^i$ agree. In what follows, we need to consider not only the Zelevinsky image of Mat but also the variety of all block upper-left triangular matrices.

Definition 17.28 The **opposite big cell** is the variety Y inside M_d obtained by setting $\mathbf{x}^{ji} = 0$ for $i < j$ and $\mathbf{x}^{ii} = \mathbf{1}$ for all i. Denote the remaining nonconstant coordinates on Y by $\mathbf{y} = \{y_{\alpha\beta}^{ji} \mid i > j\}$, so $\Bbbk[Y] = \Bbbk[\mathbf{y}]$.

Using language at the end of Section 17.3, Y is the opposite cell in the Schubert subvariety of $P \backslash GL_d$ consisting of the whole space. Note that Y is actually a cell—that is, isomorphic to an affine space. The $\mathbb{Z}^d$-grading of $\Bbbk[\mathbf{x}]$ descends to the $\mathbb{Z}^d$-grading of $\Bbbk[\mathbf{y}]$, which is positive (check this!).

Example 17.29 In the situation of Example 17.27, the coordinate ring $\Bbbk[\mathbf{y}]$ has only the variables $y_{\alpha\beta}^{ji}$ that appear in the following matrices:

	t_1^2	t_1^1	t_2^1	t_3^1	t_1^0	t_2^0
t_1^0	y_{11}^{02}	y_{11}^{01}	y_{12}^{01}	y_{13}^{01}	1	
t_2^0	y_{21}^{02}	y_{21}^{01}	y_{22}^{01}	y_{23}^{01}		1
t_1^1	y_{11}^{12}	1				
t_2^1	y_{21}^{12}		1			
t_3^1	y_{31}^{12}			1		
t_1^2	1					

$=$

	c_1	b_1	b_2	b_3	a_1	a_2
a_1	y_{11}^{02}	y_{11}^{01}	y_{12}^{01}	y_{13}^{01}	1	
a_2	y_{21}^{02}	y_{21}^{01}	y_{22}^{01}	y_{23}^{01}		1
b_1	y_{11}^{12}	1				
b_2	y_{21}^{12}		1			
b_3	y_{31}^{12}			1		
c_1	1					

In this case, the variable y_{23}^{01} has ordinary weight $t_2^0 - t_3^1 = a_2 - b_3$. $\diamond$

Definition 17.30 The **double quiver polynomial** $\mathcal{Q}_\mathbf{r}(\mathbf{t} - \mathring{\mathbf{s}})$ is the ratio

$$\mathcal{Q}_\mathbf{r}(\mathbf{t} - \mathring{\mathbf{s}}) = \frac{\mathfrak{S}_{v(\mathbf{r})}(\mathbf{t} - \mathring{\mathbf{s}})}{\mathfrak{S}_{v_*}(\mathbf{t} - \mathring{\mathbf{s}})}$$

of double Schubert polynomials in the concatenations of the two sequences of finite alphabets described in (17.11) and (17.12).

The denominator $\mathfrak{S}_{v_*}(\mathbf{t} - \mathring{\mathbf{s}})$ should be regarded as a fudge factor, being simply the product of all ordinary weights $(t_* - s_*)$ of variables lying strictly above the block superantidiagonal. These variables lie in locations corresponding to $*$ entries in the diagram of every Zelevinsky permutation, so $\mathfrak{S}_{v_*}$ obviously divides $\mathfrak{S}_{v(\mathbf{r})}$ (see Corollary 16.30).

The simple relation between double and ordinary quiver polynomials, to be presented in Theorem 17.34, justifies the notation $\mathcal{Q}_\mathbf{r}(\mathbf{t} - \mathring{\mathbf{t}})$ for the ordinary case: quiver polynomials are the specializations of double quiver polynomials obtained by setting $\mathbf{s}^i = \mathbf{t}^i$ for all i. For this purpose, write

$$\mathring{\mathbf{t}} = \mathbf{t}^n, \ldots, \mathbf{t}^0$$

to mean the reverse of the finite list $\mathbf{t}$ of alphabets from (17.11). Consequently, setting $\mathbf{s}^i = \mathbf{t}^i$ for all i is simply setting $\mathring{\mathbf{s}} = \mathring{\mathbf{t}}$. In the case where every block has size 1, so $P = B$ is the Borel subgroup of lower triangular matrices in GL_d, each alphabet in the list $\mathbf{t}$ consists of just one variable (as opposed to there being only one alphabet in the list), so the reversed list $\mathring{\mathbf{t}}$ is really just a globally reversed alphabet in that case.

Our goal is to relate double quiver polynomials to ordinary quiver polynomials. At first, we work with K-polynomials, for which we need a lemma.

Proposition 17.31 *Let $\mathcal{F}_{\bullet}$ be a $\mathbb{Z}^{2d}$-graded free resolution of $\Bbbk[\mathbf{x}]/I_{v(\mathbf{r})}$ over $\Bbbk[\mathbf{x}]$. If $\mathfrak{m}_\mathbf{y}$ is the ideal of Y in $\Bbbk[\mathbf{x}]$, then the complex $\mathcal{F}_{\bullet} \otimes_{\Bbbk[\mathbf{x}]} \Bbbk[\mathbf{y}] = \mathcal{F}_{\bullet}/\mathfrak{m}_\mathbf{y}\mathcal{F}_{\bullet}$ is a $\mathbb{Z}^d$-graded free resolution of $\Bbbk[\mathcal{Z}(\Omega_\mathbf{r})]$ over $\Bbbk[\mathbf{y}]$.*

Proof. Note that $\mathcal{F}_{\bullet}/\mathfrak{m}_\mathbf{y}\mathcal{F}_{\bullet}$ is complex of $\mathbb{Z}^d$-graded free modules over $\Bbbk[\mathbf{y}]$. Indeed, coarsening the $\mathbb{Z}^{2d}$-grading on $\Bbbk[\mathbf{x}]$ to the grading by $\mathbb{Z}^d$ in which $x_{\alpha\beta}^{ji}$ has ordinary weight $t_\alpha^j - t_\beta^i$ (by setting $s_\beta^i = t_\beta^i$) makes the generators of $\mathfrak{m}_\mathbf{y}$ homogeneous, because the variables set equal to 1 have degree zero.

The $\mathbf{x}$ variables in blocks strictly above the block superantidiagonal already lie inside $I_{v(\mathbf{r})}$, so $I(\mathcal{Z}(\Omega_\mathbf{r})) = I_{v(\mathbf{r})} + \mathfrak{m}_\mathbf{f} = I_{v(\mathbf{r})} + \mathfrak{m}_\mathbf{y}$ by Theorem 17.17. What we would like is for the generators of $\mathfrak{m}_\mathbf{y}$ to form a regular sequence on $\Bbbk[\mathbf{x}]/I_{v(\mathbf{r})}$, because then repeated application of Lemma 8.27 would complete the proof. What we will actually show is almost as good: we will check that the generators of $\mathfrak{m}_\mathbf{y}$ form a regular sequence on the localization of $\Bbbk[\mathbf{x}]/I_{v(\mathbf{r})}$ at every maximal ideal $\mathfrak{p}$ of $\Bbbk[\mathbf{x}]$ containing $I_{v(\mathbf{r})} + \mathfrak{m}_\mathbf{y}$.

This suffices because (i) the complex $\mathcal{F}_\bullet/\mathfrak{m}_\mathbf{y}\mathcal{F}_\bullet$ is acyclic if and only if its localization at every maximal ideal of $\Bbbk[\mathbf{y}]$ is acyclic [Eis95, Lemma 2.8], and (ii) if $\mathfrak{p}$ does not contain $I_{v(\mathbf{r})}$ then $(\mathcal{F}_\bullet)_\mathfrak{p}$—and hence also $(\mathcal{F}_\bullet/\mathfrak{m}_\mathbf{y}\mathcal{F}_\bullet)_\mathfrak{p}$—is a free resolution of 0, which is split exact.

For the local regular sequence property, we use [BH98, Theorem 2.1.2]: If N is a Cohen–Macaulay module over a local ring, and $z_1, \ldots, z_r$ is any sequence of elements, then $N/\langle z_1, \ldots, z_r \rangle N$ has dimension $\dim(N) - r$ if and only if $z_1, \ldots, z_r$ is a regular sequence on N. Noting that $\mathfrak{m}_\mathbf{y}$ is generated by $\dim(\overline{X}_{v(\mathbf{r})}) - \dim(\Omega_\mathbf{r})$ elements, we are done by Corollary 16.44. $\square$

If $\Bbbk[\mathbf{x}]/I(\overline{X})$ is a $\mathbb{Z}^{2d}$-graded coordinate ring of a subvariety $\overline{X}$ inside M_d, write $\mathcal{K}_M(\overline{X}; \mathbf{t}, \mathring{\mathbf{s}})$ for its K-polynomial. Similarly, write $\mathcal{K}_Y(Z; \mathbf{t})$ for the K-polynomial of a $\mathbb{Z}^d$-graded quotient $\Bbbk[\mathbf{y}]/I(Z)$ if $Z \subseteq Y$. The geometry in Corollary 17.20 has the following interpretation in terms of K-polynomials.

Corollary 17.32 $\mathcal{K}_Y(\mathcal{Z}(\Omega_\mathbf{r}); \mathbf{t}) = \mathcal{K}_M(\overline{X}_{v(\mathbf{r})}; \mathbf{t}, \mathring{\mathbf{t}})$.

Proof. This is immediate from Proposition 17.31, by Definition 8.32. $\square$

Lemma 17.33 *The K-polynomial* $\mathcal{K}_{\text{Mat}}(\Omega_\mathbf{r}; \mathbf{t})$ *of* $\Omega_\mathbf{r}$ *inside* Mat *is*

$$\mathcal{K}_{\text{Mat}}(\Omega_\mathbf{r}; \mathbf{t}) = \frac{\mathcal{K}_Y(\mathcal{Z}(\Omega_\mathbf{r}); \mathbf{t})}{\mathcal{K}_Y(\mathcal{Z}(\text{Mat}); \mathbf{t})}.$$

Proof. The equality $H(\Omega_\mathbf{r}; \mathbf{t}) = \mathcal{K}_{\text{Mat}}(\Omega_\mathbf{r}; \mathbf{t}) H(\text{Mat}; \mathbf{t})$ of Hilbert series (which are well-defined by positivity of the grading of $\Bbbk[\mathbf{f}]$ by $\mathbb{Z}^d$) follows from Theorem 8.20. For the same reason, we have

$$H(\text{Mat}; \mathbf{t}) = \mathcal{K}_Y(\mathcal{Z}(\text{Mat}); \mathbf{t}) H(Y; \mathbf{t}),$$

and also

$$H(\Omega_\mathbf{r}; \mathbf{t}) = \mathcal{K}_Y(\mathcal{Z}(\Omega_\mathbf{r}); \mathbf{t}) H(Y; \mathbf{t}).$$

Thus $\mathcal{K}_Y(\mathcal{Z}(\Omega_\mathbf{r}); \mathbf{t}) H(Y; \mathbf{t}) = \mathcal{K}_{\text{Mat}}(\Omega_\mathbf{r}; \mathbf{t}) \mathcal{K}_Y(\mathcal{Z}(\text{Mat}); \mathbf{t}) H(Y; \mathbf{t})$. $\square$

Theorem 17.34 *The ordinary quiver polynomial* $\mathcal{Q}_\mathbf{r}(\mathbf{t} - \mathring{\mathbf{t}})$ *is the* $\mathring{\mathbf{s}} = \mathring{\mathbf{t}}$ *specialization of the double quiver polynomial* $\mathcal{Q}_\mathbf{r}(\mathbf{t} - \mathring{\mathbf{s}})$. *In other words, the quiver polynomial* $\mathcal{Q}_\mathbf{r}(\mathbf{t} - \mathring{\mathbf{t}})$ *for a rank array* $\mathbf{r}$ *equals the ratio*

$$\mathcal{Q}_\mathbf{r}(\mathbf{t} - \mathring{\mathbf{t}}) = \frac{\mathfrak{S}_{v(\mathbf{r})}(\mathbf{t} - \mathring{\mathbf{t}})}{\mathfrak{S}_{v_*}(\mathbf{t} - \mathring{\mathbf{t}})}$$

of double Schubert polynomials in the two alphabets $\mathbf{t}$ *and* $\mathring{\mathbf{t}}$.

Proof. After clearing denominators in Lemma 17.33, substitute using Corollary 17.32 to get

$$\mathcal{K}_{\text{Mat}}(\Omega_\mathbf{r}; \mathbf{t}) \mathcal{K}_M(\overline{X}_{v_*}; \mathbf{t}, \mathring{\mathbf{t}}) = \mathcal{K}_M(\overline{X}_{v(\mathbf{r})}; \mathbf{t}, \mathring{\mathbf{t}}).$$

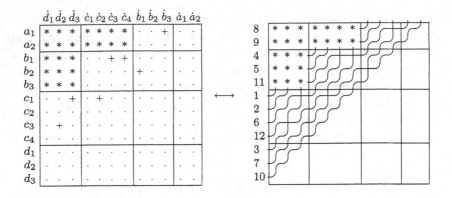

Figure 17.2: Zelevinsky pipe dream

Now substitute $1-t$ for every occurrence of each variable t, and take lowest degree terms to get

$$\mathcal{Q}_{\mathbf{r}}(\mathbf{t} - \overset{\circ}{\mathbf{t}})\,\mathfrak{S}_{v_*}(\mathbf{t} - \overset{\circ}{\mathbf{t}}) \;\; = \;\; \mathfrak{S}_{v(\mathbf{r})}(\mathbf{t} - \overset{\circ}{\mathbf{t}}).$$

The polynomial $\mathfrak{S}_{v_*}(\mathbf{t} - \overset{\circ}{\mathbf{t}})$ is nonzero, being simply the product of the $\mathbb{Z}^d$-graded ordinary weights $t_\alpha^j - t_\beta^i$ of the $\mathbf{y}$ variables $y_{\alpha\beta}^{ji}$ with $i > j$. Therefore we may divide through by $\mathfrak{S}_{v_*}(\mathbf{t} - \overset{\circ}{\mathbf{t}})$. $\square$

17.5   Pipes to laces

Having a formula for quiver polynomials in terms of Schubert polynomials produces a formula in terms of pipe dreams, given the simplicity of the denominator polynomial $\mathfrak{S}_{v_*}$. Let us begin unraveling the structure of reduced pipe dreams for Zelevinsky permutations with an example.

Example 17.35 A typical reduced pipe dream for the Zelevinsky permutation v in Example 17.14 looks like the one in Fig. 17.2, when we leave the $*$'s as they are in the diagram $D(v(\mathbf{r}))$. Although each $*$ represents a $+$ in every pipe dream for $v(\mathbf{r})$, the $*$'s will be just as irrelevant here as they were for the diagram of $v(\mathbf{r})$. The left pipe dream in Fig. 17.2 is labeled on the side and top with the row and column variables for ordinary weights. $\diamond$

Given a set D of $+$ entries in the square $d \times d$ grid, let $(\mathbf{t} - \overset{\circ}{\mathbf{s}})^D$ be its *monomial*, defined as the product over all $+$ entries in D of $(t_+ - s_+)$, where t_+ sits at the left end of the row containing $+$ and s_+ sits atop the column containing $+$.

Theorem 17.36 *The double quiver polynomial for ranks* $\mathbf{r}$ *equals the sum*

$$\mathcal{Q}_{\mathbf{r}}(\mathbf{t} - \overset{\circ}{\mathbf{s}}) \;\; = \;\; \sum_{D \in \mathcal{RP}(v(\mathbf{r}))} (\mathbf{t} - \overset{\circ}{\mathbf{s}})^{D \smallsetminus D(v_*)}$$

of the monomials for the complement of $D(v_*)$ in all reduced pipe dreams for the Zelevinsky permutation $v(\mathbf{r})$.

Proof. This follows from Definition 17.30 and Corollary 16.30, using the fact that every pipe dream $D \in \mathcal{RP}(v(\mathbf{r}))$ contains the subdiagram $D(v_*)$ and that $\mathcal{RP}(v_*)$ consists of the single pipe dream $D(v_*)$. $\square$

Double quiver polynomials $\mathcal{Q}_{\mathbf{r}}(\mathbf{t} - \overset{\circ}{\mathbf{s}})$ are thus sums of all monomials for "skew reduced pipe dreams" $D \smallsetminus D(v_*)$ with $D \in \mathcal{RP}(v(\mathbf{r}))$. That is why we only care about crosses in D occupying the block antidiagonal and superantidiagonal. The monomial $(\mathbf{t} - \overset{\circ}{\mathbf{s}})^{D \smallsetminus D(v_*)}$ for the pipe dream in Fig. 17.2 is

$$(a_1 - \dot{b}_3)(b_1 - \dot{c}_3)(b_1 - \dot{c}_4)(b_2 - \dot{b}_1)(c_1 - \dot{d}_3)(c_1 - \dot{c}_2)(c_3 - \dot{d}_2),$$

ignoring all $*$ entries as required. Removing the dots yields this pipe dream's contribution to the ordinary quiver polynomial:

Corollary 17.37 *The ordinary quiver polynomial for ranks* $\mathbf{r}$ *equals the sum*

$$\mathcal{Q}_{\mathbf{r}}(\mathbf{t} - \overset{\circ}{\mathbf{t}}) \;=\; \sum_{D \in \mathcal{RP}(v(\mathbf{r}))} (\mathbf{t} - \overset{\circ}{\mathbf{t}})^{D \smallsetminus D(v_*)}.$$

Recall that we started in Section 17.1 analyzing quiver representations by decomposing them as direct sums of laces, as in Example 17.1. Although we have by now taken a long detour, here we come back again to some concrete combinatorics: pipe dreams for Zelevinsky permutations give rise to lacing diagrams.

Definition 17.38 The jth *antidiagonal block* is the block of size $r_j \times r_j$ along the main antidiagonal in the jth block row. Given a reduced pipe dream D for the Zelevinsky permutation $v(\mathbf{r})$, define the partial permutation $w_j = w_j(D)$ sending k to ℓ if the pipe entering the kth column from the *right* of the $(j-1)^{\text{st}}$ antidiagonal block enters the jth antidiagonal block in its ℓth column from the *right*. Set $\mathbf{w}(D) = (w_1, \ldots, w_n)$, so that $\mathbf{w}(D)$ is the *lacing diagram* determined by D.

Example 17.39 The partial permutations arising from the pipe dream in Example 17.35 come from the following partial reduced pipe dreams:

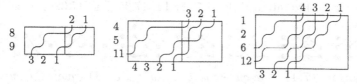

These send each number along the top either to the number along the bottom connected to it by a pipe (if such a pipe exists), or to nowhere. It is easy to see the pictorial lacing diagram $\mathbf{w}(D)$ from these pictures. Indeed, removing all segments of all pipes not contributing to one of the partial permutations leaves some pipes

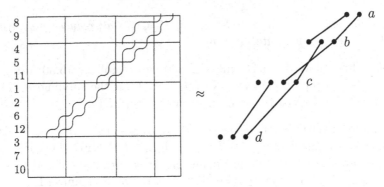

that can be interpreted directly as the desired lacing diagram

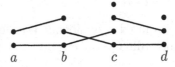

by shearing to make the rightmost dots in each row line up vertically and then reflecting through the diagonal line $\searrow$ of slope -1. $\diamondsuit$

Proposition 17.40 *Every reduced pipe dream $D \in \mathcal{RP}(v(\mathbf{r}))$ gives rise to a lacing diagram $\mathbf{w}(D)$ representing a partial permutation list with ranks $\mathbf{r}$.*

Proof. Each $\times$ entry in the permutation matrix for $v(\mathbf{r})$ corresponds to a pipe in D entering due north of it and exiting due west of it. The permutation $v(\mathbf{r})$ was specifically constructed to have exactly q_{ij} entries $\times$ (for $i \leq j$) in the intersection of the ith block row and the jth block column from the right, where $\mathbf{q}$ is the lace array from Lemma 17.5. $\square$

The fact that lacing diagrams popped out of pipe dreams for Zelevinsky permutations suggests that lacing diagrams control the combinatorics of quiver polynomials as deeply as they controlled the algebra in Section 17.1. This turns out to be true: there is a different, more intrinsic combinatorial formula for quiver polynomials in terms of Schubert polynomials. To state it, define the *length* of a lacing diagram $\mathbf{w} = (w_1, \ldots, w_n)$ to be the sum $l(\mathbf{w}) = l(w_1) + \cdots + l(w_n)$ of the lengths of its constituent partial permutations. For an irreducible rank array $\mathbf{r}$, we are interested in the set $W(\mathbf{r})$ of *minimal* lacing diagrams for $\mathbf{r}$—that is, with minimal length. For instance, with $\mathbf{r}$ as in Examples 17.1, 17.7, 17.14, 17.35, and 17.39, the set $W(\mathbf{r})$ is:

Theorem 17.41 *The quiver polynomial $\mathcal{Q}_{\mathbf{r}}(\mathbf{t} - \overset{\circ}{\mathbf{t}})$ equals the sum*

$$\mathcal{Q}_{\mathbf{r}}(\mathbf{t} - \overset{\circ}{\mathbf{t}}) = \sum_{\mathbf{w} \in W(\mathbf{r})} \mathfrak{S}_{w_1}(\mathbf{t}^0 - \mathbf{t}^1) \mathfrak{S}_{w_2}(\mathbf{t}^1 - \mathbf{t}^2) \cdots \mathfrak{S}_{w_n}(\mathbf{t}^{n-1} - \mathbf{t}^n)$$

of products of double Schubert polynomials indexed by minimal lacing diagrams $\mathbf{w} = (w_1, \ldots, w_n)$ with rank array $\mathbf{r}$.

This statement was discovered by Knutson, Miller, and Shimozono, who at first proved only that the expansion on the right-hand side has positive coefficients. After publicizing their weaker statement and conjecturing the above precise statement, independent (and quite different) proofs of the conjecture were given by the conjecturers [KMS04] and by Rimányi [BFR03]. For information on the motivation, consequences, and variations that have appeared and could in the future appear, see the Notes.

Exercises

17.1 Given the rank array $\mathbf{r} = \begin{array}{ccc|c} 2 & 1 & 0 & i/j \\ & 3 & & 0 \\ 4 & 2 & & 1 \\ 2 & 2 & 0 & 2 \end{array}$ with $n = 2$, compute the lace array $\mathbf{q}$

and rectangle array $\mathbf{R}$. Find all the minimal lacing diagrams with rank array $\mathbf{r}$.

17.2 Prove Lemma 17.5, Lemma 17.8, and Lemma 17.13.

17.3 What conditions on a dimension vector (r_0, r_1, r_2, r_3) and a rank ρ guarantee that the minors of size $\rho+1$ in the product $\Phi_1\Phi_2\Phi_3$ generate a prime ideal, where Φ_i is a generic matrix of size $r_{i-1} \times r_i$?

17.4 For the general data in Example 17.11 (and the particular case in (17.3) and Example 17.18), show that the Zelevinsky permutation has essential set of size 2. Use the Binet–Cauchy formula to prove Theorem 17.17 directly in this case.

17.5 Work out the lace array $\mathbf{q}$, rank array $\mathbf{r}$, rectangle diagram $\mathbf{R}$, and Zelevinsky permutation $v(\mathbf{r})$ for the data in Example 17.25. Check the degree calculations there. Find all six minimal lacing diagrams sharing the rank array $\mathbf{r}$. Verify the pipe dream and lacing diagram formulas in Theorems 17.36 and 17.41 for $\mathbf{r}$.

17.6 Set $d = r_0 + \cdots + r_n$ as usual, and fix an irreducible rank array $\mathbf{r}$. Consider the set $S_d(\mathbf{r})$ of permutations in S_d whose permutation matrices have the same number of nonzero entries as $v(\mathbf{r})$ does in every $r_j \times r_i$ block. Prove that $v = v(\mathbf{r})$ if and only if $v \in S_d$ and that every other permutation $v' \in S_d$ satisfies $l(v') > l(v)$.

17.7 Interpret (17.2) as a statement about the rectangles in the rectangle array $\mathbf{R}$.

17.8 A **variety of complexes** is a quiver locus $\Omega_{\mathbf{r}}$ such that for all $\phi \in \Omega_{\mathbf{r}}$, $\phi_{i-1}\phi_i = 0$ for $i = 2, \ldots, n$. Which varieties of complexes $\mathbb{k}^{r_0} \to \cdots \to \mathbb{k}^{r_n}$ are irreducible as varieties? What is the multidegree of a variety of complexes?

17.9 Pick a random quiver representation ϕ with dimension vector $(2, 3, 3, 1)$, and compute an isomorphism $\phi \cong \mathbf{w}$ with a lacing diagram $\mathbf{w}$. Could you have predicted the lace array $\mathbf{q}$ and the rank array $\mathbf{r}$ of $\mathbf{w}$? What is $\mathcal{E}ss(v(\mathbf{r}))$?

17.10 Calculate the dimension of $\Omega_{\mathbf{r}}$ in terms of the rectangle array $\mathbf{R}$ of $\mathbf{r}$.

17.11 Suppose that $\mathbf{r}$ is a rank array that is *not* irreducible. Must it always be the case that $\Omega_{\mathbf{r}}$ has more than one component? Can $\Omega_{\mathbf{r}}$ be nonreduced?

17.12 Let $\overline{P}$ be the closure of P in M_d. Verify that (17.7) and (17.8) correspond to a morphism $\overline{P} \times Mat \to \overline{X}_{v_*}$ that happens to take the subset $P \times Mat$ to the subset $P \cdot \mathcal{Z}(Mat) \subset \overline{X}_{v_*}$. Use Proposition 17.22 to show that (17.7) and (17.8) define the *only* algebra map $\mathbb{k}[\mathbf{x}_{v_*}] \to \mathbb{k}[\mathbf{p}, \mathbf{f}]$ inducing the morphism $P \times Mat \to P \cdot \mathcal{Z}(Mat)$.

17.13 Let $1 + \mathbf{r}$ be obtained by adding 1 to every entry of a rank array $\mathbf{r}$. Compare the lace arrays of $\mathbf{r}$ and $1 + \mathbf{r}$. What is the difference between the Zelevinsky permutations $v(\mathbf{r})$ and $v(1 + \mathbf{r})$? How about the rectangle arrays of $\mathbf{r}$ and $1 + \mathbf{r}$?

17.14 Let w be a permutation matrix of size $n + 1$, and consider the rank array $\mathbf{r}_w$ in Example 17.26.

(a) Prove that the Zelevinsky permutation $v(\mathbf{r}_w)$ has as many diagonal $\times$ entries as will fit in each superantidiagonal block.

(b) Show that every rectangle in the rectangle array $\boldsymbol{R}$ has size 1×1, and explain how $\boldsymbol{R}$ can be naturally identified with the diagram $D(w)$.

(c) If $\mathbf{r} = \mathbf{r}_w$, then the ordinary quiver polynomial $\mathcal{Q}_\mathbf{r}$ takes $2n$ alphabets for its argument. Suppose that the first n of these alphabets are specialized to $\{t_1\}, \{t_1, t_2\}, \ldots, \{t_1, \ldots, t_n\}$ and that the last n of these alphabets are specialized to $\{s_1, \ldots, s_n\}, \ldots, \{s_1, s_2\}, \{s_1\}$. Prove that $\mathcal{Q}_\mathbf{r}$ evaluates at these alphabets to the double Schubert polynomial $\mathfrak{S}_w(\mathbf{t} - \mathbf{s})$.

17.15 Give a direct proof of Lemma 17.16, without using exterior powers.

17.16 Let $\mathbf{y}$ be a set of variables and fix a $\Bbbk$-algebra R. Using any definition of Cohen–Macaulay that suits this generality, prove that for any nonzero $f \in \Bbbk[\mathbf{y}]$, the localization $R[\mathbf{y}][f^{-1}]$ is Cohen–Macaulay if and only if $R[\mathbf{y}]$ is Cohen–Macaulay.

17.17 Prove that the minimum length for a lacing diagram with rank array $\mathbf{r}$ is the difference $l(v(\mathbf{r})) - l(v_*)$. Hint: Given a minimal lacing diagram $\mathbf{w}$, exhibit a reduced pipe dream D for $v(\mathbf{r})$ such that $\mathbf{w}(D) = \mathbf{w}$.

17.18 Show by example that Theorem 17.41 fails for double quiver polynomials when all $\mathbf{t}$ alphabets with minus signs are replaced by corresponding $\mathbf{s}$ alphabets.

Notes

The use of laces to denote indecomposable quiver representations as in Definition 17.2 is due to Abeasis and Del Fra [AD80], who identified unordered sets of laces (called *strands* there) as giving rank conditions. The refinement of this notion to include the partial permutations between columns in a lacing diagram is due to Knutson, Miller, and Shimozono [KMS04], who needed it for the statement of Theorem 17.41. Quiver ideals, quiver loci, (indecomposable) quiver representations, and Proposition 17.9 are part of a much larger theory of representations of finite type quivers; see below. The rectangle arrays in Definition 17.6 were invented by Buch and Fulton [BF99].

The Zelevinsky map originated in Zelevinsky's two-page article [Zel85], where he proved the set-theoretic (as opposed to the scheme-theoretic) version of Corollary 17.20. Zelevinsky's original big block matrix, being essentially the inverse matrix of (17.4), visibly contained all of the consecutive products $\Phi_{i+1} \cdots \Phi_j$ for $i < j$. Theorem 17.17 and the concept of Zelevinsky permutation appeared in [KMS04], from which much of Section 17.2 has been lifted with few changes. The primality in Theorem 17.23 is due to Lakshmibai and Magyar [LM98], as is the Cohen–Macaulayness of quiver loci over fields of arbitrary characteristic, although earlier, Abeasis, Del Fra, and Kraft had proved (without primality) that the underlying reduced variety is Cohen–Macaulay in characteristic zero [ADK81].

Quiver polynomials were defined by Buch and Fulton [BF99]. Double quiver polynomials as ratios of Schubert polynomials, as well as the subsequent ratio and pipe dream formulas for ordinary quiver polynomials in Theorem 17.34 and Theorem 17.36, were discovered by Knutson, Miller, and Shimozono [KMS04]. That article also contains the combinatorial connections between lacing diagrams and reduced pipe dreams for Zelevinsky permutations in Proposition 17.40. Attributions for Theorem 17.41 appear in the text, after its statement.

In contrast to the situation for minors in Chapters 15 and 16, it is not known whether there is a term order under which the generators of I_r in Definition 17.3 form a Gröbner basis. Although the degeneration to pipe dreams at the level of matrix Schubert varieties, which results in Theorem 17.36, descends to a degeneration of the Zelevinsky image $\mathcal{Z}(\Omega_r)$, this degeneration fails to be Gröbner. Indeed, some of the variables are set equal to 1, so the resulting flat family of ideals in $\Bbbk[\mathbf{x}]$ is not obtained by scaling the variables. On the other hand, there is still a *partial* Gröbner degeneration [KMS04, Section 4 and Theorem 6.16]; the components in its special fiber are indexed by lacing diagrams, so it gives rise to the positive formula in Theorem 17.41 in the manner of Corollary 16.1.

We have drawn Exercise 17.6 from [Yon03]. Exercise 17.16 was used in the proof of Theorem 17.23; it follows from [BH98, Theorems 2.1.3 and 2.1.9]. What we have called Fulton polynomials in Example 17.26 and Exercise 17.14 were originally called *universal Schubert polynomials* by Fulton because they specialize to quantum and double Schubert polynomials [Ful99]. Treatments of combinatorial aspects of Fulton polynomials and their K-theoretic analogues appear in [BKTY04a, BKTY04b].

The topics in this chapter have historically developed in the contexts of algebraic geometry and representation theory. On the algebraic geometry side, the direct motivation comes from [BF99] and its predecessors, which deal with *degeneracy loci* for vector bundle morphisms; see [FP98, Man01] for background on the long history of this perspective. In particular, the three formulas for $\mathcal{Q}_r$ in this chapter (Theorems 17.34, 17.36, and 17.41) were originally aimed at a solution in [KMS04] of the main conjecture in [BF99], which is a positive combinatorial formula for $\mathcal{Q}_r$ as a sum of products of double Schur polynomials. Further topics in this active area of research include new proofs of Theorem 17.41 or steps along the way [BFR03, Yon03], relations between quiver polynomials and symmetric functions [Buc01, BSY03], and K-theoretic versions [Buc02, Buc03, Mil03b].

The representation theory motivation comes from general quivers. The term *quiver* is a synonym for *directed graph*. In our *equioriented type A* case, the quiver is a directed path. The definition of quiver representation makes sense for arbitrary quivers (attach a vector space to each vertex and a matrix of variables to each directed edge), and the notion of quiver locus can be extended, as well (to orbit closures for the general linear group that acts by changing bases); see [ARS97] or [GR97] for background. The extent to which we understand the multidegrees of quiver loci for orientations of *Dynkin diagrams* of type A, D, or E comes from the topological perspective of Fehér and Rimányi [FR02], but as yet, there are no known analogues of the positivity in Theorem 17.41 for other types. This open problem is only a sample of the many relations of quiver representations with combinatorial commutative algebra. Other connections include the work of Bobiński and Zwara on normality and rational singularities [BZ02] as well as Derksen and Weyman on *semi-invariants* [DW00].

Chapter 18

Hilbert schemes of points

Hilbert schemes are algebraic varieties that parametrize families of ideals in polynomial rings. They are fundamental in algebraic geometry and its applications. A simple instance of a Hilbert scheme is the Grassmannian of r-planes in $\mathbb{C}^n$, written $\mathrm{Gr}^r(\mathbb{C}^n)$ in this chapter: it parametrizes all ideals generated by r linearly independent forms in $\mathbb{C}[x_1, \ldots, x_n]$. In more general cases, Hilbert schemes are still often defined by determinantal conditions. The rings arising in the study of Hilbert schemes provide an ample supply of good research problems for combinatorial commutative algebra.

We begin this chapter with an introduction to Hilbert schemes of points in the plane, which are shown to be smooth and irreducible. This leads us to introduce the work of Haiman that relates the geometry of these Hilbert schemes to the theory of symmetric functions (the $n!$ Theorem). Then we discuss Hilbert schemes of points in $\mathbb{C}^d$ for $d \geq 3$. In the final section we present *multigraded Hilbert schemes*, which parametrize ideals having a fixed Hilbert function with respect to an arbitrary multigrading on the polynomial ring. Sections 18.1, 18.2, and 18.4 are elementary in nature, in the sense that we prove (almost) everything we state. The remaining two sections are intended more as an overview. Our purpose is to present some recent advances to nonexperts and to indicate possible future directions.

Note: Our conventions regarding the uses of n and d as the number of variables and the rank of the grading group are overridden in this chapter by notation from the literature that is too standard to warrant alteration.

18.1 Ideals of points in the plane

Consider the polynomial ring $\mathbb{C}[x, y]$ in two variables over the complex numbers. As a set, the *Hilbert scheme* $H_n = \mathrm{Hilb}^n(\mathbb{C}^2)$ of n points in the plane consists of those ideals $I \subseteq \mathbb{C}[x, y]$ for which the quotient $\mathbb{C}[x, y]/I$ has dimension n as a vector space over $\mathbb{C}$. Our goal is to see that this set can be considered naturally as a smooth algebraic variety of dimension $2n$.

To begin, let us get a feeling for what an ideal I of colength n can look like. If $P_1, \ldots, P_n \in \mathbb{C}^2$ are distinct points, for example, then the ideal of functions vanishing on these n points has colength n. Ideals of this form are the *radical* colength n ideals.

At the opposite end of the spectrum, a point I in H_n could be an ideal whose (reduced) zero set consists of only one point $P \in \mathbb{C}^2$. In this case, $\mathbb{C}[x, y]/I$ is a local ring with abundant nilpotent elements. In geometric terms, this means that P carries a nonreduced scheme structure. Such a nonreduced scheme structure on the point P is far from unique; in other words, there are many length n local rings $\mathbb{C}[x, y]/I$ supported at P. In fact, we will see in Theorem 18.26 that they come in a family parametrized by an algebraic variety of dimension $(n - 1)$.

Among the ideals supported at single points, the monomial ideals are the most special. These ideals have the form $I = \langle x^{a_1}y^{b_1}, \ldots, x^{a_m}y^{b_m} \rangle$ for some nonnegative integers $a_1, b_1, \ldots, a_m, b_m$ and are supported at $(0, 0) \in \mathbb{C}^2$. As in Part I of this book, we draw the monomials outside of I as boxes under a staircase. If the diagram of monomials outside I is a Ferrers shape with λ_i boxes in row i, then $\sum_i \lambda_i = n$ is by definition a partition λ whose parts sum to n. We write $I = I_\lambda$ and say that λ is a partition of n.

Example 18.1 Consider the partition $2 + 1 + 1$ of $n = 4$. The ideal I_{2+1+1} equals $\langle x^2, xy, y^3 \rangle$. The four boxes under the staircase form an L-shape:

The monomial x^2 would be the first box after the bottom row, whereas xy would nestle in the nook of the L, and y^3 would lie atop the first column. $\diamond$

Interpolating between the above two extreme cases, if I is an arbitrary colength n ideal, then the quotient $\mathbb{C}[x, y]/I$ is a product of local rings with maximal ideals corresponding to a finite set $\{P_1, \ldots, P_r\}$ of distinct points in $\mathbb{C}^2$. The lengths $\ell_1, \ldots, \ell_r$ of these local rings (as modules over themselves) satisfy $\ell_1 + \cdots + \ell_r = n$. (Do not confuse this partition of n with the partitions obtained from monomial ideals, where $r = 1$.) When $r = n$, it must be that $\ell_i = 1$ for all i, so the ideal I is radical.

If all colength n ideals were radical, then the Hilbert scheme H_n would be easy to describe, as follows. Every unordered list of n distinct points in $\mathbb{C}^2$ corresponds to a set of $n!$ points in $(\mathbb{C}^2)^n$, or alternatively, to a single point in the nth *symmetric product* $S^n\mathbb{C}^2$, defined as the quotient $(\mathbb{C}^2)^n/S_n$ by the symmetric group S_n. Of course, not every point of $S^n\mathbb{C}^2$ corresponds to an unordered list of *distinct* points; for that, one needs to remove the *diagonal locus*

$$\{(P_1, \ldots, P_n) \in (\mathbb{C}^2)^n \mid P_i = P_j \text{ for some } i \neq j\} \tag{18.1}$$

of $(\mathbb{C}^2)^n$ before quotienting by S_n. Since S_n acts freely on the complement $((\mathbb{C}^2)^n)^\circ$ of the diagonal locus, the complement $(S^n\mathbb{C}^2)^\circ$ of the image of the diagonal locus in the quotient $S^n\mathbb{C}^2$ is smooth. Therefore, whatever variety structure we end up with, H_n will contain $(S^n\mathbb{C}^2)^\circ$ as a smooth open subvariety. This subvariety has dimension $2n$ and parametrizes the radical ideals.

The variety structure on H_n arises by identifying it as an algebraic subvariety of a familiar variety: the Grassmannian. For each nonnegative integer m, consider the vector subspace V_m inside of $\mathbb{C}[x,y]$ spanned by the $\binom{m+2}{2}$ monomials of degree at most m.

Lemma 18.2 *Given any colength n ideal I, the image of V_m spans the quotient $\mathbb{C}[x,y]/I$ as a vector space whenever $m \geq n$.*

Proof. The n monomials outside any initial monomial ideal of I span the quotient $\mathbb{C}[x,y]/I$, and these monomials must lie inside V_m. □

The intersection $I \cap V_m$ is a vector subspace of codimension n in V_m. Furthermore, the reduced Gröbner basis of I for any term order refining the partial order by total degree consists of polynomials of degree at most n (see the proof of Lemma 18.2). In particular, I is generated by $I \cap V_m$ when $m \geq n$. Thus the Hilbert scheme H_n is—as a set, at least—contained inside the Grassmannian $\mathrm{Gr}^n(V_m)$ of codimension n subspaces of V_m.

Definition 18.3 For a partition λ of n, let $U_\lambda \subset H_n$ be the set of ideals I such that the monomials outside I_λ map to a vector space basis for $\mathbb{C}[x,y]/I$.

The set of codimension n subspaces $W \subset V_m$ for which the monomials outside I_λ span V_m/W constitutes a standard open affine subvariety of $\mathrm{Gr}^n(V_m)$. This open set is defined by the nonvanishing of the corresponding Plücker coordinate (Chapter 14). This means that W has a unique $\mathbb{C}$-basis consisting of polynomials of the form

$$x^r y^s - \sum_{hk \in \lambda} c_{hk}^{rs} x^h y^k. \tag{18.2}$$

Here, we write $hk \in \lambda$ to mean $x^h y^k \notin I_\lambda$, so the box labeled (h,k) lies under the staircase for I_λ. The affine open chart of $\mathrm{Gr}^n(V_m)$ is the affine space whose coordinate ring is the polynomial ring in the coefficients c_{hk}^{rs} from (18.2).

The intersection of each ideal $I \in U_\lambda$ with V_m is a codimension n subspace of V_m spanned by polynomials of the form (18.2), by definition of U_λ. Of course, if $W \subset V_m$ is to be expressible as the intersection of V_m with some ideal I, then the coefficients c_{hk}^{rs} cannot be chosen completely at will. Indeed, the fact that I is an ideal imposes relations on the coefficients that say "multiplication by x, which takes $x^r y^s$ to $x^{r+1} y^s$, preserves I; and similarly for multiplication by y."

Explicitly, if $x^{r+1} y^s \in V_m$ and $m \geq n$, then multiplying (18.2) by x yields another polynomial $x^{r+1} y^s - \sum_{hk \in \lambda} c_{hk}^{rs} x^{h+1} y^k$ inside $I \cap V_m$. Some

of the terms $x^{h+1}y^k$ no longer lie outside I_λ, so we have to expand them
again using (18.2) to get

$$x^{r+1}y^s - \left(\sum_{h+1,k\in\lambda} c_{hk}^{rs}x^{h+1}y^k + \sum_{h+1,k\notin\lambda} c_{hk}^{rs} \sum_{h'k'\in\lambda} c_{h'k'}^{h+1,k}x^{h'}y^{k'} \right) \in I. \quad (18.3)$$

Equating the coefficients on $x^h y^k$ in (18.3) to those in

$$x^{r+1}y^s - \sum_{hk\in\lambda} c_{hk}^{r+1,s}x^h y^k$$

from (18.2) yields relations in the polynomial ring $\mathbb{C}[\{c_{hk}^{rs}\}]$. These relations,
taken along with their counterparts that result by switching the roles of x
and y, characterize the set U_λ in Definition 18.3. Although we have yet to
see that these relations generate a radical ideal, we can at least conclude
that U_λ is an algebraic subset of an open cell in the Grassmannian.

Theorem 18.4 *The affine varieties U_λ form an open cover of the subset
$H_n \subset \mathrm{Gr}^n(V_m)$ for $m \geq n+1$, thereby endowing H_n with the structure of
a quasiprojective variety (i.e., an open subvariety of a projective variety).*

Proof. The sets U_λ cover H_n by Lemma 18.2, and each set U_λ is locally
closed in $\mathrm{Gr}^n(V_m)$ by the above discussion. (We will explain near the
beginning of Section 18.2 why we assumed $m \geq n+1$ instead of $m \geq n$.) $\square$

In summary, we have constructed the Hilbert scheme H_n as a quasiprojective variety because it is locally obtained by the intersection of a Zariski
open condition (certain monomials span modulo I) and a Zariski closed
condition ($W \subset V_m$ is closed under multiplication by x and y).
    The number of coordinates c_{hk}^{rs} used in our description of the affine vari-
eties U_λ is $n \cdot \left(\binom{m+2}{2} - n \right)$. This number can be made considerably smaller,
even when $m = n + 1$. For instance, it suffices to take those coordinates
c_{hk}^{rs} where either $(r-1,s)$ or $(r,s-1)$ is in the shape obtained from λ by
adding a strip of width 1 along its boundary. All other coordinates are
polynomial functions in these special coordinates. Moreover, the map that
projects away from the other coordinates is an isomorphism of varieties; see
the paragraph after the statement of Theorem 18.7. Sometimes it even suffices to take only those coordinates c_{hk}^{rs} where $x^r y^s$ is a minimal generator
of I_λ. We present one example where these minimal-generator coordinates
suffice and one example where they do not.

Example 18.5 Take $n = 4$ and λ the partition $2+1+1$ of Example 18.1.
Every ideal I in U_{2+1+1} is generated by three of the polynomials in (18.2):

$$\langle \underline{x^2} - ay^2 - bx - py - q, \ \underline{xy} - cy^2 - dy - ex - r, \ \underline{y^3} - fy^2 - gy - hx - s \rangle.$$

Here, we abbreviate $a = c_{02}^{20}$, $p = c_{01}^{20}$, and so on. This ideal lies in U_{2+1+1}
if and only if its three generators are a Gröbner basis with the underlined

leading terms. Buchberger's s-pair criterion implies that this happens if and only if

$$
\begin{aligned}
p &= fc^2 + ec^2 - fa + ae - bc + 2cd, \\
q &= fec^2 - c^3h - fae + gc^2 + ae^2 + ach - bec + 2ecd - ga - bd + d^2, \\
r &= -e^2c - c^2h + ah - ed, \\
s &= -fe^2 + e^3 + 2ech - ge - bh + dh
\end{aligned}
$$

all hold. Thus the affine chart U_{2+1+1} of the Hilbert scheme H_4 is the 8-dimensional affine space with coordinate ring $\mathbb{C}[a, b, c, d, e, f, g, h]$. ◇

Example 18.6 Take $n = 4$ and λ the partition $2 + 2$. Every ideal I in U_{2+2} is generated by four of the polynomials in (18.2), namely

$$
\begin{aligned}
I = \langle &x^2 - axy - ey - px - t, \ x^2y - bxy - fy - qx - u, \\
&y^2 - cxy - gx - ry - v, \ xy^2 - dxy - hx - sy - w \rangle.
\end{aligned}
$$

The quotient ring $\mathbb{C}[x, y]/I$ has the $\mathbb{C}$-basis $\{1, x, y, xy\}$ if and only if

$$
\begin{aligned}
&p = b - ad - ec, \quad q = ah + eg, \quad r = d - ag - bc, \quad s = cf + eg, \\
&t = f - ed - acf + bce, \quad u = aw + adeg - aceh - beg + eh, \\
&v = h - bg - ach + adg, \quad \text{and} \quad w = cu - bceg - acfd + deg + fg.
\end{aligned}
$$

Eliminating the parameters $\{p, q, r, s, t, u, v\}$ leaves us with one equation

$$
w(1 - ac) = \text{a polynomial in } a, b, c, d, e, f, g, h.
$$

The affine chart U_{2+2} of the Hilbert scheme H_4 is the smooth hypersurface in $\mathbb{C}^9 = \mathrm{Spec}(\mathbb{C}[a, b, c, d, e, f, g, h, w])$ defined by this equation. ◇

18.2 Connectedness and smoothness

In this section we prove the following theorem.

Theorem 18.7 *The Hilbert scheme H_n is a smooth and irreducible complex algebraic variety of dimension $2n$.*

The variety structure in Theorem 18.7 is the same as the one from Theorem 18.4, although it is not obvious from the latter that this structure is independent of m. This important fact can be deduced using the smoothness of H_n along with the fact that projection $V_{m+1} \to V_m$ maps H_n to itself by sending $I \cap V_{m+1} \mapsto I \cap V_m$. If we had allowed $m = n$ in Theorem 18.4, then one of the results in this section, namely Proposition 18.14, would sometimes fail, so the variety structure would be different. In any case, we fix $m \geq n + 1$ for the duration of this section.

Our first aim is to prove that the complex variety H_n is connected. In the next lemma, a *rational curve* inside a variety is a subvariety of dimension 1 expressible as the image of a map from the affine line. These subvarieties are the curves parametrized by polynomials in a single variable.

Lemma 18.8 *Every point $I \in H_n$ is connected to a monomial ideal by a rational curve.*

Proof. Choosing a term order and taking a Gröbner basis of I yields a family of ideals parametrized by the coordinate variable t on the affine line. Such a Gröbner degeneration is a flat family I_t over the affine line by Proposition 8.26. When $t = 1$ we get I back, and when $t = 0$ we get the initial ideal of I, which is a monomial ideal. $\qquad\square$

Example 18.9 Consider the ideal $I = \langle x^2 - xy, y^2 - xy, x^2 y, xy^2 \rangle$, which lies in the chart U_{2+2} discussed in Example 18.6. Now replace y by ty in every polynomial $f \in I$, and observe what happens as t goes to 0. Finding polynomials in I such that applying this procedure to them yields generators for the resulting ideal at $t = 0$ is the same as computing the lexicographic Gröbner basis of I. Our rational curve in H_4 is given by

$$I_t = \langle x^2 - txy, xy - t^2 y^2, x^2 y, xy^2, y^3 \rangle.$$

This represents a flat family because the quotient ring $\mathbb{C}[x, y][t]/I_t$ is a free module of rank 4 over $\mathbb{C}[t]$. The initial monomial ideal is $I_0 = \langle x^2, xy, y^3 \rangle$ from Example 18.1. Note that I_0 does not lie in the chart U_{2+2}, but it lies in the chart U_{2+1+1} discussed in Example 18.5. $\qquad\diamond$

The previous lemma shows that every point in H_n connects to a monomial ideal. The next lemma shows that monomial ideals all connect to the locus of radical ideals.

Lemma 18.10 *For every partition λ of n, the point $I_\lambda \in H_n$ lies in the closure of the locus $(S^n \mathbb{C}^2)^\circ$ of all radical ideals in the Hilbert scheme H_n.*

Proof. Consider the set of exponent vectors (h, k) on monomials $x^h y^k$ outside I_λ. This set constitutes a collection of n points in $\mathbb{N}^2 \subset \mathbb{C}^2$. The radical ideal of these points is denoted by I'_λ and called the *distraction* of I_λ. Suppose $I_\lambda = \langle x^{a_1} y^{b_1}, \ldots, x^{a_m} y^{b_m} \rangle$ and consider the polynomials

$$f_i = x(x-1)(x-2)\cdots(x - a_i + 1) y(y-1)\cdots(y - b_i + 1).$$

We have $\langle f_1, \ldots, f_m \rangle \subseteq I'_\lambda$ because the polynomials f_i vanish at the given points (a_j, b_j), and we have $\text{colength}(\langle f_1, \ldots, f_m \rangle) \leq n$ because the leading terms of the f_i are the generators of the colength n ideal I_λ. Therefore

$$I'_\lambda = \langle f_1, \ldots, f_m \rangle.$$

Moreover, I_λ is forced to be the initial monomial ideal of I'_λ with respect to every term order. The ideal $(I'_\lambda)_t$ constructed as in the proof of Lemma 18.8 is radical for each $t \neq 0$. Hence $I_\lambda = (I'_\lambda)_0$ lies in the closure of $(S^n \mathbb{C}^2)^\circ$. $\qquad\square$

Example 18.11 The distraction of $I_{2+1+1} = \langle x^2, xy, y^3 \rangle$ is the ideal

$$I'_{2+1+1} \quad = \quad \langle x(x-1),\ xy,\ y(y-1)(y-2) \rangle.$$

The zero set of each generator is a union of lines, namely integer translates of one of the two coordinate axes in $\mathbb{C}^2$. The zero set of our ideal I'_{2+1+1} is

The groups of lines on the right-hand side are the zero sets of $x(x-1)$, xy, and $y(y-1)(y-2)$, respectively. $\diamond$

Lemma 18.8 allows us to derive half of Theorem 18.7.

Proposition 18.12 *The Hilbert scheme H_n is connected.*

Proof. We connect any two points I and J in H_n by a path as follows. Go from I to any initial monomial ideal I_λ and then to its distraction I'_λ. Go from J to any initial monomial ideal I_ν and then to its distraction I'_ν. Now I'_λ and I'_ν are the radical ideals of n points in $\mathbb{C}^2$. Connect these two ideals by continuously moving one point configuration into the other. $\square$

Remark 18.13 Proposition 18.12 holds for Hilbert schemes of n points in $\mathbb{C}^d$ even when d is arbitrary, with the same proof. The connectedness theorem of Hartshorne [Har66a] implies that it holds more generally for Hilbert schemes of $\mathbb{Z}$-graded ideals in the standard grading. In Theorem 18.52 we will see that Hilbert schemes of $\mathbb{Z}^n$-graded ideals can be disconnected. In Section 18.4 we will see that Hilbert schemes of n points in $\mathbb{C}^d$ are neither smooth nor irreducible for $n \gg d \geq 3$.

Our eventual goal is to prove that H_n is smooth. This is a local property which amounts to checking that the maximal ideal of each local ring does not have more than the smallest possible number of minimal generators.

Proposition 18.14 *For each partition λ of n, the local ring $(H_n)_{I_\lambda}$ of the Hilbert scheme H_n at I_λ has embedding dimension at most $2n$; that is, the maximal ideal $\mathfrak{m}_{I_\lambda}$ satisfies $\dim_{\mathbb{C}}(\mathfrak{m}_{I_\lambda}/\mathfrak{m}_{I_\lambda}^2) \leq 2n$.*

Proof. Identify each variable c_{hk}^{rs} with an arrow pointing from the box $hk \in \lambda$ to the box $rs \notin \lambda$ (see Example 18.16). Allow arrows starting in boxes with $h < 0$ or $k < 0$, but set them equal to zero. The arrows lie inside—and in fact generate—the maximal ideal $\mathfrak{m}_{I_\lambda}$ at the point $I_\lambda \in H_n$. As each term in the double sum in (18.3) has two c's in it, the double sum lies inside $\mathfrak{m}_{I_\lambda}^2$. Moving both the tail and head of any given arrow one box to the right therefore does not change the arrow's residue class modulo $\mathfrak{m}_{I_\lambda}^2$, as long as the tail of the original arrow does not lie in the last box in a row of λ. Switching the roles of x and y, we conclude that an arrow's residue class modulo $\mathfrak{m}_{I_\lambda}^2$ is unchanged by moving vertically or horizontally, as long as the tail stays under the staircase and the head stays above it. This analysis includes the case where the tail of the arrow crosses either axis, in which case the arrow is zero.

Every arrow can be moved horizontally and vertically until one of the following occurs:

(i) The tail crosses an axis.

(ii) There is a box $hk \in \lambda$ such that the tail lies just inside row k of λ while the head lies just above column h outside λ.

(iii) There is a box $hk \in \lambda$ such that the tail lies just under the top of column h in λ while the head lies in the first box to the right outside row k of λ.

Arrows of the first sort do not contribute at all to $\mathfrak{m}_{I_\lambda}/\mathfrak{m}_{I_\lambda}^2$. On the other hand, there are exactly n northwest-pointing arrows of the second sort and exactly n southeast-pointing arrows of the third sort. Therefore the cotangent space $\mathfrak{m}_{I_\lambda}/\mathfrak{m}_{I_\lambda}^2$ has dimension at most $2n$.   □

Example 18.15 In Examples 18.5 and 18.6, the basis of $\mathfrak{m}_{I_\lambda}/\mathfrak{m}_{I_\lambda}^2$ described above consists of the parameters $\{a, b, c, d, e, f, g, h\}$. Note that four of them are northwest arrows and the other four are southeast arrows. ◇

Example 18.16 All of the following three staircase diagrams depict the same partition λ: $8 + 8 + 5 + 3 + 3 + 3 + 3 + 2 = 35$. In the left diagram, the middle of the five arrows represents $c_{31}^{54} \in \mathfrak{m}_{I_\lambda}$. As in the proof of Proposition 18.14, all of the arrows in the left diagram are equal modulo $\mathfrak{m}_{I_\lambda}^2$. Since the bottom one is manifestly zero as in item (i) from the proof of Proposition 18.14, all of the arrows in the left diagram represent zero in $\mathfrak{m}_{I_\lambda}/\mathfrak{m}_{I_\lambda}^2$.

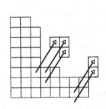 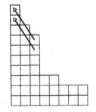 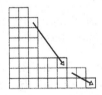

The two arrows in the middle diagram are equal, and the bottom one, c_{25}^{08}, is an example of a regular parameter in $\mathfrak{m}_{I_\lambda}$ as in (ii). Finally, the two arrows in the rightmost diagram represent unequal regular parameters as in (iii). ◇

Now we finally have enough prerequisites to prove the main result.

Proof of Theorem 18.7. Lemma 18.10 implies that the dimension of the local ring of H_n at any monomial ideal I_λ is at least $2n$, because the radical locus has dimension $2n$. On the other hand, Proposition 18.14 shows that the maximal ideal of that local ring can be generated by $2n$ polynomials. Therefore H_n is regular in a neighborhood of any point I_λ.

The two-dimensional torus $(\mathbb{C}^*)^2$ acting on $\mathbb{C}^2$ by scaling the coordinates has an induced action on H_n. The proof of Lemma 18.8 says that every orbit on H_n contains a monomial ideal (= torus-fixed point) in its closure. A point is smooth if and only if every point of its orbit under $(\mathbb{C}^*)^2$ is smooth. Since every smooth point has an open neighborhood that is smooth, the singular locus of H_n must contain a closed orbit. Since H_n is regular at every torus fixed point, the singular locus must be empty.

We now know that H_n is smooth and connected (by Proposition 18.12). This implies that H_n is irreducible: if H_n had more than one irreducible component, then any point in the intersection of two distinct components would be a singular point. $\square$

The argument using the torus action can be replaced by a completely algebraic one. The torus action on each open affine U_λ results in a positive grading of its coordinate ring. Presenting this coordinate ring by relations as in Section 18.1, we see that the singular locus is the zero set of the Jacobian ideal $J(U_\lambda)$ of these relations [Eis95, Section 16.6], which is graded. To check that the singular locus is empty, we need only check that $J(U_\lambda)$ is the unit ideal, and for this it is enough to check that no homogeneous maximal ideals contain it (because the grading is positive). All of the homogeneous maximal ideals have the form $\mathfrak{m}_{I_\lambda}$ for some partition λ, and Proposition 18.14 shows that these do not contain $J(U_\lambda)$.

Remark 18.17 A key tool in studying the topology of the manifold H_n is its *Białynicki-Birula decomposition* [Bia76] with respect to some fixed term order. Each partition λ defines one affine cell in the Białynicki-Birula decomposition. It consists of all colength n ideals I whose initial monomial ideal equals I_λ. This cell is always contained in the affine chart U_λ. Sometimes they are equal (for instance, in Example 18.5), but U_λ is generally much larger than the Białynicki-Birula cell (for instance, in Example 18.6).

18.3 Haiman's theory

An important connection between the Hilbert scheme H_n and the theory of symmetric functions was developed by Haiman, in his proof of the *n! Theorem* and the $(n+1)^{n-1}$ *Theorem*. This section provides an introduction to these results, with a view toward combinatorial commutative algebra.

Consider the following two morphisms of complex algebraic varieties:

$$(\mathbb{C}^2)^n \atop \downarrow \atop H_n \to S^n \mathbb{C}^2 \tag{18.4}$$

The elements of the $2n$-dimensional affine space $(\mathbb{C}^2)^n$ are ordered n-tuples of points (x_i, y_i) in $\mathbb{C}^2$. The coordinate ring of $(\mathbb{C}^2)^n$ is the polynomial ring

$$\mathbb{C}[\mathbf{x}, \mathbf{y}] \;=\; \mathbb{C}[x_1, x_2, \ldots, x_n, y_1, y_2, \ldots, y_n].$$

The symmetric group S_n acts on $(\mathbb{C}^2)^n$ by permuting the points (x_i, y_i). The variety $S^n\mathbb{C}^2$ is the quotient of $(\mathbb{C}^2)^n$ modulo the action of the symmetric group S_n. The coordinate ring of $S^n\mathbb{C}^2$ is the invariant ring

$$\mathbb{C}[\mathbf{x}, \mathbf{y}]^{S_n} \;=\; \{ f \in \mathbb{C}[\mathbf{x}, \mathbf{y}] \mid f(x_{w_1}, \ldots, x_{w_n}, y_{w_1}, \ldots, y_{w_n})$$
$$= f(x_1, \ldots, x_n, y_1, \ldots, y_n) \text{ for all } w \in S_n \}.$$

The vertical arrow in (18.4) is induced by the inclusion $\mathbb{C}[\mathbf{x}, \mathbf{y}]^{S_n} \hookrightarrow \mathbb{C}[\mathbf{x}, \mathbf{y}]$. The next result, due to Hermann Weyl, describes its image explicitly.

Theorem 18.18 *The invariant ring* $\mathbb{C}[\mathbf{x}, \mathbf{y}]^{S_n}$ *is generated by power sums*

$$p_{r,s}(\mathbf{x}, \mathbf{y}) \;=\; x_1^r y_1^s + \cdots + x_n^r y_n^s \quad \text{for } 1 \le r + s \le n.$$

The image of an element in $(\mathbb{C}^2)^n$ is given by the values of the power sums $p_{r,s}$ at this n-tuple of points (x_i, y_i). The horizontal map in (18.4) is understood similarly. The image of an ideal $I \in H_n$ in the symmetric product $S^n\mathbb{C}^2$ is the unordered multiset of its n zeros, counting multiplicity. It is determined algebraically by evaluating each power sum $p_{r,s}$ at this unordered multiset. This value is computed as the trace of (any $n \times n$ matrix representing) the $\mathbb{C}$-linear map from $\mathbb{C}[x, y]/I$ to itself given by multiplication with $x^r y^s$.

The diagonal locus (18.1) in $(\mathbb{C}^2)^n$ is the union of $\binom{n}{2}$ linear spaces of codimension 2. Let $I_{\mathrm{diag}} \subset \mathbb{C}[\mathbf{x}, \mathbf{y}]$ be the radical ideal of the diagonal locus.

Theorem 18.19 *The radical ideal* I_{diag} *is generated by the polynomials*

$$\Delta_D(\mathbf{x}, \mathbf{y}) \;=\; \det \begin{bmatrix} x_1^{i_1} y_1^{j_1} & x_2^{i_1} y_2^{j_1} & \cdots & x_n^{i_1} y_n^{j_1} \\ \vdots & \vdots & \ddots & \vdots \\ x_1^{i_n} y_1^{j_n} & x_2^{i_n} y_2^{j_n} & \cdots & x_n^{i_n} y_n^{j_n} \end{bmatrix}.$$

where $D = \{(i_1, j_1), \ldots, (i_n, j_n)\}$ *runs over all n-element subsets of* $\mathbb{N}^2$.

This theorem is due to Haiman, who found it in the course of proving Theorem 18.21. No elementary proof of Theorem 18.19 is known. It is also an open problem to identify a finite set of polynomials Δ_D that minimally generates I_{diag}. Each partition λ corresponds to the subset of pairs $(i, j) \in \mathbb{N}^2$ such that $x^i y^j$ lies outside of I_λ, and it is known that the minimal generating set includes the determinants Δ_λ for all partitions λ of n.

Example 18.20 For $n = 3$, the ideal I_{diag} has five minimal generators Δ_D:

$$I_{\text{diag}} = \langle x_1 - x_2, y_1 - y_2 \rangle \cap \langle x_1 - x_3, y_1 - y_3 \rangle \cap \langle x_2 - x_3, y_2 - y_3 \rangle$$

$$= \left\langle \det \begin{bmatrix} 1 & 1 & 1 \\ x_1 & x_2 & x_3 \\ y_1 & y_2 & y_3 \end{bmatrix}, \det \begin{bmatrix} 1 & 1 & 1 \\ x_1 & x_2 & x_3 \\ x_1^2 & x_2^2 & x_3^2 \end{bmatrix}, \det \begin{bmatrix} 1 & 1 & 1 \\ y_1 & y_2 & y_3 \\ y_1^2 & y_2^2 & y_3^2 \end{bmatrix}, \right.$$

$$\left. \det \begin{bmatrix} 1 & 1 & 1 \\ x_1 & x_2 & x_3 \\ x_1 y_1 & x_2 y_2 & x_3 y_3 \end{bmatrix}, \det \begin{bmatrix} 1 & 1 & 1 \\ y_1 & y_2 & y_3 \\ x_1 y_1 & x_2 y_2 & x_3 y_3 \end{bmatrix} \right\rangle .$$

The last two generators are Δ_D and $\Delta_{D'}$ for $D = \{(0,0), (1,0), (1,1)\}$ and $D' = \{(0,0), (0,1), (1,1)\}$, neither of which is a partition. $\diamond$

We now state the main results, albeit in their most basic versions.

Theorem 18.21 (Haiman's $n!$ Theorem and $(n+1)^{n-1}$ Theorem)

1. *If λ is a partition of n, then the set of all polynomials obtained from Δ_λ by applying linear partial differential operators with constant coefficients span a vector space of dimension $n!$ over the complex numbers.*

2. *The quotient of $\mathbb{C}[\mathbf{x}, \mathbf{y}]$ modulo the ideal $\langle p_{r,s} \mid 1 \leq r + s \leq n \rangle$ generated by all nonconstant homogeneous S_n-invariants is a vector space of dimension $(n+1)^{n-1}$ over $\mathbb{C}$.*

Part 1 of Theorem 18.21 can be reformulated in ideal-theoretic terms as follows. A *linear partial differential operator with constant coefficients* is by definition a polynomial

$$p(\partial \mathbf{x}, \partial \mathbf{y}) = p\left(\frac{\partial}{\partial x_1}, \ldots, \frac{\partial}{\partial x_n}, \frac{\partial}{\partial y_1}, \ldots, \frac{\partial}{\partial y_n} \right)$$

in the symbols $\frac{\partial}{\partial x_i}$ and $\frac{\partial}{\partial y_j}$. The following vector space is an ideal:

$$J_\lambda = \{ p \in \mathbb{C}[\mathbf{x}, \mathbf{y}] \mid p(\partial \mathbf{x}, \partial \mathbf{y}) \text{ annihilates } \Delta_\lambda \}. \qquad (18.5)$$

Moreover, the quotient ring $\mathbb{C}[\mathbf{x}, \mathbf{y}] / J_\lambda$ is a zero-dimensional Gorenstein ring (see Exercise 13.12 for the definition). It is isomorphic to the $\mathbb{C}$-vector space described in part 1 of Theorem 18.21. Hence the $n!$ Theorem states that J_λ has colength $n!$ for every partition λ of n.

The two parts of Theorem 18.21 are related by the observation that

$$\langle p_{r,s} \mid 1 \leq r + s \leq n \rangle \subseteq J_\lambda \text{ for all partitions } \lambda \text{ of } n.$$

Example 18.22 Let $n = 3$ and $\lambda = 2 + 1$. Then

$$\Delta_\lambda = x_1 y_2 - x_2 y_1 + x_3 y_1 - x_1 y_3 - x_3 y_2 + x_2 y_3.$$

By differentiating Δ_λ, we first get the differences $x_i - x_j$ and $y_i - y_j$, and next the constants. Together they span a vector space of dimension $3! = 6$. The annihilating ideal of Δ_λ is

$$J_\lambda = \langle x_1+x_2+x_3, \; y_1+y_2+y_3, \; y_3^2, \; y_2 y_3,$$
$$y_2^2, \; x_3^2, \; x_2 x_3, \; x_2^2, \; x_2 y_2, \; x_3 y_3, \; x_3 y_2 + x_2 y_3 \rangle.$$

Thus the ring $\mathbb{C}[\mathbf{x},\mathbf{y}]/J_\lambda$ is Gorenstein of length 6. $\diamond$

The main player in the proof of Theorem 18.21 is the *isospectral Hilbert scheme* X_n. It is defined as the reduced fiber product of the two maps in (18.4). Hence X_n is the reduced subscheme of $(\mathbb{C}^2)^n \times H_n$ consisting of all pairs $(((x_1,y_1),\ldots,(x_n,y_n)),I)$ such that the points (x_i,y_i) are the zeros of I appearing with the correct multiplicity. The two projections define the left and top morphisms in the "reduced fiber square"

$$\begin{array}{ccc} X_n & \to & (\mathbb{C}^2)^n \\ \downarrow & & \downarrow \\ H_n & \to & S^n \mathbb{C}^2 \end{array} \qquad (18.6)$$

that completes the diagram in (18.4).

It is instructive to compute the local equations of the isospectral Hilbert scheme. By this we mean the ideal in the polynomial ring $\mathbb{C}[x_i, y_i, c_{hk}^{rs}]$ defining the intersection of X_n with $(\mathbb{C}^2)^n \times U_\lambda$. To do this, let L_λ denote the ideal generated by all the incidence relations

$$x_i^r y_i^s - \sum_{hk \in \lambda} c_{hk}^{rs} x_i^h y_i^k \quad \text{for } i = 1, \ldots, n$$

together with the polynomials in the variables c_{hk}^{rs} that define U_λ. The latter were described right before Theorem 18.4. The desired radical ideal equals

$$\text{radical}(L_\lambda : I_{\text{diag}}^\infty). \qquad (18.7)$$

This is the ideal of the isospectral Hilbert scheme X_n over the chart U_λ.

Example 18.23 Let $n = 3$ and $\lambda = 2 + 1$. The ideal L_λ is generated by

$$\begin{array}{lll} x_1^2 - ax_1 - by_1 - c, & x_1 y_1 - dx_1 - ey_1 - f, & y_1^2 - gx_1 - hy_1 - i, \\ x_2^2 - ax_2 - by_2 - c, & x_2 y_2 - dx_2 - ey_2 - f, & y_2^2 - gx_2 - hy_2 - i, \\ x_3^2 - ax_3 - by_3 - c, & x_3 y_3 - dx_3 - ey_3 - f, & y_3^2 - gx_3 - hy_3 - i \end{array}$$

and the three compatibility relations

$$bd - ae + e^2 - bh - c, \quad bg - de - f, \quad d^2 - ag + eg - dh - i. \qquad (18.8)$$

The radical ideal (18.7) of X_3 over U_{2+1} is the colon ideal $(L_\lambda : I_{\text{diag}})$ with respect to the ideal I_{diag} in Example 18.20. In addition to the three polynomials in (18.8), this quotient has 11 minimal generators:

$y_1+y_2+y_3-d-h$, $x_1+x_2+x_3-a-e$,

$y_3^2-x_3g-y_3h-i$, $y_2y_3-y_2d-y_3d+x_2g+x_3g-eg+dh+i$, $y_2^2-x_2g-y_2h-i$,

$x_3^2-x_3a-y_3b-c$, $x_2x_3+y_2b+y_3b-x_2e-x_3e+e^2-bh$, $x_2^2-x_2a-y_2b-c$,

$x_2y_2-x_2d-y_2e-f$, $x_3y_3-x_3d-y_3e-f$,

$x_3y_2+x_2y_3-y_2a-y_3a+x_2d+x_3d+y_2e+y_3e-x_2h-x_3h-bg+ah+2f$.

The fiber of the isospectral Hilbert scheme X_3 over the point $I_{2+1} \in H_3$ is gotten by setting $a=b=c=d=e=f=g=h=i=0$ in these 11 polynomials. What results is precisely the ideal J_{2+1} from Example 18.22. ◇

The key result implying Theorem 18.21 is a statement in commutative algebra: the ideals in (18.7) are Gorenstein (Exercise 13.12).

Theorem 18.24 (Haiman) *The isospectral Hilbert scheme is Gorenstein.*

Consider now the morphism $X_n \to H_n$ in (18.6). The base is smooth by Theorem 18.7. The generic fiber is reduced of length $n!$. It is given by all permutations of n distinct points (x_i, y_i) in $\mathbb{C}^2$. Theorem 18.24 implies that all special fibers have the same length $n!$ (that is, the family of fibers is flat) and that they are all Gorenstein. Part 1 of Theorem 18.21 is now a consequence of Theorem 18.24 and Theorem 13.37.5, by the next lemma.

Lemma 18.25 *The fiber of the morphism $X_n \to H_n$ over the torus-fixed point $I_\lambda \in H_n$ is the zero-dimensional scheme defined by the ideal J_λ (18.5).*

For $n = 3$ and $\lambda = 2+1$, this lemma was confirmed computationally in Example 18.23. The derivation of the $(n+1)^{n-1}$ Theorem requires one more geometric ingredient. The *zero-fiber* Z_n is the scheme-theoretic fiber of the origin under the morphism $H_n \to S^n\mathbb{C}^2$ in (18.4). The equations of Z_n over an affine open U_λ are obtained from the ideal of X_n by setting all variables x_i and y_i to zero.

Theorem 18.26 (Briançon and Haiman) *The zero-fiber Z_n is reduced, irreducible, and Cohen–Macaulay of dimension $n-1$.*

Example 18.27 The ideal of the zero-fiber Z_3 over U_{2+1} is obtained from the ideal of X_3 in Example 18.23 by setting $x_i = y_i = 0$ for $i = 1, 2, 3$. It equals $\langle c, f, i, a+e, d+h, eg+h^2, bg+eh, e^2-bh \rangle$. ◇

Let P be the sheaf on the Hilbert scheme H_n obtained by pushing down the sheaf of regular functions on the isospectral Hilbert scheme X_n. The ring of global sections of this sheaf is our polynomial ring in $2n$ unknowns:

$$H^0(H_n, P) = \mathbb{C}[\mathbf{x}, \mathbf{y}].$$

The $n!$ Theorem tells us that P is a vector bundle of rank $n!$. The fiber of P over the point $I_\lambda \in H_n$ is the Gorenstein ring $\mathbb{C}[\mathbf{x}, \mathbf{y}]/J_\lambda$ by Lemma 18.25.

Now consider the restriction of the sheaf P to the zero-fiber Z_n. The ring of global sections of this restricted sheaf turns out to be

$$H^0(Z_n, P) \; = \; \mathbb{C}[\mathbf{x}, \mathbf{y}]/\langle p_{r,s} \mid 1 \leq r + s \leq n \rangle. \qquad (18.9)$$

Using this fact, Haiman derived the second part of Theorem 18.21 from Theorem 18.26 and the Bridgeland–King–Reid Theorem on the generalized McKay correspondence [BKR01].

We close this section by explaining what all of this has to do with the theory of symmetric functions. The rings $R^{(\lambda)} = \mathbb{C}[\mathbf{x}, \mathbf{y}]/J_\lambda$ carry two natural structures resulting from group actions: they are $\mathbb{Z}^2$-graded, and the symmetric group S_n acts on each $\mathbb{Z}^2$-graded component $R^{(\lambda)}_{(i,j)}$. The *formal character* of the S_n-module $R^{(\lambda)}_{(i,j)}$ is a symmetric function $F^\lambda_{ij}(\mathbf{z})$ in an infinite alphabet $\mathbf{z}$; namely $F^\lambda_{ij}(\mathbf{z})$ is the sum (with multiplicity) of all Schur functions $s_\mu(\mathbf{z})$ for irreducible S_n-modules indexed by μ appearing in $R^{(\lambda)}_{(i,j)}$. The *Hilbert–Frobenius series* of the ring $R^{(\lambda)}$ is

$$\sum_{i=1}^{\binom{n}{2}} \sum_{j=1}^{\binom{n}{2}} F^\lambda_{ij}(\mathbf{z}) \, q^i t^j. \qquad (18.10)$$

Likewise, we can define the Hilbert–Frobenius series of the ring (18.9). These expressions are symmetric functions that depend on two parameters q and t. The punch line of the $n!$ Theorem for algebraic combinatorialists who know and cherish Macdonald polynomials [Macd95] is that the symmetric functions (18.10) arise from those introduced by Macdonald.

Corollary 18.28 *The Hilbert–Frobenius series (18.10) of* $\mathbb{C}[\mathbf{x}, \mathbf{y}]/J_\lambda$ *is the transformed Macdonald polynomial* $\tilde{H}_\lambda(\mathbf{z}; q, t)$. *In particular,* $\tilde{H}_\lambda(\mathbf{z}; q, t)$ *is an* $\mathbb{N}[q, t]$-*linear combination of Schur functions* $s_\mu(\mathbf{z})$.

18.4   Ideals of points in d-space

In the first three sections, we studied the Hilbert scheme $H_n = \mathrm{Hilb}^n(\mathbb{C}^2)$ of n points in the affine plane $\mathbb{C}^2$. In this section, we consider the Hilbert scheme $H_n^d = \mathrm{Hilb}^n(\mathbb{C}^d)$ of n points in affine d-space $\mathbb{C}^d$. Its points are the ideals I of colength n in $\mathbb{C}[\mathbf{x}] = \mathbb{C}[x_1, x_2, \ldots, x_d]$. The construction of Section 18.1 extends in a straightforward manner to this new situation: if we define V_m to be the $\mathbb{C}$-vector space of all polynomials of degree at most m in $\mathbb{C}[\mathbf{x}]$, then H_n^d is a subscheme of the Grassmannian $\mathrm{Gr}^n(V_{n+1})$.

The role of the partitions λ is now played by *order ideals* of cardinality n in $\mathbb{N}^d$. An order ideal is a subset $\lambda \subset \mathbb{N}^d$ such that $\mathbf{u} \in \lambda$ and $\mathbf{v} \leq \mathbf{u}$ implies $\mathbf{v} \in \lambda$. Equivalently, an order ideal is the set of exponents on monomials outside of a monomial ideal. When $d = 3$, for example, these order ideals are the staircases discussed in Chapter 3 (in the combinatorics literature these are also known as *plane partitions*). As earlier, we write I_λ for the monomial ideal spanned by all monomials $\mathbf{x}^\mathbf{u}$ with $\mathbf{u} \notin \lambda$, and we let $U_\lambda \subset H_n^d$ denote the affine open subscheme consisting of all ideals $I \in H_n^d$ such that $\{\mathbf{x}^\mathbf{u} \mid \mathbf{u} \in \lambda\}$ is a $\mathbb{C}$-basis of $\mathbb{C}[\mathbf{x}]/I$. The equations defining U_λ are expressed in local coordinates $c_\mathbf{v}^\mathbf{u}$, where $\mathbf{v}$ runs over λ and $\mathbf{u}$ runs over monomials not in λ having degree at most n. These equations need not generate a radical ideal, which is why we refer to U_λ as an "affine subscheme" rather than an "affine subvariety".

Many of the nice properties of the Hilbert scheme of points in the plane no longer hold for H_n^d. To see that H_n^d is generally not smooth and to study its singularities, one uses the following formula for the tangent space.

Theorem 18.29 *The tangent space to the Hilbert scheme H_n^d at any point $I \in H_n^d$ is isomorphic as $\mathbb{C}$-vector space to the module $\mathrm{Hom}_{\mathbb{C}[\mathbf{x}]}(I, \mathbb{C}[\mathbf{x}]/I)$.*

This theorem is derived from the universal property of the Hilbert scheme, a topic we will only briefly mention in Section 18.5. If $I = I_\lambda$ is a monomial ideal, then the image of the parameter $c_\mathbf{v}^\mathbf{u}$ in the tangent space corresponds to the unique $\mathbb{C}$-linear map $I \to \mathbb{C}[\mathbf{x}]/I$ that maps a monomial $\mathbf{x}^\mathbf{w}$ to $\mathbf{x}^{\mathbf{w}+\mathbf{v}-\mathbf{u}}$ if $\mathbf{w} + \mathbf{v} \succeq \mathbf{u}$ and to 0 otherwise. Since I is a monomial ideal, this $\mathbb{C}$-linear map is a $\mathbb{C}[\mathbf{x}]$-module homomorphism, and hence it is an element of the module appearing in Theorem 18.29.

Corollary 18.30 *The Hilbert scheme H_n^d is not smooth if $n > d \geq 3$. In fact, the square of the maximal ideal in $\mathbb{C}[\mathbf{x}]$ is a singular point of H_{d+1}^d.*

Proof. As before, the Hilbert scheme H_n^d contains the locus $(S^n \mathbb{C}^d)^\circ$ of radical ideals as an open subvariety. This subvariety is smooth of dimension dn. It parametrizes unordered configurations of n distinct points in $\mathbb{C}^d$, or equivalently, radical ideals of colength n in $\mathbb{C}[\mathbf{x}]$. Every monomial ideal I_λ is in the closure of $(S^n \mathbb{C}^d)^\circ$, as can be seen using distractions as in Lemma 18.10. Therefore, a necessary condition for H_n^d to be smooth is that the tangent space of H_n^d at all monomial ideals has dimension dn. However, it can be checked, using Theorem 18.29, that this dimension is greater than dn if the points of $\lambda \subset \mathbb{N}^d$ do not lie in a hyperplane in $\mathbb{R}^d$. Specifically, if $\lambda = \{\mathbf{0}, \mathbf{e}_1, \ldots, \mathbf{e}_d\}$, so that $I_\lambda = \langle x_1, \ldots, x_d \rangle^2$, then a basis of the tangent space is given by the images of the parameters $c_\mathbf{v}^\mathbf{u}$, where $\mathbf{u}$ and $\mathbf{v}$ run over vectors in $\mathbb{N}^d$ having coordinate sum 2 and 1, respectively. The number of these parameters is

$$\binom{d+1}{2} \cdot d \;>\; (d+1) \cdot d.$$

We illustrate this derivation for $d = 3$ in the following example. $\qquad\qquad\square$

Example 18.31 Consider the Hilbert scheme H_4^3 of four points in affine 3-space. One of the monomial ideals in H_4^3 is the square

$$I_\lambda \;=\; \langle x, y, z \rangle^2 \;=\; \langle x^2, xy, xz, y^2, yz, z^2 \rangle$$

of the maximal ideal $\langle x, y, z \rangle$. The affine chart $U_\lambda \subset H_4^3$ consists of all colength 4 ideals of the form

$$\langle x^2 - c_1 x - c_2 y - c_3 y - d_1, \; xy - c_4 x - c_5 y - c_6 y - d_2,$$
$$xz - c_7 x - c_8 y - c_9 y - d_3, \; y^2 - c_{10} x - c_{11} y - c_{12} y - d_4,$$
$$yz - c_{13} x - c_{14} y - c_{15} y - d_5, \; z^2 - c_{16} x - c_{17} y - c_{18} y - d_6 \rangle.$$

The defining equations of U_λ are obtained by enforcing Buchberger's criterion for these six polynomials to form a Gröbner basis with respect to the total degree order. From this we find that each of the constant coefficients d_i can be expressed as a quadratic polynomial in the c_j. For instance,

$$d_2 \;=\; c_3 c_{10} + c_2 c_{10} - c_4 c_6 - c_4 c_5.$$

The remaining equations in the 18 parameters c_j are all quadratic. They generate a prime ideal of dimension 12. Hence the Hilbert scheme H_4^3 is irreducible of dimension 12, but its tangent space at I_λ has dimension 18. $\Diamond$

The Hilbert scheme H_n^d is connected. This is seen by the same Gröbner-path argument as in the case $d = 2$. However, it is generally not irreducible.

Theorem 18.32 (Iarrobino) *If $d \geq 3$ and $n \gg d$ then the Hilbert scheme H_n^d has more than one irreducible component and its dimension exceeds dn.*

Proof. The radical locus $(S^n \mathbb{C}^d)^\circ$ is an open subvariety of H_n^d. Let R_n^d denote its closure in H_n^d. Since $(S^n \mathbb{C}^d)^\circ$ is smooth and irreducible of dimension dn, we know that R_n^d is a dn-dimensional irreducible component of H_n^d. What we are claiming is that $R_n^d \neq H_n^d$ for $n \gg d \geq 3$.

The idea of the proof is to construct a family of colength n ideals whose dimension exceeds dn. This is done as follows. Determine the unique integer $r = r(d, n)$ such that

$$\binom{d + r - 1}{d} \;<\; n \;\leq\; \binom{d + r}{d}. \qquad (18.11)$$

Let W be any $\mathbb{C}$-vector space spanned by $\binom{d+r}{d} - n$ homogeneous polynomials of degree r in $\mathbb{C}[\mathbf{x}]$. Then the ideal

$$J_W \;=\; \langle W \rangle + \langle x_1, x_2, \ldots, x_n \rangle^{r+1}$$

has colength n, so J_W is a point in H_n^d. The assignment $W \mapsto J_W$ defines an injective algebraic map from the Grassmannian $\mathrm{Gr}^{\binom{d+r}{d} - n}(\mathbb{C}[\mathbf{x}]_r)$ into the Hilbert scheme H_n^d. The dimension of this Grassmannian is

$$\left(\binom{d + r}{d} - n \right) \cdot \left(n - \binom{d + r - 1}{d} \right). \qquad (18.12)$$

Hence the image of $W \mapsto J_W$ lies in an irreducible component of H_n^d whose dimension is at least the number (18.12). If n is large enough and chosen right in the middle of the bounds (18.11), then (18.12) is larger than dn. If that happens, then $R_n^d \neq H_n^d$ and the dimension of H_n^d is larger than dn.

The proof is completed by noting that $R_n^d \neq H_n^d$ implies $R_{n+1}^d \neq H_{n+1}^d$; namely, if $I \in H_n^d \smallsetminus R_n^d$ and $P = (p_1, \ldots, p_d) \in \mathbb{C}^d$ is not a zero of I, then $I \cap \langle x_1 - p_1, \ldots, x_d - p_d \rangle$ is a point in $H_{n+1}^d \smallsetminus R_{n+1}^d$. □

Example 18.33 For $d = 3$, the smallest value of n for which (18.12) exceeds the dimension $3n$ of the radical locus $(S^n \mathbb{C}^3)^\circ$ is $n = 102$. For that value, we have $r = 7$ and the lower and upper bounds in (18.11) are 84 and 120. Hence (18.12) is $18^2 = 324$ while $3n = 306$. In concrete terms: there exist ideals J_W of colength 102 in $\mathbb{C}[x, y, z]$ that are not in the closure of the locus of the ideals of 102 distinct points in affine 3-space $\mathbb{C}^3$. ◇

We call R_n^d the *radical component* of the Hilbert scheme H_n^d. The study of the radical component and all of the other components of H_n^d is a widely open problem. We do not even know what goes on for points in 3-space.

Problem 18.34 *Determine the smallest integer* n *such that* $H_n^3 \neq R_n^3$.

Another open problem is to identify the most singular point on the Hilbert scheme H_n^d. By this we mean an ideal I such that the vectorspace dimension of the tangent space $\mathrm{Hom}_{\mathbb{C}[\mathbf{x}]}(I, \mathbb{C}[\mathbf{x}]/I)$ is as large as possible. Since this dimension can only increase if we pass from I to an initial monomial ideal, this is really a combinatorial question about monomial ideals.

Problem 18.35 *Among all monomial ideals* I *of colength* n *in* $\mathbb{C}[\mathbf{x}]$, *find one that maximizes the vector space dimension of* $\mathrm{Hom}_{\mathbb{C}[\mathbf{x}]}(I, \mathbb{C}[\mathbf{x}]/I)$.

A first guess is that the most singular monomial ideal is the one with the most generators. The following example shows that this is not the case.

Example 18.36 Consider the case $d = 3$ and $n = 8$. There are 160 monomial ideals of colength 8 in $\mathbb{C}[x, y, z]$. These 160 ideals come in 33 types modulo permutations of the three variables. The ideals with the most generators are

$$\langle xy, xz, yz^2, y^2z, x^2, y^3, z^4 \rangle \quad \text{and} \quad \langle xy, xz^2, yz^2, y^2z, x^2, z^3, y^3 \rangle.$$

The tangent space of the Hilbert scheme H_8^3 at these singular points has dimension 32 in both cases. On the other hand, the point in H_8^3 given by

$$I = \langle x, y, z^2 \rangle^2 = \langle x^2, xy, y^2, xz^2, yz^2, z^4 \rangle$$

has only 6 minimal generators, but $\dim_{\mathbb{C}} \mathrm{Hom}_{\mathbb{C}[x,y,z]}(I, \mathbb{C}[x, y, z]/I) = 36$. ◇

The radical component R_n^d is generally also singular (for instance, for $d = 3$ and $n = 4$, as Example 18.31 shows). However, it is plausible that some version of Haiman's theory for $H_n^2 = R_n^2$ will extend to the irreducible variety R_n^d.

A natural object to study for commutative algebraists would be the ideal

$$I_{\text{diag}} \quad = \quad \bigcap_{1 \leq i < j \leq n} \langle y_{i1} - y_{j1}, y_{i2} - y_{j2}, \ldots, y_{id} - y_{jd} \rangle.$$

Here, the variables y_{ij} are the coordinate functions on the configuration space $(\mathbb{C}^d)^n$ of ordered n-tuples of points in $\mathbb{C}^d$. The zero set of the radical ideal I_{diag} is the *diagonal locus* consisting of all configurations with repeated points. For any n-element subset D of $\mathbb{N}^d$ consider the $n \times n$ determinant

$$\Delta_D \quad = \quad \det \left[\prod_{j=1}^{d} y_{ij}^{u_j} \right] \tag{18.13}$$

as in Theorem 18.21. The rows are indexed by $i = 1, \ldots, n$ and the columns are indexed by the boxes $(u_1, \ldots, u_d)$ in D. It is plausible that Theorem 18.19 still holds.

Conjecture 18.37 *The ideal I_{diag} is generated by the determinants Δ_D.*

The point of departure for a d-dimensional version of Haiman's theory would be the maps to the symmetric product and their reduced fiber product

$$\begin{array}{ccc}
X_n^d & \to & (\mathbb{C}^d)^n \\
\downarrow & & \downarrow \\
R_n^d & \to & S^n\mathbb{C}^d
\end{array}$$

as earlier. Thus X_n^d parametrizes pairs consisting of an n-tuple of points in d-space and an ideal I in the radical component whose zeros are the given points. The hope is that the arrows at the top and left pointing away from X_n^d represent morphisms with some of the good properties we have seen for $d = 2$. Generalizing what Haiman has proved for $d = 2$, it is natural to conjecture the following.

Conjecture 18.38 *The variety X_n^d is Cohen–Macaulay and coincides with the blowup of $(\mathbb{C}^d)^n$ along the diagonal locus. The radical component R_n^d of the Hilbert scheme is the blowup of $S^n\mathbb{C}^d$ along the ideal $I_{\text{diag}} \cap \mathbb{C}[\mathbf{x}]^{S_n}$.*

Conjecture 18.38 states in concrete terms that the determinants Δ_D are the natural coordinate functions on the radical component of the Hilbert scheme. The object in commutative algebra corresponding to the blowup of an affine space along a polynomial ideal is the *Rees algebra* of that ideal.

Problem 18.39 *Study the Rees algebra of the diagonal ideal $I_{\text{diag}} \subset \mathbb{C}[y_{ij}]$.*

The problem includes the question of finding minimal generators of I_{diag}, a question that is quite challenging even for $d = 2$; see our discussion after Theorem 18.19. The Rees algebra is the polynomial ring over $\mathbb{C}[\mathbf{x}]$ with one generator for each minimal generator Δ_D of I_{diag} and the relations are the homogeneous algebraic relations in these Δ_D with coefficients in $\mathbb{C}[\mathbf{x}]$. Thus Problem 18.39, a key issue in the study of Hilbert schemes, boils down to the following concrete question in combinatorial commutative algebra.

Problem 18.40 *Besides the quadratic Plücker relations, what are all the algebraic relations that hold among the $n \times n$ determinants Δ_D in (18.13)?*

18.5 Multigraded Hilbert schemes

In this section we present the multigraded Hilbert scheme that parametrizes all ideals in a polynomial ring $\mathbb{k}[\mathbf{x}] = \mathbb{k}[x_1, \ldots, x_n]$ with fixed Hilbert function for an arbitrary grading. Here, $\mathbb{k}$ need not be a field, but we allow $\mathbb{k}$ to be an arbitrary commutative ring. The multigraded Hilbert scheme generalizes both the Hilbert schemes of points in affine space, which we studied in previous sections, and the classical Hilbert scheme in algebraic geometry.

Assume that the polynomial ring $\mathbb{k}[\mathbf{x}] = \mathbb{k}[x_1, \ldots, x_n]$ is multigraded by an abelian group A as in the beginning of Section 8.1. Additionally, assume that A is generated as a group by the image of $\mathbb{Z}^n$. Let $A_+ = \deg(\mathbb{N}^n)$ denote the subsemigroup of A generated by $\mathbf{a}_1, \ldots, \mathbf{a}_n$.

A homogeneous ideal I in $\mathbb{k}[\mathbf{x}]$ is *admissible* if $(\mathbb{k}[\mathbf{x}]/I)_\mathbf{a} = \mathbb{k}[\mathbf{x}]_\mathbf{a}/I_\mathbf{a}$ is a locally free $\mathbb{k}$-module of finite rank for all $\mathbf{a} \in A$. Its *Hilbert function* is

$$h_I : A \to \mathbb{N} \quad \text{with} \quad h_I(\mathbf{a}) = \text{rank}_\mathbb{k}(\mathbb{k}[\mathbf{x}]/I)_\mathbf{a}.$$

Note that the support of the Hilbert function h_I must be contained in A_+.

If the grading by A is positive, then $(\mathbb{k}[\mathbf{x}]/I)_\mathbf{a}$ is a finitely generated $\mathbb{k}$-module. Hence $(\mathbb{k}[\mathbf{x}]/I)_\mathbf{a}$ is locally free if and only if it is flat over $\mathbb{k}$ [Eis95, Exercise 6.2]. Therefore, when the grading is positive, I is admissible if and only if $\mathbb{k}[\mathbf{x}]/I$ is flat over $\mathbb{k}$. In contrast, when the grading is not positive, admissibility is a stronger criterion than flatness. For example, let $\mathbb{k} = \mathbb{C}[y]$ be the polynomial ring over the complex numbers, let $\mathbb{k}[x]$ have the zero grading, and let $I = \langle 1 - xy \rangle$. Then $\mathbb{k}[x]/I$ is a flat $\mathbb{k}$ module, but it is not locally free at $y = 0$.

Example 18.41 Let $n = 3$ and $A = \mathbb{Z}^2$, and multigrade $\mathbb{C}[x, y, z]$ by

$$\deg(x) = (1, 0), \quad \deg(y) = (1, 1), \quad \deg(z) = (0, 1).$$

Here, $A_+ = \mathbb{N}^2$. Every A-homogeneous ideal I in $\mathbb{C}[x, y, z]$ is admissible and we can encode its Hilbert function by the coefficients of the *Hilbert series*

$$H_I(s, t) = \sum_{(a,b) \in \mathbb{N}^2} h_I(a, b) \cdot s^a t^b.$$

For instance, the Hilbert series of the zero ideal $I = \{0\}$ is

$$H_{\{0\}}(s,t) \;=\; \frac{1}{(1-s)(1-st)(1-t)} \;=\; 1 + s + t + s^2 + 2st + t^2 + \cdots.$$

Later, we will be interested in the artinian monomial ideal

$$M \;=\; \langle x^3, xy^2, x^2y, y^3, x^2z, xyz, y^2z, z^2 \rangle. \tag{18.14}$$

It has colength 9 and its Hilbert series is

$$H_M(s,t) \;=\; 1 + s + t + s^2 + 2st + st^2 + s^2t + s^2t^2. \tag{18.15}$$

Are there any other (monomial) ideals with the same Hilbert series? $\diamond$

Returning to the general discussion, let us fix a numerical function $h : A \to \mathbb{N}$. We wish to construct a scheme over $\Bbbk$ that parametrizes all admissible ideals I in $\Bbbk[\mathbf{x}]$ with $h_I = h$. To describe precisely what we mean by "parametrizes", we use the notion of *admissible family*. Over an affine scheme $\mathrm{Spec}(R)$ for a $\Bbbk$-algebra R, this extends the notion of admissible ideal in $\Bbbk[\mathbf{x}]$. For the definition, let us write $R[\mathbf{x}] = R \otimes_{\Bbbk} \Bbbk[\mathbf{x}]$ to mean the polynomial ring over R, which is graded in such a way that $R[\mathbf{x}]_{\mathbf{a}} = R \otimes_{\Bbbk} \Bbbk[\mathbf{x}]_{\mathbf{a}}$. Also, let $\mathbb{A}^n_{\Bbbk} = \mathrm{Spec}(\Bbbk[\mathbf{x}])$ be affine n-space over $\Bbbk$, and for a $\Bbbk$-scheme X, write $X \times \mathbb{A}^n_{\Bbbk}$ for the fiber product of X with $\mathbb{A}^n_{\Bbbk}$ over $\mathrm{Spec}(\Bbbk)$.

Definition 18.42 Fix a numerical function $h : A \to \mathbb{N}$. An **admissible ideal** over a $\Bbbk$-algebra R is a homogeneous ideal I in $R[\mathbf{x}]$ such that $R_{\mathbf{a}}/I_{\mathbf{a}}$ is a locally free R-module of rank $h(\mathbf{a})$ for each $\mathbf{a} \in A$. An **admissible family** over a $\Bbbk$-scheme X is a subscheme of $X \times \mathbb{A}^n_{\Bbbk}$ whose ideal sheaf restricts to an admissible ideal over R for every open affine subscheme $\mathrm{Spec}(R)$ of X.

Given two $\Bbbk$-algebras R and S along with an admissible ideal $I \subseteq S[\mathbf{x}]$, the image of I in $R[\mathbf{x}]$ under the map $S[\mathbf{x}] \to R[\mathbf{x}]$ generates an admissible ideal over R (see Exercise 18.10). More generally, if $X \to Y$ is a morphism of schemes, then every admissible family over Y pulls back to an admissible family over X. The scheme parametrizing the admissible ideals in $\Bbbk[\mathbf{x}]$ will be the one "best" admissible family, from which all others are pulled back.

Theorem 18.43 (Haiman and Sturmfels) *There is a quasiprojective scheme $\mathcal{H}^h_{\Bbbk[\mathbf{x}]}$ over $\Bbbk$ and an admissible family $\mathcal{U}^h_{\Bbbk[\mathbf{x}]}$ over $\mathcal{H}^h_{\Bbbk[\mathbf{x}]}$ such that for every $\Bbbk$-scheme X, the admissible families over X are in bijection with the morphisms $X \to \mathcal{H}^h_{\Bbbk[\mathbf{x}]}$, the bijection being given by pulling back $\mathcal{U}^h_{\Bbbk[\mathbf{x}]}$.*

Definition 18.44 $\mathcal{H}^h_{\Bbbk[\mathbf{x}]}$ is called the **multigraded Hilbert scheme**, and the admissible family $\mathcal{U}^h_{\Bbbk[\mathbf{x}]}$ over it is the **universal admissible family**.

Theorem 18.43 implies that both the multigraded Hilbert scheme and the universal family over it are unique up to canonical isomorphism. Thus $\mathcal{H}^h_{\mathbb{k}[\mathbf{x}]}$ really is the *one* best family. It parametrizes the admissible ideals in $\mathbb{k}[\mathbf{x}]$ in the sense that (by Theorem 18.43) they are in bijection with the $\mathbb{k}$-points of $\mathcal{H}^h_{\mathbb{k}[\mathbf{x}]}$, by which we mean the morphisms $\mathrm{Spec}(\mathbb{k}) \to \mathcal{H}^h_{\mathbb{k}[\mathbf{x}]}$.

There is an explicit (but quite complicated) algorithm to derive polynomial equations that locally describe the scheme $\mathcal{H}^h_{\mathbb{k}[\mathbf{x}]}$ [HS04]. The algorithm generalizes the derivation of the equations in the parameters $c^{\mathbf{u}}_{\mathbf{v}}$ for the charts U_λ in the previous sections. The construction has the following important consequence. Recall from Definition 8.7 what it means for the grading of $\mathbb{k}[\mathbf{x}]$ to be *positive*.

Corollary 18.45 *If the grading of the polynomial ring $\mathbb{k}[\mathbf{x}]$ is positive then the multigraded Hilbert scheme $\mathcal{H}^h_{\mathbb{k}[\mathbf{x}]}$ is projective over the ground ring $\mathbb{k}$.*

The remainder of this section is devoted to examples of multigraded Hilbert schemes, with the ground ring $\mathbb{k}$ being the complex numbers $\mathbb{C}$.

Based on the results of Section 18.2, we propose the following conjecture.

Conjecture 18.46 *The multigraded Hilbert scheme $\mathcal{H}^h_{\mathbb{C}[x,y]}$ is smooth and irreducible for any multigrading on $\mathbb{C}[x,y]$ and any Hilbert function h.*

Example 18.47 The following examples illustrate the range of Conjecture 18.46.

(i) If $A = \{0\}$ is the one-element group, then $\mathcal{H}^h_{\mathbb{C}[x,y]}$ is the Hilbert scheme of $h(0)$ points in the affine plane $\mathbb{C}^2$. We saw in Theorem 18.7 that this Hilbert scheme is smooth and irreducible of dimension $2n$.

(ii) Let $A = \mathbb{Z}$. If $\deg(x)$ and $\deg(y)$ are both positive integers and h has finite support, then $\mathcal{H}^h_{\mathbb{C}[x,y]}$ is an irreducible component in the fixed-point set of a $\mathbb{C}^*$-action on the Hilbert scheme of points. It was proved by Evain [Eva02] that this scheme is smooth and irreducible.

(iii) If $A = \mathbb{Z}$, $\deg(x) = \deg(y) = 1$, and $h(a) \equiv 1$, then $\mathcal{H}^h_{\mathbb{C}[x,y]} = \mathbb{P}^1$.

(iv) More generally, set $A = \mathbb{Z}$ and $\deg(x) = \deg(y) = 1$, but instead let $h(a) = \min(m, a+1)$ for some $m \in \mathbb{N}$. Then $\mathcal{H}^h_{\mathbb{C}[x,y]}$ is the Hilbert scheme of m points on $\mathbb{P}^1$.

(v) If $A = \mathbb{Z}$, $\deg(x) = -\deg(y) = 1$, and $h(a) \equiv 1$, then $\mathcal{H}^h_{\mathbb{C}[x,y]} = \mathbb{C}^1$.

(vi) If $A = \mathbb{Z}^2$, $\deg(x) = (1,0)$, and $\deg(y) = (0,1)$, then $\mathcal{H}^h_{\mathbb{C}[x,y]}$ is either empty or a point. In the latter case it consists of one monomial ideal.

(vii) If $A = \mathbb{Z}/2\mathbb{Z}$, $\deg(x) = \deg(y) = 1$, and $h(0) = h(1) = 1$, then $\mathcal{H}^h_{\mathbb{C}[x,y]}$ is isomorphic to the cotangent bundle of the projective line $\mathbb{P}^1$. ◇

We have seen in Section 18.4 that the statement of Conjecture 18.46 does not extend to 3 variables. The counterexample from Iarrobino's Theorem is $\mathcal{H}^{102}_{\mathbb{C}[x,y,z]}$, the Hilbert scheme of 102 points in affine 3-space. What follows is an example with only 9 points but using a two-dimensional grading. It is the smallest known example of a reducible multigraded Hilbert scheme.

Example 18.48 Let $n = 3$, fix the $\mathbb{Z}^2$-grading in Example 18.41, and let $h = h_M$ be the Hilbert function with Hilbert series (18.15). The multigraded Hilbert scheme $H^h_{\mathbb{C}[x,y,z]}$ is the reduced union of two projective lines $\mathbb{P}^1$ that intersect in the common torus fixed point M. The universal family equals

$$\langle x^3, xy^2, x^2y, y^3, a_0x^2z - a_1xy, b_0xyz - b_1y^2, y^2z, z^2 \rangle \quad \text{with} \quad a_1b_1 = 0.$$

Here, $(a_0 : a_1)$ and $(b_0 : b_1)$ are coordinates on two projective lines. This Hilbert scheme has three torus fixed points, namely the three monomial ideals in the family. The ideal M in (18.14) is the singular point on $H^h_{\mathbb{C}[x,y,z]}$. $\diamond$

Example 18.49 In algebraic geometry, there are classical examples of Hilbert schemes with multiple components. Let $n = 4$ and take $h(m) = 2m + 2$ for $m \geq 1$, but $h(0) = 1$. The corresponding Hilbert scheme has two components. A generic point of the first component corresponds to a pair of skew lines in projective space. A generic point of the second component corresponds to a conic in projective space and a point outside the plane of the conic. The two meet along a component, a generic point of which corresponds to two crossing lines in $\mathbb{P}^3$ with some nonreduced scheme structure at the crossing point in the direction normal to the plane spanned by the lines. (There are several other types of ideals in this family as well: their schemes are double lines with nonreduced structure at one point and plane conics with the extra point in the plane of the conic and not reduced.) $\diamond$

We will present two more classes of multigraded Hilbert schemes that have appeared in the commutative algebra literature. These are the classical *Grothendieck Hilbert scheme* and the *toric Hilbert scheme*.

Example 18.50 Let $A = \mathbb{Z}$ and give $\mathbb{C}[\mathbf{x}]$ the standard grading with $\deg(x_i) = 1$ for $i = 1, \ldots, n$. Consider the following family of Hilbert functions h. Let $p(t)$ be any univariate polynomial with $p(\mathbb{N}) \subseteq \mathbb{N}$. Fix a sufficiently large integer $g \gg 0$. (For experts: the number g has to exceed the Gotzmann number.) These data define a Hilbert function $h : A \to \mathbb{N}$ by

$$h(a) = \binom{n + a - 1}{a} \text{ if } a < g \quad \text{and} \quad h(a) = p(a) \text{ if } a \geq g.$$

The multigraded Hilbert scheme $H^h_{\mathbb{C}[\mathbf{x}]}$ parametrizes all subschemes of projective space $\mathbb{P}^{n-1}$ with Hilbert polynomial p. This is the **classical Hilbert scheme** due to Grothendieck. It is known to be connected [Har66a]. $\diamond$

Example 18.51 Fix any grading by an abelian group A on the polynomial ring $\mathbb{C}[\mathbf{x}] = \mathbb{C}[x_1, \ldots, x_n]$. The **toric Hilbert scheme** is defined as the multigraded Hilbert scheme $H^1_{\mathbb{C}[\mathbf{x}]}$, where $\mathbf{1}$ denotes the characteristic function of the semigroup A_+. This means that $\mathbf{1}(\mathbf{a}) = 1$ if $\mathbf{a} \in A_+$

and $\mathbf{1}(\mathbf{a}) = 0$ if $\mathbf{a} \in A \smallsetminus A_+$. There is distinguished point on the toric Hilbert scheme $H^1_{\mathbb{C}[\mathbf{x}]}$, namely the lattice ideal I_L studied in Chapter 7, for $L = \{\mathbf{u} \in \mathbb{Z}^n \mid \deg(\mathbf{u}) = \mathbf{0}\}$. To see this, note that $\mathbb{C}[\mathbf{x}]/I_L$ is isomorphic to the semigroup ring $\mathbb{C}[A_+]$, and obviously the Hilbert function of $\mathbb{C}[A_+]$ is the characteristic function $\mathbf{1}$ of A_+. The toric Hilbert scheme $H^1_{\mathbb{C}[\mathbf{x}]}$ parametrizes all A-homogeneous ideals with the same Hilbert function as the lattice ideal I_L. $\diamond$

The toric Hilbert scheme is a combinatorial object whose study is closely related to triangulations of polytopes. Using this connection to polyhedral geometry, Santos recently established the following result [San04].

Theorem 18.52 (Santos) *There exists a grading of the polynomial ring $\mathbb{C}[\mathbf{x}] = \mathbb{C}[x_1, \ldots, x_{26}]$ in 26 variables by the 6-dimensional lattice $A = \mathbb{Z}^6$ such that the toric Hilbert scheme $H^1_{\mathbb{C}[\mathbf{x}]}$ is disconnected.*

Exercises

18.1 For each of the seven partitions λ of $n = 5$, determine the equations of the affine chart U_λ of the Hilbert scheme H_5 of five points in $\mathbb{C}^2$.

18.2 For each of the eleven partitions λ of $n = 6$, find an explicit basis for the vector space $\mathfrak{m}_{I_\lambda}/\mathfrak{m}^2_{I_\lambda}$ in Proposition 18.14.

18.3 Generalizing Example 18.20, compute a minimal generating set of the ideal I_{diag} for $n = 4$ and $n = 5$.

18.4 Compute the equations of the isospectral Hilbert scheme Z_4 over U_{2+1+1}.

18.5 For each of the seven partitions λ of $n = 5$, compute the ideal J_λ.

18.6 Does the analogue of Theorem 18.26 hold for the zero-fiber of the Hilbert scheme H^3_n of n points in 3-space?

18.7 What is the Hilbert–Frobenius series of the polynomial ring $\mathbb{C}[\mathbf{x}, \mathbf{y}]$ with respect to the diagonal action of the symmetric group S_n?

18.8 Prove that H^3_5 is irreducible.

18.9 Prove Conjecture 18.37 for $n \leq 5$.

18.10 Fix a homomorphism $S \to R$ of $\Bbbk$-algebras, and let $I \subseteq S[\mathbf{x}]$ be an admissible ideal. Prove that $I_{\mathbf{a}} \otimes_S R$ maps injectively to $R[\mathbf{x}]_{\mathbf{a}}$ for every $\mathbf{a} \in A$. Conclude that the image of I in $R[\mathbf{x}]$ generates an admissible ideal.

18.11 Show that, for any grading on $\mathbb{C}[\mathbf{x}]$ and any given Hilbert function h, there are only finitely many monomial ideals having Hilbert function h.

18.12 Find a toric Hilbert scheme with exactly three irreducible components.

Notes

The result that the Hilbert scheme of points in the affine plane is smooth is due to Fogarty [Fog68]. Our proof of smoothness here is lifted with few changes from the appendix to [Hai04], which is, in turn, based on the introductory parts of [Hai98]. The smoothness holds more generally for the Hilbert scheme of points on any smooth surface, and the study of such Hilbert schemes is an active area of current research; see [Nak99, Göt02] for background and references. By contrast, there are few articles on the Hilbert scheme of points in affine 3-space or on a smooth threefold, and we hope that some readers of Section 18.4 might be interested in becoming pioneers. A noteworthy exception is the article [Iar72] by Iarrobino, which proves Theorem 18.32. Example 18.36 is drawn from [Stu00]. Recent work related to Conjecture 18.38 has been done by Ekedahl and Skjelnes [ES04].

Haiman developed his theory of Hilbert schemes and their relation to Macdonald polynomials in the two seminal articles [Hai01] and [Hai02]. Weyl's invariant theory result in Theorem 18.18 can be found in [Wey97]. In Theorem 18.26, the result that the zero-fiber is reduced, irreducible, and $(n-1)$-dimensional is due to Briançon [Bri77]. Haiman proved the Cohen–Macaulayness in [Hai02]. There are still many fascinating open questions regarding Hilbert schemes and symmetric functions. For an excellent survey of the field, see Haiman's article [Hai03].

Multigraded Hilbert schemes were introduced by Haiman and Sturmfels in [HS04], and most of the material in Section 18.5 is taken from that article. Toric Hilbert schemes were studied by Peeva and Stillman [PS02]. They are not to be confused with Hilbert schemes of subschemes of toric varieties. The latter are also multigraded Hilbert schemes, as shown by Maclagan and Smith [MS04b] using the notion of *multigraded regularity* [MS04a].

For more about universal properties of Hilbert schemes, we recommend the textbook by Eisenbud and Harris [EH00]. In particular, the general *functor of points* perspective naturally yields results such as Theorem 18.29.

References

Numbers in square brackets at the end of each entry indicate the pages in the text where that entry is cited.

[AD80]   S. Abeasis and A. Del Fra, *Degenerations for the representations of an equioriented quiver of type A_m*, Boll. Univ. Mat. Ital. Suppl. (1980), no. 2, 157–171.   [352]

[ADK81]   S. Abeasis, A. Del Fra, and H. Kraft, *The geometry of representations of A_m*, Math. Ann. **256** (1981), no. 3, 401–418.   [352]

[AB04]   Valery Alexeev and Michel Brion, *Toric degenerations of spherical varieties*, preprint, 2004. arXiv:math.AG/0403379   [288]

[AH99]   Klaus Altmann and Lutz Hille, *Strong exceptional sequences provided by quivers*, Algebr. Represent. Theory **2** (1999), no. 1, 1–17. [208]

[AH00]   Annetta Aramova and Jürgen Herzog, *Almost regular sequences and Betti numbers*, Amer. J. Math., **122** (2000), no. 4, 689–719.   [106]

[AHH00]   Annetta Aramova, Jürgen Herzog, and Takayuki Hibi, *Shifting operations and graded Betti numbers*, J. Algebr. Combin. **12** (2000), no. 3, 207–222.   [40]

[Aud91]   Michèle Audin, *The topology of torus actions on symplectic manifolds*, Progress in Mathematics Vol. 93, Birkhäuser Verlag, Basel, 1991, translated from the French by the author.   [208]

[ARS97]   Maurice Auslander, Idun Reiten, and Sverre O. Smalø, *Representation theory of Artin algebras*, Cambridge Studies in Advanced Mathematics Vol. 36, Cambridge University Press, Cambridge, 1997, corrected reprint of the 1995 original.   [353]

[AGHSS04] L. Avramov, M. Green, C. Huneke, K. Smith, and B. Sturmfels (eds.), *Lectures in Contemporary Commutative Algebra*, Mathematical Sciences Research Institute Publications, Cambridge University Press, Cambridge, 2004.   [viii]

[BNT02]   Eric Babson, Isabella Novik, and Rekha Thomas, *Symmetric iterated Betti numbers*, J. Combin. Theory, Ser. A **105** (2004), 233–254. [40]

[BaS96]   Imre Barany and Herbert Scarf, *Matrices with identical sets of neighbors*, Math. Oper. Res. **23** (1998), no. 4, 863–873.   [189, 190]

[Bar94]   Alexander I. Barvinok, *A polynomial time algorithm for counting integral points in polyhedra when the dimension is fixed*, Math. Oper. Res. **19** (1994), no. 4, 769–779.   [246]

[BP99]   Alexander Barvinok and James E. Pommersheim, *An algorithmic theory of lattice points in polyhedra*, New perspectives in algebraic combinatorics (Berkeley, CA, 1996–97), Mathematical Sciences Research Institute Vol. 38, Cambridge University Press, Cambridge, 1999, pp. 91–147.   [246]

[BW03]   Alexander Barvinok and Kevin Woods, *Short rational generating functions for lattice point problems*, J. Amer. Math. Soc. **16** (2003), no. 4, 957–979 (electronic).   [246]

[Bay96]   Dave Bayer, *Monomial ideals and duality*, Lecture notes, Berkeley 1995–96, available online at http://math.columbia.edu/~bayer/Duality_B96/.   [86]

[BCP99]   Dave Bayer, Hara Charalambous, and Sorin Popescu, *Extremal Betti numbers and applications to monomial ideals*, J. Algebra **221** (1999), no. 2, 497–512.   [19, 106]

[BPS98]   Dave Bayer, Irena Peeva, and Bernd Sturmfels, *Monomial resolutions*, Math. Res. Lett. **5** (1998), no. 1–2, 31–46.   [80, 126, 190]

[BPS01]   Dave Bayer, Sorin Popescu, and Bernd Sturmfels, *Syzygies of unimodular Lawrence ideals*, J. Reine Angew. Math. **534** (2001), 169–186.   [190]

[BS87]   David Bayer and Michael Stillman, *A criterion for detecting m-regularity*, Invent. Math. **87** (1987), no. 1, 1–11.   [40, 45]

[BS98]   Dave Bayer and Bernd Sturmfels, *Cellular resolutions of monomial modules*, J. Reine Angew. Math. **502** (1998), 123–140. [79, 80, 190]

[BDDPS04]   M. Beck, J. A. De Loera, M. Develin, J. Pfeifle, and R. P. Stanley, *Coefficients and roots of Ehrhart polynomials*, Contemp. Math., to appear, 2004. arXiv:math.CO/0402148   [246]

[BB93]   Nantel Bergeron and Sara Billey, *RC-graphs and Schubert polynomials*, Exp. Math. **2** (1993), no. 4, 257–269.   [329]

[BGG73]   I. N. Bernšteĭn, I. M. Gelfand, and S. I. Gelfand, *Schubert cells, and the cohomology of the spaces G/P*, Usp. Mat. Nauk **28** (1973), no. 3(171), 3–26.   [309]

[Bia76]   A. Białynicki-Birula, *Some properties of the decompositions of algebraic varieties determined by actions of a torus*, Bull. Acad. Polon. Sci. Sér. Sci. Math. Astronom. Phys. **24** (1976), no. 9, 667–674. [363]

[Big93]   Anna Maria Bigatti, *Upper bounds for the Betti numbers of a given Hilbert function*, Commun. Algebra **21** (1993), no. 7, 2317–2334.   [40]

[BP79]   Louis J. Billera and J. Scott Provan, *A decomposition property for simplicial complexes and its relation to diameters and shellings*, Second International Conference on Combinatorial Mathematics (New York, 1978), New York Academy of Sciences, New York, 1979, pp. 82–85.   [330]

[BiS96] Louis J. Billera and A. Sarangarajan, *The combinatorics of permutation polytopes*, Formal power series and algebraic combinatorics (New Brunswick, NJ, 1994), American Mathematical Society, Providence, RI, 1996, pp. 1–23. [80]

[BJS93] Sara C. Billey, William Jockusch, and Richard P. Stanley, *Some combinatorial properties of Schubert polynomials*, J. Algebr. Combin. **2** (1993), no. 4, 345–374. [329]

[BL00] Sara Billey and V. Lakshmibai, *Singular loci of Schubert varieties*, Birkhäuser, Boston, MA, 2000. [309]

[Bj00] Anders Björner, *Face numbers of Scarf complexes*, Discrete Comput. Geom. **24** (2000) no. 2–3, 185–196. [190]

[BB04] Anders Björner and Francesco Brenti, *Combinatorics of Coxeter groups*, Graduate Texts in Mathematics, Springer–Verlag, 2004, to appear. [viii, 309, 330]

[BK88] Anders Björner and Gil Kalai, *An extended Euler–Poincaré theorem*, Acta Math. **161** (1988), no. 3–4, 279–303. [40]

[BK89] Anders Björner and Gil Kalai, *On f-vectors and homology*, Combinatorial Mathematics: Proceedings of the Third International Conference (New York, 1985) (New York), Annals of the New York Academy of Science Vol. 555, New York Academy of Science, 1989, pp. 63–80. [40]

[BLSWZ99] Anders Björner, Michel Las Vergnas, Bernd Sturmfels, Neil White, and Günter M. Ziegler, *Oriented matroids*, second ed., Cambridge University Press, Cambridge, 1999. [72, 329]

[BZ02] Grzegorz Bobiński and Grzegorz Zwara, *Schubert varieties and representations of Dynkin quivers*, Colloq. Math. **94** (2002), no. 2, 285–309. [353]

[BB82] Walter Borho and Jean-Luc Brylinski, *Differential operators on homogeneous spaces. I. Irreducibility of the associated variety for annihilators of induced modules*, Invent. Math. **69** (1982), no. 3, 437–476. [172]

[BB85] Walter Borho and Jean-Luc Brylinski, *Differential operators on homogeneous spaces. III. Characteristic varieties of Harish-Chandra modules and of primitive ideals*, Invent. Math. **80** (1985), no. 1, 1–68. [172]

[Bot57] Raoul Bott, *Homogeneous vector bundles*, Ann. Math. (2) **66** (1957), 203–248. [288]

[Bri77] Joël Briançon, *Description de Hilb^{n}C{x, y}*, Invent. Math. **41** (1977), no. 1, 45–89. [378]

[BKR01] Tom Bridgeland, Alastair King, and Miles Reid, *The McKay correspondence as an equivalence of derived categories*, J. Amer. Math. Soc. **14** (2001), no. 3, 535–554 (electronic). [368]

[Bri88] Michel Brion, *Points entiers dans les polyèdres convexes*, Ann. Sci. École Norm. Sup. (4) **21** (1988), no. 4, 653–663. [246]

[BV97]   Michel Brion and Michèle Vergne, *Residue formulae, vector partition functions and lattice points in rational polytopes*, J. Amer. Math. Soc. **10** (1997), no. 4, 797–833.   [246]

[BrS98]   M. P. Brodmann and R. Y. Sharp, *Local cohomology: an algebraic introduction with geometric applications*, Cambridge Studies in Advanced Mathematics Vol. 60, Cambridge University Press, Cambridge, 1998.   [269]

[BC01]   Winfried Bruns and Aldo Conca, *KRS and determinantal ideals*, Geometric and combinatorial aspects of commutative algebra (Messina, 1999), Lecture Notes in Pure and Applied Mathematics Vol. 217, Marcel Dekker, New York, 2001, pp. 67–87.   [329]

[BC03]   Winfried Bruns and Aldo Conca, *Gröbner bases and determinantal ideals*, Commutative algebra, singularities and computer algebra (Sinaia, 2002), NATO Science Series II Mathematics, Physics, and Chemistry Vol. 115, Kluwer Academic, Dordrecht, 2003, pp. 9–66.
                                                                        [329]

[BG99]   Winfried Bruns and Joseph Gubeladze, *Normality and covering properties of affine semigroups*, J. Reine Angew. Math. **510** (1999), 161–178.   [148]

[BG05]   Winfried Bruns and Joseph Gubeladze, *Polytopes, rings, and K-theory*, in preparation, 2005.   [148, 172, 208]

[BH98]   Winfried Bruns and Jürgen Herzog, *Cohen–Macaulay rings*, revised edition, Cambridge Studies in Advanced Mathematics Vol. 39, Cambridge University Press, Cambridge, 1998.   [vii, 19, 80, 100, 106, 227, 251, 258, 264, 265, 266, 269, 270, 329, 342, 347, 353]

[BV88]   Winfried Bruns and Udo Vetter, *Determinantal rings*, Lecture Notes in Mathematics Vol. 1327, Springer–Verlag, Berlin, 1988.   [329]

[Buc01]   Anders Skovsted Buch, *Stanley symmetric functions and quiver varieties*, J. Algebra **235** (2001), no. 1, 243–260.   [353]

[Buc02]   Anders Skovsted Buch, *Grothendieck classes of quiver varieties*, Duke Math. J. **115** (2002), no. 1, 75–103.   [309, 353]

[Buc03]   Anders Skovsted Buch, *Alternating signs of quiver coefficients*, preprint, 2003. arXiv:math.CO/0307014   [353]

[BFR03]   Anders S. Buch, László M. Fehér, and Richárd Rimányi, *Positivity of quiver coefficients through Thom polynomials*, preprint, 2003. http://home.imf.au.dk/abuch/papers/   [351, 353]

[BF99]   Anders Skovsted Buch and William Fulton, *Chern class formulas for quiver varieties*, Invent. Math. **135** (1999), no. 3, 665–687.   [288, 352, 353]

[BKTY04a]   Anders S. Buch, Andrew Kresch, Harry Tamvakis, and Alexander Yong, *Schubert polynomials and quiver formulas*, Duke Math. J. **122** (2004), no. 1, 125–143.   [353]

[BKTY04b]   Anders S. Buch, Andrew Kresch, Harry Tamvakis, and Alexander Yong, *Grothendieck polynomials and quiver formulas*, Amer. J. Math., to appear, 2004. arXiv:math.CO/0306389   [353]

[BSY03]   Anders Skovsted Buch, Frank Sottile, and Alexander Yong, *Quiver coefficients are Schubert structure constants*, preprint, 2003. arXiv: math.CO/0311390   [353]

[BP02]   Victor M. Buchstaber and Taras E. Panov, *Torus actions and their applications in topology and combinatorics*, University Lecture Series Vol. 24, American Mathematical Society, Providence, RI, 2002. [208]

[BCKV00]   Daniel Bump, Kwok-Kwong Choi, Pär Kurlberg, and Jeffrey Vaaler, *A local Riemann hypothesis. I*, Math. Zeit. **233** (2000), no. 1, 1–19.
                                                                    [246]

[Cal02]   Philippe Caldero, *Toric degenerations of Schubert varieties*, Transform. Groups **7** (2002), no. 1, 51–60.   [288]

[Chi00]   R. Chirivì, *LS algebras and application to Schubert varieties*, Transform. Groups **5** (2000), no. 3, 245–264.   [288]

[CG97]   Neil Chriss and Victor Ginzburg, *Representation theory and complex geometry*, Birkhäuser, Boston, MA, 1997.   [172]

[CoC]   CoCoATeam, *CoCoA: a system for doing computations in commutative algebra*, available at http://cocoa.dima.unige.it.   [20]

[Con95]   Aldo Conca, *Ladder determinantal rings*, J. Pure Appl. Algebra **98** (1995), no. 2, 119–134.   [309]

[CH97]   Aldo Conca and Jürgen Herzog, *Ladder determinantal rings have rational singularities*, Adv. Math. **132** (1997), no. 1, 120–147. [309]

[CS04]   Aldo Conca and Jessica Sidman, *Generic initial ideals of points and curves*, preprint, 2004. arXiv:math.AC/0402418   [40]

[Cox95]   David Cox, *The homogeneous coordinate ring of a toric variety*, J. Algebr. Geom. **4** (1995), 17–50.   [208]

[CLO97]   David Cox, John Little, and Donal O'Shea, *Ideals, varieties, and algorithms: An introduction to computational algebraic geometry and commutative algebra*, second ed., Undergraduate Texts in Mathematics, Springer–Verlag, New York, 1997.   [viii, 24]

[CLO98]   David Cox, John Little, and Donal O'Shea, *Using algebraic geometry*, Graduate Texts in Mathematics Vol. 185, Springer–Verlag, New York, 1998.   [viii]

[DEP82]   Corrado De Concini, David Eisenbud, and Claudio Procesi, *Hodge algebras*, Astérisque Vol. 91, Société Mathématique de France, Paris, 1982, with a French summary.   [329]

[DH³SY03] Jesus De Loera, David Haws, Raymond Hemmecke, Peter Huggins, Bernd Sturmfels, Ruriko Yoshida, *Short rational functions for toric algebra and applications*, preprint, 2003. arXiv:math.CO/0307350
                                                                    [246]

[DH³TY03] J. A. De Loera, D. Haws, R. Hemmecke, P. Huggins, J. Tauzer, and R. Yoshida, *A user guide for LattE v1.1 and Software package*, 2003, available at http://www.math.ucdavis.edu/~latte.   [246]

[DRS04]   J. A. De Loera, J. Rambau, and F. Santos, *Triangulations of point sets: Applications, structures and algorithms*, Algorithms and Computation in Mathematics, Springer–Verlag, Heidelberg, to appear.
[78, 144, 232]

[Dem74]   Michel Demazure, *Désingularisation des variétés de Schubert généralisées*, Ann. Sci. École Norm. Sup. (4) **7** (1974), 53–88.   [309]

[DW00]   Harm Derksen and Jerzy Weyman, *Semi-invariants of quivers and saturation for Littlewood-Richardson coefficients*, J. Amer. Math. Soc. **13** (2000), no. 3, 467–479 (electronic).   [353]

[DRS74]   Peter Doubilet, Gian-Carlo Rota, and Joel Stein, *On the foundations of combinatorial theory. IX. Combinatorial methods in invariant theory*, Studies Appl. Math. **53** (1974), 185–216.   [288, 329]

[DK00]   J. J. Duistermaat and J. A. C. Kolk, *Lie groups*, Universitext, Springer–Verlag, Berlin, 2000.   [288]

[ER98]   John A. Eagon and Victor Reiner, *Resolutions of Stanley–Reisner rings and Alexander duality*, J. Pure Appl. Algebra **130** (1998), no. 3, 265–275.   [106]

[EG87]   Paul Edelman and Curtis Greene, *Balanced tableaux*, Adv. Math. **63** (1987), no. 1, 42–99.   [330]

[EG98]   Dan Edidin and William Graham, *Equivariant intersection theory*, Invent. Math. **131** (1998), no. 3, 595–634.   [172, 309]

[Eis95]   David Eisenbud, *Commutative algebra, with a view toward algebraic geometry*, Graduate Texts in Mathematics Vol. 150, Springer–Verlag, New York, 1995.   [viii, 12, 24, 26, 32, 40, 133, 135, 147, 152, 154, 155, 156, 159, 165, 230, 231, 263, 264, 301, 347, 363, 373]

[Eis04]   David Eisenbud, *Geometry of Syzygies*, Graduate Texts in Mathematics, Springer–Verlag, New York, 2004, to appear.   [viii]

[EH00]   David Eisenbud and Joe Harris, *The geometry of schemes*, Graduate Texts in Mathematics Vol. 197, Springer–Verlag, New York, 2000.
[viii, 378]

[EMS00]   David Eisenbud, Mircea Mustaţă, and Michael Stillman, *Cohomology of sheaves on toric varieties*, J. Symbolic Comp. **29** (2000), 583–600.   [269]

[Ehr62a]   Eugène Ehrhart, *Sur les polyèdres rationnels homothétiques à n dimensions*, C. R. Acad. Sci. Paris **254** (1962), 616–618.   [246]

[Ehr62b]   Eugène Ehrhart, *Sur les polyèdres homothétiques bordésà n dimensions*, C. R. Acad. Sci. Paris **254** (1962), 988–990.   [246]

[Ehr67a]   Eugène Ehrhart, *Sur un problème de géométrie diophantienne linéaire. I. Polyèdres et réseaux*, J. Reine Angew. Math. **226** (1967), 1–29.   [246]

[Ehr67b]   Eugène Ehrhart, *Sur un problème de géométrie diophantienne linéaire. II. Systèmes diophantiens linéaires*, J. Reine Angew. Math. **227** (1967), 25–49.   [246]

[Ehr67c]   Eugène Ehrhart, *Démonstration de la loi de réciprocité pour un polyèdre entier*, C. R. Acad. Sci. Paris Sér. A-B **265** (1967), A5–A7. [246]

[ES04]   Torsten Ekedahl and Roy Skjelnes, *Recovering the good component of the Hilbert scheme*, preprint, 2004. arXiv:math.AG/0405073 [378]

[EK90]   Shalom Eliahou and Michel Kervaire, *Minimal resolutions of some monomial ideals*, J. Algebra **129** (1990), no. 1, 1–25. [40]

[EGM98]   J. Elias, J. M. Giral, and R. M. Miró-Roig (eds.), *Six lectures on commutative algebra*, Progress in Mathematics Vol. 166, Birkhäuser Verlag, Basel, 1998. [viii]

[Eva02]   Laurent Evain, *Incidence relations among the Schubert cells of equivariant punctual Hilbert schemes*, Math. Zeit. **242** (2002), no. 4, 743–759. [375]

[Ewa96]   Günter Ewald, *Combinatorial convexity and algebraic geometry*, Graduate Texts in Mathematics Vol. 168, Springer–Verlag, New York, 1996. [viii, 208]

[FR02]   László Fehér and Richárd Rimányi, *Classes of degeneracy loci for quivers: the Thom polynomial point of view*, Duke Math. J. **114** (2002), no. 2, 193–213. [353]

[Fel01]   Stefan Felsner, *Convex drawings of planar graphs and the order dimension of 3-polytopes*, Order **18** (2001), no. 1, 19–37. [60]

[Fel03]   Stefan Felsner, *Geodesic embeddings and planar graphs*, Order **20** (2003), no. 2, 135–150. [60]

[Fog68]   John Fogarty, *Algebraic families on an algebraic surface*, Amer. J. Math. **90** (1968), 511–521. [377]

[FK96]   Sergey Fomin and Anatol N. Kirillov, *The Yang–Baxter equation, symmetric functions, and Schubert polynomials*, Discrete Math. **153** (1996), no. 1–3, 123–143. [329]

[FS94]   Sergey Fomin and Richard P. Stanley, *Schubert polynomials and the nil-Coxeter algebra*, Adv. Math. **103** (1994), no. 2, 196–207. [329]

[Ful92]   William Fulton, *Flags, Schubert polynomials, degeneracy loci, and determinantal formulas*, Duke Math. J. **65** (1992), no. 3, 381–420. [309, 330]

[Ful93]   William Fulton, *Introduction to toric varieties*, Princeton University Press, Princeton, NJ, 1993. [208]

[Ful97]   William Fulton, *Young tableaux*, London Mathematical Society Student Texts Vol. 35, Cambridge University Press, Cambridge, 1997. [288]

[Ful99]   William Fulton, *Universal Schubert polynomials*, Duke Math. J. **96** (1999), no. 3, 575–594. [353]

[FM05]   William Fulton and Mircea Mustaţă, book on toric varieties, in preparation. [207, 208, 256]

[FP98]   William Fulton and Piotr Pragacz, *Schubert varieties and degeneracy loci*, Springer–Verlag, Berlin, 1998. [309, 353]

[GR97]   P. Gabriel and A. V. Roiter, *Representations of finite-dimensional algebras*, Springer–Verlag, Berlin, 1997, translated from the Russian. [353]

[Gal74] André Galligo, *À propos du théorème de-préparation de Weierstrass*, Fonctions de plusieurs variables complexes, Lecture Notes in Mathematics Vol. 409, Springer, Berlin, 1974, pp. 543–579. [40]

[GPW99] Vesselin Gasharov, Irena Peeva, and Volkmar Welker, *The lcm-lattice in monomial resolutions*, Math. Res. Lett. **6** (1999), no. 5–6, 521–532. [80]

[GM88]   Rüdiger Gebauer and H. Michael Möller, *On an installation of Buchberger's algorithm*, J. Symbolic Comput. **6** (1988), no. 2–3, 275–286. [60]

[GT50]   I. M. Gelfand and M. L. Tsetlin, *Finite-dimensional representations of the group of unimodular matrices*, Dokl. Akad. Nauk SSSR (N.S.) **71** (1950), 825–828. [288]

[Gia04] G. Z. Giambelli, *Ordine di una varietà più ampia di quella rappresentata coll'annullare tutti i minori di dato ordine estratti da una data matrice generica di forme*, Mem. R. Ist. Lombardo **3** (1904), no. 11, 101–135. [310]

[GL96]   N. Gonciulea and V. Lakshmibai, *Degenerations of flag and Schubert varieties to toric varieties*, Transform. Groups **1** (1996), no. 3, 215–248. [288]

[GL97]   N. Gonciulea and V. Lakshmibai, *Schubert varieties, toric varieties, and ladder determinantal varieties*, Ann. Inst. Fourier (Grenoble) **47** (1997), no. 4, 1013–1064. [309]

[GL00]   N. Gonciulea and V. Lakshmibai, *Singular loci of ladder determinantal varieties and Schubert varieties*, J. Algebra **229** (2000), no. 2, 463–497. [309]

[GM00]   Nicolae Gonciulea and Claudia Miller, *Mixed ladder determinantal varieties*, J. Algebra **231** (2000), no. 1, 104–137. [309]

[GW78]   Shiro Goto and Keiichi Watanabe, *On graded rings, II ($\mathbb{Z}^n$-graded rings)*, Tokyo J. Math. **1** (1978), no. 2, 237–261. [228, 269, 270]

[Göt02] L. Göttsche, *Hilbert schemes of points on surfaces*, Proceedings of the International Congress of Mathematicians, Vol. II (Beijing, 2002), Higher Education Press, Beijing, 2002, pp. 483–494. [378]

[Grä84] Hans-Gert Gräbe, *The canonical module of a Stanley–Reisner ring*, J. Algebra **86** (1984), 272–281. [269]

[GS04]   Daniel R. Grayson and Michael E. Stillman, *Macaulay 2, a software system for research in algebraic geometry*, available by ftp at http://www.math.uiuc.edu/Macaulay2/ [20, 75, 106]

[GM92]   John P. C. Greenlees and J. Peter May, *Derived functors of I-adic completion and local homology*, J. Algebra **149** (1992), no. 2, 438–453. [106, 270]

[GP02]   Gert-Martin Greuel and Gerhard Pfister, *A singular introduction to commutative algebra*, Springer–Verlag, Berlin, 2002. [viii]

[GPS01] G.-M. Greuel, G. Pfister, and H. Schönemann, SINGULAR 2.0, *A Computer Algebra System for Polynomial Computations*, Centre for Computer Algebra, University of Kaiserslautern (2001), available at http://www.singular.uni-kl.de [20]

[Gru93] Peter Gruber, *Geometry of numbers*. Handbook of convex geometry, North-Holland, Amsterdam, 1993, Vol. A, B, pp. 739–763. [243]

[Grü03] Branko Grünbaum, *Convex polytopes*, second ed., Graduate Texts in Mathematics Vol. 221, Springer–Verlag, New York, 2003. [viii]

[GS83] Victor Guillemin and Shlomo Sternberg, *The Gel'fand–Cetlin system and quantization of the complex flag manifolds*, J. Funct. Anal. **52** (1983), no. 1, 106–128. [288]

[Hai92] Mark D. Haiman, *Dual equivalence with applications, including a conjecture of Proctor*, Discrete Math. **99** (1992), no. 1–3, 79–113. [330]

[Hai98] Mark Haiman, *t, q-Catalan numbers and the Hilbert scheme*, Discrete Math. **193** (1998), no. 1–3, 201–224. [378]

[Hai01] Mark Haiman, *Hilbert schemes, polygraphs and the Macdonald positivity conjecture*, J. Amer. Math. Soc. **14** (2001), no. 4, 941–1006. (electronic). [266, 378]

[Hai02] Mark Haiman, *Vanishing theorems and character formulas for the Hilbert scheme of points in the plane*, Invent. Math. **149** (2002), no. 2, 371–407. [266, 378]

[Hai03] Mark Haiman, *Combinatorics, symmetric functions, and Hilbert schemes*, Current developments in mathematics, 2002, International Press, Somerville, MA, 2003, pp. 39–111. [378]

[Hai04] Mark Haiman, *Commutative algebra of N points in the plane*, Lectures in Contemporary Commutative Algebra (L. Avramov, M. Green, C. Huneke, K. Smith, and B. Sturmfels, eds.), Mathematical Sciences Research Institute Publications, Cambridge University Press, Cambridge, 2004. [378]

[HS04] Mark Haiman and Bernd Sturmfels, *Multigraded Hilbert schemes*, J. Alg. Geom. **13** (2004), no. 4, 725–769. [375, 378]

[Har66a] Robin Hartshorne, *Connectedness of the Hilbert scheme*, Inst. Hautes Études Sci. Publ. Math. **29** (1966), 5–48. [40, 361, 376]

[Har66b] Robin Hartshorne, *Residues and duality*, Lecture Notes in Mathematics Vol. 20, Springer–Verlag, Berlin, 1966. [233, 246, 265, 270]

[Har70] Robin Hartshorne, *Affine duality and cofiniteness*, Invent. Math. **9** (1969/1970), 145–164. [269]

[Har77] Robin Hartshorne, *Algebraic geometry*, Graduate Texts in Mathematics Vol. 52, Springer–Verlag, New York, 1977. [viii, 172, 301]

[Hat02] Allen Hatcher, *Algebraic topology*, Cambridge University Press, Cambridge, 2002. [9, 19, 106]

[HM03] David Helm and Ezra Miller, *Bass numbers of semigroup-graded local cohomology*, Pacific J. Math. **209**, no. 1 (2003), 41–66. [270]

[HM04]   David Helm and Ezra Miller, *Algorithms for graded injective resolu-
            tions and local cohomology over semigroup rings*, J. Symbolic Com-
            put., to appear, 2004. arXiv:math.CO/0309256   [228, 270]

[HT92]   Jürgen Herzog and Ngô Viêt Trung, *Gröbner bases and multiplicity
            of determinantal and Pfaffian ideals*, Adv. Math. **96** (1992), no. 1,
            1–37.   [309, 310, 329]

[Hib86]   Takayuki Hibi, *Every affine graded ring has a Hodge algebra struc-
            ture*, Rend. Sem. Mat. Univ. Politec. Torino **44** (1986), no. 2, 277–
            286 (1987).   [329]

[Hib87]   Takayuki Hibi, *Distributive lattices, affine semigroup rings and al-
            gebras with straightening laws*, Commutative algebra and combina-
            torics (Kyoto, 1985), Advanced Studies in Pure Mathematics Vol. 11,
            North-Holland, Amsterdam, 1987, pp. 93–109.   [288]

[Hib92]   Takayuki Hibi, *Algebraic combinatorics on convex polytopes*, Cars-
            law Publications, Glebe, Australia, 1992.   [viii, 19, 329]

[Hil98]   Lutz Hille, *Toric quiver varieties*, Algebras and modules, II (Gei-
            ranger, 1996), CMS Conference Proceedings Vol. 24, American
            Mathematical Society, Providence, RI, 1998, pp. 311–325.   [208]

[Hoc72]   M. Hochster, *Rings of invariants of tori, Cohen–Macaulay rings gen-
            erated by monomials, and polytopes*, Ann. Math. (2) **96** (1972), 318–
            337.   [270]

[Hoc77]   Melvin Hochster, *Cohen–Macaulay rings, combinatorics, and sim-
            plicial complexes*, Ring theory, II (Proc. Second Conf., Univ. Okla-
            homa, Norman, Okla., 1975) (B. R. McDonald and R. Morris, eds.),
            Lecture Notes in Pure and Applied Mathematics Vol. 26, Marcel
            Dekker, New York, 1977, pp. 171–223.   [19, 105, 269]

[Hof79]   Douglas R. Hofstadter, *Gödel, Escher, Bach: An eternal golden
            braid*, Basic Books, New York, 1979.   [208]

[HM99]   S. Hoşten and W. Morris, Jr. *The order dimension of the complete
            graph*, Discrete Math. **201** (1999), 133–139.   [121, 122, 126]

[HoS02]   Serkan Hoşten and Gregory G. Smith, *Monomial ideals*, Compu-
            tations in algebraic geometry with Macaulay 2, Algorithms and
            Computation in Mathematics Vol. 8, Springer–Verlag, Berlin, 2002,
            pp. 73–100.   [106]

[Hul93]   Heather A. Hulett, *Maximum Betti numbers of homogeneous ideals
            with a given Hilbert function*, Commun. Algebra **21** (1993), no. 7,
            2335–2350.   [40]

[Hum90]   James E. Humphreys, *Reflection groups and Coxeter groups*, Cam-
            bridge University Press, Cambridge, 1990.   [309]

[Iar72]   Anthony Iarrobino, *Reducibility of the families of 0-dimensional
            schemes on a variety*, Invent. Math. **15** (1972), 72–77.   [378]

[Ish80]   Masa-Nori Ishida, *Torus embeddings and dualizing complexes*, Tô-
            hoku Math. J. (2) **32** (1980), no. 1, 111–146.   [246, 270]

[Ish87]   Masa-Nori Ishida, *The local cohomology groups of an affine semi-group ring*, Algebraic geometry and commutative algebra in Honor of Masayaoshi Nagata, Vol. I, Kinokuniya, Tokyo, 1987, pp. 141–153.
[148, 246, 270]

[Jos84]   Anthony Joseph, *On the variety of a highest weight module*, J. Algebra **88** (1984), no. 1, 238–278.
[172]

[KK79]   Bernd Kind and Peter Kleinschmidt, *Schälbare Cohen–Macauley-Komplexe und ihre Parametrisierung*, Math. Zeit. **167** (1979), no. 2, 173–179.
[270]

[KMS04]   Allen Knutson, Ezra Miller, and Mark Shimozono, *Four positive formulae for type A quiver polynomials*. arXiv:math.AG/0308142   [172, 330, 351, 352, 353]

[KnM04a] Allen Knutson and Ezra Miller, *Subword complexes in Coxeter groups*, Adv. Math. **184** (2004), 161–176.
[106, 330]

[KnM04b] Allen Knutson and Ezra Miller, *Gröbner geometry of Schubert polynomials*, Ann. Math. (2), to appear, 2004. arXiv:math.AG/0110058
[172, 309, 329, 330]

[Kog00]   Mikhail Kogan, *Schubert geometry of flag varieties and Gel'fand–Cetlin theory*, Ph.D. thesis, Massachusetts Institute of Technology, 2000.
[288, 330]

[KoM04]   Mikhail Kogan and Ezra Miller, *Toric degeneration of Schubert varieties and Gelfand–Tsetlin polytopes*, Adv. Math., to appear, 2004. arXiv:math.AG/0303208
[288, 330]

[KP99]   C. Krattenthaler and M. Prohaska, *A remarkable formula for counting nonintersecting lattice paths in a ladder with respect to turns*, Trans. Amer. Math. Soc. **351** (1999), no. 3, 1015–1042.   [309]

[KR00]   Martin Kreuzer and Lorenzo Robbiano, *Computational commutative algebra. 1*, Springer–Verlag, Berlin, 2000.   [viii]

[Lak03]   V. Lakshmibai, *The development of standard monomial theory. II*, A tribute to C. S. Seshadri (Chennai, 2002), Birkhäuser, Basel, 2003, pp. 283–309.   [288]

[LM98]   V. Lakshmibai and Peter Magyar, *Degeneracy schemes, quiver schemes, and Schubert varieties*, Int. Math. Res. Notices (1998), no. 12, 627–640.   [352]

[LS82a]   Alain Lascoux and Marcel-Paul Schützenberger, *Polynômes de Schubert*, C. R. Acad. Sci. Paris Sér. I Math. **294** (1982), no. 13, 447–450.
[309]

[LS82b]   Alain Lascoux and Marcel-Paul Schützenberger, *Structure de Hopf de l'anneau de cohomologie et de l'anneau de Grothendieck d'une variété de drapeaux*, C. R. Acad. Sci. Paris Sér. I Math. **295** (1982), no. 11, 629–633.   [309]

[LS85]   Alain Lascoux and Marcel-Paul Schützenberger, *Schubert polynomials and the Littlewood–Richardson rule*, Lett. Math. Phys. **10** (1985), no. 2–3, 111–124.   [330]

[LS89]   Alain Lascoux and Marcel-Paul Schützenberger, *Tableaux and non-commutative Schubert polynomials*, Funct. Anal. Appl. **23** (1989), 63–64.   [330]

[Lit98a]   Peter Littelmann, *Cones, crystals, and patterns*, Transform. Groups **3** (1998), no. 2, 145–179.   [288]

[Lyu88]   Gennady Lyubeznik, *A new explicit finite free resolution of ideals generated by monomials in an R-sequence*, J. Pure Appl. Algebra **51** (1988), no. 1–2, 193–195.   [80]

[Mac27]   Francis S. Macaulay, *Some properties of enumeration in the theory of modular systems*, Proc. London Math. Soc. **26** (1927), 531–555.   [34, 40]

[Macd63]   Ian G. Macdonald, *The volume of a lattice polyhedron*, Proc. Cambridge Philos. Soc. **59** (1963), 719–726.   [246]

[Macd71]   I. G. Macdonald, *Polynomials associated with finite cell-complexes*, J. London Math. Soc. (2) **4** (1971), 181–192.   [246]

[Macd91]   Ian G. Macdonald, *Notes on Schubert polynomials*, Publications du LACIM, Universitè du Québec à Montréal, Montréal, 1991.   [309]

[Macd95]   Ian G. Macdonald, *Symmetric functions and Hall polynomials*, second ed., Clarendon Press/Oxford University Press, New York, 1995.   [305, 368]

[MS04a]   Diane Maclagan and Gregory G. Smith, *Multigraded Castelnuovo-Mumford regularity*, J. Reine Angew. Math. **571** (2004), 179–212.   [378]

[MS04b]   Diane Maclagan and Gregory G. Smith, *Uniform bounds on multi-graded regularity*, J. Alg. Geom., to appear, 2004. arXiv:math.AG/0305215   [378]

[MacL95]   Saunders Mac Lane, *Homology*, Classics in Mathematics, Springer-Verlag, Berlin, 1995, reprint of the 1975 edition.   [20, 269]

[MacL98]   Saunders Mac Lane, *Categories for the working mathematician*, second ed., Graduate Texts in Mathematics Vol. 5, Springer-Verlag, New York, 1998.   [viii, 183]

[Man01]   Laurent Manivel, *Symmetric functions, Schubert polynomials and degeneracy loci*, SMF/AMS Texts and Monographs Vol. 6, American Mathematical Society, Providence, RI, 2001, translated from the 1998 French original by John R. Swallow, Cours Spécialisés [Specialized Courses], 3.   [309, 353]

[Mar03]   Jeremy L. Martin, *Geometry of graph varieties*, Trans. Amer. Math. Soc. **355** (2003), no. 10, 4151–4169 (electronic).   [330]

[Mar03]   Jeremy Martin, *The slopes determined by n points in the plane*, preprint, 2003. arXiv:math.AG/0302106   [172, 330]

[Mil98]   Ezra Miller, *Alexander duality for monomial ideals and their resolutions*. arXiv:math.AG/9812095   [80, 126, 270]

[Mil00a]   Ezra Miller, *The Alexander duality functors and local duality with monomial support*, J. Algebra **231** (2000), 180–234.   [20, 106, 126, 228, 269, 270]

[Mil00b]   Ezra Miller, *Resolutions and duality for monomial ideals*, Ph.D. thesis, University of California at Berkeley, 2000.   [106, 262]

[Mil02a]   Ezra Miller, *Graded Greenlees–May duality and the Čech hull*, Local cohomology and its applications (Guanajuato, 1999), Lecture Notes in Pure and Applied Mathematics Vol. 226, Marcel Dekker, New York, 2002, pp. 233–253.   [106, 270]

[Mil02b]   Ezra Miller, *Planar graphs as minimal resolutions of trivariate monomial ideals*, Documenta Math. **7** (2002), 43–90 (electronic).   [60, 80, 106]

[Mil02c]   Ezra Miller, *Cohen–Macaulay quotients of normal semigroup rings via irreducible resolutions*, Math. Res. Lett. **9** (2002), no. 1, 117–128.   [228]

[Mil03a]   Ezra Miller, *Mitosis recursion for coefficients of Schubert polynomials*, J. Combin. Theory, Ser. A **103** (2003), 223–235.   [329]

[Mil03b]   Ezra Miller, *Alternating formulas for K-theoretic quiver polynomials*, Duke Math J., to appear. arXiv:math.CO/0312250   [353]

[MP01]   Ezra Miller and David Perkinson, *Eight lectures on monomial ideals*, COCOA VI: Proceedings of the International School, Villa Gualino—May–June, 1999 (Anthony V. Geramita, ed.), Queens Papers in Pure and Applied Mathematics Vol. 120, Queen's University, Kingston, Ontario, Canada, 2001, pp. 3–105.   [vii]

[MS99]   Ezra Miller and Bernd Sturmfels, *Monomial ideals and planar graphs*, Applied Algebra, Algebraic Algorithms and Error-Correcting Codes (M. Fossorier, H. Imai, S. Lin, and A. Poli, eds.), Springer Lecture Notes in Computer Science Vol. 1719, Springer–Verlag, Berlin, 1999, pp. 19–28.   [60, 75]

[MSY00]   Ezra Miller, Bernd Sturmfels, and Kohji Yanagawa, *Generic and cogeneric monomial ideals*, J. Symbolic Comput. **29** (2000), 691–708.   [80, 126]

[MS96]   J. V. Motwani and M. A. Sohoni, *Divisor class groups of ladder determinantal varieties*, J. Algebra **186** (1996), no. 2, 338–367. [309]

[Mul89]   S. B. Mulay, *Determinantal loci and the flag variety*, Adv. Math. **74** (1989), no. 1, 1–30.   [309]

[MFK94]   D. Mumford, J. Fogarty, and F. Kirwan, *Geometric invariant theory*, third ed., Ergebnisse der Mathematik und ihrer Grenzgebiete (2) [Results in Mathematics and Related Areas (2)] Vol. 34, Springer–Verlag, Berlin, 1994.   [208]

[Mun84]   James R. Munkres, *Elements of algebraic topology*, Addison–Wesley, Menlo Park, CA, 1984.   [9, 19, 106]

[Mus03]   C. Musili, *The development of standard monomial theory. I*, A tribute to C. S. Seshadri (Chennai, 2002), Birkhäuser, Basel, 2003, pp. 385–420.   [288]

[Mus94]   Ian M. Musson, *Differential operators on toric varieties*, J. Pure Appl. Algebra **95** (1994), no. 3, 303–315.   [208]

[Mus00]   Mircea Mustaţă, *Local cohomology at monomial ideals*, J. Symbolic
            Comput. **29** (2000), 709–720.   [269, 270]

[Mus02]   Mircea Mustaţă, *Vanishing theorems on toric varieties*, Tohoku
            Math. J. (2) **54** (2002), no. 3, 451–470.   [208, 269]

[Nak99]   Hiraku Nakajima, *Lectures on Hilbert schemes of points on surfaces*,
            University Lecture Series Vol. 18, American Mathematical Society,
            Providence, RI, 1999.   [378]

[NPS02]   Isabella Novik, Alexander Postnikov, and Bernd Sturmfels, *Syzygies
            of oriented matroids*, Duke Math. J. **111** (2002), no. 2, 287–317. [80]

[Oda88]   Tadao Oda, *Convex bodies and algebraic geometry*, Ergebnisse der
            Mathematik und ihrer Grenzgebiete (3) [Results in Mathematics and
            Related Areas (3)] Vol. 15, Springer–Verlag, Berlin, 1988.   [208]

[Par94]   Keith Pardue, *Nonstandard Borel-fixed ideals*, Ph.D. thesis, Brandeis
            University, 1994.   [40]

[PS98a]   Irena Peeva and Bernd Sturmfels, *Generic lattice ideals*, J. Amer.
            Math. Soc. **11** (1998), no. 2, 363–373.   [190]

[PS02]   Irena Peeva and Mike Stillman, *Toric Hilbert schemes*, Duke Math.
            J. **111** (2002), no. 3, 419–449.   [378]

[PS98b]   Irena Peeva and Bernd Sturmfels, *Syzygies of codimension 2 lattice
            ideals*, Math. Zeit. **229** (1998), no. 1, 163–194.   [190]

[PS04]   Alexander Postnikov and Boris Shapiro, *Trees, parking functions,
            syzygies, and deformations of monomial ideals*, Trans. Amer. Math.
            Soc. **356** (2004), no. 8, 3109–3142 (electronic).   [126]

[PSS99]   Alexander Postnikov, Boris Shapiro, and Mikhail Shapiro, *Algebras
            of curvature forms on homogeneous manifolds*, Differential topology,
            infinite-dimensional Lie algebras, and applications, American Math-
            ematical Society Translations Series 2 Vol. 194, American Mathe-
            matical Society, Providence, RI, 1999, pp. 227–235.   [80]

[Ram85]   A. Ramanathan, *Schubert varieties are arithmetically Cohen–Ma-
            caulay*, Invent. Math. **80** (1985), no. 2, 283–294.   [330]

[RS95]   Victor Reiner and Mark Shimozono, *Plactification*, J. Algebr. Com-
            bin. **4** (1995), no. 4, 331–351.   [330]

[Rei76]   Gerald Allen Reisner, *Cohen–Macaulay quotients of polynomial
            rings*, Adv. Math. **21** (1976), no. 1, 30–49.   [106, 270]

[RS90]   Lorenzo Robbiano and Moss Sweedler, *Subalgebra bases*, Commu-
            tative algebra (Salvador, 1988), Lecture Notes in Mathematics
            Vol. 1430, Springer–Verlag, Berlin, 1990, pp. 61–87.   [288]

[Ros89]   W. Rossmann, *Equivariant multiplicities on complex varieties III:
            Orbites unipotentes et représentations*, Astérisque **11** (1989), no.
            173–174, 313–330.   [172]

[Rot88]   Joseph J. Rotman, *An introduction to algebraic topology*, Graduate
            Texts in Mathematics Vol. 119, Springer–Verlag, New York, 1988.
                                                                 [viii, 9, 19, 94, 106]

[San04]   Francisco Santos, *Non-connected toric Hilbert schemes*, Math. Ann.,
            to appear, 2004. arXiv:math.CO/0204044   [377]

[Sca86]   Herbert Scarf, *Neighborhood systems for production sets with indi-
            visibilities*, Econometrica **54** (1986), no. 3, 507–532.   [126, 190]

[SS90]   Uwe Schäfer and Peter Schenzel, *Dualizing complexes of affine semi-
            group rings*, Trans. Amer. Math. Soc. **322** (1990), no. 2, 561–582.
                                                                        [270]

[Sch03]   Hal Schenck, *Computational algebraic geometry*, London Mathemati-
            cal Society Student Texts Vol. 58, Cambridge University Press, Cam-
            bridge, 2003.   [viii]

[Sch86]   Alexander Schrijver, *Theory of linear and integer programming*,
            Wiley-Interscience Series in Discrete Mathematics, John Wiley &
            Sons, Chichester, 1986.   [148]

[Ses95]   C. S. Seshadri, *The work of P. Littelmann and standard monomial
            theory*, Current Trends in Mathematics and Physics, Narosa, New
            Delhi, 1995, pp. 178–197.   [288]

[Sta78]   Richard P. Stanley, *Hilbert functions of graded algebras*, Adv. Math.
            28 (1978), no. 1, 57–83.   [264]

[Sta84]   Richard P. Stanley, *On the number of reduced decompositions of ele-
            ments of Coxeter groups*, Eur. J. Combin. **5** (1984), no. 4, 359–372.
                                                                        [330]

[Sta96]   Richard P. Stanley, *Combinatorics and commutative algebra*, second
            ed., Progress in Mathematics Vol. 41, Birkhäuser, Boston, MA, 1996.
                                               [vii, 8, 19, 190, 266, 269, 270, 406]

[Sta97]   Richard P. Stanley, *Enumerative combinatorics. Vol. 1*, Cambridge
            Studies in Advanced Mathematics Vol. 49, Cambridge University
            Press, Cambridge, 1997.   [232]

[Stu90]   Bernd Sturmfels, *Gröbner bases and Stanley decompositions of de-
            terminantal rings*, Math. Zeit. **205** (1990), no. 1, 137–144.   [329]

[Stu93]   Bernd Sturmfels, *Algorithms in invariant theory*. Texts and Mono-
            graphs in Symbolic Computation, Springer–Verlag, Vienna, 1993.
                                                                        [288]

[Stu95]   Bernd Sturmfels, *On vector partition functions*, J. Combin. Theory
            Ser. A **72** (1995), no. 2, 302–309.   [246]

[Stu96]   Bernd Sturmfels, *Gröbner bases and convex polytopes*, AMS Univer-
            sity Lecture Series Vol. 8, American Mathematical Society, Provi-
            dence, RI, 1996.   [viii, 148, 187, 286]

[Stu99]   Bernd Sturmfels, *The co-Scarf resolution*, Commutative algebra, al-
            gebraic geometry, and computational methods (Hanoi, 1996) (David
            Eisenbud, ed.), Springer–Verlag, Singapore, 1999, pp. 315–320. [126]

[Stu00]   Bernd Sturmfels, *Four counterexamples in combinatorial algebraic
            geometry*, J. Algebra **230** (2000), no. 1, 282–294.   [378]

[SWZ95]   Bernd Sturmfels, Robert Weismantel and Günter Ziegler, *Gröbner
            bases of lattices, corner polyhedra, and integer programming*, Beit.
            Alg. und Geom. **36** (1995), 281–298.   [148]

[SW89]   Bernd Sturmfels and Neil White, *Gröbner bases and invariant theory*, Adv. Math. **76** (1989), no. 2, 245–259. [288]

[Tay60]   Diana Taylor, *Ideals generated by monomials in an R-sequence*, Ph.D. thesis, University of Chicago, 1960. [80]

[Ter99a] Naoki Terai, *Alexander duality theorem and Stanley–Reisner rings*, Free resolutions of coordinate rings of projective varieties and related topics (Kyoto, 1998), Sūrikaisekikenkyūsho Kōkyūroku Vol. 1078, 1999, pp. 174–184 (Japanese). [106]

[Ter99b] Naoki Terai, *Local cohomology modules with respect to monomial ideals*, preprint, 1999. [269, 270]

[Tho02]   Howard Thompson, *On toric log schemes*, Ph.D. thesis, University of California at Berkeley, 2002. [148]

[Tot99]   Burt Totaro, *The Chow ring of a classifying space*, Algebraic K-theory (Seattle, WA, 1997), American Mathematical Society, Providence, RI, 1999, pp. 249–281. [172, 309]

[TH86]   Ngô Viêt Trung and Lê Tuấn Hoa, *Affine semigroups and Cohen–Macaulay rings generated by monomials*, Trans. Amer. Math. Soc. **298** (1986), no. 1, 145–167. [269, 270]

[Vas98]   Wolmer V. Vasconcelos, *Computational methods in commutative algebra and algebraic geometry*, Algorithms and Computation in Mathematics Vol. 2, Springer–Verlag, Berlin, 1998. [viii, 172, 228]

[Ver03]   Michèle Vergne, *Residue formulae for Verlinde sums, and for number of integral points in convex rational polytopes*, European women in mathematics (Malta, 2001), World Scientific Publishing, River Edge, NJ, 2003, pp. 225–285. [246]

[Vil01]   Rafael H. Villarreal, *Monomial algebras*, Monographs and Textbooks in Pure and Applied Mathematics Vol. 238, Marcel Dekker, New York, 2001. [viii, 148]

[Wag96]   David G. Wagner, *Singularities of toric varieties associated with finite distributive lattices*, J. Algebr. Combin. **5** (1996), no. 2, 149–165. [287]

[Wei94]   Charles A. Weibel, *An introduction to homological algebra*, Cambridge Studies in Advanced Mathematics Vol. 38, Cambridge University Press, Cambridge, 1994. [15, 17, 20, 252, 269]

[Wei92]   Volker Weispfenning, *Comprehensive Gröbner bases*, J. Symbolic Comput. **14** (1992), no. 1, 1–29. [25]

[Wes01]   Douglas B. West, *Introduction to graph theory*, second ed., Prentice–Hall, Upper Saddle River, NJ, 2001. [53]

[Wey97]   Hermann Weyl, *The classical groups, Their invariants and representations*, Princeton Landmarks in Mathematics, Princeton University Press, Princeton, NJ, 1997 (reprint of the 1946 second edition). [378]

[Woo04a] Alexander Woo, *Multiplicities of the most singular point on Schubert varieties in GL_n/B for $n = 5, 6$*, preprint, 2004. arXiv:math.AG/0407158 [330]

[Woo04b] Alexander Woo, *Catalan numbers and Schubert polynomials for w =* $1(n+1)...2$, preprint, 2004. arXiv:math.CO/0407160   [330]

[Yan00] Kohji Yanagawa, *Alexander duality for Stanley–Reisner rings and squarefree* $\mathbb{N}^n$*-graded modules*, J. Algebra **225** (2000), no. 2, 630–645.   [106]

[Yan01] Kohji Yanagawa, *Sheaves on finite posets and modules over normal semigroup rings*, J. Pure Appl. Algebra **161** (2001), no. 3, 341–366.   [269]

[Yan02] Kohji Yanagawa, *Squarefree modules and local cohomology modules at monomial ideals*, Local cohomology and its applications (Guanajuato, 1999), Lecture Notes in Pure and Applied Mathematics Vol. 226, Marcel Dekker, New York, 2002, pp. 207–231.   [270]

[Yon03] Alexander Yong, *On combinatorics of quiver component formulas*, preprint, 2003. arXiv:math.CO/0307019   [353]

[You77] Alfred Young, *The collected papers of Alfred Young (1873–1940)*, University of Toronto Press, Toronto, Ontario, Buffalo, NY, 1977.   [288]

[Zel85] A. V. Zelevinskiĭ, *Two remarks on graded nilpotent classes*, Usp. Mat. Nauk **40** (1985), no. 1(241), 199–200.   [352]

[Zie95] Günter M. Ziegler, *Lectures on polytopes*, Graduate Texts in Mathematics Vol. 152, Springer–Verlag, New York, 1995.   [viii, 62, 73, 77, 119, 134, 199, 205, 235]

Glossary of notation

We use the standard arithmetic, algebraic, and logical symbols, including: "=" and "≅" for equality and isomorphism; "∅" and "{...}" for the empty set and the set consisting of "..."; "∩" and "∪" for intersection and union; "⊕" and "∏" for direct sum and product; ⊗ for tensor product; "∈" and "⊆" for set membership and containment (allowing equality; we use "⊂" if strict containment is intended); "∧" and "∨" for meet and join; "M/N" for the quotient of M by N; and "⟨...⟩" for the ideal generated by "...".

We use square brackets [...] to delimit matrices appearing "as is", whereas we use parentheses (...) to delimit column vectors written horizontally in the text. Thus, column vectors represented vertically in displayed equations or figures are delimited by square brackets.

Our common symbols beyond the very standard ones above are defined in the following table. The notations listed are those that span more than one chapter. If the notation has a specific definition, we have given the page number for it; otherwise, we simply list the page number of a typical (often not the first) usage.

symbol	typical usage or definition	page
$\succeq$	partial order on $\mathbb{N}^n$	11
$\mathbf{0}$	the zero vector	63, 133
$\mathbf{1}$	$(1,\ldots,1) \in \mathbb{N}^n$	76
A	abelian group with distinguished elements $\mathbf{a}_1,\ldots,\mathbf{a}_n$	150
$\mathbf{A}$	integer matrix whose columns $\mathbf{a}_1,\ldots,\mathbf{a}_n$ generate A	133
$\mathbf{a}$	vector $(a_1,\ldots,a_n)$ in $\mathbb{N}^n$	3
	element in A (often, a vector $(a_1,\ldots,a_d)$ in $\mathbb{Z}^d$)	133
$\mathbf{a}_F$	vector label on face F of labeled cell complex	62
$\mathbf{a}_i$	$\deg(x_i)$, one of the distinguished elements $\mathbf{a}_1,\ldots,\mathbf{a}_n \in A$	149
$\mathbf{a}_\sigma$	$\deg(m_\sigma) = \bigvee_{i \in \sigma} \mathbf{a}_i$	107
$\mathbf{a} \smallsetminus \mathbf{b}$	complementation of $\mathbf{b}$ in $\mathbf{a}$, for Alexander duality	88
$\langle \mathbf{a}, \mathbf{t} \rangle$	linear form $a_1 t_1 + \cdots + a_d t_d$	166
$\mathbf{b}$	analogous to $\mathbf{a}$	4, 129
$|\mathbf{b}|$	$b_1 + \cdots + b_n$	30
$\beta_{i,\mathbf{a}}(M)$	The ith Betti number of M in degree $\mathbf{a}$	157
Buch(I)	Buchberger graph of I	48
C	a real polyhedral cone (usually a rational polyhedral cone in $\mathbb{R}^d$)	134

symbol	typical usage or definition	page
$\mathrm{in}(f)$	initial term of f	24
$\mathrm{in}(I)$	initial ideal of I	24
$\mathrm{in}(M)$	initial submodule of M	27
J	an ideal	44
$\mathbb{K}.$	Koszul complex	13
$K^{\mathbf{b}}(I)$	upper Koszul simplicial complex	16
$\mathcal{K}(M;\mathbf{t})$	K-polynomial of M in variables $\mathbf{t}$	157
$\Bbbk$	field (sometimes with chapter-wide hypotheses)	3
$\Bbbk[\mathbf{x}]$	polynomial ring in variables $\mathbf{x}$	3
$\Bbbk[Q]$	semigroup ring for semigroup Q over $\Bbbk$ (sometimes $\Bbbk = \mathbb{Z}$)	129
$\Bbbk\{T\}$	vector space $\bigoplus_{\mathbf{a}\in T} \Bbbk \cdot \mathbf{t}^{\mathbf{a}}$, usually as $\Bbbk[Q]$-module	133
L	lattice in $\mathbb{Z}^n$ (often the kernel of $\mathbb{Z}^n \to A$)	130
$L_{\mathbb{R}}^{\perp}$	orthogonal complement in $\mathbb{R}^n$ of the real span of L	144
$\mathbf{L}$	integer matrix with cokernel A (so the rows generate L)	131
lcm	least common multiple	42
$\mathrm{link}_\Delta(\sigma)$	link of σ in Δ	17
$l(w)$	length of partial permutation w	294
λ	a real number	177
	a partition	285
λ_{qp}	scalar entries in monomial matrix	12, 217
M	a module	11
$M^{\vee}$	Matlis dual of module M	216
$M_{\mathbf{a}}$	graded component of M in degree $\mathbf{a}$	153
$M(\mathbf{a})$	graded translate of M satisfying $M(\mathbf{a})_{\mathbf{b}} = M_{\mathbf{a}+\mathbf{b}}$	153
$M_{k\ell}$	matrices with k rows and ℓ columns over the field $\Bbbk$	290
m_i	minimal generator of monomial ideal $\langle m_1, \ldots, m_r \rangle$	28
m_σ	least common multiple of $\{m_i \mid i \in \sigma\}$	107
$\mathfrak{m}$	graded maximal ideal	257
$\mathfrak{m}^{\mathbf{b}}$	irreducible monomial ideal $\langle x_i^{b_i} \mid b_i \geq 1 \rangle$	87
$\mathbb{N}$	the natural numbers $\{0, 1, 2, \ldots\}$	3
n	number of variables in polynomial ring S	3
$n!$	n factorial $= n(n-1)\cdots 3\cdot 2\cdot 1$	356
$[n]$	the set $\{1, \ldots, n\}$	81, 274
$\binom{n}{k}$	binomial coefficent $\frac{n!}{k!(n-k)!}$	48
ν	a normal vector	77, 199
$\Omega_Q^{\bullet}$	dualizing complex for affine semigroup Q	233
ω_Q	canonical module for semigroup ring $\Bbbk[Q]$	233
P_F	monomial prime ideal of semigroup ring	134
$\mathbb{P}^r$	projective space of dimension r	198
$\mathcal{P}$	a polytope or polyhedron	62, 197
$\mathcal{P}_\lambda$	hull polyhedron for real number $\lambda \gg 0$	177
$\mathfrak{p}$	a prime ideal	165
Q	subsemigroup of A generated by $\mathbf{a}_1, \ldots, \mathbf{a}_n$	150
Q_{sat}	saturation of semigroup Q	140
$\mathcal{Q}$	a polytope	62
R	a ring	159

symbol	typical usage or definition	page
$\mathbb{R}$	field of real numbers	41
$\mathbb{R}^n_{\geq 0}$	orthant of all nonnegative real vectors	72
$\mathbb{R}_{\geq 0} Q$	real cone generated by affine semigroup Q	134
$\mathcal{RP}(w)$	set of reduced pipe dreams for partial permutation w	312
$r_{pq}(w)$	rank of submatrix $w_{p \times q}$ of partial permutation w	290
S	polynomial ring $\Bbbk[\mathbf{x}]$	3
S^G	ring of invariants in S under action of group G	193, 364
S_n	symmetric group of permutations of $\{1, \ldots, n\}$	291
$\mathrm{supp}(\mathbf{a})$	support $\{i \in \{1, \ldots, n\} \mid a_i \neq 0\}$	7
$\mathbf{s}$	auxiliary symbol/variables analogous to $\mathbf{t}$	164
σ	squarefree vector or face of simplicial complex	4–5
$\overline{\sigma}$	complement $\{1, \ldots, n\} \smallsetminus \sigma$	5
σ_i	transposition switching i and $i+1$	298
$\mathfrak{S}_w(\mathbf{t})$	Schubert polynomial	304
$\mathfrak{S}_w(\mathbf{t} - \mathbf{s})$	double Schubert polynomial	304
Tor^S_i	i^{th} Tor module	15
$\mathbf{t}$	dummy variable for monomials in semigroup rings	129
	dummy variable for Hilbert series and K-polynomials	154
	variables $t_1, \ldots, t_d$ for K-polynomials and multidegrees	166
τ	analogous to σ	4
$\mathbf{u}$	vector $(u_1, \ldots, u_n)$ in $\mathbb{Z}^n$	130
$v \leq w$	Bruhat and weak orders on partial permutations	295, 299
$\mathbf{v}$	vector $(v_1, \ldots, v_n)$ in $\mathbb{Z}^n$	130
w	weight vector in $\mathbb{R}^n_{\geq 0}$	142
	partial permutation (matrix)	290
w_0	long word (permutation), reversing the order of $1, \ldots, n$	291
$\mathbf{w}$	vector $(w_1, \ldots, w_n)$ in $\mathbb{Z}^n$	179
X	cell complex, often labeled	62
$\underline{X}$	underlying unlabeled cell complex	92
$X_{\prec \mathbf{b}}$	subcomplex of X on face with labels $\prec \mathbf{b}$	64
$X_{\preceq \mathbf{b}}$	subcomplex of X on face with labels $\preceq \mathbf{b}$	64
$\overline{X}_w$	matrix Schubert variety for partial permutation w	290
$\mathbf{x}$	variables $x_1, x_2, \ldots$ in polynomials rings	3
	coordinates $x_1, x_2, \ldots$ on affine space	192
	variables $x_{\alpha\beta}$ in a square or rectangular array	290
$\mathbf{x}^{\mathbf{a}}$	monomial $x_1^{a_1} \ldots x_n^{a_n}$	3
$\mathbf{x}^{\mathbf{a}} < \mathbf{x}^{\mathbf{b}}$	comparison of monomials under term order $<$	24
$\mathbf{x}_{p \times q}$	upper-left $p \times q$ submatrix of matrix $\mathbf{x}$	290
$\mathbf{y}$	auxiliary variables analogous to $\mathbf{x}$	25, 139
$\mathbb{Z}$	ring of integers	6
$\mathbb{Z}F$	group generated by face F of affine semigroup	134
$Z_{p \times q}$	upper-left $p \times q$ submatrix of matrix Z	290
$\mathbf{z}$	Laurent variables $z_1, \ldots, z_n$; coordinates on $(\mathbb{C}^*)^n$	192

Index

3-connected, *53**

abelian group, 129, 149
  cyclic, 194
  divisible, *see* divisible group
  finite, 172, 194
  finitely generated, 129
  free, 131, 133
  sequence defines multigrading, 149, 191
  torsion, 152, 161
  torsion-free, 151, 152, 187
acyclic cover, *94*
additive identity, 129
additivity, *166*, 169, 172, 306, 311
  yields Schubert polynomials, 323–324
adjacent transposition, *291*, *298*, 303, 325
adjointness, *see* functor, adjoint
admissible family, *374*
  universal, *374*
admissible ideal, 373, *374*, 377
Alexander duality, 81, 105–106
  as planar map duality, 99–100, 106
  on antidiagonal ideals, 318
  on arbitrary ideals, *88*, 89–91, 226, 269
  on cogeneric ideals, 123
  on free and injective resolutions, 106
  on free and irreducible resolutions, 225
  on free resolutions, *see* duality for resolutions
  on generic ideals, 122
  on homological invariants, 100, 102–104
  on irrelevant ideal, 199
  on $\mathbb{N}^n$-graded modules, *228*
  on upper bound problems, 125
  principle behind, 96, 126
  simplicial, *16*, 17, 81, 85, 98, 105
  squarefree, *16*, 81–82, 89, 102, 226, 318
  tight, *104*
  topological, 83, 84
Alexander inversion formula, 86, 106
algebraic geometry, viii, 21, 41, 106, 193, 353, 355, 376
algebraic shifting, 40, 45, 106

algebraic torus, 21, 172, 191, 197, 200, 363
  coordinates on, 192
almost 3-connected, *53*, 54
almost n-connected, *59*
antidiagonal complex, *318*, 319–323, 329
  from matrix Schubert variety, 323
  is ball or sphere, 329
  is shellable, 327
  is subword complex, 327
antidiagonal term, *280*, *318*
  caused by rank condition, *321*
associated prime
  multigraded, 133, 152, 166
  of Borel-fixed ideal, 39
  of initial ideal, 145
  of local cohomology, 254, 256, 270
  of principal ideal, 147, 269
  of $\mathbb{Z}$-graded module, 263
Auslander–Buchsbaum formula, 100, 264

ball, 145, 329, 330
Barvinok's algorithm, 229, 244
Barvinok's Theorem, 241
barycenter, *112*
basis weights, *158*
Bass number, 104, 106, *223*, 224, 228, 265
  of local cohomology, 255, 270
Bayer, Dave, 86, 106
Bender–Knuth involution, 329
`betti` diagram, *102*, 103
Betti number, *see also* syzygy
  characteristic dependence, 18, 58, 80
  dual to Bass number, 104
  duality for, 76, 98
  extremal, *see* extremal Betti number
  from cellular resolution, 65–66
  multigraded, *157*
  $\mathbb{N}^n$-graded, *14*, 15–18
  of Borel-fixed ideal, 30–33, 38
  of generic ideal, 53, 112
  of generic lattice ideal, 190
  of lattice ideal, 174, 175
  of lex-segment ideal, 35
  of local cohomology, 255
  of monomial ideal, 16, 85

---

*Italic page numbers refer to definitions

Graduate Texts in Mathematics

(continued from page ii)